Carpenter's Guide to Innovative SAS® Techniques

Art Carpenter

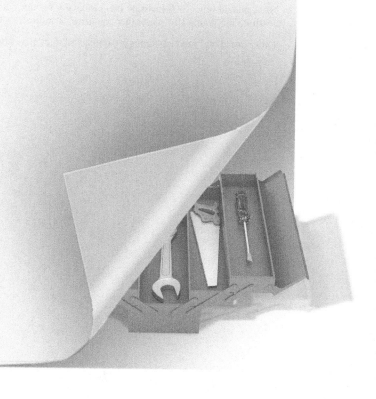

support.sas.com/publishing

The correct bibliographic citation for this manual is as follows: Carpenter, Art. 2012. *Carpenter's Guide to Innovative SAS® Techniques*. Cary, NC: SAS Institute Inc.

Carpenter's Guide to Innovative SAS® Techniques

Copyright © 2012, SAS Institute Inc., Cary, NC, USA

ISBN 978-1-61290-202-9 (electronic book)
ISBN 978-1-60764-991-5

All rights reserved. Produced in the United States of America.

SAS Institute Inc., SAS Campus Drive, Cary, North Carolina 27513-2414

1st printing, March 2012

SAS® Publishing provides a complete selection of books and electronic products to help customers use SAS software to its fullest potential. For more information about our e-books, e-learning products, CDs, and hard-copy books, visit the SAS Publishing Web site at **support.sas.com/publishing** or call 1-800-727-3228.

For the ancient history buffs - as the Mamas and Papas used to say, "This is dedicated to the one I love." That would be my wife Marilyn who supported me (sometimes quite literally) during "one more book project," and who suggested the word 'Innovative' for the title.

iv

Contents

Part 2 Data Summary, Analysis, and Reporting 231

About This Book

The Intent of this Book

The goal of this book is to broaden the usage of a number of SAS programming tools and techniques. This is a very eclectic collection of ideas and tips that have been advanced over the years by any number of users. Some are quite advanced; however, most require only an intermediate understanding of the general concepts surrounding the tip. For instance if the technique involves the use of a double SET statement, you should have a decent understanding of the DATA step and how it is compiled and executed. Many of the techniques are even simple and are essentially suggestions along the lines of "Did you know that you can . . . .?"

What this Book is NOT

As is the case with any book that deals with a very broad range of topics, no single topic can be covered with all possible detail. For example SAS Formats are discussed in this book in several places; however, if you want more information on SAS formats, a full book has been written on that subject alone (Bilenas, 2005), consequently the content of that book will not be repeated in this one.

Except for a few of the especially advanced topics (I get to decide which ones), for most topics, this book makes no attempt to explain the basics. There are several very good "getting started" books on various aspects of SAS, this book is NOT one of them. If you want the basic how-to for a procedure or technique consult one of these other books. Of course, the reality is that some of the readers of this book will have more, or less, experience than others. I have made some attempt at offering brief explanations on most topics. Hopefully the depth of this book will be enough to get you started in the right direction for any given topic, even if it does not cover that topic thoroughly.

By its very nature this book is not designed to be read linearly, front to back, instead I anticipate that the reader will use it either as a reference for a specific technique, an exploration tool for learning random new 'tidbits', or perhaps most effectively as a sleeping aid. The MORE INFORMATION and SEE ALSO sections, as well as, the topical index in Appendix A, and the usage index in Appendix B should help you find and navigate to related topics.

What this Book is . . .

There are literally hundreds of techniques used on a daily basis by the users of SAS software as they perform analyses and generate reports. Although sometimes obscure, most of these techniques are relatively easy to learn and generally do not require any specialized training before they can be implemented. Unfortunately a majority of these techniques are used by only a very small minority of the analysts and programmers. They are not used more frequently, simply because a majority of SAS users have not been exposed to them. Left to ourselves it is often very difficult to '*discover*' the intricacies of these techniques and then to sift through them for the nuggets that have immediate value. Certainly this is true for myself as I almost daily continue to learn new techniques. I regret that the nugget that I learn tomorrow will not make it into this book.

This book introduces and demystifies a series of those nuggets. It covers a very broad range of mostly Base SAS topics that have proven to be useful to the intermediate or advanced SAS programmer who is involved with the analysis and reporting of data. The intended audience is expected to have a firm grounding in Base SAS. For most of the covered topics, the book will introduce useful techniques and options, but will not 'teach the procedure'.

I have purposefully avoided detailed treatment of advanced topics that are covered in other books. These include, but are not limited to: statistical graphics (Friendly, 1991), advanced ODS topics (Haworth et al, 2009), the macro language (Carpenter, 2004), PROC REPORT (Carpenter, 2007a), PROC TABULATE (Haworth, 1999), SAS/GRAPH (Carpenter and Shipp, 1995), and the annotate facility (Carpenter, 1999).

The more advanced users may find that they are already using some of these techniques, and I hope that this is the case for you. However I believe that the range of topics is broad enough that there will be something for everyone. It may only take a single nugget to 'pay for the book'.

Intended Audience

This book is intended to be used by intermediate and advanced SAS programmers and SAS users who are faced with large or complex reporting and analysis tasks. It is especially for those that have a desire to learn more about the sometimes obscure options and techniques used when writing code for the advanced analysis and reporting of data. SAS is complex enough that it can be very difficult, even for an advanced user, to have a knowledge base that is diverse enough to cover all the necessary topics. Covering, at least at a survey level, as many of these diverse topics as possible is the goal of this book.

This book has not been written for the user who is new to SAS. While this book contains a great deal that the new user will find valuable, unlike an introductory book that goes into great detail, most of the topics in this book are fairly brief and are intended more to spark the reader's interest rather than to provide a complete reference. The assumption is that most readers of this book will have sufficient background to 'dig deeper' for the details of the topics that most interest them.

Overview of Chapters

Part 1 Data Preparation

Most tasks involving the use of SAS revolve around the data. The analyst is often responsible for bringing the data into the SAS world, manipulating it so that it can be analyzed, and for the analysis preparation itself. Although not all phases of data preparation are necessary for every project or task, the analyst must be prepared for a wide variety of *variations on the theme*.

Chapter 1: Moving, Copying, Importing and Exporting Data

The issues surrounding the movement of data into and out of the SAS environment are as diverse as the types and sources of data.

Chapter 2: Working with Your Data

Once the data is available to SAS there are a number of ways that it can be manipulated and prepped for analysis. In addition to the DATA step, SAS contains a number of tools to assist in the process of data preparation.

Chapter 3: Just in the DATA Step

There are a number of tools and techniques that apply only to the DATA step.

Chapter 4: Sorting the Data

The order of the rows in a data table can affect not only how the data are analyzed, but also how it is presented.

Chapter 5: Working with Data Sets

Very often there are things that we can do to the data tables that will assist with the analysis and reporting process.

Chapter 6: Table Lookup Techniques

The determination of a value for a variable based on another variable's value requires a lookup for the desired value. As our tables become complex lookup techniques can become quite specialized.

Part 2 Data Summary, Analysis, and Reporting

The use of SAS for the summarization and analysis of data is at the heart of what SAS does best. And of course, since there is so much that you can do, it is very hard to know of all the techniques that are available. This part of the book covers some of the more useful techniques, as well as a few that are underutilized, either because they are relatively new, or because they are somewhat obscure.

Several of these techniques apply to a number of different procedures. And the discussion associated with them can be found in various locations within this book. In all cases these are techniques of which I believe the SAS power user should be aware.

Chapter 7: MEANS and SUMMARY Procedures

Although almost all SAS programs make use of these procedures, there are a number of options and techniques that are often overlooked.

Chapter 8: Other Reporting and Analysis Procedures

Several commonly used procedures have new and/or underutilized options, which when used, can greatly improve the programmer's efficiency.

**Chapter 9: SAS/GRAPH Elements You Should Know – Even If You Don't Use
            SAS/GRAPH**

A number of statements, options, and techniques that were developed for use with SAS/GRAPH can also be taken advantage of outside of SAS/GRAPH.

Chapter 10: Presentation Graphics – More than Just SAS/GRAPH

A number of Base SAS procedures as well as procedures from products other than SAS/GRAPH produce presentation-quality graphics. Some of the highlights and capabilities of those procedures are discussed in this chapter.

Chapter 11: Output Delivery System

Most reporting takes advantage of the Output Delivery System. A great deal has been written about ODS; in this chapter a few specialized techniques are discussed.

Part 3 Techniques, Tools, and Interfaces

In addition to the coding nuts and bolts of SAS, there are a number of tools and techniques, many of which transcend SAS that can be especially helpful to the developer. This part of the book is less about DATA and PROC steps and more about how they work together and how they interface with the operating environment.

Chapter 12: Taking Advantage of Formats

There is a great deal more that you can do with formats in addition to the control of the display of values.

Chapter 13: Interfacing with the Macro Language

When building advanced macro language applications there are a number of things of which the developer should be aware.

Chapter 14: Operating System Interface and Environmental Control

While not necessarily traditional SAS, application programmers must be able to interface with the operating system, and there is a great deal more than one would anticipate at first glance.

Chapter 15: Miscellaneous Topics

There are a number of isolated topics that, while they do not fit into the other chapters, do indeed still have value.

Software Used to Develop the Book's Content

This book is based on SAS 9.3. Although every effort has been made to include the latest information available at the time of printing, new features will be made available in later releases. Be sure to check out the SAS Web site for current updates and check the SAS OnlineDoc for enhancements and changes in new releases of SAS.

Using this Book

Initial publication of this book will be the traditional hard copy paper. As time and technology permits, it is hoped that the book will also be made available in various forms electronically.

Display of SAS Code and Output

The type face for the bulk of the text is Times New Roman.

The majority of the code will appear in a shaded box and will appear in the `Courier New` font.

```
SAS Code appears in a
shaded box.
```

```
LOG Window in a box with a
dotted border.
```

Text written to the SAS LOG will appear in a box with a dotted border, and like the code box the text will be in the `Courier New` font.

The Output Delivery System, ODS, has been used to present the output generated by SAS procedures. Throughout the book it is common to show only portions of the output from a given procedure. Output written to the LISTING destination will appear in an un-shaded solid bordered box using the `Courier New` font. The output written to other ODS destinations will be presented as screen shot graphics appropriate to that destination.

```
SAS OUTPUT Window in an
unshaded box.
```

Although color is included in most ODS styles, color will not be presented in this book. If you want to see the color output, you are encouraged to execute the sample code associated with the appropriate section so that you can see the full output. Occasionally raw data will also be presented in an unshaded box with a solid border.

SAS terms, keywords, options and such are capitalized, as are data set and variable names. Terms that are to be emphasized are written in italics, as are nonstandard English words (such as *fileref*) that are common in the SAS vernacular.

References and Links

Throughout the book references are included so that the reader can find more detail on various topics. Most, but not all, of these references are shown in the MORE INFORMATION and SEE ALSO sections.

MORE INFORMATION Sections

Related topics that are discussed further within this book are pointed out in the MORE INFORMATION section that follows most sections of the book. Locations are identified by section number.

SEE ALSO Sections

References to sources outside of this book are made in the SEE ALSO section. Citations refer to a variety of sources. Usually the citation will include the author's name and the year of publication. Additional detail for each citation, including a live link, can be found in the References section.

There are also a number of references to SAS Institute's support site (support.sas.com). Unfortunately internal addressing on this site is changing constantly, and while every effort has been made to make all links as current as possible, any links to this site should be considered to be suspect until verified.

Locating References

If you are reading this book using an electronic device, you will notice that most of the links cited in the SEE ALSO sections are live. Each of the papers or books listed also has a live link in the References section of this book. Every attempt has been made to ensure that these links are current; however, it is the very nature of the Web that links change, and this will be especially true throughout the life of this book.

Whether you are reading this book using the traditional paper format or if you are using an electronic device all of the links in this book, including the links to all the cited papers in the References section, as well as the links embedded within the text of the book, have been made available to you as live links on sasCommunity.org under a category named using the title of this book. As I discover that links have gone stale or have changed they will be updated at this location whenever possible. Please let me know if you discover a stale or bad link.

Navigating the Book

In addition to the standard word index at the back of the book two appendixes have been provided that will help you navigate the book and to find related topics:

- Appendix A – Topical Index Find related items by technique or topic
- Appendix B – Usage Index Find statements, options, and keywords as they are used in examples.

The **MORE INFORMATION** sections will also guide you to related topics elsewhere within the book.

Using the Sample Programs and Sample Data

A series of sample programs and data sets from this book are available for your use. These are available in a downloadable ZIP file, either from the author page for this book at support.sas.com/authors or from sasCommunity.org. The sample programs are organized by chapter, and named according to the section in which they are described. They can be used 'out of the box'; however, you may need to establish some macro variables and libraries. This is done automatically for you if you use the suggested AUTOEXEC.SAS program and the assumed folder structure.

The ZIP file will contain the primary folder \InnovativeTechniques and the three subfolders \SASCode, \Results, and \Data. To use the SAS programs you will want to first set up a SAS environment as described in Chapter 14, "Operating System Interface and Environmental Control." The \SASCode directory contains an AUTOEXEC.SAS program that you will want to take advantage of by following the instructions in Section 14.2 and 14.1.1. As it is currently written the autoexec program expects that the SAS session initialization will include an &SYSPARM definition (see Section 14.1.1).

The following SAS catalogs and data tables are used by the sample programs, and are made available through the use of the ADVRPT *libref,* which is automatically established by the AUTOEXEC.SAS program.

The clinical trial study data has been fabricated for this book and does not reflect any real or actual study. Although the names of drugs and symptoms are nominally factual, data values do not necessarily reflect real-world situations. Careful inspection of the data tables will surface a number of data issues that are, in part, discussed throughout the book. Although I have introduced some data errors for use in this book, the bulk of the ADVRPT.DEMOG data set was created by Kirk Lafler, Software Intelligence Corporation, and has been used with his permission.

The manufacturing data is nominally actual data, but it has been highly edited for use in this book. I would suggest that you do not adjust any process controls based on this data.

Data Group	Data Group	Member Name	Description
Clinical Trial	Study Data	AE	Adverse events
		CLINICNAMES	Clinic names and locations
		CONMED	Concomitant Medications
		DEMOG	Demographic Information
		LAB_CHEMISTRY	Laboratory Chemistry results
	Study Metadata	DATAEXCEPTIONS	Data exclusion criteria
		DSNCONTROL	Data set level metadata
		FLDCHK	Automated data field check metadata (see Section 13.5.2)
Manufacturing	Manufacturing Data	MFGDATA	Manufacturing process test data
Miscellaneous	Function Definitions	FUNCTIONS	User-defined functions using PROC FCMP (see Section 15.2)
	Password Control	PASSTAB	See Section 5.4.2
		PWORD	This is a simplified version of the PASSTAB data set (See Section 2.1.2)

Catalog Name	Member Type	Description
FONTS	Fonts, graphical	User-defined SAS/GRAPH font
PROJFMT	Formats	User-defined format library
SASMACR	Stored Compiled Macros	Stored compiled macro library (see Section 13.9)

Corrections, Typos, and Errors

Although every effort has been made by numerous reviewers and editors to catch my typos and technical errors, it is conceivable – however unlikely – that one still remains in the book. Any *errata* that are discovered after publication will be collected and published on sasCommunity.org. Please visit the category dedicated to this book on sasCommunity.org. There you can get the latest updates and corrections, and you can let me know of anything that you discover. Will you be the first to report something?

Author Page

You can access the author page for this book at http://support.sas.com/authors. This page includes several features that relate to this specific book, including more information about the book and author, book reviews, and book updates; book extras such as example code and data; and contact information for the author and SAS Press.

Additional Resources

SAS offers a rich variety of resources to help build your SAS skills and explore and apply the full power of SAS software. Whether you are in a professional or academic setting, we have learning products that can help you maximize your investment in SAS.

Bookstore	http://support.sas.com/publishing/
Training	http://support.sas.com/training/
Certification	http://support.sas.com/certify/
Higher Education Resources	http://support.sas.com/learn/
SAS OnDemand for Academics	http://support.sas.com/ondemand/
Knowledge Base	http://support.sas.com/resources/
Support	http://support.sas.com/techsup/
Learning Center	http://support.sas.com/learn/
Community	http://support.sas.com/community/
SAS Forums	http://communities.sas.com/index.jspa
User community wiki	http://www.sascommunity.org/wiki/Main_Page

Comments or Questions?

If you have comments or questions about this book, you can contact the author through SAS as follows:

Mail: SAS Institute Inc.
SAS Press
Attn: Art Carpenter
SAS Campus Drive
Cary, NC   27513

Email: saspress@sas.com

Fax: (919) 677-4444

Please include the title of this book in your correspondence.

SAS Publishing News

Receive up-to-date information about all new SAS publications via e-mail by subscribing to the SAS Publishing News monthly eNewsletter. Visit support.sas.com/subscribe.

Acknowledgments

Writing this book has been a fun adventure, and it has provided me the opportunity to work with a number of talented and helpful folks. From small conversations to extended dialogs I would like to thank all of those who helped to make this book possible. I would especially like to thank the following members of the SAS community who gave so freely to this endeavor.

Bob Virgle and Peter Eberhardt both contributed several suggestions for the title of this book. Because their suggestions were judged by others to be too humorous, their ideas were ultimately not used; however, I liked them nonetheless and wish they could have been incorporated.

During the writing of this book it was my privilege to learn from the contributions and help of several of the world-class SAS programming legends, including, but not limited to, Peter Crawford, Paul Dorfman, John King, Art Trabachneck, and Ian Whitlock.

I'd like to thank the following technical reviewers at SAS who contributed substantially to the overall quality of the book's content: Amber Elam, Kim Wilson, Scott McElroy, Kathryn McLawhorn, Ginny Piechota, Russ Tyndall, Grace Whiteis, Chevell Parker, Jan Squillace, Ted Durie, Jim Simon, Kent Reeve, and Kevin Russell. In addition to the SAS reviewers, helpful comments and suggestions were received from William Benjamin Jr. and Peter Crawford. Rick Langston at SAS contributed a number of his examples and explanations to the sections on formats and functions. The reviewers and editors were thorough; therefore, any mistakes or omissions that remain are mine alone and in no way reflect a lack of effort on the part of the reviewers to guide me. To paraphrase Merle Haggard, "they tried to guide me better, but their pleadings I denied, I have only me to blame, because they tried."

About the Author

This is Art Carpenter's fifth book and his publications list includes numerous papers and posters presented at SAS Global Forum, SUGI, and other user group conferences. Art is a SAS Silver Circle member and has been using SAS® since the mid 1970's, and he has served in various leadership positions in local, regional, national, and international user groups. He is a SAS Certified Base Programmer for SAS 9, SAS Certified Clinical Trials Programmer Using SAS 9 and a SAS Certified Advanced Programmer for SAS 9. Through California Occidental Consultants he teaches SAS courses and provides contract SAS programming support nationwide.

Author Contact

Arthur L. Carpenter
California Occidental Consultants
10606 Ketch Circle
Anchorage, AK 99515

(907) 865-9167
art@caloxy.com
http://www.caloxy.com
http://www.sascommunity.org/wiki/User:ArtCarpenter
http://support.sas.com/publishing/authors/carpenter.html

Part 1

Data Preparation

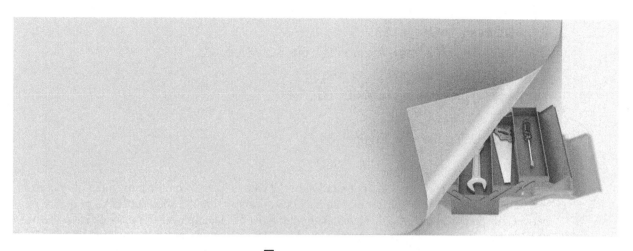

Chapter 1

Moving, Copying, Importing, and Exporting Data

A great deal of the process of the preparation of the data is focused on the movement of data from one table to another. This transfer of data may be entirely within the control of SAS or it may be between disparate data storage systems. Although most of the emphasis in this book is on the use of SAS, not all data are either originally stored in SAS or even ultimately presented in SAS. This chapter discusses some of the aspects associated with moving data between tables as well as into and out of SAS.

When moving data into and out of SAS, Base SAS allows you only limited access to other database storage forms. The ability to directly access additional databases can be obtained by licensing one or more of the various SAS/ACCESS products. These products give you the ability to utilize the SAS/ACCESS engines described in Section 1.1 as well as an expanded list of databases that can be used with the IMPORT and EXPORT procedures (Section 1.2).

SEE ALSO

Andrews (2006) and Frey (2004) both present details of a variety of techniques that can be used to move data to and from EXCEL.

1.1 LIBNAME Statement Engines

In SAS®9 a number of engines are available for the LIBNAME statement. These engines allow you to read and write data to and from sources other than SAS. These engines can reduce the need to use the IMPORT and EXPORT procedures.

The number of available engines depends on which products your company has licensed from SAS. One of the most popular is SAS/ACCESS® Interface to PC Files.

You can quickly determine which engines are available to you. An easy way to build this list is through the NEW LIBRARY window.

From the SAS Explorer right click on LIBRARIES and select NEW. Available engines appear in the ENGINE pull-down list.

Pulling down the engine list box on the 'New Library' dialog box shown to the right, indicates the engines,

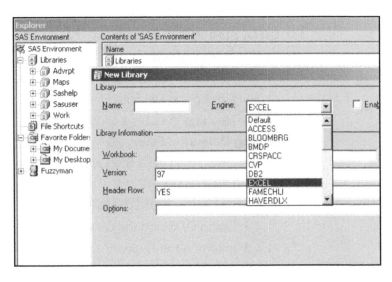

including the EXCEL engine, among others, which are available to this user.

PROC SETINIT can also be used to determine which products have been licensed.

The examples in this section show various aspects of the EXCEL engine; however, most of what is demonstrated can be applied to other engines as well.

SEE ALSO
Choate and Martell (2006) discuss the EXCEL engine on the LIBNAME statement in more detail. Levin (2004) used engines to write to ORACLE tables.

1.1.1 Using Data Access Engines to Read and Write Data

In the following example, the EXCEL engine is used to create an EXCEL workbook, store a SAS data set as a sheet in that workbook, and then read the data back from the workbook into SAS.

```
libname toxls excel "&path\data\newwb.xls"; ❶

proc sort data=advrpt.demog
          out=toxls.demog; ❷
   by clinnum;
   run;

data getdemog;
   set toxls.demog; ❸
   run;

libname toxls clear; ❹
```

❶ The use of the EXCEL engine establishes the TOXLS *libref* so that it can be used to convert to and from the Microsoft Excel workbook NEWWB.XLS. If it does not already exist, the workbook will be created upon execution of the LIBNAME statement.

For many of the examples in this book, the macro variable &PATH is assumed to have been defined. It contains the upper portion of the path appropriate for the installation of the examples on your system. See the book's introduction and the AUTOEXEC.SAS in the root directory of the example code, which you may download from support.sas.com/authors.

❷ Data sets that are written to the TOXLS *libref* will be added to the workbook as named sheets. This OUT= option adds a sheet with the name of DEMOG to the NEWWB.XLS workbook.

❸ A sheet can be read from the workbook, and brought into the SAS world, simply by naming the sheet.

❹ As should be the case with any *libref*, when you no longer need the association, the *libref* should be cleared. This can be especially important when using data engines, since as long as the *libref* exists, access to the data by applications other than SAS is blocked. Until the *libref* is cleared, we are not able to view or work with any sheets in the workbook using Excel.

MORE INFORMATION
LIBNAME statement engines are also discussed in Sections 1.1.2 and 1.2.6. The XML engine is discussed in Section 1.6.2.

1.1.2 Using the Engine to View the Data

Once an access engine has been established by a *libref,* we are able to do almost all of the things

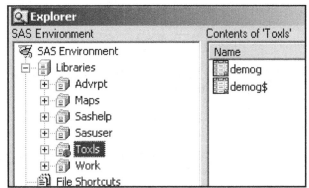

that we typically do with SAS data sets that are held in a SAS library.

The SAS Explorer shows the contents of the workbook with each sheet appearing as a data table.

When viewing an EXCEL workbook through a SAS/ACCESS engine, each sheet appears as a data set. Indeed you can use the VIEWTABLE or View Columns tools against what are actually sheets. Notice in this image of the SAS

Explorer, that the DEMOG sheet shows up twice. Sheet names followed by a $ are actually named ranges, which under EXCEL can actually be a portion of the entire sheet. Any given sheet can have more than one named range, so this becomes another way to filter or subset what information from a given sheet will be brought into SAS through the SAS/ACCESS engine.

1.1.3 Options Associated with the Engine

The SAS/ACCESS engine is acting like a translator between two methods of storing information, and sometimes we need to be able to control the interface. This can often be accomplished through the use of options that modify the translation process. Many of these same options appear in the PROC IMPORT/EXPORT steps as statements or options.

It is important to remember that not all databases store information in the same relationship as does SAS. SAS, for instance, is column based - an entire column (variable) will be either numeric or character. EXCEL, on the other hand, is cell based – a given cell can be considered numeric, while the cell above it in the same column stores text. When translating from EXCEL to SAS we can use options to establish guidelines for the resolution of ambiguous situations such as this.

Connection Options

For database systems that require user identification and passwords these can be supplied as options on the LIBNAME statement.

- USER   User identification
- PASSWORD   User password
- *others*   Other connection options vary according to the database to which you are connecting

LIBNAME Statement Options

These options control how information that is passed through the interface is to be processed. Most of these options are database specific and are documented in the sections dealing with your database.

When working with EXCEL typical LIBNAME options might include:

- HEADER   Determines if a header row exists or should be added to the table.
- MIXED   Some columns contain both numeric and character information.
- VER   Controls which type (version) of EXCEL is to be written.

Data Source Options

Some of the same options associated with PROC IMPORT (see Section 1.2.3) can also be used on the LIBNAME statement. These include:

- GETNAMES   Incoming variable names are available in the *first row* of the incoming data.
- SCANTEXT   A length is assigned to a character variable by scanning the incoming column and determining the maximum length.

1.1.4 Replacing EXCEL Sheets

While the EXCEL engine allows you to establish, view, and use a sheet in an Excel workbook as a SAS data set, you cannot update, delete or replace the sheet from within SAS. It is possible to replace the contents of a sheet, however, with the help of PROC DATASETS and the SCAN_TEXT=NO option on the LIBNAME statement. The following example shows how to replace the contents of an EXCEL sheet.

In the first DATA step the programmer has 'accidently' used a WHERE clause ❶ that writes the incorrect data, in this case 0 observations, to the EXCEL sheet. Simply correcting and rerunning the DATA step ❷ will not work because the sheet already exists.

```
libname toxls excel "&path\data\newwb.xls";

data toxls.ClinicNames;
   set advrpt.clinicnames;
   where clinname>'X'; ❶
   run;

* Running the DATA step a second time
* results in an error;
data toxls.ClinicNames; ❷
   set advrpt.clinicnames;
   run;
```

We could step out of SAS and use EXCEL to manually remove the bad sheet; however, we would rather do it from within SAS. First we must reestablish the *libref* using the SCAN_TEXT=NO option ❸. PROC DATASETS can then be used to *delete* the sheet. In actuality the sheet

```
libname toxls excel
             "&path\data\newwb.xls"
             scan_text=no ❸;
proc datasets library=toxls nolist;
   delete ClinicNames;
   quit;
```

has not truly been deleted, but merely cleared of all contents. Since the sheet is now truly empty and the SCAN_TEXT option is set to NO, we can now replace the empty sheet with the desired contents.

```
data toxls.ClinicNames; ❹
   set advrpt.clinicnames;
   run;

libname toxls clear; ❺
```

The DATA step can now be rerun ❹, and the sheet contents will now be correct. When SAS has completed its work with the workbook, and before you can use the workbook using EXCEL you will need to clear the *libref*. This can be done using the CLEAR option on the LIBNAME statement ❺.

MORE INFORMATION

See Section 1.2 for more information on options and statements in PROC IMPORT and PROC EXPORT. In addition to PROC DATASETS, Section 5.4 discusses other techniques that can be used to delete tables. Section 14.4.5 also has an example of deleting data sets using PROC DATASETS.

SEE ALSO

Choate and Martell (2006) discuss this and numerous other techniques that can be used with EXCEL.

1.1.5 Recovering the Names of EXCEL Sheets

Especially when writing automated systems you may need to determine the names of workbook sheets. There are a couple of ways to do this.

If you know the *libref*(s) of interest, the automatic view SASHELP.VTABLE can be used in a DATA step to see the sheet names. This view contains one observation for every SAS data set in every SAS library in current use, and for the TOXLS *libref* the sheet names will be shown as data set names.

```
data sheetnames;
set sashelp.vtable;
where libname = 'TOXLS';
run;
```

```
proc sql;
create table sheetnames as
   select * from dictionary.members
      where engine= 'EXCEL' ;
   quit ;
```

When there are a number of active libraries, the process of building this table can be lengthy. As a general rule using the DICTIONARY.MEMBERS table in a PROC SQL step has a couple of advantages. It is usually quicker than the SASHELP.VTABLE view, and it also has an ENGINE column which allows you to search without knowing the specific *libref*.

The KEEP statement or the preferred KEEP= data set option could have been used in these examples to reduce the number of variables (see Section 2.1.3).

MORE INFORMATION

SASHELP views and DICTIONARY tables are discussed further in Section 13.8.1.

SEE ALSO

A thread in the SAS Forums includes similar examples.
http://communities.sas.com/thread/10348?tstart=0

1.2 PROC IMPORT and EXPORT

Like the SAS/ACCESS engines discussed in Section 1.1, the IMPORT and EXPORT procedures are used to translate data into and out of SAS from a variety of data sources. The SAS/ACCESS product, which is usually licensed separately through SAS (but may be bundled with Base SAS), controls which databases you will be able to move data to and from. Even without SAS/ACCESS you can still use these two procedures to read and write text files such as comma separated variables (CSV), as well as files using the TAB and other delimiters to separate the variables.

1.2.1 Using the Wizard to Build Sample Code

The import/export wizard gives you a step-by-step guide to the process of importing or exporting data. The wizard is easy enough to use, but like all wizards does not lend itself to automated or batch processing. Fortunately the wizard is actually building a PROC IMPORT/EXPORT step in the background, and you can capture the completed code. For both the import and export process the last screen prompts you to 'Create SAS Statements.'

```
PROC EXPORT DATA= WORK.A ❶
            OUTFILE= "C:\temp\junk.xls"❷
            DBMS=EXCEL❸
            REPLACE❹;
      SHEET="junk"; ❺
RUN;
```

The following PROC EXPORT step was built using the EXPORT wizard. A simple inspection of the code indicates what needs to be changed for a future application of the EXPORT procedure. Usually this means that the wizard itself needs to be run infrequently.

❶ The DATA= option identifies the data set that is to be converted.

❷ In this case, since we are writing to EXCEL ❸ the OUTFILE= identifies the workbook.

❹ If the sheet already exists, it will be replaced.

❺ The sheet name can also be provided.

Converting the previous generic step to one that creates a CSV file is very straightforward.

```
PROC EXPORT DATA= sashelp.class
            OUTFILE= "&path\data\class.csv"
            DBMS=csv
            REPLACE;
    RUN;
```

SEE ALSO
Raithel (2009) discusses the use of the EXPORT wizard to generate code in a sasCommunity.org tip.

1.2.2 Control through the Use of Options

There are only a few options that need to be specified. Of these most of the interesting ones are used when the data are being imported (clearly SAS already knows all about the data when it is being exported).

- DBMS= Identifies the incoming database structure (including .CSV and .TXT). Since database structures change with versions of the software, you should know the database version. Specific engines exist at the version level for some databases (especially Microsoft's EXCEL and ACCESS). The documentation discusses which engine is optimized for each software version.

- REPLACE Determines whether or not the destination target (data set, sheet, table) is replaced if it already exists.

1.2.3 PROC IMPORT Data Source Statements

These statements give you additional control over how the incoming data are to be read and interpreted. Availability of any given source statement depends on the type (DBMS=) of the incoming data.

- DATAROW First incoming row that contains data.
- GETNAMES The names of the incoming columns are available in the *first row* of the incoming data. Default column names when none are available on the incoming table are VAR1, VAR2, etc.

- GUESSINGROWS Number of rows SAS will scan before determining if an incoming column is numeric or character. This is especially important for mixed columns and early rows are all numeric. In earlier versions of SAS modifications to the SAS Registry were needed to change the number of rows used to determine the variable's type, which is fortunately no longer necessary.

- RANGE and SHEET For spreadsheets a specific sheet name, named range, or range within a sheet can be specified.
- SCANTEXT and TEXTSIZE PROC IMPORT assigns a length to a character variable by scanning the incoming column and determining the maximum.

When using GETNAMES to read column names from the source data, keep in mind that most databases use different naming conventions than SAS and may have column names that will cause problems when imported. By default illegal characters are replaced with an underscore (_) by PROC IMPORT. When you need the original column name, the system option VALIDVARNAME=ANY (see Section 14.1.2) allows a broader range of acceptable column names.

In the contrived data for the following example we have an EXCEL file containing a subject number and a response variable (SCALE). The import wizard can be used to generate a PROC

	A	B
1	subject	scale
2	200	1
3	200	2
4	200	3
5	200	4
6	200	5
7	200	6
8	200	7
9	200	8
10	200	9

IMPORT step that will read the XLS file (MAKESCALE.XLS) and create the data set WORK.SCALEDATA. This PROC IMPORT step creates two numeric variables.

```
PROC IMPORT OUT= WORK.scaledata
          DATAFILE= "C:\Temp\makescale.xls"
                        DBMS=EXCEL REPLACE;
     RANGE="MAKESCALE";
     GETNAMES=YES;  ❶
     MIXED=NO;  ❷
     SCANTEXT=YES;
     USEDATE=YES;
     SCANTIME=YES;
   RUN;
```

Notice that the form of the supporting statements is different than form most procedures. They look more like options (option=value;) than like statements. The GETNAMES= statement ❶ is used to determine the variable names from the first column.

When importing data SAS must determine if a given column is to be numeric or character. A number of clues are utilized to make this determination. SAS will scan a number of rows for each column to try to determine if all the values are numeric. If a non-numeric value is found, the column will be read as a character variable; however, only some of the rows are scanned and consequently an incorrect determination is possible. ❷ The MIXED= statement is used to specify that the values in a given column are always of a single type (numeric or character). When set to YES, the IMPORT procedure will tend to create character variables in order to accommodate mixed types.

In this contrived example it turns out that starting with subject 271 the variable SCALE starts

706	270	5
707	270	6
708	270	7
709	270	8
710	270	9
711	270	10
712	271	a
713	271	b
714	271	c
715	271	d

GUESSINGROWS statements.

taking on non-numeric values. Using the previous PROC IMPORT step does not detect this change, and creates SCALE as a numeric variable. This, of course, means that data will be lost as SCALE will be missing for the observations starting from row 712.

For PROC IMPORT to correctly read the information in SCALE it needs to be a character variable. We can encourage IMPORT to create a character variable by using the MIXED and

```
PROC IMPORT OUT= WORK.scaledata
          DATAFILE= "C:\Temp\makescale.xls"
               DBMS=excel REPLACE;
     GETNAMES=YES;
     MIXED=YES;  ❸
   RUN;
```

Changing the MIXED= value to YES ❸ is not necessarily sufficient to cause SCALE to be a character value; however, if the value of the DBMS option is changed from EXCEL to XLS ❹, the MIXED=YES statement ❺ is honored and SCALE is written as a character variable in the data set SCALEDATA.

```
PROC IMPORT OUT= WORK.scaledata
            DATAFILE= "C:\Temp\makescale.xls"
            DBMS=xls REPLACE;  ❹
     GETNAMES=YES;  ❺
     GUESSINGROWS=800;  ❻
  RUN;
```

When MIXED=YES is not practical the GUESSINGROWS= statement can sometimes be used to successfully determine the type for a variable.

GUESSINGROWS cannot be used when DBMS=EXCEL, however it can be used when DBMS=XLS. Since GUESSINGROWS ❻ changes the number of rows that are scanned prior to determining if the column should be numeric or character, its use can increase the time and resources required to read the data.

SEE ALSO
The SAS Forum thread http://communities.sas.com/thread/12743?tstart=0 has a PROC IMPORT using NAMEROW= and STARTROW= data source statements. The thread http://communities.sas.com/thread/30405?tstart=0 discusses named ranges, and it and the thread http://communities.sas.com/thread/12293?tstart=0 show the use of several data source statements.

1.2.4 Importing and Exporting CSV Files

Comma Separated Variable, CSV, files have been a standard file type for moving data between systems for many years. Fortunately we now have a number of superior tools available to us so that we do not need to resort to CSV files as often. Still they are commonly used and we need to understand how to work with them.

Both the IMPORT and EXPORT procedures can work with CSV files (this capability is a part of the Base SAS product and a SAS/ACCESS product is not required). Both do the conversion by first building a DATA step, which is then executed.

Building a DATA Step
When you use the import/export wizard to save the PROC step (see Section 1.2.1), the resulting DATA step is not saved. Fortunately you can still get to the generated DATA step by recalling the last submitted code.

1.   Execute the IMPORT/EXPORT procedure.

2.   While in the Display Manager, go to RUN→Recall Last Submit.

Once the code generated by the procedure is loaded into the editor, you can modify it for other purposes or simply learn from it. For the simple PROC EXPORT step in Section 1.2.1, the following code is generated:

```
/**************************************************************
 *    PRODUCT:   SAS
 *    VERSION:   9.1
 *    CREATOR:   External File Interface
 *    DATE:      11APR09
 *    DESC:      Generated SAS Datastep Code
 *    TEMPLATE SOURCE:  (None Specified.)
 ***********************************************************/
    data _null_;
    set  SASHELP.CLASS                              end=EFIEOD;
    %let _EFIERR_ = 0; /* set the ERROR detection macro variable */
    %let _EFIREC_ = 0;     /* clear export record count macro variable */
    file 'C:\InnovativeTechniques\data\class.csv' delimiter=','
         DSD DROPOVER lrecl=32767;
      format Name $8. ;
      format Sex $1. ;
      format Age best12. ;
      format Height best12. ;
      format Weight best12. ;
    if _n_ = 1 then         /* write column names */
     do;
       put
       'Name'
       ','
       'Sex'
       ','
       'Age'
       ','
       'Height'
       ','
       'Weight'
       ;
     end;
     do;
       EFIOUT + 1;
       put Name $ @;
       put Sex $ @;
       put Age @;
       put Height @;
       put Weight ;
       ;
     end;
    if _ERROR_ then call symputx('_EFIERR_',1);  /*set ERROR detection
                                                 macro variable*/
    if EFIEOD then call symputx('_EFIREC_',EFIOUT);
    run;
```

Headers are Not on Row 1

The ability to create column names based on information contained in the data is very beneficial. This is especially important when building a large SAS table from a CSV file with lots of columns. Unfortunately we do not always have a CSV file with the column headers in row 1. Since GETNAMES=YES assumes that the headers are in row 1 we cannot use GETNAMES=YES. Fortunately this is SAS, so there are alternatives.

The CSV file created in the PROC EXPORT step in Section 1.2.1 has been modified so that the column names are on row 3. The first few lines of the file are:

```
Class Data from SASHELP,,,,
Comma Separated rows; starting in row 3,,,,
Name,Sex,Age,Height,Weight
Alfred,M,14,69,112.5
Alice,F,13,56.5,84
Barbara,F,13,65.3,98
Carol,F,14,62.8,102.5
                        . . . . data not shown . . . .
```

The DATA step generated by PROC IMPORT (E1_2_3c_ImportWO.SAS), simplified somewhat for this example, looks something like:

```
data WORK.CLASSWO                                        ;
infile "&path\Data\classwo.csv" delimiter = ','
        MISSOVER DSD lrecl=32767 firstobs=4 ;
    informat VAR1 $8. ;
    informat VAR2 $1. ;
    informat VAR3 best32. ;
    informat VAR4 best32. ;
    informat VAR5 best32. ;
    format VAR1 $8. ;
    format VAR2 $1. ;
    format VAR3 best12. ;
    format VAR4 best12. ;
    format VAR5 best12. ;
input
            VAR1 $
            VAR2 $
            VAR3
            VAR4
            VAR5
;
run;
```

Clearly SAS has substituted VAR1, VAR2, and so on for the unknown variable names. If we knew the variable names, all we would have to do to fix the problem would be to rename the variables. The following macro reads the header row from the appropriate row in the CSV file, and uses that information to rename the columns in WORK.CLASSWO.

```
%macro rename(headrow=3, rawcsv=, dsn=);
%local lib ds i;
data _null_        ;
    infile "&path\Data\&rawcsv"
           scanover lrecl=32767 firstobs=&headrow;
    length temp $ 32767;
    input temp $;
    i=1;
    do while(scan(temp,i,',') ne ' ');
      call symputx('var'||left(put(i,4.)),scan(temp,i,','),'l');
      i+1;
    end;
    call symputx('varcnt',i-1,'l');
    stop;
    run;

    %* Determine the library and dataset name;
    %if %scan(&dsn,2,.) =  %then %do;
      %let lib=work;
      %let ds = %scan(&dsn,1,.);
    %end;
    %else %do;
      %let lib= %scan(&dsn,1,.);
      %let ds = %scan(&dsn,2,.);
    %end;

    proc datasets lib=&lib nolist;
      modify &ds;
      rename
      %do i = 1 %to &varcnt;
         var&i = &&var&i
      %end;
      ;
      quit;
%mend rename;

%rename(headrow=3, rawcsv=classwo.csv, dsn=work.classwo)
```

SEE ALSO
McGuown (2005) also discusses the code generated by PROC IMPORT when reading a CSV file. King (2011) uses arrays and hash tables to read CSV files with unknown or varying variable lists. These flexible and efficient techniques could be adapted to the type of problem described in this section.

1.2.5 Preventing the Export of Blank Sheets

PROC EXPORT does not protect us from writing a blank sheet when our exclusion criteria excludes all possible rows from a given sheet ❶. In the following example we have inadvertently asked to list all students with SEX='q'. There are none of course, and the resulting sheet is blank, except for the column headers.

```
proc export data=sashelp.class(where=(sex='q'❶))
            outfile='c:\temp\classmates.xls'
            dbms=excel2000
            replace;
   SHEET='sex: Q';
   run;
```

We can prevent this from occurring by first identifying those levels of SEX that have one or more rows. There are a number of ways to generate a list of values of a variable; however, an SQL step is ideally suited to place those values into a macro variable for further processing.

The name of the data set that is to be exported, as well as the classification variable, are passed to the macro %MAKEXLS as named parameters.

```
%macro makexls(dsn=,class=);
%local valuelist listnum i value;
proc sql noprint;
select distinct &class ❷
   into :valuelist separated by ' ' ❸
      from &dsn;
%let listnum = &sqlobs;
quit;

%* One export for each sheet;
%do i = 1 %to &listnum; ❹
    %let value = %scan(&valuelist,&i,%str( )); ❺
    proc export data=&dsn(where=(&class="&value")) ❻
                outfile="c:\temp\&dsn..xls"
                dbms=excel2000
                replace;
       SHEET="&class:&value";
       run;
%end;
%mend makexls;
%makexls(dsn=sashelp.class,class=sex)
```

❷ An SQL step is used to build a list of distinct values of the classification variable.

❸ These values are saved in the macro variable &VALUELIST.

❹ A %DO loop is used to process across the individual values, which are extracted ❺ from the list using the %SCAN function.

❻ The PROC EXPORT step then creates a sheet for the selected value. ❼

SEE ALSO

A similar example which breaks a data set into separate sheets can be found in the article "Automatically_Separating_Data_into_Excel_Sheets" on sasCommunity.org.
http://www.sascommunity.org/wiki/Automatically_Separating_Data_into_Excel_Sheets

1.2.6 Working with Named Ranges

By default PROC IMPORT and the LIBNAME statement's EXCEL engine expect EXCEL data to be arranged in a certain way (column headers, if present, on row one column A; and data starting on row two). It is not unusual, however, for the data to be delivered as part of a report or as a subset of a larger table. One solution is to manually cut and paste the data onto a blank sheet so that it conforms to the default layout. It can often be much easier to create a named range.

	A	B	C	D	E	F	G
1			Data From SASHELP.CLASS				
2				Data Columns			
3	Variable Names		Name	Sex	Age	Height	Weight
4			Alfred	M	14	69	112.5
5			Alice	F	13	56.5	84
6			Barbara	F	13	65.3	98
7			Carol	F	14	62.8	102.5
8			Henry	M	14	63.5	102.5
9			James	M	12	57.3	83
10			Jane	F	12	59.8	84.5
11			Janet	F	15	62.5	112.5
12			Jeffrey	M	13	62.5	84
13			John	M	12	59	99.5

The EXCEL spreadsheet shown here contains the SASHELP.CLASS data set (only part of which is shown here); however, titles and columns have been added. Using the defaults PROC IMPORT will not be able to successfully read this sheet.

To facilitate the use of this spreadsheet, a named range was created for the rectangle defined by C3-G22 . This range was given the name 'CLASSDATA'. This named range can now be used when reading the data from this sheet.

When reading a named range using the EXCEL engine on the LIBNAME statement, the named range (CLASSDATA) is used just as you would the sheet name ❶.

```
libname seexls excel "&path\data\E1_2_6classmates.xls";

data class;
   set seexls.classdata; ❶
   run;
libname seexls clear; ❷
```

❷ When using an engine on the LIBNAME statement be sure to clear the *libref* so that you can use the spreadsheet outside of SAS.

When using PROC IMPORT to read a named range, the RANGE= statement ❸ is used to designate the named range of interest. Since the name of the named range is unique to the workbook, a sheet name is not required.

```
proc import out=work.classdata
          datafile= "&path\data\E1_2_6classmates.xls"
          dbms=xls replace;
   getnames=yes;
   range='classdata'; ❸
   run;
```

MORE INFORMATION
The EXCEL LIBNAME engine is introduced in Section 1.1.

1.3 DATA Step INPUT Statement

The INPUT statement is loaded with options that make it extremely flexible. Since there has been a great deal written about the basic INPUT statement, only a few of the options that seem to be under used have been collected here.

SEE ALSO
An overview about reading raw data with the INPUT statement can be found in the SAS documentation at http://support.sas.com/publishing/pubcat/chaps/58369.pdf. Schreier (2001) gives a short overview of the automatic _INFILE_ variable along with other information regarding the reading of raw data.

1.3.1 Format Modifiers for Errors

Inappropriate data within an input field can cause input errors that prevent the completion of the data set. As the data are read, a great many messages can also be generated and written to the LOG. The (?) and (??) format modifiers control error handling. Both the ? and the ?? suppress error messages in the LOG; however, the ?? also resets the automatic error variable (_ERROR_) to 0. This means that while both of these operators control what is written to the LOG only the ?? will necessarily prevent the step from terminating when the maximum error count is reached.

In the following step, the third data row contains an invalid value for AGE. AGE is assigned a missing value, and because of the ?? operator no 'invalid data' message is written to the LOG.

```
data base;
input age ?? name $;
datalines;
15   Fred
14   Sally
x    John
run;
```

MORE INFORMATION
The ?? modifier is used with the INPUT function in Sections 2.3.1 and 3.6.1.

SEE ALSO
The SAS Forum thread found at http://communities.sas.com/message/48729 has an example that uses the ?? format modifier.

1.3.2 Format Modifiers for the INPUT Statement

Some of the most difficult input coding occurs when combining the use of informats with LIST style input. This style is generally required when columns are not equally spaced so informats can't be easily used, and the fields are delimited with blanks. LIST is also the least flexible input style. Informat modifiers include:

    &   allows embedded blanks in character variables

    :   allows the use of informats for non-aligned columns

    ~   allows the use of quotation marks within data fields

Because of the inherent disadvantages of LIST input (space delimited fields), when it is possible, consider requesting a specific unique delimiter. Most recently generated files of this type utilize a non-blank delimiter, which allows you to take advantage of some of the options discussed in Section 1.3.3. Unfortunately many legacy files are space delimited, and we generally do not have the luxury of either requesting a specific delimiter or editing the existing file to replace the spaces with delimiters.

There are two problems in the data being read in the following code. The three potential INPUT statements (two of the three are commented) highlight how the ampersand and colon can be used to help read the data. Notice that DOB does not start in a consistent column and the second last name has an embedded blank.

```
title '1.3.2a List Input Modifiers';
data base;
length lname $15;
input fname $ dob mmddyy10. lname $ ; ❶
*input fname $ dob :mmddyy10. lname $ ; ❷
*input fname $ dob :mmddyy10. lname $ &; ❸
datalines;
Sam 12/15/1945 Johnson
Susan   10/10/1983 Mc Callister
run;
```

Using the first INPUT statement without informat modifiers ❶ shows, that for the second data line, both the date and the last name have been read incorrectly.

```
1.3.2a List Input Modifiers

Obs    lname        fname         dob

 1     Johnson      Sam       12/15/1945
 2     83           Susan     10/10/2019
```

Assuming the second INPUT statement ❷ was commented and used, the colon modifier is placed in front of the date informat. The colon allows the format to essentially float to the appropriate starting point by using LIST input and then applying the informat once the value is found.

```
1.3.2a List Input Modifiers

Obs    lname        fname         dob

 1     Johnson      Sam       12/15/1945
 2     Mc           Susan     10/10/1983
```

The birthdays are now being read correctly; however, Susan's last name is being split because the embedded blank is being interpreted as a field delimiter. The ampersand ❸ can be used to allow embedded spaces within a field.

By placing an ampersand after the variable name (LNAME) ❸, the blank space becomes part of

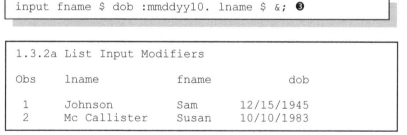

```
input fname $ dob :mmddyy10. lname $ &; ❸
```

```
1.3.2a List Input Modifiers

Obs    lname          fname         dob

 1     Johnson        Sam       12/15/1945
 2     Mc Callister   Susan     10/10/1983
```

the variable rather than a delimiter. We are now reading both the date of birth and the last name correctly.

While the ampersand is also used as a macro language trigger, this will not be a problem when it is used as an INPUT statement modifier as long as it is not immediately followed by text that could be interpreted as a macro variable name (letter or underscore). In this example the ampersand is followed by the semicolon so there will be no confusion with the macro language.

While the trailing ampersand can be helpful it can also introduce problems as well. If the data had been slightly more complex, even this solution might not have worked. The following data also contains a city name. Even though the city is not being read, the trailing & used with the last name (LNAME) causes the city name to be confused with the last name.

```
title '1.3.2b List Input Modifiers';
data base;
length lname $15;
input fname $ dob :mmddyy10. lname $ &;
format dob mmddyy10.;  ❹
datalines;
Sam 12/15/1945 Johnson    Seattle
Susan   10/10/1983 Mc Callister New York
;  ❺
run;
```

Because of the trailing & and the length of LNAME ($15) a portion of the city (New York) has been read into the LNAME for the second observation. On the first observation the last name is correct because more than one space separates Johnson and Seattle. Even with the trailing &, more than one space is still successfully seen as a field delimiter.

```
1.3.2b List Input Modifiers

Obs    lname            fname        dob

 1     Johnson          Sam      12/15/1945
 2     Mc Callister Ne  Susan    10/10/1983
```

On the second observation the city would not have been confused with the last name had there been two or more spaces between the two fields.

❹ Placing the FORMAT statement within the DATA step causes the format to be associated with the variable DOB in subsequent steps. The INFORMAT statement is only used when reading the data.

❺ The DATALINES statement causes subsequent records to be read as data up to, but not including, the first line that contains a semicolon. In the previous examples the RUN statement doubles as the end of data marker. Many programmers use a separate semicolon to perform this task. Both styles are generally considered acceptable (as long as you are using the RUN statement to end your step).

With only a single space between the last name and the city, the trailing & alone is not sufficient to help the INPUT statement distinguish between these two fields. Additional variations of this example can be found in Section 1.3.3.

MORE INFORMATION
LIST input is a form of delimited input and as such these options also apply to the examples discussed in Section 1.3.3. When the date form is not consistent one of the *any date* informats may be helpful. See Section 12.6 for more information on the use of these specialized informats.

SEE ALSO
The SAS Forum thread http://communities.sas.com/message/42690 discusses the use of list input modifiers.

1.3.3 Controlling Delimited Input

Technically LIST input is a form of delimited input, with the default delimiter being a space. This means that the modifiers shown in Section 1.3.2 apply to other forms of delimited input, including comma separated variable, CSV, files.

INFILE Statement Options

Options on the INFILE statement are used to control how the delimiters are to be interpreted.

- **DELIMITER** Specifies the character that delimits fields (other than the default - a space). This option is often abbreviated as DLM=.
- **DLMSTR** Specifies a single multiple character string as a delimiter.
- **DLMOPT** Specifies parsing options for the DLMSTR option.
- **DSD** Allows character fields that are surrounded by quotes (by setting the comma as the delimiter). Two successive delimiters are interpreted as individual delimiters, which allow missing values to be assigned appropriately. DSD also removes quotation marks from character values surrounded by quotes. If the comma is not the delimiter you will need to use the DLM= option along with the DSD option.

Some applications, such as Excel, build delimiter separated variable files with quotes surrounding the fields. This can be critical if a field's value can contain the field separator. For default list input, where a space is a delimiter, it can be very difficult to successfully read a field with an embedded blank (see Section 1.3.2 which discusses the use of trailing & to read embedded spaces). The DSD option alerts SAS to the *potential* of quoted character fields. The following example demonstrates simple comma-separated data.

```
data base;
length lname $15;
infile datalines ❶ dlm=','; ❷
*infile datalines dlm=',' dsd; ❸
input fname $ lname $ dob :mmddyy10.;
datalines;
'Sam','Johnson',12/15/1945
'Susan','Mc Callister',10/10/1983
run;
```

❶ Although the INFILE statement is often not needed when using the DATALINES, CARDS, or CARDS4 statements, it can be very useful when the options associated with the INFILE statement are needed. The *fileref* can be DATALINES or CARDS.

The DLM= option is used to specify the delimiter. In this example the field delimiter is specified as a comma ❷.

```
1.3.3a Delimited List Input Modifiers

Obs    lname             fname          dob

 1     'Johnson'         'Sam'       12/15/1945
 2     'Mc Callister'    'Susan'     10/10/1983
```

The fields containing character data have been quoted. Since we do not actually want the quote marks to be a part of the data fields, the DSD option ❸ alerts the parser to this possibility and the quotes themselves become a part of the field delimiting process.

```
infile datalines dlm=',' dsd; ❸
```

Using the DSD option results in data fields without the quotes.

```
1.3.3a Delimited List Input Modifiers

Obs    lname           fname         dob

 1     Johnson         Sam        12/15/1945
 2     Mc Callister    Susan      10/10/1983
```

On the INPUT Statement

The tilde (~) ❹can be used to modify a format, much the same way as a colon (:); however, the two modifiers are not exactly the same.

```
title '1.3.3b Delimited List Input Modifiers';
title2 'Using the ~ Format Modifier';
data base;
length lname $15;
infile datalines dlm=',' dsd;
input fname $ lname $ birthloc $~❹15. dob :mmddyy10. ;
datalines;
'Sam','Johnson', 'Fresno, CA','12/15/1945'
'Susan','Mc Callister','Seattle, WA',10/10/1983
run;
```

The tilde format modifier correctly reads the BIRTHLOC field; however, it preserves the quote marks that surround the field. Like the colon, the tilde can either precede or follow the $ for character variables. As an aside notice that for this example quote marks surround the numeric date value for the first row. The field is still processed correctly as a numeric SAS date value.

```
1.3.3b Delimited List Input Modifiers
Using the ~ Format Modifier

Obs    lname           fname      birthloc           dob

 1     Johnson         Sam        'Fresno, CA'    12/15/1945
 2     Mc Callister    Susan      'Seattle, WA'   10/10/1983
```

Replacing the tilde ❹ with a colon (:) would cause the BIRTHLOC value to be saved without the quote marks. If instead we supply a length for BIRTHLOC ❺, neither a format nor the tilde will be needed.

```
title '1.3.3c Delimited List Input Modifiers';
title2 'BIRTHLOC without a Format Modifier';
title3 'BIRTHLOC Length Specified';
data base;
length lname birthloc $15; ❺
infile datalines dlm=',' dsd;
input fname $ lname $ birthloc $ dob :mmddyy10. ;
datalines;
'Sam','Johnson', 'Fresno, CA',12/15/1945
'Susan','Mc Callister','Seattle, WA',10/10/1983
run;
```

```
1.3.3c Delimited List Input Modifiers
BIRTHLOC without a Format Modifier
BIRTHLOC Length Specified

Obs    lname           birthloc        fname           dob

 1     Johnson         Fresno, CA      Sam        12/15/1945
 2     Mc Callister    Seattle, WA     Susan      10/10/1983
```

Multiple Delimiters

It is possible to read delimited input streams that contain more than one delimiter. In the following small example two delimiters, a comma and a slash are both used to delimit the data values.

```
data imports;
infile cards dlm='/,';
input id importcode $ value;
cards;
14,1,13
25/Q9,15
6,D/20
run;
```

Obs	id	importcode	value
1	14	1	13
2	25	Q9	15
3	6	D	20

```
data imports;
retain dlmvar '/,'; ❻
infile cards dlm=dlmvar;
input id importcode $ value;
cards;
14,1,13
25/Q9,15
6,D/20
run;
```

Notice that the DLM option causes either the comma or the slash to be used as field delimiters, but not the slash comma together as a single delimiter (see the DLMSTR option below to create a single multiple character delimiter).

❻ Because the INFILE statement is executed for each observation, the value assigned to the DLM option does not necessarily need to be a constant. It can also be a variable or can be changed using IF-THEN/ELSE logic. In the simplest form this variable could be assigned in a retain statement.

```
data imports;
infile cards;
input dlmvar $1. @;
infile cards dlm=dlmvar; ❼
input @2 id importcode $ value;
cards;
,14,1,13
/25/Q9/15
~6~D~20
run;
```

❼ This simple example demonstrates a delimiter that varies by observation. Here the first character of each line is the delimiter that is to be used in that line. The delimiter is read, stored, and then used on the INFILE statement. Here we are taking advantage of the executable nature of the INFILE statement.

Using DLMSTR

Unlike the DLM option, which designates one or more delimiters, the DLMSTR option declares a specific list of characters to use as a delimiter. Here the delimiter is the sequence of characters comma-comma-slash (,,/). Notice in the LISTING of the IMPORT data set, that extra commas and slashes are read as data.

```
data imports;
infile cards dlmstr=',,/';
input  id importcode $ value;
cards;
14,,/1/,,/13
25,,/Q9,,,/15
6,,/,D,,/20
run;
```

1.3.3g Use a delimiter string			
Obs	id	importcode	value
1	14	1/	13
2	25	Q9,	15
3	6	,D	20

SEE ALSO
The following SAS Forum thread discussed the use of the DLM and DLMSTR options
http://communities.sas.com/message/46192. The use of the tilde when writing data was discussed
on the following forum thread: http://communities.sas.com/message/57848. The INFILE and
FILE statements are discussed in more detail by First (2008).

1.3.4 Reading Variable-Length Records

For most raw data files, including the small ones shown in most of the preceding examples, the
number of characters on each row has not been consistent. Inconsistent record length can cause
problems with lost data and incomplete fields. This is especially true when using the formatted
style of input. Fortunately there are several approaches to reading this kind of data successfully.

The Problem Is

Consider the following data file containing a list of patients. Unless it has been built and defined
as a fixed-length file, which is very unlikely on most operating systems including Windows, each
record has a different length. The individual records physically stop after the last non-blank
character. When we try to read the last name on the third row (Rachel's last name is unknown),
we will be attempting to read past the end of the physical record and there will almost certainly be
an error.

```
F       Linda       Maxwell
M       Ronald      Mercy
F       Rachel
M       Mat         Most
M       David       Nabers
F       Terrie      Nolan
F       June        Olsen
M       Merv        Panda
M       Mathew      Perez
M       Robert      Pope
M       Arthur      Reilly
M       Adam        Robertson
```

The following code attempts to read the above data. However, we have a couple of problems.

```
filename patlist "&path\data\patientlist.txt";
data patients;
   infile patlist;
   input @2  sex $1.
         @8  fname $10.
         @18 lname $15.;
   run;
title '1.3.4a Varying Length Records';
proc print data=patients;
   run;
```

The LOG shows two notes; there is a LOST CARD and the INPUT statement reached past the
end of the line.

```
NOTE: LOST CARD.
sex=M fname=Adam lname=   _ERROR_=1 _N_=6
NOTE: 12 records were read from the infile PATLIST.
      The minimum record length was 13.
      The maximum record length was 26.
NOTE: SAS went to a new line when INPUT statement reached past the end of a line.
```

The resulting data set has a number of data problems. Even a quick inspection of the data shows that the data fields have become confused.

```
1.3.4a Varying Length Records

Obs     sex     fname           lname

 1       F      Linda        M     Ronald
 2       F      M     Mat    M     David
 3       F      Terrie       F     June
 4       M      Merv         M     Mathew
 5       M      Robert       M     Arthur
```

Our INPUT statement requests SAS to read 15 spaces starting in column 18; however, there are never 15 columns available (the longest record is the last – Robertson – with a last name of 9 characters. To fill our request, it skips to column 1 of the next physical record to read the last name. When this happens the notes mentioned in the LOG are generated.

INFILE Statement Options (TRUNCOVER, MISSOVER)

Two INFILE statement options can be especially useful in controlling how SAS handles *short* records.

- MISSOVER   Assigns missing values to variables beyond the end of the physical record. Partial variables are set to missing.

- TRUNCOVER   Assigns missing values to variables beyond the end of the physical record. Partial variables are truncated, but not necessarily set to missing.

- FLOWOVER   SAS finishes the logical record using the next physical record. This is the default.

```
title '1.3.4b Varying Length Records';
title2 'Using TRUNCOVER';
data patients(keep=sex fname lname);
   infile patlist truncover;
   input @2  sex $1.
         @8  fname $10.
         @18 lname $15.;
   run;
```

The TRUNCOVER option is specified and as much information as possible is gathered from each record; however, SAS does not go to the next physical record to complete the observation.

```
1.3.4b Varying Length Records
Using TRUNCOVER

Obs    sex     fname        lname

 1      F      Linda        Maxwell
 2      M      Ronald       Mercy
 3      F      Rachel
 4      M      Mat          Most
 5      M      David        Nabers
 6      F      Terrie       Nolan
 7      F      June         Olsen
 8      M      Merv         Panda
 9      M      Mathew       Perez
10      M      Robert       Pope
11      M      Arthur       Reilly
12      M      Adam         Robertson
```

Generally the TRUNCOVER option is easier to apply than the $VARYING informat, and there is no penalty for including a TRUNCOVER option on the INFILE statement even when you think that you will not need it.

By including the TRUNCOVER option on the INFILE statement, we have now correctly read the data without skipping a record, while correctly assigning a missing value to Rachel's last name.

Using the $VARYING Informat

The $VARYING informat was created to be used with variable-length records. This informat allows us to determine the record length and then use that length for calculating how many columns to read. As a general rule, you should first attempt to use the more flexible and easier to apply TRUNCOVER option on the INFILE statement, before attempting to use the $VARYING informat.

Unlike other informats $VARYING utilizes a secondary value to determine how many bytes to read. Very often this value depends on the overall length of the record. The record length can be retrieved with the LENGTH= option ❶ and a portion of the overall record length is used to read the field with a varying width.

The classic use of the $VARYING informat is shown in the following example, where the last field on the record has an inconsistent width from record to record. This is also the type of data read for which the TRUNCOVER option was designed.

```
title2 'Using the $VARYING Informat';
data patients(keep=sex fname lname);
   infile patlist length=len ❶;
   input @; ❷
   namewidth = len-17; ❸
   input @2  sex $1.
         @8  fname $10.
         @18 lname $varying15. namewidth ❹;
   run;
```

❶ The LENGTH= option on the INFILE statement specifies a temporary variable (LEN) which holds the length of the current record.

```
1.3.4c Varying Length Records
Using the $VARYING Informat

Obs    sex     fname        lname

 1      F      Linda        Maxwell
 2      M      Ronald       Mercy
 3      F      M      Mat  ❺
 4      M      David        Nabers
 5      F      Terrie       Nolan
 6      F      June         Olsen
 7      M      Merv         Panda
 8      M      Mathew       Perez
 9      M      Robert       Pope
10      M      Arthur       Reilly
11      M      Adam         Robertson
```

❷ An INPUT statement with just a trailing @ is used to load the record into the input buffer. Here the length is determined and loaded into the variable LEN. The trailing @ holds the record so that it can be read again.

❸ The width of the last name is calculated (total length less the number of characters to the left of the name). The variable NAMEWIDTH holds this value for use by the $VARYING informat.

❹ The width of the last name field for this particular record follows the $VARYING15. informat. Here the width used with the $VARYING informat is the widest possible value for LNAME and also establishes the variable's length.

Inspection of the resulting data shows that we are now reading the correct last name; however, we still have a data issue ❺ for the third and fourth input lines. Since the third data line has *no* last name, the $VARYING informat jumps to the next data record. The TRUNCOVER option on the INFILE statement discussed above addresses this issue successfully.

In fact for the third record the variable FNAME, which uses a $10 informat, reaches beyond the end of the record and causes the data to be misread.

```
data patients(keep=sex fname lname namewidth ❿);
   length sex $1 fname $10 lname $15; ❻
   infile patlist length=len;
   input @;
   if len lt 8 then do; ❼
      input @2  sex $;
   end;
   else if len le 17 then do; ❽
      namewidth = len-7;
      input @2  sex $
            @8  fname $varying. namewidth;
   end;
   else do; ❾
      namewidth = len-17;
      input @2  sex $
            @8  fname $
            @18 lname $varying. namewidth; ❿
   end;
   run;
```

❻ Using a LENGTH statement to declare the variable lengths avoids the need to add a width to the informats.

❼ Neither a first or last name is included. This code assumes that a gender (SEX) is always present.

❽ The record is too short to have a last name, but must contain a first name of at least one letter.

❾ The last name must have at least one letter.

❿ The variable NAMEWIDTH will contain the width of the rightmost variable. The value of this variable is generally of no interest, but it is kept here so that you can see its values change for each observation.

It is easy to see that the $VARYING informat is more difficult to use than either the TRUNCOVER or the MISSOVER options. However, the $VARYING informat can still be helpful. In the following simplified example suggested by John King there is no delimiter and yet the columns are not of constant width. To make things more interesting the variable with the inconsistent width is not on the end of the input string.

```
data datacodes;
   length dataname $15;
   input @1 width 2.
         dataname $varying. width
         datacode :2.;
   datalines;
5 Demog43
2 AE65
13lab_chemistry32
   run;
```

The first field (WIDTH) contains the number of characters in the second field (DATANAME). This value is used with the $VARYING informat to correctly read the data set name while not reading past the name and into the next field (DATACODE).

SEE ALSO

Cates (2001) discusses the differences between MISSOVER and TRUNCOVER. A good comparison of these options can also be found in the SAS documentation http://support.sas.com/documentation/cdl/en/basess/58133/HTML/default/viewer.htm#a00264581 2.htm .

SAS Technical Support example #37763 uses the $VARYING. informat to write a zero-length string in a REPORT example http://support.sas.com/kb/37/763.html.

1.4 Writing Delimited Files

Most modern database systems utilize metadata to make the data itself more useful. When transferring data to and from Excel, for instance, SAS can take advantage of this metadata. Flat files do not have the advantage of metadata and consequently more information must be transferred through the program itself. For this reason delimited data files should not be our first choice for transferring information from one database system to another. That said we do not always have that choice. We saw in Section 1.3 a number of techniques for reading delimited data.

Since SAS already knows all about a given SAS data set (it has access to the metadata), it is much more straightforward to write delimited files.

MORE INFORMATION

Much of the discussion on reading delimited data also applies when writing delimited data (see Section 1.3).

1.4.1 Using the DATA Step with the DLM= Option

When reading delimited data using the DATA step, the INFILE statement is used to specify a number of controlling options. Writing the delimited file is similar; however, the FILE statement is used. Many of the same options that appear on the INFILE statement can also be used on the FILE statement. These include:

- DLM=
- DLMSTR=
- DSD

While the DSD option by default implies a comma as the delimiter, there are differences between the uses of these two options. The DSD option will cause values which contain an embedded delimiter character to be double quoted. The DSD option also causes missing values to appear as two consecutive delimiters, while the DLM= alone writes the missing as either a period or a blank.

In the following example three columns from the ADVRPT.DEMOG data set are to be written to the comma separated variable (CSV) file. The FILE statement is used to specify the delimiter

```
filename outspot "&path\data\E1_4_1demog.csv";

data _null_;
   set advrpt.demog(keep=fname lname dob);
   file outspot dlm=',' ❶
                dsd;  ❷
   if _n_=1 then put 'FName,LName,DOB'; ❸
   put fname lname dob mmddyy10.; ❹
   run;
```

using the DLM= ❶ option. Just in case one of the fields contains the delimiter (a comma in this example), the Delimiter Sensitive Data option, DSD ❷, is also included. Using the DSD option is a good general practice.

When you also want the first row to contain the column names, a conditional PUT ❸ statement can be used to write them. The data itself is also written using a PUT statement ❹.

MORE INFORMATION
The example in Section 1.4.4 shows how to insert the header row without explicitly naming the variables.

All the variables on the PDV can be written by using the statement `PUT (_ALL_)(:);` (see Section 1.4.5).

1.4.2 PROC EXPORT

Although a bit less flexible than the DATA step, the EXPORT procedure is probably easier to use for simple cases. However, it has some characteristics that make it 'not so easy' when the data are slightly less straightforward.

The EXPORT step shown here is intended to mimic the output file generated by the DATA step in Section 1.4.1; however, it is not successful and we need to understand why.

```
filename outspot "&path\data\E1_4_2demog.csv";

proc export data=advrpt.demog(keep=fname lname dob) ❶
            outfile=outspot ❷
            dbms=csv ❸ replace;
   delimiter=','; ❹
   run;
```

❶Three variables have been selected from ADVRPT.DEMOG and EXPORT is used to create a CSV file.

❷ The OUTFILE= option points to the *fileref* associated with the file to be created. Notice that the extension of the file's name matches the selected database type ❸.

❸ The DBMS= option is used to declare the type for the generated file. In this case a CSV file. Other choices include TAB and DLM (and others if one of the SAS/ACCESS products has been licensed).

❹ The DELIMITER= option is used to designate the delimiter. It is not necessary in this example as the default delimiter for a CSV file is a comma. This option is most commonly used when DBMS is set to DLM and something other than a space, the default delimiter for DBMS=DLM, is desired as the delimiter.

A quick inspection of the file generated by the PROC EXPORT step shows that **all** the variables from the ADVRPT.DEMOG data set have been included in the file; however, only those variables in the KEEP= data set option have values. Data set options ❶ cannot be used with the incoming data set when EXPORT creates delimited data. Either you will need to write all the variables or the appropriate variables need to be selected in a previous step (see Section 1.4.3). This behavior is an artifact of the way that PROC EXPORT writes the delimited file. PROC EXPORT writes a DATA step and builds the variable list from the metadata, ignoring the data set options. When the data are actually read into the constructed DATA step; however, the KEEP= data set option is applied, thus resulting in the missing values.

```
subject,clinnum,lname,fname,ssn,sex,dob,death,race,edu,wt,ht,symp,death2
,,Adams,Mary,,,,12AUG51,,,,,,,
,,Adamson,Joan,,,,,,,,,,,
,,Alexander,Mark,,,15JAN30,,,,,,,
,,Antler,Peter,,,15JAN34,,,,,,,
,,Atwood,Teddy,,,14FEB50,,,,,,,
        . . . . data not shown . . . .
```

1.4.3 Using the %DS2CSV Macro

The DS2CSV.SAS file is a macro that ships with Base SAS, and is accessed through the SAS autocall facility. Its original authorship predates many of the current capabilities discussed elsewhere in Section 1.4. The macro call is fairly straightforward; however, the macro code itself utilizes SCL functions and lists and is outside the scope of this book.

The macro is controlled through the use of a series of named or keyword parameters. Only a small subset of this list of parameters is shown here.

```
data part;
    set advrpt.demog(keep=fname lname dob);  ❶
    run;

%ds2csv(data=part,  ❷
        runmode=b,  ❸
        labels=n,  ❹
        csvfile=&path\data\E1_4_3demog.csv)  ❺
```

❶ As was the case with PROC EXPORT in Section 1.4.2, if you need to eliminate observations or columns a separate step is required.

❷ The data set to be processed is passed to the macro.

❸ The macro can be executed on a server by using RUNMODE=Y.

❹ By default the variable labels are used in the column header. Generally you will want the column names to be passed to the CSV file. This is done using the LABELS= parameter.

❺ The CSVFILE= parameter is used to name the CSV file. This parameter does not accept a *fileref.*

SEE ALSO

A search of SAS documentation for the macro name, DS2CSV, will surface the documentation for this macro.

1.4.4 Using ODS and the CSV Destination

The Output Delivery System, ODS, and the CVS tagset can be used to generate CSV files. When you want to create a CSV file of the data, complete with column headers, the CSV destination can be used in conjunction with PROC PRINT.

```
ods csv file="&path\data\E1_4_4demog.csv)" ❶
        options(doc='Help' ❷
              delimiter=";");❸
proc print data=advrpt.demog
           noobs;❹
   var fname lname dob; ❺
   run;
ods csv close; ❻
```

❶ The new delimited file is specified using the FILE= option.

❷ TAGSET options are specified in the OPTIONS list. A list of available options can be seen using the DOC='HELP' option.

```
"fname";"lname";"dob"
"Mary";"Adams";"12AUG51"
"Joan";"Adamson";"."
"Mark";"Alexander";"15JAN30"
"Peter";"Antler";"15JAN34"
"Teddy";"Atwood";"14FEB50"
       .... data not shown ....
```

❸ The delimiter can be changed from a comma with the DELIMITER= option.

❹ The OBS column is removed using the NOOBS option.

❺ Select variables and variable order using the VAR statement in the PROC PRINT step.

❻ As always be sure to close the destination.

MORE INFORMATION

Chapter 11 discusses a number of aspects of the Output Delivery System.

SEE ALSO

There have been several SAS forum postings on the CSV destination.
http://communities.sas.com/message/29026#29026
http://communities.sas.com/message/19459

1.4.5 Inserting the Separator Manually

When using the DATA step to create the delimited file, the techniques shown in Section 1.4.1 will generally be sufficient. However you may occasionally require more control, or you may want to take control of the delimiter more directly.

One suggestion that has been seen in the literature uses the PUT statement to insert the delimiter.

```
data _null_;
   set advrpt.demog(keep=fname lname dob);
   file csv_a;
   if _n_=1 then put 'FName,LName,DOB';
   put (_all_)(','); ❶
   run;
```

Here the _ALL_ variable list shortcut has been used to specify that all variables are to be written. This shortcut list requires a corresponding text, format, or other modifier for each of the variables. In this case we have specified a comma, e.g., (',') ❶.

This approach will work to some extent, but it is not perfect in that a comma precedes each line of data.

The DSD option on the FILE statement ❷ implies a comma as the delimiter, although the DLM= option can be used to specify a different option (see Section 1.4.1). The _ALL_ list abbreviation can still be used; however, a neutral modifier must also be selected. Either the colon (:) or the question mark (?) ❸, will serve the purpose.

```
data _null_;
   set advrpt.demog(keep=fname lname dob);
   file csv_b dsd;❷
   if _n_=1 then put 'FName,LName,DOB';
   put (_all_)(?) ❸;
   run;
```

Because the DSD option has been used, an approach such as this one will also work when one or more of the variables contain an embedded delimiter.

1.5 SQL Pass-Through

SQL pass through allows the user to literally pass instructions through a SAS SQL step to the server of another database. Passing code or SQL instructions out of the SQL step to the server can have a number of advantages, most notably significant efficiency gains.

1.5.1 Adding a Pass-Through to Your SQL Step

The pass-through requires three elements to be successful:

- A connection must be formed to the server/database. ❶
- Code must be passed to the server/database. ❷
- The connection must be closed. ❸

These three elements will be formulated as statements (❶ CONNECT and ❸ DISCONNECT) or as a clause within the FROM CONNECTION phrase ❷.

```
proc sql noprint;
   connect to odbc (dsn=clindat uid=Susie pwd=pigtails); ❶

   create table stuff as select * from connection to odbc (
      select * from q.org ❷
         for fetch only
      );

   disconnect from odbc; ❸
   quit;
```

The connection that is established using the CONNECT statement ❶ and is then referred to in the FROM CONNECTION TO phrase.

Notice that the SQL code that is being passed to the database, not a SAS database, ❷ is within the parentheses. This code must be appropriate for the receiving database. In this case the pass through is to a DB2 table via an ODBC connection.

There are a number of types of connections and while ODBC connections, such as the one established in this example, are almost universally available in the Microsoft/Windows world, they are typically slower than SAS/ACCESS connections.

1.5.2 Pass-Through Efficiencies

When using PROC SQL to create and pass database-specific code to a database other than SAS, such as Oracle or DB2, it is important that you be careful with how you program the particular problem. Depending on how it is coded SQL can be very efficient or very inefficient, and this can be an even more important issue when you use pass-through techniques to create a data subset.

Passing information back from the server is usually slower than processing on the server. Design the pass-through to minimize the amount of returned information. Generally the primary database will be stored at a location with the maximum processing power. Take advantage of that power. At the very least minimizing the amount of information that has to be transferred back to you will help preserve your bandwidth.

In SQL, data sets are processed in memory. This means that large data set joins should be performed where available memory is maximized. When a join becomes memory bound subsetting the data before the join can be helpful. Know and understand your database and OS, some WHERE statements form clauses that are applied to the result of the join rather than to the incoming data set.

Even when you do not intend to write to the primary database that is being accessed using an SQL pass-through, extra process checking may be involved against that data table. These checks, which can be costly, can potentially be eliminated by designating the incoming data table as read-only. This can be accomplished in a number of ways. In DB2 using the clause `for fetch only` in the code that is being passed to the database eliminates write checks against the incoming table. In the DB2 pass-through example in Section 1.5.1 we only want to extract or fetch data. We speed up the process by letting the database know that we will not be writing any data – only fetching it.

MORE INFORMATION
An SQL step using pass-through code can be found in Section 5.4.2.

1.6 Reading and Writing to XML

Extensible Markup Language, XML, has a hierarchical structure while SAS data sets are record or observation based. Because XML is fast becoming a universal data exchange format, it is incumbent for the SAS programmer to have a working knowledge of how to move information from SAS to XML and from XML to SAS.

The XML engine (Section 1.6.2) was first introduced in Version 8 of SAS. Later the ODS XML destination was added; however, currently the functionality of the XML destination has been built into the ODS MARKUP destination (see Section 1.6.1).

Because XML is text based and each row contains its own metadata, the files themselves can be quite large.

SEE ALSO
A very nice overview of XML and its relationship to SAS can be found in (Pratter, 2008). Other introductory discussions on the relationship of XML to SAS include: Chapal (2003), Palmer (2003 and 2004), and in the SAS documentation on "XML Engine with DATA Step or PROC COPY".

1.6.1 Using ODS

You can create an XML file using the ODS MARKUP destination. The file can contain procedure output in XML form, and this XML file can then be passed to another application that utilizes / reads XML. By default the MARKUP destination creates a XML file.

```
title1 '1.6.1 Using ODS MARKUP';
ods markup file="&path\data\E1_6_1Names.xml"; ❶

* create a xml file of the report; ❷
proc print data=advrpt.demog;
   var lname fname sex dob;
   run;
ods markup close; ❸
```

❶ The FILE= option is used to designate the name of the file to be created. Notice the use of the XML extension.

❷ The procedure must be within the ODS 'sandwich.'

❸ The destination must be closed before the file ❶ can be used outside of SAS.

MORE INFORMATION
If the application that you are planning to use with the XML file is Excel, the EXCELXP tagset is a superior choice (see Section 11.2).

SEE ALSO
The LinkedIn thread
http://www.linkedin.com/groupItem?view=&srchtype=discussedNews&gid=70702&item=74453221&type=member&trk=eml-anet_dig-b_pd-ttl-cn&ut=34c4-P0gjofkY1
follows a discussion of the generation of XML using ODS.

1.6.2 Using the XML Engine

The use of the XML engine is a process similar to the one shown in Section 1.6.1, and can be used to write to the XML format. XML is a markup language and XML code is stored in a text file that can be both read and written by SAS. As in the example above, an engine is used on the LIBNAME statement to establish the link with SAS that performs the conversion. A *fileref* is established and it is used in the LIBNAME statement.

```
filename xmllst "&path\data\E1_6_2list.xml";

libname toxml xml xmlfileref=xmllst; ❶

* create a xml file (E1_6_2list.xml);
data toxml.patlist; ❷
   set advrpt.demog(keep=lname fname sex dob);
   run;

* convert xml to sas;
data fromxml;
   set toxml.patlist; ❷
   run;
```

❶ On the LIBNAME statement that has the XML engine, the XMLFILEREF= option is used to point to the *fileref* either containing the XML file or, as is the case in this example, the file that is to be written.

```
<?xml version="1.0" encoding="windows-1252" ?>
- <TABLE>
- <PATLIST> ❸
  <lname>Adams</lname>
  <fname>Mary</fname>
  <sex>F</sex>
  <dob>1951-08-12</dob>
  </PATLIST>
- <PATLIST>
  <lname>Adamson</lname>
  <fname>Joan</fname>
  <sex>F</sex>
  <dob missing="." />
  </PATLIST>
  . . . . the remaining observations are not shown . . . .
```

❷ The *libref* TOXML can be used to both read and write the XML file. The name of the data set (PATLIST) is recorded as a part of the XML file ❸. This means that multiple SAS data sets can be written to the same XML file.

The selected variables are written to the XML file. Notice that the variables are named on each line and that the date has been re-coded into a YYYY-MM-DD form, and that the missing DOB for 'Joan Adamson' has been written using the `missing=` notation.

SEE ALSO
Hemedinger and Slaughter (2011) briefly describe the use of XML and the XML Mapper.

Chapter 2

Working with Your Data

In SAS the data set is central to most of our analyses and reporting. This means that it is crucial that we have the power to know all sorts of things about our data. The power that we need comes from a multitude of SAS tools and techniques. This chapter is a fairly random collection of these tools and techniques that can help us accomplish our goals of working with our data sets.

2.1 Data Set Options

Data set options can be used to modify how a data set is either read or written. There are over three dozen of these options, and while you will generally only make use of a hand full of them, you should have a good idea of their scope.

To use these option(s) place them in parentheses immediately following the name of the data set to which they are to be applied. While data set options can be used virtually anytime the data set is named, some of the options are situation dependent, which means that you will have to understand what an option does before applying it. For instance, options that control how a data set is to be read would not be used on a DATA statement.

In the following example the KEEP data set option is applied to the data set being used by PROC SORT.

```
proc sort data=advrpt.demog(keep= lname fname ssn)
              out=namesonly;
by lname fname;
run;
```

Regardless of how many variables are in ADVRPT.DEMOG the SORT procedure will only have to deal with the three variables named in the KEEP= option. For a SORT procedure this can substantially speed up the processing.

Data set options apply only to the data set to which they are associated. Here all the variables from YEAR2006 will be included. Only the variables from YEAR2007 will be limited by the KEEP= data set option.

```
data yr6_7;
   set year2006
       year2007(keep=subject visit labdt);
   run;
```

```
data yr6_7;
   set year:(keep=subject visit labdt);
   run;
```

This is not the case when a data set list abbreviation is used. Only these three variables will be read from the incoming data sets (WORK.YEARxxxxxx). The variable list applies to each incoming data set; consequently, an error is generated if a variable is not present on one or more of the incoming data sets.

For a number of the data set options, similar functionality can be obtained through the use of DATA step statements or through the use of system options. System options are the most general (they would apply to all data sets); DATA step statements will only apply to data sets within that DATA step, and data set options are the most specific as they apply only to a specific data set.

MORE INFORMATION

Additional information on the use of data set options with PROC SORT can be found in Section 4.2. The INDSNAME= option, along with the IN= data set options, are discussed in Section 3.8.2.

2.1.1 REPLACE and REPEMPTY

Since it is possible to create an empty (zero observation) data set, we may want to control whether

```
data advrpt.class;
   set sashelp.class;
   where age > 25;
   run;
```

or not the new table will replace an existing table of the same name. For this example, assume that the data set ADVRPT.CLASS already exists. Because there are no ages > 25 in the SASHELP.CLASS data set, the WHERE clause in this DATA step will always be false, no observations will be written, and the data set ADVRPT.CLASS will be replaced with an empty data set.

The REPLACE and REPEMPTY data set options allow us to control the conditions under which the data set is replaced.

- REPLACE   REPLACE=NO prevents the replacement of a *permanent* data set. This data set option overrides the system option of the same name. REPLACE=YES is the default.

- REPEMPTY   Determines whether or not an empty data set can overwrite an existing data set.

For full protection these two options are usually used together. Normally we want to be able to replace permanent data sets, unless the new version is empty. In the following DATA step there are no observations where AGE is greater than 25, so zero observations will be returned.

```
data advrpt.class(replace=yes repempty=no);
   set sashelp.class;
   where age > 25;
   run;
```

However, since REPEMPTY=NO, the data set ADVRPT.CLASS will not be replaced.

Traditionally the issue of overwriting a data set which has observations with an empty one has been especially problematic when the semicolon has been left off of the DATA statement. In the following DATA step, because of the missing semicolon ❶ the SET statement is masked and *three* empty data sets are created (ADVRPT.VERYIMPORTANT, WORK.SET, and SASHELP.CLASS).

```
options DATASTMTCHK=NONE;  ❷
data advrpt.VeryImportant  ❶
   set sashelp.class;
   run;
```

❶ The missing semicolon causes SAS to see the SET statement as part of the DATA statements. The result is that there is no incoming data set; consequently, the created data sets will have no variables or observations.

❷ The DATASTMTCHK system option protects us from this very problem by not allowing data sets to be created with names, such as SET and MERGE. Setting DATASTMTCHK to NONE removes this protection.

```
options DATASTMTCHK=NONE;  ❷
data advrpt.VeryImportant(replace=yes repempty=no)  ❸
   set sashelp.class;
   run;
```

❸ The REPEMPTY=NO option will protect our very important data set, but unfortunately not the SASHELP.CLASS data set.

Without a compelling reason to do so, it is my opinion that the value of DATASTMTCHK should not be set to NONE. If you must change the option, use the REPLACE and REPEMPTY data set options ❸ to provide some protection.

2.1.2 Password Protection

Data sets can be both encrypted and password protected. Password and encryption data set options include:

▪	ALTER	Password to alter the data set
▪	ENCRYPT	Encrypt the data set
▪	PW	Specify the password
▪	PWREQ	Password request window
▪	READ	Password to read the data set
▪	WRITE	Password to write to the data set

The following DATA step creates a data set that is not only encrypted, but requires different passwords for both reading and writing.

```
data advrpt.pword(encrypt=yes   pwreq=yes
                  read=readpwd write=writepwd);
   DB='DEApp'; UID='MaryJ'; pwd='12z3'; output;
   DB='p127';  UID='Mary';  pwd='z123'; output;
   run;
proc print data=advrpt.pword;
   run;
```

Before PROC PRINT can display the protected data set, the following dialogue box will appear requesting the READ password.

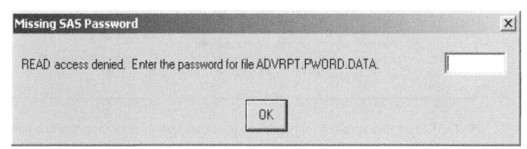

While these password protections can be useful within SAS, the protected files are still vulnerable to deletion or manipulation using tools outside of SAS. Including the ENCRYPT option adds another layer of protection from tools other than SAS.

MORE INFORMATION

More about password protection including the use of a data set containing passwords, such as the one used in this section, is used in Section 5.4.2.

2.1.3 KEEP, DROP, and RENAME Options

When using the KEEP, DROP, or RENAME in a DATA step, you can choose between using data set options or DATA step statements. When multiple data sets are either created or read, DATA step statements apply to all created data sets, while data set options can be applied to specific data sets. As a general rule, when you have a choice, data set options are preferred over statements, as the data set options give you more control and the code is generally clearer.

The KEEP statement is the functional opposite of the DROP statement. The following discussion for the KEEP statement and the KEEP= data set option could just as easily been applied to the DROP statement and DROP= data set option. Here I am showing the KEEP statement and KEEP= data set option because I have an admitted bias against the DROP statement and DROP= data set option. The DROP statement and DROP= data set option work fine, both do exactly what they are supposed to do, and both can save the programmer typing when the list of variables to keep is long, and the list of variables to drop is short. However, the KEEP conveys more information to the programmer by documenting which variables the programmer *does* need to continue to think about.

The following examples highlight the differences between the KEEP statement and the KEEP= data set option. The KEEP statement below ❶ is only applied to the new data set (WORK.LABS) and in no way affects the Program Data Vector or what variables will be read from the incoming data set (ADVRPT.LAB_CHEMISTRY).

```
data labs;
   set advrpt.lab_chemistry;
   keep subject visit labdt;  ❶
   if sodium>'142';  ❷
   run;
```

❶ The KEEP statement variable list is applied to the new outgoing data set.

❷ The IF statement is executed after the entire observation is read and loaded into the PDV.

Using the KEEP statement ❶ is exactly the same as specifying the KEEP= option on the data set in the DATA statement ❸. The KEEP= option on the SET statement ❹, however, is applied before the PDV is built. Only those variables listed will be read from the incoming data set and included on the PDV.

```
data labs(keep=subject visit labdt);  ❸
   set advrpt.lab_chemistry(keep=subject visit labdt sodium  ❹
                            where=(sodium>'142'));  ❺
   run;
```

❸ The KEEP= data set option only impacts which variables will be written to the new data set.

❹ On the SET statement the KEEP= data set option is applied to the incoming data set; therefore, this variable list affects which variables will appear on the PDV. Because SODIUM is used in the WHERE= clause ❺ in this example, the KEEP= data set option must also include SODIUM, even though it is not written to the new data set.

❺ The WHERE= filter is specified as a data set option and is applied before observations are read.

The RENAME option allows you to change the name of a variable either as it is read or as it is written. Like the RENAME statement, the syntax is of the form *oldname=newname*. Placing the RENAME= option on the SET statement ❻ causes the name to be changed before it is written to the PDV.

```
data labs(keep=subject visit labdate)❼;
   set advrpt.lab_chemistry(rename=(labdt=labdate))❻;
   if sodium>'142';
   run;
```

❻ The original name (LABDT) is changed to the new name (LABDATE).

❼ The new name would be used in any programming statements and it also appears on the KEEP= variable list.

When the RENAME= and KEEP= options are both used on the same data set, it is important to understand which is applied first. In the following DATA step the incoming data set has both a RENAME= and KEEP= option.

```
data labs(keep=subject visit labdate)❼;
   set advrpt.lab_chemistry(rename=(labdt=labdate)❻
                            keep=subject visit labdt❽ sodium
                            );
   if sodium>'142';
   run;
```

❽ Since the KEEP= option is applied before the RENAME= option, the original variable name is used on the incoming KEEP= variable list.

MORE INFORMATION
When combined with PROC SORT these data set options can have a huge impact on processing efficiencies, see Section 4.2.

2.1.4 Observation Control Using FIRSTOBS and OBS Data Set Options

The data set options FIRSTOBS and OBS can be used separately or in conjunction with one another to limit which observations are read and/or written. Their operation is similar to the system options with corresponding names; however, as data set options their application can be more refined.

- FIRSTOBS   specifies the number of the first observation to be read
- OBS   specifies the last observation that is to be read (when FIRSTOBS is not also used, this corresponds to the number of observations that will be read).

In the following PROC PRINT step we have requested that only the first 6 observations be printed ❶.

```
proc print data=sashelp.class(obs=6);  ❶
   run;
```

```
2.1.4a obs=6

Obs       Name        Sex      Age     Height      Weight

 1        Alfred       M        14      69.0        112.5
 2        Alice        F        13      56.5         84.0
 3        Barbara      F        13      65.3         98.0
 4        Carol        F        14      62.8        102.5
 5        Henry        M        14      63.5        102.5
 6        James        M        12      57.3         83.0
```

The FIRSTOBS= option ❷ identifies the first observation that is to be printed.

```
title1 '2.1.4b firstobs=4';
proc print data=sashelp.class(firstobs=4); ❷
   run;
```

```
2.1.4b firstobs=4

Obs       Name        Sex      Age     Height      Weight
 4        Carol        F        14      62.8        102.5
 5        Henry        M        14      63.5        102.5
 6        James        M        12      57.3         83.0
 7        Jane         F        12      59.8         84.5
 8        Janet        F        15      62.5        112.5
 9        Jeffrey      M        13      62.5         84.0
10        John         M        12      59.0         99.5
11        Joyce        F        11      51.3         50.5
12        Judy         F        14      64.3         90.0
13        Louise       F        12      56.3         77.0
14        Mary         F        15      66.5        112.0
15        Philip       M        16      72.0        150.0
16        Robert       M        12      64.8        128.0
17        Ronald       M        15      67.0        133.0
18        Thomas       M        11      57.5         85.0
19        William      M        15      66.5        112.0
```

When these two options are used together ❸ they work independently of each other. It is important to remember that the OBS= option counts from observation 1 regardless of the value of FIRSTOBS. This is demonstrated in the following example.

```
title1 '2.1.4c firstobs=4 obs=6';
proc print data=sashelp.class(firstobs=4 obs=6); ❸
   run;
```

Only the first 6 observations are available to be printed (OBS=6); however, the first to be printed is the fourth observation (FIRSTOBS=4). As a result only three observations are actually printed.

```
2.1.4c firstobs=4 obs=6

Obs       Name        Sex      Age     Height      Weight

 4        Carol        F        14      62.8        102.5
 5        Henry        M        14      63.5        102.5
 6        James        M        12      57.3         83.0
```

Adding a WHERE clause ❹ changes the counting process. The WHERE clause is applied first; consequently, the counts and the selection of the observations is based on the subset of observations.

```
title1 '2.1.4d firstobs=4 obs=6 where (sex=m)';
proc print data=sashelp.class(firstobs=4 obs=6
                    where=(sex='M')); ❹
   run;
```

The resulting LISTING shows that only the fourth, fifth, and sixth male patients have been displayed.

```
2.1.4d firstobs=4 obs=6 where (sex=m)

Obs    Name     Sex    Age    Height    Weight

  9    Jeffrey   M      13     62.5      84.0
 10    John      M      12     59.0      99.5
 15    Philip    M      16     72.0     150.0
```

2.2 Evaluating Expressions

A SAS expression contains a combination of operators, constants, functions, and variables. Expressions are used in a number of ways, both within the DATA step and in the PROC step. Very often when we encounter the term '*expression*', we most commonly think of comparison expressions; however, they are actually much more general and can also appear in other statements such as assignment statements. It is important to remember that, regardless of their use, the evaluation of an expression will follow a series of steps or rules. Understanding these rules can lead us to a more expansive use of expressions.

2.2.1 Operator Hierarchy

Operators are like verbs in the expression. They tell SAS what to do with things like constants and variables. In order to avoid confusion and ambiguity, operators are assigned a hierarchy or order in which they are applied. The hierarchy is formed by seven groups of operators, and within a group, operators of equal rank are applied from left to right (except Group 1 which is applied right to left).

At a simple level, we need to understand why the expression (5+6*2) is equal to 17 and not 22. But as we encounter expressions in non-standard form, such as some of those in Sections 2.2.2 and 2.2.3, we need to have a solid understanding of this hierarchy, if we are to understand why the expressions evaluate the way that they do.

Group	Operators
Parentheses	Operations within parentheses are performed first
Group 1 (performed right to left)	Exponentiation (**) Prefix operators, such as, positive (+), negative (-), and negation Minimum (MIN,><) and maximum (MAX, <>)
Group 2	Multiplication (*) and division (/)
Group 3	Addition (+) and subtraction (-)
Group 4	Concatenation (‖)
Group 5	Comparisons such as equal (=) and less than (<)
Group 6	AND - Boolean comparison (&)
Group 7	OR - Boolean comparison (|)

Since any of these operators can appear in any expression, whether in an assignment statement or an IF statement, we need to expand our perception of what an expression *should* contain. The

```
season = 1*(1 le month(dob) le 3)
       + 2*(4 le month(dob) le 6)
       + 3*(7 le month(dob) le 9)
       + 4*(10 le month(dob) le 12);
```

assignment statement shown here creates the variable SEASON which can contain one of the numeric values 0 thru 4 depending on the month of the date of birth. The 'less-than-or-equal-to' comparison operators (Group 5) return a zero or one which is multiplied against the constants. The comparison operators are just another form of expression operators and are perfectly suited to assignment statements as well as to logical expressions.

Although it is important to understand the logic of the previous assignment statement, it could

```
season= ceil(month(dob)/3);
```

have been more simply written using the CEIL function. However the two statements are not equivalent. When DOB is missing, the first assignment statement returns a zero, while this one returns a missing value.

MORE INFORMATION
There are some additional comparison operators that are unique to the WHERE statement (see Section 2.7.1). The MIN and MAX operators are further discussed in Section 2.2.5.

2.2.2 Using the Colon as a Comparison Modifier

The colon (:) can be used as an operator modifier when character values are being compared.

```
data Mar;
   set advrpt.demog (keep=lname fname);
   if lname =: 'Mar';
   run;
```

The colon permits the comparison of two strings of unequal length, and the colon follows the comparison operator of choice. In the example to the left, the subsetting IF statement will select all observations which have a LNAME starting with 'Mar'. Since it does not matter whether the value with the smaller length is on the left or right side of the equal sign, a last name of 'Ma' would also be selected. The lengths of the values on both sides are determined and the smaller length is selected and applied to both sides. An IF statement with the two values reversed would produce the same result.

```
if 'Mar' =: lname;
```

The colon comparison operator modifier can also be used with the IN operator. In this example

```
where lname in:('Me', 'Mar', 'Adams');
```

the WHERE statement will select a variety of last names that start with the indicated strings. Notice that the target strings are not all of the same length. The number of compared characters in LNAME will be appropriate for each of the individual values.

Trailing blanks are counted and are used to determine matches. The following WHERE statement

```
where 'Adamso'=: trim(lname);
```

will return both 'Adams' and 'Adamson'; however, if the TRIM function had not been used only 'Adamson' would have been found.

Similar functionality can be achieved in an SQL step; however, the syntax can vary. The colon operator can be used in a WHERE= data set option whenever you are importing a SAS data set (see Section 2.1).

```
proc sql;
   title2 'Used in SQL data set WHERE=';
   select lname, fname, dob
      from advrpt.demog(where=(lname=:'Adams')); ❶

   title2 'Used in SQL WHERE Clause';
   select lname, fname, dob
      from advrpt.demog
         where lname=:'Adams'; ❷
proc sql;
   title2 'Using the EQT operator';
   select lname, fname, dob
      from advrpt.demog
         where lname eqt 'Adams'; ❸
   quit;
```

❶ The =: can be used within the WHERE= data set option even when it is used within the SQL step.

❷ Although it worked in the WHERE= data set option, the =: will **not** work in an SQL WHERE clause. This SELECT statement will fail.

❸ In SQL the EQT operator is similar to the =:, and because it is an SQL operator it can be used in the SQL WHERE clause. In addition to the EQT operator, SQL also supports the LET (<=:) and GET (>=:) operators.

SEE ALSO
The colon comparison modifier is not available in the macro language; however, macros have been written to provide similar functionality. See Carpenter (2004, Section 7.6.3 pg. 196).

2.2.3 Logical and Comparison Operators in Assignment Statements

In Section 2.2.1 the following assignment statement is briefly introduced. This is an example of a value look-up where the value of month determines the value of SEASON (Chapter 6 goes into detail on a variety of table look-up

```
season = 1*(1 le month(dob) le 3)
       + 2*(4 le month(dob) le 6)
       + 3*(7 le month(dob) le 9)
       + 4*(10 le month(dob) le 12);
```

coding strategies). In this case the process could have been simplified by the use of a format. More importantly, however, it demonstrates the use of a comparison operator (LE) in an assignment statement.

In fact, as was mentioned in Section 2.2.1 there is no reason why any of the logical and comparison operators cannot appear in an assignment statement. The key to their use is to

```
season = 1*(0)
       + 2*(1)
       + 3*(0)
       + 4*(0);
```

remember that logical expressions will yield either TRUE or FALSE, which is represented by 1 or 0 respectively. For a date of birth in May the previous equation is evaluated as is shown on the left. The expression results in a value of 2 for SEASON. When you are generating a numeric value based on a logical determination, such as this one, you should be able to write the assignment statement in a form similar to the one above. However, it is not unusual to see this type of assignment made through a series of less efficient IF-THEN/ELSE statements.

Although the previous example could have also been made using a user-defined format and a PUT function (see Section 6.5), the assignment of a value to GROUP using the series of IF-THEN/ELSE statements, such as the one shown here, does not so easily lend itself to a solution

```
if sex = 'M' and year(dob) >  1949 then group=1;
else if sex = 'M' and year(dob) le  1949 then group=2;
else if sex = 'F' and year(dob) >  1949 then group=3;
else if sex = 'F' and year(dob) le  1949 then group=4;
```

involving a format. The value can, however, be determined with an assignment

```
group = 1*(sex = 'M' and year(dob) >  1949)
      + 2*(sex = 'M' and year(dob) le  1949)
      + 3*(sex = 'F' and year(dob) >  1949)
      + 4*(sex = 'F' and year(dob) le  1949);
```

statement containing the same logic as was used in these IF-THEN/ELSE statements. Since assignment statements tend to be processed faster than IF-THEN/ELSE

statements, it is likely that the use of assignment statements can decrease processing time. This type of assignment statement will also generally out perform a PUT function.

Since True/False determinations always result in either a 0 or a 1, this same approach can be especially useful if assigning a numeric 0,1 value to a variable. In the following DATA step we would like to create a flag that indicates whether or not the date of birth is before 1950. Three equivalent flags have been created to demonstrate three different methods.

```
data flags;
   set advrpt.demog (keep=lname fname dob sex);
   if year(dob) >  1949 then boomer=1; ❶
   else boomer=0;
   boomer2 = year(dob) >  1949; ❷
   boomer3 = ifn(year(dob) >  1949, 1, 0); ❸
   run;
```

❶ Very often IF-THEN/ELSE statements are used. These statements tend to process slower than assignment statements.

❷ The logical expression appears on the right of the equal sign.

❸ The IFN function can be used to assign the result. This function has added value when a result other than just 0 or 1 is to be returned.

A similar coding structure is used when you need to create a flag that depends on whether or not an item is in a list of values. Here we need to determine if the number 3 is contained in one or more of the variables X1 through X4.

```
data _null_;
input x1-x4;
array a {*} x1-x4;
flag1 = (3 in a);  ❹
flag2 = ^^whichn(3,of x:);  ❺
put flag1= flag2=;
datalines;
1 2 3 4
5 6 7 8
run;
```

more of the variables X1 through X4.

❹ FLAG1 is created by determining if the value 3 is in the array A. If it is, a 1 is stored.

❺ The WHICHN function returns an item number if the value in the first argument is found (otherwise a 0 is returned). This value is then converted to a 0 or a 1 by use of a double negation (see Section 2.2.6 for more on the use of the double negation).

MORE INFORMATION
The assignment statements discussed in this section are simple table lookups. Chapter 6 discusses a variety of different table lookup techniques in more detail. The IFN and IFC functions are discussed more in Section 3.6.6.

SEE ALSO
The sasCommunity.org tip
http://www.sascommunity.org/wiki/Tips:Creating_a_flag_avoiding_the_If_..._Then_Structure discusses the use of this type of expression in an assignment statement. The discussion tab includes alternative forms that can be used in an SQL step. The example of flags used to indicate presence of a value in a list was suggested by Chang Chung and Mike Rhoads. Their examples can be found at:
http://www.sascommunity.org/wiki/Tips:Double_negatives_to_normalize_a_boolean_value and http://www.listserv.uga.edu/cgi-bin/wa?A2=ind1101c&L=sas-l&D=1&O=D&P=9693, respectively.

2.2.4 Compound Inequalities

When a compound inequality is used in an expression it is important to understand how the expression is evaluated. An expression with a compound inequality very often contains a variable

```
if 13 le edu le 16;
```

between two values, which form the upper and lower limits of a range. This compound expression is effectively interpreted as two distinct inequalities which

```
if (13 le edu) and (edu le 16);
```

are joined with an AND. The value of EDU must meet both conditions for the overall expression to evaluate as true.

Misplacing the parentheses totally changes the way that the expression is evaluated. Notice the

```
where (13 le edu) le 16 ;
```

placement of the parentheses in this WHERE statement. The inequality inside the parentheses is evaluated to True or False (0 or 1), and the result compared to 16. This expression will be true for all values of EDU (both 0 and 1 are less than 16).

You may be thinking 'OK, so what, I will be careful with parentheses. Why should I care?' Of course I think that the way that SAS evaluates these expressions is both interesting and important, but there are also practical benefits.

In the macro language compound inequalities are not evaluated the same way as we might think

```
%let x = 5;
%let y = 4;
%let z = 3;
%if &x lt &y lt &z %then
        %put &x < &y < &z;
```

based on our knowledge of DATA step expressions. The resolved macro variables in the expression in this %IF-%THEN statement show that the expression should be evaluated as false; however, it evaluates as TRUE and the %PUT executes.

```
%if 5 lt 4 lt 3 %then %put 5 < 4 < 3;
```

This happens because the compound

```
%if (&x lt &y) lt &z %then %put &x < &y < &z;
```

expression in the macro language is not broken up into

two distinct inequalities like it would be in the DATA step. Instead it is evaluated as if there were parentheses around the first portion of the comparison. The expression is evaluated left to right, and (&x lt &y) will be either TRUE or FALSE (1 or 0). Either way as long as &Z is > 1, the overall expression will be TRUE.

In actual applications compound inequalities such as the ones shown above are quite common. It is much less likely that you will use other forms of compound expressions; however, as was shown in the example with the %IF statement, it is important to understand how the expression is evaluated.

2.2.5 The MIN and MAX Operators

CAVEAT

It is my opinion that these two operators, and their mnemonics (see Section 2.2.1), should *never* be used. They are included in this section primarily so that this warning can be given. When you want to return either the minimum or maximum, always use either the MIN or MAX function. These functions have none of the problems associated with these the two operators.

The MIN and MAX operators return the minimum or maximum of *two* values respectively.

```
where x = (y min z);
```

```
where x = min(y, z);
```

In this WHERE statement the smaller value of the variables Y and Z will be compared to the value of the variable X. The MIN and MAX operators allow the comparison of exactly two variables, while the preferred MIN and MAX functions can work with any number of variables.

These two operators do not behave the same in WHERE statement expressions as they do in an IF statement expression, and the differences are important. Actually they are only important if you intend to ignore the caveat and go ahead and use these operators. Assuming that you like living on the edge, read on.

Potential Problem #1 - Mnemonics

In the IF statement these two operators can be replaced with their mnemonics; however, these mnemonics do not work in the same way in the WHERE statement.

Operator	IF Statement Mnemonic	WHERE Statement Caveat for the Mnemonic	IF and WHERE Clause	Results

MIN	><	Mnemonic is not supported in WHERE, although MIN is converted to >< in the LOG	where x=(y><z);
			
			 if x=(y><z);
				WHERE → error
				
				 IF works as expected

MAX	<>	Mnemonic is interpreted as not equal in the WHERE	where x=(y<>z);	True when X=1 and Y not equal to Z or when X=0 and Y=Z

Potential Problem #2 – Negative Values

When the negative sign is used with the first value associated with these operators, the expressions are not interpreted the same in the WHERE and IF statements.

Expression	Result
`if -2 = (-5 min 2);`	True – the minus sign is applied after the MIN operator, essentially the same as: `if -2 = -(5 min 2);`
`where -2 = (-5 min 2);`	False – the minus sign is applied before the MIN operator, comparison is the same as: `where -2 = -5;`

2.2.6 Numeric Expressions and Boolean Transformations

It is sometimes necessary to transform numeric values to Boolean (0 or 1) values. There is no one function that does this, and indeed the exact transformation may be too situational for a specific function.

True / False

Partly because of the way that SAS handles TRUE/FALSE, i.e., false is 0 or missing and all else is true, the missing values must also map to 0. The double negation (NOT) is used to perform the transformation. Since negation is a Boolean operator, it converts the original value to either a zero or a one.

```
x = ^^dob;
```

MORE INFORMATION

Double negation is used to convert a number to a binary 0/1 operator in an example in Section 12.6.2.

Replace Missing with 0

For reporting purposes missing values can be replaced by a 0 using a simple assignment statement. The COALESCE function returns the first non-missing value. In this example if DOB is missing a 0 is returned. Prior to the inclusion of the COALESCE function, this same operation was sometimes accomplished using the SUM function. Be careful when working with dates as was done here. Remember that, although both are false, a date of missing and a date of 0 have different meanings.

```
z = coalesce(dob,0);
```

```
y = sum(dob,0);
```

These two expressions do not result in a Boolean value. If you want to convert all missing values to 0 and all other values to 1 (including 0) you can use the negation of the MISSING function. In this expression MVAL will be 1 for all numbers except missing.

```
mval=^missing(val);
```

Determine Positive or Negative Values

Because of missing values and the 0 value, we have four distinct possibilities when separating positive and negative values. The groups of positive and non-negative values are not necessarily the same. Fortunately we can build the Boolean flag for each of these four possibilities, with the use of the SIGN function, which returns -1 for values < 0, 0 for values=0, 1 for values > 0, and missing for missing values.

```
data posneg;
   do v=.,-2 to 2;
      *if positive;
      pos = sign(v)=1;
      * Not positive;
      notpos = (sign(v) in(-1,0));
      * Negative;
      neg = ^sign(v)=-1;
      * Not negative;
      notneg = sign(v) in (0,1);
      output posneg;
   end;
   run;
```

```
2.2.6  Boolean Conversions
Positive or Negative?

Obs  v  pos  notpos  neg  notneg

 1   .   0      0      0     0
 2  -2   0      1      0     0
 3  -1   0      1      0     0
 4   0   0      1      0     1
 5   1   1      0      0     1
 6   2   1      0      0     1
```

SEE ALSO

Several of the code examples in this section have been suggested by Howard Schreier in the sasCommunity.org article:
http://www.sascommunity.org/wiki/Numeric_transformations .

2.3 Data Validation and Exception Reporting

Although we sometimes have the opportunity to work with data that has already been scrubbed clean, often a major portion of our work is the preparation and cleaning of the data. This is especially true for data that has been hand entered or for data that comes from a source without your high standards of excellence. In the biotech/pharmaceutical industries a great deal of time and careful attention to detail is spent on the validation of the data.

For large or complex data sets the manual/visual process of finding data errors is just not practical. We need some tools that will allow us to automate the validation process.

When reporting on errors in your data, you will need to be able to communicate the specific problem and precisely where it occurs. Generally this means that you will have to identify the specific row and column (observation and variable). To identify a particular row you need to know the values of all the variables that form the data set's primary key (the BY variables that identify the data down to the row level).

SEE ALSO

Wright (2006), and the related papers by Nelson (2004a, 2004b), discusses an approach to the validation of SAS programs. Bahler (2001) provides a summary of a variety of data cleaning techniques. Ron Cody (2008b) has written an entire book on data cleaning techniques.

2.3.1 Date Validation

Since we will work with dates, we need to know how to detect and work with incomplete or inappropriate date values. There are several issues that we need to understand.

Using Formats to Check Date Strings

Formats, whether they are user defined or provided by SAS, can be powerful tools when searching the data for values that are inappropriate, incomplete, missing, or out of range.

Dates can be particularly problematic for some types of data, such as survey response data. Often the dates are collected as character strings and then checked before conversion to a SAS date.

```
data visitdates;
   visit=1; v_date_ = '05/25/2004'; output;
   visit=2; v_date_ = '06/XX/2004'; output;
   run;

proc print data=visitdates;
   where input(v_date_,mmddyy10.) = .;
   run;
```

Collecting them as character strings is important so that we can record partial dates. In the second observation shown to the left, the date might be from the response: "My second visit was in June of 2004. I do not remember the exact day."

```
2.3.1 Date Validation Using Formats
Date Errors


Obs     visit      v_date_

 2        2        06/XX/2004
```

The INPUT function in the WHERE statement attempts to convert the character string into a date. Invalid values will result in a missing value, which is selected by the WHERE.

```
9     proc print data=visitdates;
10       where input(v_date_,mmddyy10.) eq .;
11       run;


ERROR: INPUT function reported 'ERROR: Invalid date value'
while processing WHERE clause.
```

When the INPUT function is used with the format MMDDY. for an incorrectly structured

value, as it is in the second observation, an error is issued in the LOG. This error message can be suppressed by using either the ?? format modifier (see Section 1.3.1) or the NOFMTERR system option.

The ?? format modifier suppresses format error messages and can be used with the INFORMAT in the INPUT function, and is more specific than the NOFMTERR system option as it can be

```
data dateerrors;
   set visitdates;
   if input(v_date_,?? mmddyy10.) eq .;
   run;
proc print data=dateerrors;
   run;
```

applied to a specific informat. The subsetting IF statement in this step replaces the WHERE statement in the previous example, as the ?? informat modifier cannot be used in the WHERE statement.

Because we can suppress the error messages, we have more latitude in the use of formats to rebuild dates with missing components. As before, let's assume that our incoming dates are received as character strings with either the day or month potentially coded as 'XX'. This example shows the subject's response to the question: "When did you stop smoking?" Very typically the patient is unable to remember the date let alone the day of the month. In this example we have decided to replace missing days with the 15th and if the month is missing with the year midpoint July 1. This code assumes that at least the year is present.

```
data quit_dates;
   subject=201; q_date_ = '05/25/1975'; output;
   subject=205; q_date_ = '10/XX/2001'; output;
   subject=208; q_date_ = 'XX/XX/1966'; output;
   run;
data Qdates(keep=subject q_date_ q_date);
   set quit_dates;
   format q_date date9.;
   q_date = input(q_date_,??mmddyy10.);  ❶
   if missing(q_date) then do;  ❷
      * Substitute missing day of month with 15;
      if substr(q_date_,4,2)='XX' then substr(q_date_,4,2)='15';  ❸
      q_date = input(q_date_,??mmddyy10.);
   end;
   if missing(q_date) then do;  ❹
      * Substitute missing month with 07;
      if substr(q_date_,1,2)='XX' then do;  ❺
         substr(q_date_,1,2)='07';
         * reset day of month also;
         substr(q_date_,4,2)='01';
      end;
      q_date = input(q_date_,??mmddyy10.);  ❻
   end;
   run;
```

❶ The first attempt is made to build the SAS date (Q_DATE) using the INPUT function.

❷ If the SAS date is missing, one or more elements are probably coded as XX. First we check the day of the month.

❸ When the day of the month is coded as XX, the value of 15 is substituted. Notice the use of the SUBSTR function on the left side of the = sign to perform the substitution (see Section 3.6.6 for more on the use of the SUBSTR function to substitute characters into the string).

❹ The date is again checked now for a potentially missing month value.

❺ A month coded as XX is detected, and the month and day are substituted.

❻ The corrected date string is converted to a SAS date.

Checking for Missing Date Values

In the ADVRPT.DEMOG data set the variable DEATH records the date of death for those patients that died while under the care of a clinic (remember this is made up data and no patients were harmed for the writing of this book). We would like to filter the patients for those with a date of death. We need to remember that a SAS date contains the number of elapsed days since January, 1, 1960 (day 0). Dates before this date are negative numbers.

The WHERE statement can be used and a couple alternative forms of the expression could be suggested. We need to keep in mind that we are filtering for valid dates.

```
where death>0;
```

One possible expression might be to filter for values that are greater than zero; however, this clearly excludes any dates before January 2, 1960, and is NOT sufficient.

Since we want to exclude missing dates, and missing values are false, another possible expression

```
where death;
```

would be to simply check to see if DEATH is false. This nearly works, but it has a subtle flaw. Here we will exclude the valid date of death January

1, 1960, which is day zero and will be interpreted as false.

What we really need to do is eliminate only the missing values and we should focus on this. Two expressions can be used. We can explicitly exclude the missing value and allow all others or we

```
where death ne .;
```
```
where death > .;
```

can accept any value that is larger than missing (all numbers positive or negative are larger than missing).

Sometimes you will be working with data that utilizes more than one type of missing value. Since numeric variables can take on up to 28 types of missing values (., .a, .b, …. , .z, and ._), you may need to account for these when testing for missing. Of the 28 types of numeric missing values the

```
where death > .z;
```

smallest is ._, the next smallest is ., and the largest is .z. Therefore, to effectively test for all these missing value types, we need to code

for the largest (.z). All missing values would also be detected by using the MISSING function.

```
where ^missing(death);
```

MORE INFORMATION
The WHERE= data set option is used in examples in Sections 2.1.3, 2.1.4d, and 2.2.2.

SEE ALSO
Hunt (2010) shows alternate methodologies for the completion of dates. Alternate solutions for the completion of partial dates can be found on the SAS Forums, see the following thread:

http://communities.sas.com/message/40648#40648.

2.3.2 Writing to an Error Data Set

When attempting to identify the observations that contain data errors, it is often helpful to build a secondary data set containing just those observations with errors. Error reporting through a data set is as simple as adding conditional logic and another OUTPUT statement.

```
data medstartdates(keep=subject mednumber drug
                        medstdt_ medstartdate)
    medstarterr(keep=subject mednumber drug medstdt_); ❶
  set advrpt.conmed(keep=subject  mednumber drug medstdt_);
  medstartdate = input(medstdt_,?? mmddyy10.); ❷
  if medstartdate = . then output medstarterr; ❸
  output medstartdates; ❹
  run;
```

❶ The variable list in the error data set should contain all the variables that form the primary key, as well as those being tested.

❷ In this example we are converting the text date to a SAS date. The use of the ?? format modifier to suppress error messages in the LOG is discussed in Section 2.3.1.

❸ Incomplete dates, those that cannot be converted to valid SAS dates, result in a missing value.

❹ In this example the data set MEDSTARTDATES will contain all observations, including those with errors. To remove observations with errors from this data set, simply start the OUTPUT statement with an ELSE statement. The statement becomes: `else output medstartdates;`

The resulting data table (WORK.MEDSTARTERR) will contain one observation for each MEDSTDT_ value that cannot be converted to a valid SAS date.

```
2.3.2a Collecting Date Errors

Obs      SUBJECT     mednumber     drug                          MEDSTDT_

 51        206           2         LUPRON                        XX/XX/1999
 52        206           3         CALCIUM CITRATE               --/--/----
 53        206           4         MVI                           --/--/----
 54        207           1         METOPROLOL                    --/--/----
 55        207           1         CLONIDINE                     --/--/----
 56        207           1         LUPRON                        XX/XX/1996
 57        207           2         DYAZIDE                       --/--/----
 58        207           2         COUMADIN                      XX/XX/1996
                        .... portions of the table are not shown ....
```

When testing more than one variable or different variables in different data sets, this form for the error reporting data set is not flexible enough. This is often the case when we build an automated system for validating our data. In an automated system, the exception data set needs to contain sufficient information for the data manager to get back to the value in question. A minimum amount of information would include:

- DSN   data set name
- *List of Variables*   key variables and their values
- ERRVAR   variable containing the exception
- ERRVAL   flagged value
- ERRTEXT   reason that the value was flagged (exception criteria)
- ERRRATING   exception rating (may be needed to identify critical variables)

This form of the error data set has one observation per error. It allows the reporting of more than one variable per observation and even errors from multiple data sets. In order to maximize flexibility, consider making the flagged value a character variable. This will require the conversion of numeric values to character, but adds flexibility. Also, since the list of key variables will likely change between data sets, many of the various key variables will be missing for observations that come from data sets for which they are not part of the key. As long as you are careful when designing your original data sets so that variables that appear in multiple data sets always have the same attributes, this will not be a problem.

```
title1 '2.3.2b Hardcoded Exception Reporting';

data errrpt(keep=dsn errvar errval errtxt errrating
               subject visit labdt); ❶
   length dsn        $25
          errvar     $15
          errval     $25
          errtxt     $20
          errrating  8;
   set advrpt.lab_chemistry; ❷
   if potassium lt 3.1 then do; ❸
      dsn = 'advrpt.lab_chemistry'; ❹
      errvar = 'potassium'; ❺
      errval = potassium; ❻
      errtxt = 'Low value(<3.1)'; ❼
      errrating= 1; ❽
      output errrpt; ❾
      end;
   if potassium gt 6.7 then do; ❿
      dsn = 'advrpt.lab_chemistry';
      errvar = 'potassium';
      errval = potassium;
      errtxt = 'High value(>6.7)';
      errrating=2;
      output errrpt;
      end;
   run;
```

❶ The list of variables in the error reporting data table includes the variables which form the primary key for this data table (SUBJECT VISIT LABDT).

❷ Read the data set to be checked.

❸ Enter the DO block when this specific data exception has been detected.

❹ Save the name of the data table being checked.

❺ Save the name of the variable being tested.

❻ Capture the data value for further inspection. When character values have been converted for the check, save the raw or character value. This allows you to see any text that might contribute to character to numeric conversion issues.

❼ The error test should describe the problem detected in the data.

❽ If required, assign an error severity code.

❾ Write this observation out to the error report data set (ERRRPT).

❿ The second exception criterion is tested in a DO block that is essentially the same as the first DO block ❸.

The resulting error report data set contains all the information necessary to locate, evaluate, and start the correction process.

```
2.3.2b Hardcoded Exception Reporting

Obs        dsn              errvar  errval    errtxt      errating SUBJECT VISIT     LABDT

 1  advrpt.lab_chemistry potassium  6.8   High value(>6.7)    2      203     4  09/29/2006
 2  advrpt.lab_chemistry potassium   3    Low value(<3.1)     1      208    10  03/09/2007
```

From a coding standpoint, the beauty of this approach is that you can have as many checks as are needed, and each one is simply implemented by the addition of another DO block. The disadvantage becomes apparent for large studies and for complex data tables. While not terribly complex, the program(s) can become large. More importantly, since each individual check is implemented in the program, new or changed criteria require that the program itself be revalidated. These problems can be addressed by storing the test criteria outside of the program. One very convenient way to do this is to store the exception criteria in a data table (see Section 2.3.3).

MORE INFORMATION
Non-numeric values are detected and written to an error data set in Section 3.6.1.

2.3.3 Controlling Exception Reporting with Macros

In the DATA step shown in Example 2.3.2b, a DO block is constructed for each data check. Since each one of these DO blocks is very similar, they are prime candidates for being written for us by the macro language. The following macro, %ERRRPT, will build this DO block.

```
%macro errrpt(dsn=,errvar=,errval=,errtxt=,errating=);
        dsn = "&dsn";
        errvar = "&errvar";
        errval = &errval;
        errtxt = "&errtxt";
        errating= &errating;
        output errrpt;
        end;
%mend errrpt;
```

When the %ERRRPT macro is called from within the DATA step, the IF THEN/DO statements become:

```
    if potassium lt 3.1 then do;
        %errrpt(dsn = advrpt.lab_chemistry,
                errvar = potassium,
                errval = potassium,
                errtxt = %str(Low value%(<3.1%)),
                errating= 1)

    if potassium gt 6.7 then do;
        %errrpt(dsn = advrpt.lab_chemistry,
                errvar = potassium,
                errval = potassium,
                errtxt = %str(High value%(>6.7%)),
                errating= 2)
    run;
```

Although we have simplified the code somewhat, there has still not been a huge savings in our coding effort. Of course we could have the macro language do even more of the lifting for us. If we could tell the macro language what and how many checks were needed, then each of the individual macro calls, including the IF THEN/DO, could be generated by a single macro call.

An easy way to store and pass along the needed information is through the use of a SAS data set. We can create a data set that contains the constraints for each data exception check. Each observation can then be used to build data exception and error trapping reports. For the previous example the data set might contain the following.

```
2.3.3b Data Set with Exception Criteria

Obs     errtst      errvar        errval          errtxt        errrating

 1     lt 3.1    potassium     potassium     Low value(<3.1)       1
 2     gt 6.7    potassium     potassium     High value(>6.7)      2
```

This data set is then used by a macro to build a series of macro variable lists. These lists are then processed (each observation in the data set becomes a test or DO block). The %ERRPT macro shown below builds these lists using an SQL step and then uses the lists to create a series of DO blocks using the macro variables in the list to 'fill in the blanks'.

```
%macro errrpt(dsn=, bylst=subject); ❶
%local i chkcnt;
proc sql noprint;
   select errtst, errvar, errval, errtxt, errrating
      into :errtst1-:errtst99, ❷
           :errvar1-:errvar99,
           :errval1-:errval99,
           :errtxt1-:errtxt99,
           :errrating1-:errrating99
        from vallab; ❸
   %let chkcnt = &sqlobs; ❹
   quit;
data errrpt(keep=dsn errvar errval errtxt errrating
               &bylst); ❺
   length dsn        $25
          errvar     $15
          errval     $25
          errtxt     $15
          errrating  8;

set &dsn ❻ ;

%do i = 1 %to &chkcnt; ❼
   %* Write as many error checks as are needed;
   if &&errvar&i &&errtst&i ❽ then do;
      dsn = "&dsn";
      errvar = "&&errvar&i"; ❾
      errval = &&errval&i;
      errtxt = "&&errtxt&i";
      errrating= &&errrating&i;
      output errrpt;
      end;
   %end;
   run;
%mend errrpt;
%errrpt(dsn=advrpt.lab_chemistry,
        bylst=subject visit labdt) ❿
```

❶ The macro %ERRRPT is used to control the error reporting process. Macro %DO loops must appear inside of a macro definition.

❷ The values that are being read from the control data set ❸ are stored in a series of macro variables. These take the form of &ERRVAL1, &ERRVAL2, etc. This code would allow up to, but no more than, 99 tests. There is no penalty for making this number too big (I like to over shoot by at least an order of magnitude); however, there is no hint in the LOG or elsewhere if the number is too small (values will just not get saved).

❹ The number of observations that are read from the control data set are saved in the macro variable &CHKCNT. This will be the total number of checks for this data set.

❺ The variables that form the primary key are included in the KEEP= list.

❻ The data set to be checked is specified on the SET statement.

❼ The macro %DO loop will iterate the number of times of the number contained in the macro variable &CHKCNT, which is once for each observation in the control data set. Each of the iterations will result in a DO block, with one DO block for each test to be performed.

❽ For the second %DO loop iteration the IF statement becomes:

```
if potassium gt 6.7 then do;
```

❾ The individual macro variables are addressed using the &&VAR&I macro variable form. For the second pass of the macro %DO loop (&i=2), the macro variable reference &&ERRVAR&I resolves to &ERRVAR2, which in turn resolves to potassium.

❿ The macro call contains the name of the data set to be checked and its list of BY variables.

MORE INFORMATION
The use of data sets to drive macros is discussed further in Section 13.5.

2.4 Normalizing - Transposing the Data

Most, but not all, SAS procedures prefer to operate against normalized data, which tends to be tall and narrow, and often contains classification variables that are used to identify individual rows. In the following presentation of the data, there is one value of SODIUM per observation and the classification variables are SUBJECT and VISIT.

```
2.4 Normalizing Data
Normal Form

Obs      SUBJECT     VISIT      sodium

  1        208         1         13.7
  2        208         2         14.1
  3        208         4         14.1
  4        208         5         14.1
  5        208         6         13.9
  6        208         7         13.9
  7        208         8         14.0
  8        208         9         14.0
  9        208        10         14.0
 10        209         1         14.0
  .... portions of the table are not shown ....
```

The same data could also be presented in a non-normal form, which tends to have one column for each level of one of the classification variables. In the following example, there is one observation per patient, with a column for each visit's sodium value.

```
2.4 Normalizing Data
Non-normal Form
       S                                                 V     V     V     V     V     V     V
       U     V     V     V     V     V     V     V     V  i     i     i     i     i     i     i
       B     i     i     i     i     i     i     i     i  s     s     s     s     s     s     s
       J     s     s     s     s     s     s     s     s  i     i     i     i     i     i     i
 O     E     i     i     i     i     i     i     i     i  t     t     t     t     t     t     t
 b     C     t     t     t     t     t     t     t     t  1     1     1     1     1     1     1
 s     T     1     2     4     5     6     7     8     9  0     1     2     3     4     5     6
 1   208  13.7  14.1  14.1  14.1  13.9  13.9   14    14  14.0   .     .     .     .     .     .
 2   209  14.0  14.0  13.9  14.2  14.5  13.8   14    .   13.8  14   14.1  14.2   14   .1   14  14.1
```

Since we often do not have control over the form of the data when we receive it, we need to be able to both convert the data from the normal to non-normal form and from non-normal to normal form. This process is known as transposing the data and the operations are commonly performed either by PROC TRANSPOSE or in the DATA step.

PROC TRANSPOSE is an efficient, powerful procedure for performing a transpose operation. The DATA step can be more flexible; however, PROC TRANSPOSE has the advantage of not requiring knowledge of how many transformed variables there will be prior to the transformation.

SEE ALSO
Toby Dunn (2010) discusses the differences between the normal and non-normal data forms and suggests programming motivations for using one form over the other. When summarizing at the same time as transposing, the MEANS and SUMMARY procedures can be very useful. King and Zdeb (2010) use the IDGROUP option on the OUTPUT statement to control the transpose process.

2.4.1 Using PROC TRANSPOSE

PROC TRANSPOSE tends to be less than intuitive for most users. The coding is not particularly difficult; however, for most users it is often hard to visualize what the resulting data set will look like. There is also a trap in this procedure that, when sprung, can cause the corruption of the data.

The following step, which creates a non-normal version of the lab chemistry data, demonstrates a simple PROC TRANSPOSE and will also be used to demonstrate the PROC TRANSPOSE trap.

```
proc transpose data=lab_chemistry(keep=subject visit sodium) ❶
               out=lab_nonnormal(keep=subject visit:) ❷
               prefix=Visit; ❸
   by subject; ❹
   var sodium; ❺
   run;
```

❶ DATA= identifies the incoming data set.

❷ The transposed data set is named with the OUT= option. Notice the use of the colon to select all variable names that start with the letters VISIT (see Section 2.6.1 for more on variable naming shortcuts).

❸ The PREFIX= option identifies text that will be used to form the new column names.

❹ The transposition process takes place within the group of variable(s) in the BY statement. In this example each distinct SUBJECT will form one row.

❺ The column SODIUM is transposed to rows.

```
Using PROC TRANSPOSE
2.4.1a Incompletely Specified Observations
    S                                           V    V    V    V    V
    U    V    V    V    V    V    V    V    V    V    i    i    i    i    i
    B    i    i    i    i    i    i    i    i    i    s    s    s    s    s
    J    s    s    s    s    s    s    s    s    s    i    i    i    i    i
O   E    i    i    i    i    i    i    i    i    i    t    t    t    t    t
b   C    t    t    t    t    t    t    t    t    t    1    1    1    1    1
s   T    1    2    3    4    5    6    7    8    9    0    1    2    3    4
1  208 13.7 14.1 14.1 14.1 13.9 13.9 14  14.0 14    .    .    .    .    .
2  209 14.0 14.0 13.9 14.2 14.5 13.8 14  13.8 14  14.1 14.2 14.1 14  14.1
```

Notice the value of SODIUM for patient 208 on visit #3 and #9. Compare this result to that shown in Section 2.4. Although it is not immediately obvious without careful inspection of the data, the TRANSPOSE trap has been sprung and the transposed data has been corrupted. The values of SODIUM have *in some cases* become associated with the wrong visit!

When using PROC TRANSPOSE there are two ways to identify the row, or rows, within which the transpose is to take place. In the previous example the BY statement is used. However, since there are multiple visits for each subject, there are multiple rows within each SUBJECT and these become the new columns (VISIT1, VISIT2, VISIT3, and so on). The problem is that there is nothing in our code that ties the VISIT variable to the new column, which in this case will be one of the variables VISIT1 through VISIT14. Patient 208 missed visits 3 and 9, consequently their ninth visit should have been classified as VISIT11. However, since it was the ninth observation for patient 208, PROC TRANSPOSE incorrectly classified the data to VISIT9. Notice that both patients 208 and 209 are showing data for VISIT3, although both actually missed visit #3.

The ID statement, which was not included in the previous example, can also be used to help identify rows. This statement names a variable that will be used to create the variable names for the new columns. More importantly it also ties a value in a specific row to a specified new column.

This type of problem is easily solved by the following rule: the combination of variables on the BY and ID statements must identify down to the row level.

In the following PROC TRANSPOSE step, an ID statement has been added. The BY variable, SUBJECT ❻, and the ID variable, VISIT ❼, form a unique key for this data set (any given combination of these two variables will identify at most one observation).

```
title2 '2.4.1b BY and ID Form a Unique Key';
proc transpose data=lab_chemistry(keep=subject visit sodium)
               out=lab_nonnormal(keep=subject visit:)
               prefix=Visit;
   by subject; ❻
   id visit; ❼
   var sodium;
   run;
```

Notice that there is no variable for VISIT3, because neither of these two patients had a visit number 3, so their third physical visit was visit number 4.

```
Using PROC TRANSPOSE
2.4.1b BY and ID Form a Unique Key
      S                                          V   V   V   V   V   V   V
      U    V    V    V    V    V    V    V   V   i   i   i   i   i   i   i
      B    i    i    i    i    i    i    i   i   s   s   s   s   s   s   s
      J    s    s    s    s    s    s    s   s   i   i   i   i   i   i   i
 O    E    i    i    i    i    i    i    i   i   t   t   t   t   t   t   t
 b    C    t    t    t    t    t    t    t   t   1   1   1   1   1   1   1
 s    T    1    2    4    5    6    7    8   9   0   1   2   3   4   5   6
 1   208  13.7 14.1 14.1 14.1 13.9 13.9 14  14  14.0  .   .    .    .   .    .
 2   209  14.0 14.0 13.9 14.2 14.5 13.8 14   .  13.8  14  14.1 14.2 14.1 14  14.1
```

SEE ALSO
The IDGROUP option in PROC MEANS and in PROC SUMMARY is used to transpose data in King and Zdeb (2010).

2.4.2 Transposing in the DATA Step

The DATA step offers a great deal of flexibility to the process of transposing data. Commonly the process of transposing will involve the use of an array and an iterative DO loop.

Rows to Columns

In order to transpose observations into columns, a series of observations must be processed for each new observation. The array statement is used to hold the values from the individual observations. Once all of the individual observations have been consolidated, the values in the array are written out to the new observation in the new data set.

```
data lab_nonnormal(keep=subject visit1-visit16);
   set lab_chemistry(keep=subject visit sodium);
   by subject;
   retain visit1-visit16 ; ❶
   array visits {16} visit1-visit16; ❷
   if first.subject then do i = 1 to 16; ❸
      visits{i} = .;
   end;
   visits{visit} = sodium; ❹
   if last.subject then output lab_nonnormal; ❺
   run;
```

❶ The array values are retained so that we can accumulate sodium values across visits.

❷ We know that there can be no more than 16 patient visits, so this becomes the dimension of the array.

❸ Since visit values are retained, the array containing the visit values are cleared at the start of each subject.

❹ The array is indexed using the visit number. This guarantees that the value of sodium will be assigned to the correct array variable.

❺ The new observation is written after all the incoming observations for each subject have been processed.

Unlike the data set generated by PROC TRANSPOSE, notice that, even though VISIT3 does not appear in the untransposed data, this data set includes a variable for VISIT3. This is a result of the implicit use of VISIT3 in the ARRAY statement ❶.

```
Transposing in the DATA step
2.4.2a Rows to Columns

     S                                                          v    v    v    v    v    v    v
     U     v    v    v    v    v    v    v    v    v    i    i    i    i    i    i    i
     B     i    i    i    i    i    i    i    i    i    s    s    s    s    s    s    s
     J     s    s    s    s    s    s    s    s    s    i    i    i    i    i    i    i
 O   E     i    i    i    i    i    i    i    i    i    t    t    t    t    t    t    t
 b   C     t    t    t    t    t    t    t    t    t    1    1    1    1    1    1    1
 s   T     1    2    3    4    5    6    7    8    9    0    1    2    3    4    5    6

 1  208  13.7 14.1   .  14.1 14.1 13.9 13.9  14   14  14.0   .    .    .    .    .    .
 2  209  14.0 14.0   .  13.9 14.2 14.5 13.8  14    .  13.8  14  14.1 14.2 14.1  14  14.1
```

Columns to Rows

You can also use the DATA step to normalize a data set. In the previous example we converted some lab data for two patients from a normal form to a non-normal form. We will now use a similar DATA step to convert it back to its original form.

```
title2 '2.4.2b Columns to Rows';
data lab_normal(keep=subject visit sodium);
   set lab_nonnormal(keep=subject visit:);   ❻
   by subject;
   array visits {16} visit1-visit16;   ❼
   do visit = 1 to 16;   ❽
      sodium = visits{visit};   ❾
      output lab_normal;   ❿
   end;
   run;
```

❻ The variables necessary for the conversion are kept on the incoming data set.

❼ The VISITS array is declared with each of the visits as an element.

❽ The DO loop index is VISIT which increments for each visit. The index variable, VISIT, is added to the Program Data Vector and will be written to the new data set.

❾ The variable SODIUM is created from the array element identified with VISIT as the index.

❿ Since the OUTPUT statement is inside the DO loop, it will write an observation for each of the iterations of the DO loop.

The resulting data set will have an observation for each patient X visit combination. This now includes combinations that did not originally exist.

In this example SUBJECT 208 now has observations for VISIT 3 and VISIT 11-VISIT16, and the value for SODIUM is appropriately missing for each of these visits. Observations with missing SODIUM values could easily be removed by adding an IF criterion to the OUTPUT statement ❿.

```
Transposing in the DATA step
2.4.2b Columns to Rows

Obs     SUBJECT     visit     sodium

 1        208          1        13.7
 2        208          2        14.1
 3        208          3          .
 4        208          4        14.1
 5        208          5        14.1
 6        208          6        13.9
 7        208          7        13.9
 8        208          8        14.0
 9        208          9        14.0
10        208         10        14.0
11        208         11          .
12        208         12          .
13        208         13          .
14        208         14          .
15        208         15          .
16        208         16          .
17        209          1        14.0
18        209          2        14.0
19        209          3          .
```
portions of the table not shown

```
If sodium gt .z then output lab_normal;
```

2.5 Filling Sparse Data

Sometimes when data are entered observations are created only when there is a specific value. Observations, which reflect only missing values or for count data counts of 0, are not created. This creates denser data as there will be fewer missing values. Depending on which observations are included entire classification levels or combinations of classification levels could be missing from the data. This means that the data itself does not reflect the true sampling scheme. Sometimes we need to show all possible levels - not just those with non-missing values.

Creating observations, with the appropriate missing values, is sometimes known as creating sparse (less dense) data.

The examples shown in this section work with the LAB_CHEMISTRY data, which has one row per patient per visit. Missed visits are not represented in the data and will not be represented in tables and reports.

Each patient (SUBJECT) in the LAB_CHEMISTRY data should have an observation for each of the first 10 visits and may or may not have subsequent follow-up for visits 11 through 16. We need to make sure that each patient has an observation for the first 10 visits.

2.5.1 Known Template of Rows

When we know the full list of values that the classification variable(s) should take on, a template can be built and merged back onto the original data. The process of merging the template containing *all possible* combinations of the classification variables will add the appropriate observations to the data set. For the following example we want each SUBJECT to have at least the first 10 visits.

```
proc sort data=advrpt.lab_chemistry ❶
        out=lab_chemistry;
   by subject visit;
   run;
proc sort data=advrpt.lab_chemistry
        out=sublist(keep=subject) ❷
        nodupkey; ❸
   by subject;
   run;

data subvislist;
   set sublist;
   do visit = 1 to 10; ❹
      output subvislist;
   end;
   run;

data sparsed;
   merge subvislist ❺
        lab_chemistry;
   by subject visit; ❻
   run;
```

❶ The data set to be filled (sparsed) is sorted by the classification variables that will be used to fill the data.

❷ A data set is created that contains only the classification variable(s) that do not need to be filled. In this example it is SUBJECT number.

❸ NODUPKEY is used to eliminate all duplicate subjects. This list of distinct subject numbers could have also been created in a simple SQL step.

❹ The list of unique subjects is read and a DO loop is used to output an observation for each SUBJECT - VISIT combination. When multiple classification variables need to be filled, nested DO loops are needed.

❺ The template data set (SUBVISLIST) is merged back onto the original data set. Any *extra* rows in the template data, rows that did not appear in the original data, will now also appear in the sparsed data set.

❻ The BY list will contain all the classification variables.

The LISTING for SUBJECT 210 shows that an observation has been added for visits 3 and 9. Even if they had not already been in the data for SUBJECT 210, visits 11-16 would not have been added.

```
2.5.1 Creating a Sparse Table
Every Patient should have the first 10 visits
Some patients have up to 16 visits

Obs     SUBJECT    visit         LABDT    potassium    sodium   chloride

114       210        1      02/19/2007       5.0        14.0      103
115       210        2      02/28/2007       4.0        14.2      103
116       210        3          .            .           .          .
117       210        4      03/14/2007       3.9        14.1      101
118       210        5      03/09/2007       4.7        14.4      105
119       210        6      03/16/2007       4.7        14.5      104
120       210        7      03/26/2007       4.7        14.3      103
121       210        8      03/28/2007       4.4        14.1      102
122       210        9          .            .           .          .
123       210       10      07/10/2007       4.3        14.2      106
124       210       11      04/06/2007       4.1        14.2      104
125       210       12      04/17/2007       4.0        13.9      103
126       210       13      04/19/2007       4.4        14.2      104
127       210       14      04/26/2007       4.1        14.1       99
128       210       15      05/22/2007       3.8        13.8       99
129       210       16      06/27/2007       5.2        14.3      104
                  . . . . Results for only SUBJECT 210 are shown . . . .
```

This process can also be done in a PROC SQL step, either way the concept is the same.

2.5.2 Double Transpose

When all possible levels of a classification variable are not known, or not easily specified in a DO loop, it is often possible to create a sparsed data set by performing two PROC TRANSPOSE steps. Rather than building a template data set as was done in Section 2.5.1, we will let the data itself determine the classification levels. It should be noted that the results of this technique and those obtained in Section 2.5.1 are not necessarily the same.

The first PROC TRANSPOSE step creates one column for each value of the variable to be sparsed (VISIT). Assuming that a given visit exists somewhere in the data, it will be represented as a column after the first PROC TRANSPOSE step.

The second PROC TRANSPOSE step reconverts the columns (one for each possible visit) into rows. The data now has the original form (as before the first PROC TRANSPOSE); however, every visit column is now represented as a row for every subject.

```
proc sort data=advrpt.lab_chemistry  ❶
          out=lab_chemistry
          nodupkey;
   by subject visit;
   run;

proc transpose data=lab_chemistry  ❷
               out=labtran
               prefix=Visit;  ❸
   by subject;  ❹
   id visit;  ❺
   var sodium potassium chloride;
   run;

proc transpose data=labtran  ❻
               out=sparsed(rename=(_name_=Visit));  ❼
   by subject;  ❽
   id _name_;  ❾
   var visit:;  ❿
   run;
```

❶ The incoming data set must be sorted at least to the level of the BY statement ❹.

❷ PROC TRANSPOSE is used to convert the rows that are to be filled into columns.

❸ The PREFIX= option specifies the text used to form the root portion of the new variable names.

❹ The BY statement lists the classification variables that do not need to be filled.

❺ The variable VISIT will be used to identify which columns were formed from which rows. In this case, the numeric variable VISIT will be combined with the prefix text ❸ to form the new column name.

❻ The data set LABTRAN will be transposed back to the original set of rows and columns.

❼ By default rows in the new data set (SPARSED) will be identified with the variable _NAME_. This variable is renamed to VISIT.

❽ The incoming data set (LABTRAN) has one observation per SUBJECT _NAME_ combination.

❾ The ID statement identifies the variable (_NAME_) that contains the names of the new columns.

❿ Each of the variables starting with VISIT is included in the transpose (see Section 2.6.1 for more on variable list abbreviations).

Prior to the first transpose, inspection of the data for SUBJECT=210 shows that this subject has missed both visits 3 and 9.

```
2.5.2 Creating a Sparse Table
Using a Double Transpose
Prior to First Transpose

 Obs    SUBJECT    VISIT        LABDT    potassium    sodium    chloride

  95      210        1     02/19/2007       5.0        14.0       103
  96      210        2     02/28/2007       4.0        14.2       103
  97      210        4     03/14/2007       3.9        14.1       101
  98      210        5     03/09/2007       4.7        14.4       105
  99      210        6     03/16/2007       4.7        14.5       104
 100      210        7     03/26/2007       4.7        14.3       103
 101      210        8     03/28/2007       4.4        14.1       102
 102      210       10     07/10/2007       4.3        14.2       106
 103      210       11     04/06/2007       4.1        14.2       104
 104      210       12     04/17/2007       4.0        13.9       103
 105      210       13     04/19/2007       4.4        14.2       104
 106      210       14     04/26/2007       4.1        14.1        99
 107      210       15     05/22/2007       3.8        13.8        99
 108      210       16     06/27/2007       5.2        14.3       104
```

After the first PROC TRANSPOSE ❷ the data (WORK.LABTRAN) looks like the following (for SUBJECT=210). There is now a VISIT9 column for this subject, even though this subject did not have a VISIT9 in the data; however, there is still no column for visit 3. This is because no subject in the entire data set had a VISIT3.

```
2.5.2 Creating a Sparse Table
Using a Double Transpose
First Transpose

Obs   SUBJECT    _NAME_       Visit1   Visit2   Visit4   Visit5   Visit6   Visit7   Visit8

 31     210      sodium          14     14.2     14.1     14.4     14.5     14.3     14.1
 32     210      potassium        5      4.0      3.9      4.7      4.7      4.7      4.4
 33     210      chloride       103    103.0    101.0    105.0    104.0    103.0    102.0


Obs   Visit9    Visit10   Visit11   Visit12   Visit13   Visit14   Visit15   Visit16

 31      .        14.2      14.2      13.9      14.2      14.1      13.8      14.3
 32      .         4.3       4.1       4.0       4.4       4.1       3.8       5.2
 33      .       106.0     104.0     103.0     104.0      99.0      99.0     104.0
```

The second PROC TRANSPOSE ❻ uses this data as input and since all visits are included on the VAR statement ❿, each visit becomes a row in the new table.

```
2.5.2 Creating a Sparse Table
Using a Double Transpose
Second Transpose

Obs    SUBJECT     Visit      sodium     potassium    chloride

151      210       Visit1      14.0         5.0         103
152      210       Visit2      14.2         4.0         103
153      210       Visit4      14.1         3.9         101
154      210       Visit5      14.4         4.7         105
155      210       Visit6      14.5         4.7         104
156      210       Visit7      14.3         4.7         103
157      210       Visit8      14.1         4.4         102
158      210       Visit9        .           .           .
159      210       Visit10     14.2         4.3         106
160      210       Visit11     14.2         4.1         104
161      210       Visit12     13.9         4.0         103
162      210       Visit13     14.2         4.4         104
163      210       Visit14     14.1         4.1          99
164      210       Visit15     13.8         3.8          99
165      210       Visit16     14.3         5.2         104
```

This subject now has a data row for VISIT9 (with missing values) even though VISIT9 was not in the original data for this subject. This implies that at least one subject had a VISIT9. Since VISIT3 still does not appear, we can infer that no subject in our study had a VISIT3. This technique requires that the row that is to be sparsed (visits 3 and 9) appear in the data set somewhere at least once.

Remember when using PROC TRANSPOSE that it is *very* important that some combination of the BY and ID variables identify down to the row level (see Section 2.4.1).

MORE INFORMATION

The TRANSPOSE procedure and some of its pitfalls (gotcha's) can be found in Section 2.4.1.

2.5.3 Using COMPLETETYPES with PROC MEANS or PROC SUMMARY

The COMPLETETYPES option can be used on the PROC MEANS or PROC SUMMARY statement to force the procedure to generate statistics for all combinations of the classification variables.

```
proc means data=advrpt.lab_chemistry
        completetypes noprint nway;
    class subject visit;
    var sodium potassium chloride;
    output out=allvisits sum=;
    run;
```

In Section 2.5.2 a double PROC TRANSPOSE is used to determine all of the combinations of SUBJECT and VISIT. This can also be accomplished using the COMPLETETYPES option on the PROC MEANS or PROC SUMMARY statement.

```
2.5.3 Creating a Sparse Table
Using COMPLETETYPES

Obs    SUBJECT   VISIT   sodium   potassium   chloride

151      210       1     14.0        5.0        103
152      210       2     14.2        4.0        103
153      210       4     14.1        3.9        101
154      210       5     14.4        4.7        105
155      210       6     14.5        4.7        104
156      210       7     14.3        4.7        103
157      210       8     14.1        4.4        102
158      210       9      .           .          .    ❶
159      210      10     14.2        4.3        106
160      210      11     14.2        4.1        104
161      210      12     13.9        4.0        103
162      210      13     14.2        4.4        104
163      210      14     14.1        4.1         99
164      210      15     13.8        3.8         99
165      210      16     14.3        5.2        104
```

There are no VISIT 9 observations for SUBJECT 210; however, since at least one subject somewhere in the LAB_CHEMISTRY data table had a VISIT 9, the report generated from the PROC MEANS results will show a VISIT 9 for all subjects ❶.

Behind the scenes PROC MEANS and PROC SUMMARY are really the same procedure, so this technique works with either procedure.

MORE INFORMATION
The COMPLETETYPES option is also discussed in Section 7.10. The COMPLETETYPES option also has implications when using preloaded formats; see Section 12.1.3.

2.5.4 Using CLASSDATA

The CLASSDATA option is used with the TABULATE, MEANS, and SUMMARY procedures to specify a data set that contains levels of one or more classification variables. If the data set contains levels that are not found in the data, those levels will be included in the resulting summary.

```
proc sort data=advrpt.demog(keep=subject)
   out=subjects nodupkey;
   by subject;
   run;

data Visits;
   set subjects;
   do visit = 1 to 16;   ❶
     output visits;
   end;
   run;
proc means data=advrpt.lab_chemistry
         classdata=visits   ❷
         noprint nway exclusive;
   class subject visit;
   var sodium potassium chloride;
   output out=allvisits   ❸ sum=;
   run;
```

The data set WORK.VISITS is constructed to have one observation for each of the potential 16 visits ❶. This data set is then used with the CLASSDATA= option ❷ and the EXCLUSIVE option in the PROC MEANS step.

Although in the LAB_CHEMISTRY data set there are no subjects that have a visit 3 and SUBJECT 210 does not have a visit 9, in the summary data set (WORK.ALLMEANS) ❸ which was created by PROC MEANS, each subject will have a summary row for all sixteen visits. Subject 210, which is shown here, now has both a visit 3 ❹, and a visit 9 ❺.

ASIDE: The CLASSDATA data set *must* contain *each* of the CLASS variables.

```
2.5.4 Using CLASSDATA
MEANS / SUMMARY

Obs   SUBJECT   VISIT   sodium   potassium   chloride

161     210       1      14.0       5.0         103
162     210       2      14.2       4.0         103
163     210       3       .          .           .    ❹
164     210       4      14.1       3.9         101
165     210       5      14.4       4.7         105
166     210       6      14.5       4.7         104
167     210       7      14.3       4.7         103
168     210       8      14.1       4.4         102
169     210       9       .          .           .    ❺
170     210      10      14.2       4.3         106
171     210      11      14.2       4.1         104
172     210      12      13.9       4.0         103
173     210      13      14.2       4.4         104
174     210      14      14.1       4.1          99
175     210      15      13.8       3.8          99
176     210      16      14.3       5.2         104
```

MORE INFORMATION

The CLASSDATA option is also discussed in Sections 7.9 (PROC MEANS and PROC SUMMARY) and 8.1.4 (PROC TABULATE).

2.5.5 Using Preloaded Formats

For the TABULATE, MEANS, SUMMARY, and REPORT procedures, preloaded formats can be used to add rows to output tables. Like the CLASSDATA option shown in Section 2.5.4, this method adds the sparsed rows to the table, not the data set. Thus, we are not required to either modify the original data or to even make a copy.

```
proc format; ❶
value visits
 1='1'
 2='2'
 3='3'
 4='4'
 5='5'
 6='6'
 7='7'
 8='8'
 9='9'
 10='10';
 run;
ods pdf file="&path\results\E2_5_5.pdf";
proc report data=advrpt.lab_chemistry nowd
           completerows; ❷
  column visit sodium potassium chloride;
  define visit / group
                f=visits. preloadfmt ❸
                'Visit' order=data;
  define sodium /analysis mean f=5.2;
  define potassium /analysis mean f=5.3;
  define chloride/analysis mean f=5.1;
  run;
ods pdf close;
```

Let's assume that we need to generate a report of mean lab chemistry values for lab visits. The report must contain the first 10 visits regardless of whether or not they appear in the data.

❶ A format is created which contains each of the first 10 visits.

❷ The COMPLETEROWS option, which is unique to PROC REPORT, is used to ensure that every row in the preloaded format will appear in the report.

❸ The PRELOADFMT option will always be present when using preloaded formats. Here the PRELOADFMT option is

associated with the format to be preloaded by placing both on the DEFINE statement. A portion of the resultant report is shown to the right.

MORE INFORMATION
Preloaded formats can also be used to exclude observations, and are introduced and discussed in more detail in Section 12.1.

2.5.5 Using Preloaded Formats
PROC REPORT with COMPLETEROWS

Vist	sodium	potassium	chloride
1	14.01	4.206	100.6
2	14.07	4.294	101.3
3	.	.	.
4	14.05	4.231	102.8
5	14.21	4.500	101.4
6	14.16	4.506	100.3
7	14.03	4.244	100.2

2.5.6 Using the SPARSE Option with PROC FREQ

By default the table generated by PROC FREQ will contain only those levels that actually exist in the data. In the first TABLE statement ❶, only the combinations of the two classification variables (EDU and SYMP) will exist in the table. The SPARSE option on the second TABLE statement ❷ will have all combinations of any value of EDU and SYMP. Notice that on the first table EDU=10 has only two levels of SYMP (04 and 10); however, on the second table ❷ each level of SYMP that exists somewhere in the data set is associated

```
proc freq data=advrpt.demog;
   table edu*symp/ list;  ❶
   table edu*symp/ list sparse;  ❷
   run;
```

2.5.6 Using SPARSE with FREQ   ❶

The FREQ Procedure

edu	symp	Frequency	Percent	Cumulative Frequency	Cumulative Percent
10	04	6	9.38	6	9.38
10	10	3	4.69	9	14.06
12	02	2	3.13	11	17.19
12	03	2	3.13	13	20.31
12	05	4	6.25	17	26.56
12	06	4	6.25	21	32.81
12	10	3	4.69	24	37.50

edu	symp	Frequency	Percent	Cumulative Frequency	Cumulative Percent
10	01	0	0.00	0	0.00
10	02	0 ❷	0.00	0	0.00
10	03	0	0.00	0	0.00
10	04	6	9.38	6	9.38
10	05	0	0.00	6	9.38
10	06	0	0.00	6	9.38
10	09	0	0.00	6	9.38
10	10	3	4.69	9	14.06

with EDU=10. Notice that, since no subject has a SYMP of either '07' or '08', those levels are not included in the SPARSED reports.

2.6 Some General Concepts

There are a number of general techniques, shortcuts, and *did you know that you can*s, of which you should be aware.

2.6.1 Shorthand Variable Naming

When creating a long list of variable names it is sometimes helpful to not actually write each name individually. Fortunately there are several ways to create lists of variables that require less coding.

These shorthand variable lists can be used wherever a list of variables is expected. This includes the VAR, KEEP, DROP, and ARRAY statements.

Common Prefix Variable Lists (Numbered Range)
Variables with a common prefix and a numeric suffix can be listed as:

```
visit1 - visit10
```

This list will include all the variables between VISIT1 and VISIT10 inclusively. As a general rule, it does not matter if all the variables are already present on the PDV, and their order on the PDV is not important. However, as with any list of variables, the usage itself can have unintended consequences.

In the following ARRAY statement only the first 10 visits will be included in the array; however, if one of these variables is not already on the PDV, it will be added.

```
array vis {10} visit1 - visit10;
```

The KEEP statement does not establish variables, so unlike the previous ARRAY statement, variables in the list that are not already on the PDV will cause an error. If there is no VISIT3 variable on the PDV, the following KEEP statement will produce a warning.

```
keep visit1 - visit10;
```

This type of list can be used wherever a variable list is expected. This includes statements and options such as: KEEP, DROP, VAR, RETAIN. Functions that accept a list of values, e.g., MIN, MAX, MEAN, require the use of the OF operator to prevent confusion with a subtraction.

```
m = max(of visit1 - visit10);
```

PDV Order Dependent Lists (Named Range)

When the order of the variables on the PDV is known, you can use the double dash to specify the list. Unlike the common prefix variable list shown above, the order of the variables on the PDV is very important and this form of variable list cannot be used to create variables. The following PROC CONTENTS step shows the variables in ADVRPT.DEMOG and their relative position on the PDV (through the use of the VARNUM option – the VARNUM option replaces the now outdated POSITION option).

```
title1 '2.6.1 Variable Shorthand Lists';
title2 'List of variables and their positions';
proc contents data=advrpt.demog varnum;
   run;
```

The resulting listing shows the names of the variables, their attributes, and their order.

```
2.6.1 Variable Shorthand Lists
List of variables and their positions

The CONTENTS Procedure

                    Variables in Creation Order

  #     Variable    Type    Len    Format    Label

  1     subject     Num      8
  2     CLINNUM     Char     6                clinic number
  3     LNAME       Char     10               last name
  4     FNAME       Char     6                first name
  5     SSN         Char     9                social security number
  6     SEX         Char     1                patient sex
  7     DOB         Num      8     DATE7.     date of birth
  8     DEATH       Num      8     DATE7.     date of death
  9     RACE        Char     1                race
 10     EDU         Num      8                years of education
 11     WT          Num      8                weight in pounds
 12     HT          Num      8                height in inches
 13     SYMP        Char     2                symptom code
 14     death2      Num      8     DATE9.
```

The variable list LNAME--SYMP includes all variables (numeric and character) in the data set ADVRPT.DEMOG except SUBJECT, CLINNUM, and DEATH2.

Inclusion of the list modifiers NUMERIC and CHARACTER can be used to restrict the list to just numeric or just character. Again the list is order dependent and includes the endpoints, assuming they are the correct type. The list DOB-numeric-HT excludes RACE, while the list SEX-character-SYMP, contains only three variables (SEX, RACE, and SYMP).

Inclusion of an incorrect type does not cause an error. The designation death-character-symp will correctly contain the two variables RACE and SYMP.

Unlike the list abbreviation with a single dash (common prefix numbered list), this list form cannot be used to create variables or add variables to the PDV. It can, however, be otherwise used where you need a list of variables.

CAVEAT
Since the order of the variables on the PDV is generally of secondary importance to most SAS programmers be very careful when using these forms of lists. If the variable order changes for some reason, the list may no longer be what you intend.

Using the Colon Operator (Name Prefix)
Variables named with a common prefix (with or without a numeric suffix) can be listed by following the prefix with a colon.

For the data set STATS generated by the following PROC SUMMARY, you could select all the statistics associated with HT by using the list HT_ : ❸.

```
proc summary data=advrpt.demog;
   class race edu;
   var ht wt; ❶
   output out=stats
          mean=
          stderr=
          min=/autoname; ❷
   run;

proc print data=stats;
   id race edu;
   var ht_: ❸;
   run;
```

❶ The analysis variables HT and WT are used to generate a series of statistics. The names of these statistics are automatically generated ❷, and are of the form of *analsysisvariable_statistic* (see Section 7.2 more details on the AUTONAME option).

❸ The list of all statistics generated for the HT variable will be printed using the name prefix list in the VAR statement.

MORE INFORMATION
This list abbreviation is used in a PROC TRANSPOSE example in Section 2.5.2.

Special Name Lists
Three name lists exist that allow you to address variables by their type. These include:

- _CHARACTER_   All character variables
- _NUMERIC_   All numeric variables
- _ALL_   All variables on the PDV

Since each of these lists pertains to the current list of variables, they will not create variables. In each case the resulting list of variables will be in the same order as they are on the Program Data Vector.

The _ALL_ list abbreviation is used in the following DATASETS step to remove the label and format attributes from a data set. This example was suggested by SAS Sample #25052.

```
proc datasets lib=work nolist;
   modify demog;
      attrib _all_ label=' '
                    format=;
      contents;
   quit;
```

The MODIFY statement opens the WORK.DEMOG data set and the ATTRIB statement is applied to all the data set's variables by listing the variables using the _ALL_ list abbreviation.

SEE ALSO
This example with further explanation can be found in the SAS Sample library at http://support.sas.com/kb/25/052.html.

2.6.2 Understanding the ORDER= Option

The ORDER= option can be used with most procedures that classify or summarize data. It allows us to control both the analysis and display order of information without physically sorting the data. Depending on the procedure the option may be applied on the PROC statement or on one or more of the supporting statements, such as the CLASS statement.

The option can take on the values of:

Option Value:	Order is based on:
INTERNAL	the unformatted values (like PROC SORT)
FORMATTED	the formatted value
FREQ	the descending frequency
DATA	the order of the data values

For most procedures the default value for ORDER= is INTERNAL.

In each of the following examples, a simple PROC MEANS with a single classification variable (SYMP) is used to demonstrate the effect of the ORDER= option.

ORDER=INTERNAL

This is typically the order of the variable if it had been sorted with PROC SORT, and is usually the procedure's default. Its alias is UNFORMATTED.

```
title2 'order=internal';
proc means data=advrpt.demog
          n mean;
   class symp;
   var ht;
   run;
```

The ORDER= option is not specified and the PROC MEANS default order for all classification variables is ORDER=INTERNAL. As a result the symptoms appear in alphabetical order (SYMP is character).

2.6.2 Understanding ORDER=
order=internal

The MEANS Procedure

Analysis Variable : ht height in inches

symptom code	N Obs	N	Mean
01	4	4	67.5000000
02	10	10	66.8000000
03	4	4	66.5000000
04	13	13	68.6923077
05	8	8	67.5000000
06	11	11	64.0000000
09	2	2	68.0000000
10	13	13	68.5384615

ORDER=FORMATTED

When the ORDER=FORMATTED option is used the values are first formatted and then ordered.

```
proc format;
    value $SYMPTOM
        '01'='Sleepiness'
        '02'='Coughing'
        '03'='Limping'
        '04'='Bleeding'
        '05'='Weak'
        '06'='Nausea'
        '07'='Headache'
        '08'='Cramps'
        '09'='Spasms'
        '10'='Shortness of Breath';
    run;

title2 'order=formatted';
proc means data=advrpt.demog
          n mean
          order=formatted;  ❶
    class symp;
    var ht;
    format symp $symptom.;  ❷
    run;
```

2.6.2 Understanding ORDER= order=formatted

The MEANS Procedure

Analysis Variable : ht height in inches

symptom code	N Obs	N	Mean
Bleeding	13	13	68.6923077
Coughing	10	10	66.8000000
Limping	4	4	66.5000000
Nausea	11	11	64.0000000
Shortness of Breath	13	13	68.5384615
Sleepiness	4	4	67.5000000
Spasms	2	2	68.0000000
Weak	8	8	67.5000000

❶ The ORDER=FORMATTED option on the PROC statement is applied to *all* classification variables.

❷ The user-defined format $SMPTOM. is applied to the classification variable SYMP.

The formatted values now determine the order of the rows for the classification variable.

ORDER=FREQ

The frequency of the levels of the classification variable is used to determine the order when ORDER=FREQ is used.

```
title2 'order=freq';
proc means data=advrpt.demog
          n mean;
    class symp / order=freq;  ❸
    var ht;
    run;
```

❸ Placing the ORDER= option on the CLASS statement instead of on the PROC statement, allows the selective application of the option to only specific classification variables (see Section 7.1.3 for more on the use of options on the CLASS statement).

The symptoms are now listed in order of decreasing frequency. The CLASS statement also supports the ASCENDING option which can be used with the ORDER=FREQ option to list the levels in ascending order.

2.6.2 Understanding ORDER= order=freq

The MEANS Procedure

Analysis Variable : ht height in inches

symptom code	N Obs	N	Mean
10	13	13	68.5384615
04	13	13	68.6923077
06	11	11	64.0000000
02	10	10	66.8000000
05	8	8	67.5000000
03	4	4	66.5000000
01	4	4	67.5000000
09	2	2	68.0000000

ORDER=DATA
The order of the classification variables will reflect their order in the data itself. The first level detected will be written first. The data do not have to be in any particular order.

```
title2 'order=data';
proc means data=advrpt.demog
        n mean
        order=data;
   class symp;
   var ht;
   run;
```

Symptom 02 (coughing) is the first symptom in the data, followed by 10 and 06.

MORE INFORMATION
Missing values of classification variables are not normally included in the table, see Section 7.1.1 to change this behavior. The ORDER= option is discussed in terms of the TABULATE procedure in Section 8.1.5.

2.6.2 Understanding ORDER= order=data

The MEANS Procedure

			Analysis Variable : ht height in inches
symptom code	N Obs	N	Mean
02	10	10	66.8000000
10	13	13	68.5384615
06	11	11	64.0000000
04	13	13	68.6923077
03	4	4	66.5000000
09	2	2	68.0000000
05	8	8	67.5000000
01	4	4	67.5000000

2.6.3 Quotes within Quotes within Quotes

The quote mark is used to identify constant text to the parser. Sometimes that quoted string of constant text will itself contain quotes. Fortunately SAS comes with both single and double quotes and either can be used within the other. But what happens if you need to call a macro variable within the interior string? Regardless of which type is used on the inside, the macro variable will be within single quotes and, therefore, will *probably* not be resolved.

Each of the following three statements has a quoted string within a quoted string. And each executes successfully.

A DEFINE routine in a REPORT step compute block:

```
call define(_col_,'style', 'style={flyover="myloc"}');
```

An X statement executing a Windows DIR command:

```
x 'dir "c:\myloc\*.sas" /o:n /b > c:\myloc\pgmlist.txt';
```

A DM statement being used to reroute the LOG file (see Section 14.4.2):

```
dm 'log; file "c:\myloc\logdump1.log"';
```

Now assume that we need to embed a macro variable in the above examples. Since macro variables tend not to be resolved when they are used inside of single quotes, we need to understand not only how the statements are parsed and executed, but how we can recode them.

```
%let temp = myloc;

call define(_col_,'style', 'style={flyover="&temp"}'); ❶

x 'dir "c:\&temp\*.sas" /o:n /b > c:\&temp\pgmlist.txt'; ❷

dm 'log; file "c:\&temp\logdump1.log"'; ❸
```

Simply substituting the macro variable into the statement does not always work. Interestingly it does not necessarily fail either.

❶ While it is generally true that macro variables will not be resolved when they occur within single quotes, this is not strictly true. The CALL DEFINE routine is only called from within a PROC REPORT compute block (not shown), and because of the way that these blocks are executed the macro variable will be resolved even though it is inside of single quotes. Nothing special needs to be done.

❷ The X statement will not work as it is currently coded; the macro variable will not resolve. When we pass a path to the OS under Windows, the path should be enclosed in double quotes. Under the current versions of SAS the X statement *generally* no longer requires the use of the quotes that surround the

```
x dir "c:\&temp\*.sas" /o:n /b > c:\&temp\pgmlist.txt;
```

quotes that surround the command that is to be passed to the OS. This simplifies the statement and eliminates the problem. When this does not work, consider one of the solutions used for the DM statement.

❸ In the DM statement the string following the keyword DM must be quoted, and the macro variable will not be resolved. The macro quoting functions can be helpful by temporarily masking the single quotes until after the macro variable has been resolved. Since the single quote has been

```
dm %unquote(%bquote(')log%bquote(;)file "c:\&temp\logdump1.log" %bquote('));
```

masked, the semicolon used to separate the two DM commands must also be temporarily masked. Prior to execution the macro quoting is removed using the %UNQUOTE function. An approach similar to this may be needed in the FILENAME statement as well. Here we are using the pipe engine to route the results of the DOS command to a virtual (piped) file.

```
filename list pipe %unquote(%bquote(')dir "&temp\*.rtf" /o:n /b %bquote('));
```

In fact, since we are delaying recognition of the quote marks, we do not even need to use both types of quote marks. In the X statement shown here, the %STR function is used to delay

```
x "%str(md %"&temp\output%"; )";
```

recognition of the inner pair of double quotes until after the outer ones have been utilized.

Within SAS it is not too unusual to be able to delay the parser's recognition of a character by doubling it. This technique was common before double quotes were introduced into the language. To show an apostrophe in a title statement two single quotes were used. This works because the parser sees the two single quotes and in a second

```
title1 'Tom''s Truck';
```

pass of the string, converts them to a single quote mark (an apostrophe). This technique still works and we can use it to our advantage in the DM statement that we have been working with.

Here only double quote marks are used. Notice a single double quote at the start, a double double quote in the middle of the string, and a triple double quote at the end. This will require three passes for the parser to resolve all the strings. In the meantime the macro variable will have been resolved.

```
dm "log; file ""c:\&temp\logdump1.log""";
```

This works because the quote marks are being used by the parser to 'mark' the strings in such a way as to tell the parser how to handle the string. Double double quote marks are resolved to a just one 'mark' in a second pass of the parser, and by then the macro variable has been resolved.

Rewriting the DM statement using single quotes would only partially be successful. The parser would handle the resolution process for the quotes the same; however, since the first pair of single quotes still resolves to a single quote, that quote would prevent the resolution of the macro variable.

```
dm 'log; file ''c:\&temp\logdump1.log''';
```

More rarely you may need a third level of quoting. For this problem what we really want is a third type of quote mark. We only have two, however, so again we can take advantage of the parsing process, and consequently expand the previous technique to additional levels. Surround the whole string with single quotes, as you have already done. Then change each interior single quote to two single quotes (not a double quote). This forces the parser to take a second pass.

2.6.4 Setting the Length of Numeric Variables

While we regularly reduce or control the length of character variables, we more rarely do so for numeric variables. In both cases reduction of variable length can be a successful strategy to reduce data set storage requirements. However, there are specific issues associated with reduction of the length of numeric variables – reducing the length of a numeric variable, especially non-integers, can drastically reduce the precision of the variable. The documentation associated with your version of SAS and your OS will cover topics such as the loss of precision and the size of integers that can be stored with a given length.

So how can you minimize storage costs by controlling length? With character variables it is easy; use the minimum length that avoids truncation. For numeric variables it is less straightforward. One of the first considerations is the value itself. If a numeric code is just a code, such as clinic number, and will not be used in calculations, it should generally be stored as a character variable. An exception would be social security numbers (SSN) and Employer Identification Numbers (EIN), which can be stored in 8 bytes as a numeric variable, but require at least 9 bytes as character variables.

While SAS dates, which are always integer values, can be safely stored in four bytes, most users and some companies (as company policy) never reduce (or never allow the reduction of) the length of numeric variables – "just in case." Given that storage is generally cheap and access is generally fast, my rule of thumb is that codes (regardless of content) are text. Only numbers are stored as numeric values. And I only rarely reduce the length of a numeric value.

While the readers of this book are probably an exception, most users are not sophisticated enough to understand the subtle implications of reducing length for numeric variables. If they do understand AND/OR they know that they are only dealing with integers, then some reduction of storage requirements can be achieved by reducing the length of numeric variables.

2.7 WHERE Specifics

The WHERE statement (DATA and PROC steps), WHERE= data set option, and PROC SQL WHERE clause provide subsetting capabilities that are not otherwise available. While the subsetting IF in the DATA step can have similar syntax and often similar results, the filter generated by a WHERE can be substantially different in efficiency, usage, syntax, and result. The differences are important.

When importing or exporting data, it is often necessary to filter portions of data that are to be transferred. There are several ways to provide this filtering, and building a WHERE clause, which can be used in a variety of data import and export situations, is a core methodology. It can be used in both procedure and DATA steps, and can be generated using statements, options, and clauses. Having a firm understanding of the capability of the WHERE can have a major impact when transferring large amounts of data.

The WHERE is a primary tool for the creation of subsets of data. It can be used as a statement, data set option, and as an SQL clause. Not only is it flexible in how it can be used, it has a number of inherently beneficial properties. The following are a few comments about the WHERE that mostly fall into the category of "Did you know that....".

When creating a data subset in a DATA step the WHERE generally tends to be more efficient than the subsetting IF statement. The selection criteria associated with the IF statement is applied to the values in the PDV, which means that every observation is read, and then potentially discarded. The WHERE clause on the other hand is applied before the observation is read, which can save resources by minimizing the I/O. The WHERE does have some additional overhead and the efficiency gains are first noticed and become more pronounced (compared to the subsetting IF statement) as the fraction of discarded data becomes larger.

Clearly the WHERE clause must evaluate observations; however, sometimes complete blocks of observations can be eliminated depending on what SAS knows about the data. When the WHERE clause is applied to an indexed data set, the WHERE clause will take advantage of the indexes to optimize the selection process.

In the DATA step and in procedures, a WHERE clause can be established either through the use of the WHERE statement ❶ or as a WHERE= data set option ❷ on the incoming data set. As a general rule best practices suggest that the WHERE statement should be used only when the WHERE= data set option is not available, as the use of the data set option tends to make the code easier to understand. The following two steps yield the same subset of the observations in the data set ADVRPT.DEMOG.

```
title1 'E1.4a WHERE Statement';
proc print data=advrpt.demog;
   var lname fname sex dob;
   where year(dob)>1960;❶
   run;
```

```
title1 'E1.4b WHERE Data Set Option';
proc print data=advrpt.demog(where=(year(dob)>1960))❷;
   var lname fname sex dob;
   run;
```

In PROC SQL there are three ways of using the WHERE clause:

- WHERE clause in a pass-through
- WHERE clause in the SAS SQL
- WHERE= data set option

Obviously if we are writing code that will be passed through SQL to a database system other than SAS, Oracle for instance, the WHERE= data set option cannot be used. The WHERE clause in SQL pass-through code has to be appropriate to the receiving database, and SAS data set options can only be applied to SAS data sets. More on the efficiency issues of an SQL pass-through is discussed in Section 1.4.

In a SAS SQL step there can be performance differences between the WHERE clause and the WHERE= data set option. As was mentioned above, the WHERE= data set option is generally optimized for use with indexes. Depending on the type of JOIN the WHERE clause will sometimes be applied after the read.

2.7.1 Operators Just for the WHERE

While the basic syntax of the WHERE statement is similar to that of the subsetting IF statement, there are several operators that can be used only with the WHERE (statement, data set option, or SQL clause). These include:

- BETWEEN   Builds an inclusive range
- CONTAINS   String search
- IS MISSING   Check for missing values
- LIKE   Pattern matching
- SAME   WHERE clause concatenation
- =*   Sounds like

The examples below all use the WHERE statement, but these operators apply to the WHERE= data set option and the SQL WHERE clause as well.

BETWEEN

The BETWEEN operator allows us to establish an inclusive range.

```
title2 'BETWEEN';
proc print data=advrpt.demog;
   var lname fname edu;
   where edu between 15 and 17;  ❶
/*   where 15 le edu le 17;*/  ❷
   run;
```

❶ The acceptable range for EDU is between 15 and 17 inclusively.

❷ The same list could have been established using this compound expression, which can also be used in an IF statement.

The negation of this range is requested by using the NOT operator with the BETWEEN operator. The following two WHERE statements are equivalent to each other and are the exact opposites of those in the previous PROC PRINT step.

```
where edu not between 15 and 17;
where edu lt 15 or edu gt 17;
```

CONTAINS

The CONTAINS operator works much like the INDEX function to determine if a text string can be found within another string. The word CONTAINS can be replaced with its mnemonic, the question mark (?).

```
title2 'CONTAINS';
proc print data=advrpt.demog;
   var lname fname edu;
   where lname contains 'son'; ❸
/*   where lname ? 'son';*/ ❹
/*   where index(lname,'son');*/ ❺
   run;
```

❸ All last names that contain the letters 'son' will be printed. Like all string comparisons the search is case sensitive.

❹ The question mark could be used to replace the CONTAINS operator.

❺ The INDEX function could also be used.

CONTAINS is negated by preceding the operator with a NOT or other negation mnemonic.

IS MISSING

The IS MISSING operator can be used to check for missing values. One advantage is that it can be used to check for either a numeric or a character missing value. Either the IS MISSING or the IS NULL operator can be used.

```
title2 'IS MISSING';
proc print data=advrpt.demog;
   var lname fname edu symp;
   where edu is missing or symp is missing; ❻
/*   where edu is null or symp is null;*/ ❼
/*   where edu = . or symp = ' ';*/ ❽
   run;
```

❻ The syntax is the same for numeric variables (EDU) and character variables (SYMP).

❼ IS NULL can be used instead of IS MISSING.

❽ When checking for missing values using the 'traditional' approach the programmer must be aware of the variable's type.

Negation is formed using NOT (or other negation operator). The NOT may appear either before the IS or between the IS and MISSING.

```
where edu is not missing or symp not is missing;
```

LIKE

The LIKE operator allows for simple pattern matching. When appropriate, more complex pattern matching can be accomplished using regular expressions in the RX() family of functions. This operator uses the percent sign (%) and the underscore (_) as wildcards. The % will stand for any number of characters and the _ is a place holder for exactly one character.

The following table shows some examples using the LIKE operator and alternate equivalent expressions.

Using the LIKE Operator	Without the LIKE Operator	What It Does
`lname like 'S%'`	`lname =: 'S'` `substr(lname,1,1) = 'S'`	Find all last names beginning with a capital S.
`lname like '%ly%'`	`index(lname,'ly') > 0` `lname contains 'ly'`	Find all last names containing an 'ly'.
`lname like 'Ch__'`	`substr(lname,1,2) = 'Ch' and` `length(lname)<5`	Any two, three, or four letter last names starting with Ch.

When using the % and _ with the LIKE operator, you need to be careful, as it is possible to return unanticipated values.

- A trailing _ will not select anything if the _ is past the length of the variable.

- Whenever the % is followed by other search characters, be sure to enclose the string in single quotes to prevent the macro parser from interpreting the % as a macro trigger.

- When the searched string contains either a _ or a % there can be confusion between the wildcards and the actual characters. Be sure that you specify what you really mean. Since the CONTAINS operator does not utilize wildcards, it can be used when your target string contains an underscore or percent sign.

Negation of the LIKE operator is achieved using the standard negation operators.

SAME

The SAME operator allows you to specify a composite clause through the use of separate WHERE statements. Primarily used in interactive or run-group processing, it has little use in other programming situations, since the clause cannot be maintained across step boundaries.

The first clause is specified as usual, and the second is appended with the SAME operator. If the

```
proc print data=advrpt.demog;
   var lname fname edu symp;
   where lname like 'S%';
   where same edu le 15;
   run;
```

SAME operator had not been included on the second WHERE statement, the second clause would have replaced the first. When joining two WHERE clauses with the SAME operator, the two clauses are effectively joined with an AND (both clauses have to be true for the overall result to be true).

```
   where lname like 'S%';
   where same and edu le 15;
```

It is common to explicitly specify the AND on the subsequent WHERE clauses. This can reduce ambiguity.

In both cases the resulting WHERE statement could have been written as:

```
   where lname like 'S%' and edu le 15;
```

Sounds like

The sounds like operator, which is coded using the mnemonic = *, uses the same algorithm as the SOUNDEX function:

- The first letter is preserved
- Vowels are eliminated
- Double letters are compressed
- The remaining letters are converted to numbers using a scheme that nominally groups letters that sound similarly in English

In theory two words with similar pronunciation will yield the same code.

```
proc print data=advrpt.demog;
   var lname fname dob;
   where lname =* 'che';
   run;
```

In this example we are searching for all patients with last names that sound something like 'che'.

```
2.7.1 Operators Just for the WHERE
Sounds like

Obs    LNAME     FNAME        DOB

 14    Chou      John      15MAY58
 15    Chu       David     18JUN51
```

The resulting listing shows two names that match the requested text string. Since vowels are dropped, as is the silent 'h', the portion that is actually used to form the comparison is the 'C'.

MORE INFORMATION
There can be serious and sometimes unanticipated consequences for using the MIN and MAX operators. These operators are available to expressions in both the WHERE and IF statements, but they behave differently depending on how they are used. Please review Section 2.2.5 before using either of these two operators or their mnemonics.

2.7.2 Interaction with the BY Statement

When a WHERE clause is created and a BY statement is also present, the groups of observations formed using the BY variables are created after the application of the WHERE clause. This means that any FIRST. or LAST. processing will be applied only to those observations that meet the WHERE criteria. The result can be quite different from using a subsetting IF statement to form the groups, as the BY groups are formed before the IF statement is applied.

The following DATA step correctly counts the number of distinct symptoms within each clinic (CLINNUM). Since we do not want to count observations without symptoms (SYMP=' '), a WHERE statement ❶ is used to exclude those observations, and the variable WSYMPCNT is used to accumulate the symptom count within each clinic (CLINNUM).

```
data WHEREcnt(keep=clinnum Wsympcnt);
   set demog(keep=clinnum symp);
   by clinnum symp;
   where symp ne ' ';  ❶
   if first.clinnum then do;  ❷
      Wsympcnt=0;
      end;
   if first.symp then Wsympcnt+1;  ❸
   if last.clinnum then output;  ❹
   run;
```

❷ The first observation for a given clinic is used to initialize the counter, WSYMCNT.

❸ Each distinct value of SYMP is counted. We could count either FIRST or LAST here.

❹ When we have processed all the rows for this clinic we know that we have the total count and we can output the result.

Using the subsetting IF statement ❺ instead of the WHERE

```
if symp ne ' ';  ❺
```

statement ❶ changes the way that the BY groups are formed. If the DATA step that uses the IF statement ❺ does not take this formation process into account, the results will be incorrect.

```
2.7.2 WHERE and BY Group Processing
Showing Counts

Obs    CLINNUM   IFsympcnt   Wsympcnt
 1     031234        2           2
 2     033476        3           1
 3     036321        2           2
 4     043320        3           1
 5     046789        1           1
 6     049060        1           1
```

```
2.7.2 WHERE and BY Group Processing

Obs    CLINNUM    SYMP    subject
 1     031234      01      127
 2     031234      01      168
 3     031234      04      110
 4     031234      04      156
 5     033476              148
 6     033476              161
 7     033476      09      116
 8     033476      09      157
 9     036321      02      128
10     036321      02      147
11     036321      06      135
12     036321      06      169
13     038362              145
14     038362              175
15     043320              132
16     043320              134
17     043320      02      124
18     043320      02      152
19     046789      10      107
20     046789      10      121
21     049060      02      101
22     049060      02      108
23     049060      02      164
24     049060      02      165
```

The table above shows the counts for these two methods for a few of the clinics. Clearly we can see that there is disagreement for two of these clinics, and the reason for this difference is at the heart of the problem.

Examination of the data used to form the counts shows that the variable SYMP has at least one missing value for each of the clinics that have an incorrect count.

When the subsetting IF statement ❺ executes, it eliminates these rows, but more importantly these rows are also the rows for which FIRST.CLINNUM is true. Since the row is eliminated, the counter cannot be reset ❷. This does not happen when the WHERE statement ❶ is used because the BY groups are formed after the elimination of rows. As a result FIRST.CLINNUM will be available for testing and will not be eliminated inappropriately.

You are not constrained to using the WHERE with BY-group processing, but you must be careful. If we remove the subsetting IF statement from the DATA step and add the same logic to the line that counts the symptoms, the counting problem is corrected.

```
if first.symp & symp ne ' ' then IFsympcnt+1;
```

Interestingly a comparison of the WHERE statement and the corrected logic using the IF statement highlight another difference between the two approaches.

```
2.7.2 WHERE and BY Group Processing
Showing Counts Using the Corrected IF

Obs    CLINNUM   IFsympcnt   Wsympcnt
 1     031234        2           2
 2     033476        1           1
 3     036321        2           2
 4     038362        0           .
 5     043320        1           1
 6     046789        1           1
 7     049060        1           1
```

Notice that a clinic number (038362), which was not in the previous report, now appears. All observations for this clinic have SYMP=' '; consequently, it was completely removed from consideration by both the WHERE and the subsetting IF statements ❶❺. Since the revised DATA step does not eliminate rows, and instead chooses which rows should be counted, this clinic now shows up with a count of 0. Of course you have to decide whether or not this clinic is appropriate for your report.

2.8 Appending Data Sets

There are several approaches that can be taken when appending two (or more) data sets. Each of these approaches has its own costs and capabilities. It is important for the programmer to understand the differences, similarities, and efficiencies of these techniques.

```
title '2.8 Appending Data Sets';
* Create a not so big data set;
data big;
   set sashelp.class
         (keep=name sex age height weight);
   where name > 'L';
   output big;
   * create a duplicate for Mary;
   if name=:'M' then output big;
   run;
data small;
   * The variable WEIGHT has been misspelled as WT;
   * The variables WT and HEIGHT are out of order;
   name='fred'; sex='m'; age=5;  wt=45; height=30;
   run;
```

The examples in this section use the data sets BIG and SMALL to build the data set BIGGER. The BIG data set, which is really only pretending to be big, is simply a portion of the familiar SASHELP.CLASS data set and SMALL is a single observation data set with nominally the same variables.

Notice, however, that the SMALL data set has the variable WT instead of WEIGHT, and the order of WT and HEIGHT on the PDV has been reversed.

2.8.1 Appending Data Sets Using the DATA Step and SQL UNION

One of the simplest, if not the most simple, approach for appending two data sets is through the use of the SET statement.

```
data bigger;
   set big small;
   run;
```

In this simplified example, a small transaction data set (SMALL) is appended onto a larger data set. The variable list in the new data set will be a union of the two data sets. Although this approach is sometimes necessary, it is unfortunately more commonly used by a programmer who does not understand the operations conducted by SAS in order to carry out the instructions. This DATA step will read and write each of the observations one at a time from the BIG data set before reading any of the observations from the SMALL

```
2.8 Appending Data Sets
Using the SET Statement

Obs    Name      Sex    Age    Height    Weight    wt

1      Louise    F      12     56.3      77        .
2      Mary      F      15     66.5      112       .
3      Mary      F      15     66.5      112       .  ❷
4      Philip    M      16     72.0      150       .
5      Robert    M      12     64.8      128       .
6      Ronald    M      15     67.0      133       .
7      Thomas    M      11     57.5      85        .
8      William   M      15     66.5      112       .
9      fred      m      5      30.0      .         45  ❶
```

data set. Since we are not doing anything with these observations (only reading and then writing), this is not very efficient. However it can be used to concatenate a very large number of data sets with a minimal amount of coding.

Variable attributes are determined from those in the left-most data set in which the variable appears, in this case the data set BIG. Variables unique to the SMALL data set will be added to the PDV on the right and subsequently to data set BIGGER. The order of the variables in data set SMALL is not important. ❶ Notice that the single observation contributed by data set SMALL is the last one in the listing, and the variable WT has been added last to the PDV for data set BIGGER. ❷ The duplicate observation for Mary correctly appears twice in data set BIGGER.

```
data bigger;
   set big
       small(rename=(wt=weight));
   run;
```

A simple RENAME= data set option for the data set SMALL corrects this naming issue and the data sets are appended correctly.

The use of the SQL UNION clause is similar to the previous DATA step, in that all observations from both data sets must be read (in the case of SQL, they are read into memory first) and then written. This SQL step is more sensitive than the DATA step to variable order and type. In fact it is the order of the variables in the second table, *and not the variable name*, that determines which column from data set SMALL is matched to which column in data set BIG. This can have disastrous consequences.

```
proc sql noprint;
create table bigger as
   select *
       from  big
   union
   select *
       from small;
   quit;
```

This SQL UNION produces almost the same data set as the SET in the previous example; however, the differences are very important. In data set SMALL the variable HEIGHT has a value of 30 and WT is 45. The SQL UNION has ignored the names of the variables in data set SMALL and has appended their values onto the BIGGER data set using position alone. ❸

Notice also that the duplicate observation for MARY has been

```
2.8 Appending Data Sets
Using SQL UNION

Obs    Name      Sex   Age   Height   Weight
 1     Louise    F     12    56.3      77
 2     Mary      F     15    66.5     112  ❷
 3     Philip    M     16    72.0     150
 4     Robert    M     12    64.8     128
 5     Ronald    M     15    67.0     133
 6     Thomas    M     11    57.5      85
 7     William   M     15    66.5     112
 8     fred      m      5    45.0      30  ❸
```

eliminated ❷. If the keyword ALL had been used with the SQL UNION operator (`UNION ALL`), the duplicate observation would not have been removed. The CORR keyword can also be used with SQL UNION. This keyword would both eliminate duplicate records and any variables that are not in common to both tables (`UNION CORR`).

In an SQL step we can duplicate the results of the DATA step's SET statement by naming the incoming variables, while also renaming WT. ❹ Variable order is determined on the SELECT clause (Height has been listed before Weight to match the order in data set BIGGER), and ❺ WT is renamed to WEIGHT. ❻ Notice that to prevent the elimination of duplicate observations, the ALL keyword has been added to the UNION statement.

```
proc sql noprint;
create table bigger as
   select *
       from  big
   union all  ❻
   select Name,Sex,Age,Height  ❹,
          wt as Weight  ❺
       from small;
   quit;
```

2.8.2 Using the DATASETS Procedure's APPEND Statement

The APPEND statement in PROC DATASETS is designed to efficiently append two data tables. The primary advantage of using PROC DATASETS' APPEND statement is that it does not read any of the observations from the data set named with the BASE= option ❶. The second data set (DATA= option ❷) is read and appended to the first. Rather than create a new data set, the BIG data set is to be replaced with the appended version.

```
proc datasets library=work nolist;
   append base=big ❶
           data=small ❷;
   quit;
```

APPEND assumes that both data tables have the same suite of variables and that those variables have the same attributes. The APPEND statement above fails because of the inconsistencies in the two PDVs:

- NAME is $8 in BIG and $4 in SMALL

- WT exists in SMALL, but is not found in BIG

Variables in the BASE= data set that are not in the DATA= data set do not create a problem.

Adding the FORCE option ❸ to the APPEND statement permits the process to take place despite the inconsistencies. The new version of the data set BIG will retain the same variables and variable attributes as were found on the first

```
proc datasets library=work nolist;
   append base=big
           data=small
           force ❸;
   quit;
```

version of data set BIG. A listing of the new version of data set BIG shows that the single observation from data set SMALL has been added; however, its value for WT has been lost. A warning is issued to the LOG for each of the inconsistencies.

```
2.8 Appending Data Sets
Using the APPEND Statement

Obs    Name      Sex   Age   Height   Weight
1      Louise    F     12    56.3     77
2      Mary      F     15    66.5     112
3      Philip    M     16    72.0     150
4      Robert    M     12    64.8     128
5      Ronald    M     15    67.0     133
6      Thomas    M     11    57.5     85
7      William   M     15    66.5     112
8      fred      m     5     30.0     .
```

The functionality of the APPEND procedure, which is no longer documented, has been incorporated into the DATASETS procedure's APPEND statement.

2.9 Finding and Eliminating Duplicates

When talking about duplicates, we need to be careful about our terminology. Duplicate observations are equivalent in all regards – the values of *all* the variables are the same. Observations with duplicate key variables (BY variables) may or may not be duplicate observations. Checks for duplicate key variables, such as the NODUPKEY option in PROC SORT, ignore all the variables that are not on the BY statement, and only compare values associated with the key variables.

The detection and elimination of duplicate observations can be very important, especially when dealing with data sets that should have no duplicates. There are several techniques for dealing with duplicate observations; however, they are not equally effective. It is also important to note that very often program authors are not as careful as they should be when distinguishing between duplicate observations and duplicate key values.

```
Obs    SUBJECT    VISIT         LABDT    potassium    sodium    chloride
  1      200        1      07/06/2006        3.7        14.0        103
  2      200        2      07/13/2006        4.9        14.4        106
  3      200        1      07/06/2006        3.7        14.0        103
  4      200        4      07/13/2006        4.1        14.0        103
```

This table shows the first few lines of the lab chemistry data. The variables SUBJECT and VISIT should form a unique key (they don't) and there should be no duplicate observations (there are – see observations 1 and 3 above). This table will be used in the examples below.

SEE ALSO
Kohli (2006) reviews some of the techniques shown in this section, as well as discussing some others.

2.9.1 Using PROC SORT

The NODUPLICATES option (*a.k.a.* NODUPREC and NODUPS) on the PROC SORT statement is often used in the mistaken belief that it removes duplicate observations. Actually it *will* eliminate duplicate observations, but only up to a point. This option only eliminates duplicate observations if they fall next to each other after sorting. This means that if your key (BY) variables are insufficient to bring two duplicate observations next to each other, the duplicate will not be eliminated.

To be absolutely certain that all duplicates are eliminated, the BY statement must contain either a sufficient key or all the variables in the data set (_ALL_). This is generally not very practical and certainly not very efficient. I have found that if your data set contains a derived variable such as a statistic, for instance a mean or variance, including that variable on the BY statement is *likely* to create a sufficient key so that the NODUPLICATES option will indeed eliminate all duplicate observations.

The data set used in the examples in this section, ADVRPT.LAB_CHEMISTRY, has 166 observations. These include three pairs of duplicate observations and two more pairs of observations with duplicate key variable values (SUBJECT VISIT).

After using the NODUPLICATES option with an insufficient key, the LOG shows that only 5 duplicate observations were eliminated.

```
proc sort data=advrpt.lab_chemistry
          out=none noduprec
          ;
   by subject;
   run;
```

```
NOTE: There were 169 observations read from
the data set ADVRPT.LAB_CHEMISTRY.
NOTE: 5 duplicate observations were deleted
```

Re-running the SORT step using all the variables in the data set shows, however, that 6 duplicate observations were eliminated.

```
proc sort data=advrpt.lab_chemistry
          out=none nodup;
   by _all_;
   run;
```

```
NOTE: There were 169 observations read from
the data set ADVRPT.LAB_CHEMISTRY.
NOTE: 6 duplicate observations were deleted.
```

The NODUPKEY option will successfully return a list of unique combinations of the BY variables, and does not suffer from any of the limitations of the NODUPLICATES option.

MORE INFORMATION
Section 4.1.1 discusses in more detail the NODUPREC option and its associated inability to remove duplicate observations. This section also presents the NODUPKEY and DUPOUT= options.

2.9.2 Using FIRST. and LAST. BY-Group Processing

In the DATA step checks for duplicate key fields can be implemented using the BY-group processing techniques known as FIRST. and LAST. processing. Because FIRST. and LAST. processing can only be used with variables listed in a BY statement, these checks are necessarily restricted to duplicates in the key fields.

```
title1 '2.9.2 Using FIRST. and LAST. Processing';
proc sort data=advrpt.lab_chemistry
                        (keep = subject visit labdt)
            out=labs;
   by subject visit; ❶
   run;
data dups;
   set labs;
   by subject visit; ❶
   if not (first.visit and last.visit); ❷
   run;
proc print data=dups;
   run;
```

The data are sorted and the same BY statement is used in both the SORT and DATA steps ❶. Inclusion of the BY statement automatically makes the FIRST. and LAST. temporary variables available for each variable on the BY statement. An observation with a unique set of key fields will necessarily be both the FIRST. and the LAST. observation for that combination of key variables. An observation that is not both FIRST. and LAST. ❷ will necessarily be non-unique.

```
Obs     SUBJECT     VISIT          LABDT

  1       200         1        07/06/2006
  2       200         1        07/06/2006
  3       200         4        07/13/2006
  4       200         4        07/13/2006
  5       200         7        08/04/2006
  6       200         7        08/04/2006
  7       200         9        09/12/2006 ❸
  8       200         9        09/13/2006 ❸
  9       200         9        09/13/2006
        portions of the table not shown
```

The listing of the duplicates shows those observations that do not have unique values of their key variables. Since we have only shown three variables, we do not know if the entire observation is duplicated or not. Certainly for SUBJECT 200 ❸ the lab date (LABDT) indicates that while the key fields are not unique, the observations are not necessarily duplicates.

Clearly this technique allows us to distinguish between unique and non-unique combinations of key variables, but does not create a data set with an exhaustive list of all possible unique combinations. However, to build a data set of all possible unique combinations of the key variables requires only minor changes to the DATA step. By changing the IF statement to a subsetting IF

```
data unique;
   set labs(keep=subject visit);
   by subject visit;
   if first.visit;
   run;
```

statement, which checks only for the first or last occurrence of the BY variable combination, we guarantee that each combination of the two BY variables will be unique.

MORE INFORMATION

The use of FIRST. and LAST. processing is described in more detail in Section 3.1.1.

SEE ALSO

The following SAS Forum thread contains examples of NODUPKEY, the DUPOUT option and the use of FIRST. and LAST. processing http://communities.sas.com/message/41965#41965.

2.9.3 Using PROC SQL

We can remove duplicate observations using an SQL step and the DISTINCT function. The asterisk in the DISTINCT * function is used to indicate that all the variables are to be considered when looking for duplicates. Since SQL holds the entire data set in memory, all observations – not just adjacent ones – are compared. The resulting data set will not contain any duplicate observations. Adding an ORDER BY clause will cause the new data set to be sorted.

```
proc SQL noprint;
create table nodups as
   select distinct *
       from advrpt.lab_chemistry
           order by subject,visit;
quit;
```

If you only want to create a list of unique key values, adding a KEEP= option to the incoming data set in the FROM clause will restrict the variables that are checked by the DISTINCT function. The SQL SET operators EXCEPT, INTERSECT, and UNION can also be used to return unique rows.

2.9.4 Using PROC FREQ

PROC FREQ can be used to build a data set of key variable combinations that are either already unique or already non-unique. It can also be used to create unique combinations of key variables.

The following code does not eliminate duplicates, but like the first example in Section 3.9.2 (FIRST. and LAST. processing), it only selects combinations that are already unique. PROC FREQ

```
proc freq data=advrpt.lab_chemistry;
   table subject*visit / noprint
                    out=unique(where=(count=1));
   run;
```

creates a data set with one row for each combination of SUBJECT and VISIT. The variable COUNT indicates how often that combination appears in the data. Using COUNT in a WHERE clause allows us to select for duplicated (COUNT>1) or unique (COUNT=1) combinations of the key variables (SUBJECT and VISIT).

The default data set that is created by the TABLE statement contains a list of unique combinations of the key variables. Using the KEEP= option is a simple way to save this list of unique

```
proc freq data=advrpt.lab_chemistry;
   table subject*visit / noprint
                    out=unique(keep=subject visit);
   run;
```

combinations and the result is sorted!

2.9.5 Using the Data Component Hash Object

The data component hash object can be used to eliminate both duplicate observations and duplicate key values. Because the hash object is loaded into memory it can be a fast alternative that does not require the incoming data set to be sorted.

An incoming data set can be loaded directly into the hash object or it can be loaded one observation at a time using one or more methods that have been designed to work with the data component objects.

Determining a Unique Key

In the following example the hash object is loaded from a SAS data set one observation at a time. Each observation is written to the object using the key variables and successive observations overwrite previous observations with the same combination of key values.

```
data _null_;
   if _n_=1 then do;  ❶
      declare hash chem (ordered:'Y') ;  ❷
      chem.definekey ('subject', 'visit');  ❸
      chem.definedata ('subject','visit','labdt');  ❹
      chem.definedone ();  ❺
   end;
   set advrpt.lab_chemistry end=eof;  ❻
   rc = chem.replace();  ❼
   if eof then chem.output(dataset:'nokeydups');  ❽
   run;
```

❶ The hash table must be defined before it can be used. This DO block will only be executed once. It is at this time that the hash object is instantiated and its structure is defined. This IF statement and its associated overhead could be eliminated if the SET statement ❻ was placed within a special type of DO loop known as a DOW loop (introduced in Section 3.9.1 and used later in this section).

❷ The CHEM hash table is established. The ORDERED: 'Y' option causes the table to be written in ascending order based on the values of the variables that have been named as key variables ❸.

❸ The key variables for the CHEM hash table are defined using the DEFINEKEY method. These variables are used to determine how an observation is to be written to the hash table when the REPLACE method ❼ is executed. Notice the use of the dot notation to form the association with the specific hash table and its key variables.

❹ The data variables that are to be transferred from the PDV to the hash table are specified. In this example we are interested in building a list of unique key variables, so LABDT is not needed. It has been included here to make it easier to see how the REPLACE method works (see ❼).

❺ The definition of the CHEM hash table is complete.

❻ Observations from the incoming data set are read sequentially using the SET statement.

```
2.9.5 Using the Hash Object
Eliminating Duplicate KEY Values

Obs     SUBJECT     VISIT          LABDT

 7        200         8        08/11/2006
 8        200         9        09/13/2006
 9        200        10        10/13/2006
        portions of the table not shown
```

❼ The REPLACE method is used to write the contents of the Program Data Vector to the CHEM hash object. If the current combination of key variable values already exists in the hash table, they will be replaced – not added. Subject 200 has three observations for VISIT 9 (lab dates were 9/12 and 9/13 – see table in Section 2.9.2). Because of the use of the

REPLACE method, the observation with a lab date of 9/13 overwrote the one for 9/12

❽ After the last observation has been read, the contents of the CHEM hash table are written to the data set WORK.NOKEYDUPS.

Eliminating Duplicate Observations

With only a slight modification, the previous example can be used to eliminate duplicate observations rather than duplicate key values. The difference is in the definition of the key variables ❾.

```
data _null_;
   if _n_=1 then do;
      declare hash chem (ordered:'Y') ;
      chem.definekey ('subject', 'visit','labdt', 'sodium', 'potassium', 'chloride'); ❾
      chem.definedata('subject', 'visit','labdt', 'sodium', 'potassium', 'chloride') ;
      chem.definedone () ;
   end;
   set advrpt.lab_chemistry end=eof;
   rc = chem.replace();
   if eof then chem.output(dataset:'nodups');
   run;
```

❾ The list of key variables has been expanded to include all the variables. Much like using the _ALL_ in the BY statement of PROC SORT (see Section 2.9.1), this forces the hash object to recognize and replace duplicate observations.

While the previous code does what we want it to do, it could be more efficient. There are two IF statements (❶ and ❽) that are executed for every observation on the incoming data set, but each is true only once. We can eliminate both IF statements by using what is commonly referred to as a DOW loop (DO-Whitlock). Named for Ian Whitlock, who popularized the approach by demonstrating its advantages, this loop places the SET statement inside of a DO UNTIL loop.

```
data _null_;
   declare hash chem (ordered:'Y') ;
   chem.definekey ('subject', 'visit','labdt', 'sodium', 'potassium', 'chloride');
   chem.definedata('subject', 'visit','labdt', 'sodium', 'potassium', 'chloride') ;
   chem.definedone () ;
   do until(eof);
      set advrpt.lab_chemistry end=eof;
      rc = chem.replace();
   end;
   chem.output(dataset:'nodups');
   stop;
   run;
```

Because all of the incoming observations are read inside of the DO UNTIL loop, there is only one pass of the DATA step during execution. Here the STOP statement is not necessary since we have read the last observation from the incoming data set. As a general rule the STOP provides insurance against infinite loops when processing using this approach.

If you want to eliminate duplicate observations you can take better advantage of the properties of the hash object. In the following DATA step the incoming data are loaded into the hash object directly using the DATASET: constructor on the DECLARE statement, and then written to the data set WORK.NODUPS using the OUTPUT method.

```
data _null_;
   length subject $3 ❿
          visit    8
          labdt    8
          sodium potassium chloride $12;
   declare hash chem (dataset:'advrpt.lab_chemistry', ordered:'Y') ;
      chem.definekey ('subject','visit','labdt','sodium','potassium','chloride');
      chem.definedata('subject','visit','labdt','sodium','potassium','chloride');
      chem.definedone () ;
   call missing(subject,visit,labdt, sodium, potassium,chloride);
   chem.output(dataset:'nodups');
   run;
```

❿ Since there is no SET statement, the attributes of variables in the data set created by the OUTPUT method must be established. In this example the attributes are defined using the LENGTH statement. The CALL MISSING routine initializes the variables and assigns them missing values.

Using the LENGTH statement to set the variable attributes requires a certain level of knowledge about the incoming data set. We can avoid the LENGTH statement and the use of the MISSING method by taking advantage of the information that SAS already knows.

In the following DATA step the SET statement will never be executed (the expression in the IF statement is false), so no observations are read; however, during DATA step compilation the attributes of the variables are loaded into the PDV. Since the last observation is not read from the incoming data set, the STOP statement is needed to close the implied loop created by the SET statement.

```
data _null_;
   if 0 then set advrpt.lab_chemistry(keep= subject visit labdt
                                       sodium potassium chloride);
   declare hash chem (dataset:'advrpt.lab_chemistry', ordered:'Y') ;
   chem.definekey ('subject', 'visit','labdt', 'sodium', 'potassium', 'chloride');
   chem.definedata('subject', 'visit','labdt', 'sodium', 'potassium', 'chloride') ;
   chem.definedone () ;
   chem.output(dataset:'nodups');
   stop;
   run;
```

MORE INFORMATION
DATA step component objects are discussed in more detail in Section 3.3.

SEE ALSO
Kohli (2006) includes a brief example of the use of the hash object to remove duplicate observations. Secosky and Bloom (2007) provide a nice introduction to DATA step component (HASH) objects.

2.10 Working with Missing Values

While even the most novice of SAS programmers is familiar with the general concept of working with missing values, most do not realize that there is a great deal more to the topic. To many programmers this seems like a simple topic; however, the ability to fully take advantage of missing values, both character and numeric, is essential when processing data within SAS.

MORE INFORMATION
The replacement of missing values in PROC TABULATE is discussed separately in Section 8.1.1.

SEE ALSO
Humphreys (2006) includes a number of nice examples and explanations on the use of missing values.

2.10.1 Special Missing Values

Although we usually think of a period (.) as the symbol for a numeric missing value, there are actually 28 different numeric missing values. In addition to the period, which is most commonly used, numeric missing values can also be designated by preceding each of the 26 letters of the alphabet (a through z) and the underscore with a period. These different values can be used to distinguish between kinds of missing values, such as a dropped sample as opposed to a sample that was not taken.

When using these special missing values we need to know how to read them, how to use them, and how they will be displayed. In the following step two different missing values are read from a flat file. These are then processed as a part of an expression. Notice that the missing values are designated by following the period with the designation letter.

```
data ages;
input name $ age;
if age=.y then note='Too Young';
else if age=.f then note='Refused';
datalines;
Fred 15
Sally .f
Joe .y
run;
```

```
2.10.1 Missing Numerics

Obs     name     age      note

 1      Fred     15
 2      Sally     F       Refused
 3      Joe       Y       Too Young
```

Interestingly, when this data set is printed, the special missing value is displayed capitalized and without the period.

```
data ages;
missing f y;
input name $ age;
if age=.y then note='Too Young';
else if age=.f then note='Refused';
datalines;
Fred 15
Sally f
Joe y
run;
```

When the data are coded without the dot in front of the letter, the MISSING statement can be used to declare specific letters as special missing values. The dot is still used when designating the missing value in code.

There is a hierarchy associated with the 28 numeric missing values, and understanding this hierarchy can become critical when comparisons between them are made. In terms of size (sort order) the traditional missing value (.) is neither the smallest nor the largest of the 28 types of numeric missing values. The ._ is smallest, and is the only missing value smaller than (.). The largest numeric missing value is .z.

Suppose we want to subset for all valid dates in a data set. The WHERE clause or subsetting IF statement might be written as `where date > .;` . However, this expression would only eliminate two of the 28 potential types of numeric missing values. In order to guarantee that all numeric missing values are eliminated, the expression should be written as `where date > .z;` . Conversely, if you are searching for the smallest numeric value, (._) is smaller than the traditional missing (.).

MORE INFORMATION
The .z missing value is used in a subsetting example in Section 2.3.1. A user-defined informat is created to import special codes that need to be mapped to special numeric missing values in Section 12.5.3.

2.10.2 MISSING System Option

The MISSING system option allows you to specify a character to display other than the period (.). Like all system option settings, once specified the replacement value remains in effect, persists, until the end of the SAS session, job, or until reset.

The data set SHOWMISS has three observations and two missing values, the special missing value .f (see Section 2.10.1) and a standard missing value. The MISSING option will not change how a missing value is read or how it is used in an expression; however, it does change how the missing value is displayed. ❶ Here the MISSING system option is given the value of 'X' (the use of the quotes is optional on the OPTIONS statement.

```
data showmiss;
input name $ age;
datalines;
Fred 15
Sally .f
Joe .
run;
options missing=X;   ❶
title2 'MISSING Text is: X';
proc print data=showmiss;
run;
```

Examination of the PROC PRINT results shows that special missing values (.f) are not replaced; however, the missing value for Joe's age ❷ has been replaced with an X.

Because you are limited to a single character when using the MISSING system option, it is often far more flexible to write and use a user-defined format to recode missing values (see Section 12.5.3).

```
2.10.2 Using the MISSING System Option
MISSING Text is: X

Obs     name        age

1      Fred         15
2      Sally        F
3      Joe          X  ❷
```

SEE ALSO
The SAS Forum thread found at http://communities.sas.com/message/57619#57619 discusses the use of the MISSING system option.

2.10.3 Using the CMISS, NMISS, and MISSING Functions

The CMISS and NMISS functions have been designed to count the number of arguments (character and numeric arguments respectively) with missing values, while the MISSING function detects whether or not its argument (numeric or character) is missing.

```
data cntmiss;
infile cards missover;
input (a b c) ($1.) x y z;
nmisscnt = nmiss(x,y,z);
cmisscnt = cmiss(a,b,c);
missval  = missing(x+y+z);
datalines;
abc 1 2 3

de  3 4 .
    . . .
    1 2 .a
ghi
run;
```

This example uses the NMISS and CMISS functions to count the number of numeric and character missing values within each observation. The expression used as the argument for the MISSING function will return a 1 if any one of the values of X, Y, or Z are missing.

```
2.10.3 Using the NMISS, CMISS and MISSING Functions
Noticing Missing Values

Obs  a  b  c  x  y  z  nmisscnt  cmisscnt  missval
 1   a  b  c  1  2  3     0          0         0
 2               .  .  .     3          3         1
 3   d  e     3  4  .     1          1         1
 4               .  .  .     3          3         1
 5            1  2  A     1          3         1
 6   g  h  i  .  .  .     3          0         1
```

When you do not know the variable names or you just do not want to list them, the NMISS and CMISS functions can still be used. A variant of the following expression was suggested on a SAS Forum thread to perform this count.

```
                  ❶        ❷                    ❶         ❸      ❹
tot_missing = nmiss( of _numeric_,1 ) + cmiss( of _character_, 'a' ) -1;
```

❶ The _NUMERIC_ and _CHARACTER_ variable list abbreviations (see Section 2.6.1) are used instead of explicit variable lists.

❷ A non-missing numeric constant has been added as an argument just in case there are no numeric variables.

❸ A non-missing character constant has been added as an argument to prevent an error if there are no character variables.

❹ The variable TOT_MISSING will always be missing at this point (unless its value is retained), therefore it will be counted by NMISS. Consequently we want to decrease the count by one.

MORE INFORMATION
The MISSING function is used in Section 2.2.6 to convert missing values to 0.

SEE ALSO
These two SAS Forum threads discuss missing value functions:
http://communities.sas.com/message/57614

http://communities.sas.com/message/57624.

2.10.4 Using the CALL MISSING Routine

Unlike the MISSING function which detects missing values, the CALL MISSING routine assigns missing values. The arguments to the MISSING routine can be numeric, character, or both. The arguments to this routine can also be variable lists, list abbreviations, and even calls to arrays.

```
data annual(keep=year q: totsales);
   set sashelp.retail(keep=sales date year);
   by year;
   retain q1-q4 .;
   array annual {4} q1-q4;
   if first.year then call missing(of annual{*});
   annual{qtr(date)}=sales;
   if last.year then do;
      totsales=sum(of q:);
      output annual;
   end;
   run;
```

In the example shown here the CALL MISSING routine is used to clear the values of an array (ANNUAL) by setting them all to missing. Rather than using a DO loop to step through the array one element at a time to assign the missing values, the MISSING routine allows us to access the array values much more efficiently.

SEE ALSO
This example is used in a sasCommunity.org tip:
http://www.sascommunity.org/wiki/Tips:Use_CALL_MISSING_to_Set_a_List_of_Variables_to_Missing.

The CALL MISSING routine is used to avoid uninitialized variable notes in the SAS Forum thread:
http://communities.sas.com/message/56784.

2.10.5 When Classification Variables Are Missing

Throughout SAS, when classification variables are missing, their associated observation is excluded from the analysis. This is true for procedures with explicit CLASS statements, such as PROC MEANS and PROC GLM, as well as for those with implicit classification variables, such as PROC FREQ and PROC REPORT. Sometimes this is the behavior that you want; however, often it is important that these observations not be removed. The MISSING option allows missing values to be valid levels of the classification variable.

The MISSING option can be used with most procedures that have either implicit or explicit classification variables. This option can be used on a CLASS statement or on the PROC statement. When used on the PROC statement the option applies to all the classification variables; however, when it is used on the CLASS statement it is only applied to those specific classification variables. In PROC FREQ the MISSING option can also be used as an option on the TABLES statement, and in PROC REPORT it can appear on the DEFINE statement.

MORE INFORMATION
The MISSING option on a CLASS statement is discussed in Section 7.1.1.

2.10.6 Missing Values and Macro Variables

The macro language does not support the concept of a missing value. While a macro variable can take on the value of a blank or a period, these values are not treated as missing values by the macro language. A macro variable can take on a null value; that is, the macro variable can store nothing. This is not possible for variables on a data set.

When working with null macro variables the syntax may at first look odd to the DATA step programmer.

```
%if &city = %then %do;
```

This %IF statement is considered to be standard syntax for comparing a macro variable (&CITY) to a null value. Since DATA step comparisons must have something on the right of the comparison operator, this form makes some macro programmers uneasy. Other methods for comparing against a null value include the use of a quoting function such as %STR. Since the macro variable can contain nothing the %LENGTH function can return a zero, and this can also be used to detect a null value in a macro variable.

```
% if &city = %str() %then %do;
```

```
% if %length(&city) = 0 %then %do;
```

2.10.7 Imputing Missing Values

There are a number of techniques that have been proposed for imputing missing values in a data set. These include various schemes using spline fitting techniques, which can be found in the SAS/GRAPH procedure G3GRID. The SAS/ETS procedure EXPAND and the SAS/STAT procedure STDIZE can also estimate missing values. Of these, however, only PROC MI (Multiple Imputation) has the primary objective of imputing missing values and is by far the most sophisticated.

This procedure works well; however, there is a caveat of which the user should be aware. Since the procedure calculates values based on the values of related variables it can be sensitive to changes in the order of the data. The PROC MI results may change, although usually not by a lot, just by changing the sort order of the data. And mere changes in the order of the variables in the VAR statement can also result in minor changes to the imputed values even with a fixed SEED value.

Neither of these situations is alarming, unless you encounter them and are not expecting them.

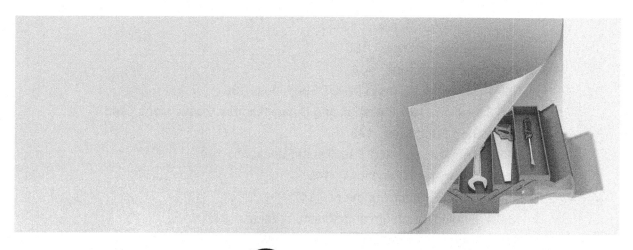

C h a p t e r **3**

Just In the DATA Step

The DATA step is the heart of the data preparation and analytic process. It is here that the true power of SAS resides. It is complex and rich in capability. A good SAS programmer must be strong in the DATA step. This chapter explores some of those things that are unique to the DATA step.

SEE ALSO

Whitlock (2008) provides a nice introduction to the process of debugging one's program.

3.1 Working across Observations

Because SAS reads one observation at a time into the PDV, it is difficult to remember the values from an earlier observation (look-back) or to anticipate the values of a future observation (look-ahead). Without doing something extra, only the current observation is available for use. This is of course not a problem when using PROC SQL or even Excel, because the entire table is loaded into memory. In the DATA step even the values of temporary or derived variables must be retained if they are to be available for future observations.

The problems inherent with single observation processing are especially apparent when we need to work with our data in groups. The BY statement can be used to define groups, but the detection and handling of group boundaries is still an issue. Fortunately there is more than one approach to this type of processing.

SEE ALSO

The sasCommunity.org article "Four methods of performing a look-ahead read" discusses a number of different methods that can be used to process across observations http://www.sascommunity.org/wiki/Four_methods_of_performing_a_look-ahead_read.

Another sasCommunity.org article "Look-Ahead and Look-Back" also presents methods for performing look-back reads. http://www.sascommunity.org/wiki/Look-Ahead_and_Look-Back.

Howard Schreier has written a number of papers and sasCommunity.org articles on look-ahead and look-back techniques, including one of the classics on the subject (Schreier, 2003). Dunn and Chung (2005) discuss additional techniques, such as interleaving, which is not covered in this book.

3.1.1 BY-Group Processing—Using FIRST. and LAST. Processing

FIRST. and LAST. processing refers to the temporary variables that are automatically available when a BY statement is used in a DATA step. For each variable in the BY statement, two temporary numeric variables will be created with the naming convention of FIRST.*varname* and LAST.*varname*. The values of these Boolean variables will either be 1 for true or 0 for false. On the first observation of the BY group FIRST.*varname*=1 and on the last observation of the BY group LAST.*varname*=1.

The data set REGIONS contains observations on subjects within clinics. The clinics are scattered across the country, which for administration purposes has been grouped into regions. The BY statement causes the FIRST. and LAST. temporary variables (temporary variables are not written to the new data set) to be created. Before the BY statement can be used, the data must be either sorted or indexed. Sorting REGIONS and clinic numbers, as is done in this example, using the BY statement `by region clinnum;` allows us to use the same BY statement in the DATA step.

```
proc sort data=regions;
   by region clinnum;
   run;

data showfirstlast;
   set regions;
   by region clinnum;
   FirstRegion = first.region;
   LastRegion = last.region;
   FirstClin = first.clinnum;
   LastClin = last.clinnum;
   run;
```

The following table demonstrates the values taken on by these temporary variables. FIRST.REGION=1 on the first observation for each value of REGION (obs.=1, 5, 11), while FIRST.CLINNUM=1 each time CLINNUM changes within REGION (obs=1, 3, 5, 7, 11). LAST.REGION and LAST.CLINNUM are set in a similar manner for the last values in a group.

Obs	REGION	CLINNUM	SSN	First Region	Last Region	First Clin	Last Clin
1	1	011234	345751123	1	0	1	0
2	1	011234	479451123	0	0	0	1
3	1	014321	075312468	0	0	1	0
4	1	014321	190473627	0	1	0	1
5	10	107211	315674321	1	0	1	0
6	10	107211	471094671	0	0	0	1
7	10	108531	366781237	0	0	1	0
8	10	108531	476587764	0	0	0	0
9	10	108531	563457897	0	0	0	0
10	10	108531	743787764	0	1	0	1
11	2	023910	066425632	1	0	1	0
12	2	023910	075345932	0	0	0	0
13	2	023910	091550932	0	0	0	1

. . . . . Portions of the output table not shown . . . .

These temporary variables can be used to detect changes of groups (group boundaries) within a data set. This is especially helpful when we want to count items within groups, which is exactly what we do in the following example. Our study was conducted in clinics across the country and the country is divided into regions. We need to determine how many subjects and how many clinics there are within each region.

```
data counter(keep=region clincnt patcnt);
   set regions(keep=region clinnum);
   by region clinnum; ❶
   if first.region then do; ❷
      clincnt=0;
      patcnt=0;
   end;

   if first.clinnum then clincnt + 1; ❸
   patcnt+1; ❹

   if last.region then output; ❺
   run;
```

The DATA step must contain a BY statement ❶ with the variables that form the groups of interest (regions and clinics).

The count accumulator variables (CLINCNT and PATCNT) must be initialized each time a new region is encountered. This group boundary is detected using FIRST.REGION ❷.

Using FIRST.CLINNUM as is done here ❸ or alternatively LAST.CLINNUM ensures that we count each clinic only once within each region.

In this incoming data set each observation represents a unique patient; consequently, each observation contributes to the patient count ❹.

After all observations within a region have been processed (counted) LAST.REGION=1, and the final counts are written to the new data set, COUNTER. ❺

Whenever you write a DATA step such as this one to count items within a group, watch to make sure that it contains the three primary elements shown in this example:

- Counter initialization ❷
- Counting of the elements of interest ❸❹
- Saving / writing the counters ❺

In this particular example, the statement at ❸ can be simplified and made more efficient at the same time by replacing the IF-THEN with a slightly modified SUM statement. The temporary variable FIRST.CLINNUM is always 1 or 0.

```
clincnt + first.clinnum;  ❸
```

A change in a higher order variable on the BY statement (FIRST. or LAST. is true) necessitates a change on any lower order variable (any variable to the right in the BY statement). This is stressed by the example shown here, where PART and UNIT are ordered using the BY statement BY PART UNIT;. Notice that whenever FIRST.UNIT=1 necessarily FIRST.PART=1. This is the case even when the same value of PART was in the previous observation (observation 3 is the first occurrence of UNIT='B', and FIRST.PART=1 although PART='x' is on observation 2 as well).

```
show lower level changes

                    First   Last    First   Last
Obs    unit   part   Unit    Unit    Part    Part

 1      A      w      1       0       1       1
 2      A      x      0       1       1       1
 3      B      x      1       0       1       0
 4      B      x      0       1       0       1
 5      C      x      1       1       1       1
```

3.1.2 Transposing to ARRAYs

Performing counts within groups, as was done in Section 3.1.1, is a fairly straightforward process because each observation is handled only one time. When more complex statistics are required, or when we need to be able to examine two or more observations at a time, temporary arrays can be used to hold the data of interest.

Moving items into temporary arrays allows us to process across observations. Moving averages, interval analysis, and other statistics are easily generated once the array has been filled. Essentially we are temporarily transposing the data using arrays in the DATA step (see Section 2.4.2 for more on transposing data in the DATA step).

In the following example an array of lab visit dates ❶ is used to determine the average number of days between scheduled lab visits. The dimension of the array VDATE is the upper bound of the number of possible visits. Since we are not interested in the dates themselves, the keyword _TEMPORARY_ is used to create a list of temporary variables.

We want to calculate the mean number of days for each subject, and FIRST.SUBJECT is used to detect the initial observation for each subject ❷. This allows us to initialize the array and other variables used to generate the statistics of interest.

```
data labvisits(keep=subject count meanlength);
   set advrpt.lab_chemistry;
   by subject;

   array Vdate {16} _temporary_;   ❶
   retain totaldays count 0;

   if first.subject then do;   ❷
      totaldays=0;
      count = 0;
      do i = 1 to 16;
         vdate{i}=.;
      end;
   end;
   vdate{visit} = labdt;   ❸
   if last.subject then do;   ❹
      do i = 1 to 15;
         between = vdate{i+1}-vdate{i};   ❺
         if between ne . then do;
            totaldays = totaldays+between;   ❻
            count = count+1;
         end;
      end;
      meanlength = totaldays/count;   ❼
      output;
   end;
   run;
```

❸ The visit number provides the index to the array and the date itself (LABDT) is loaded into the array.

Once all the visits for this subject have been loaded into the array (LAST.SUBJECT=1) ❹, we can process across the array in whatever ways we need to solve the problem at hand. In this case we are interested in determining the number of days between any two nominal visit dates. This difference is calculated ❺ and summed ❻ so that the mean number of days between visits can be determined ❼.

This solution only considers intervals between nominal visits and not between actual visits. If a subject missed visit three, the intervals between visit two and visit four would not be calculated (both are missing and do not contribute to the number of intervals because visit 3 was missed). The change to the program to use all intervals based on actual dates is simple because all the visit dates are already in the array. Although not shown here, the alternate DATA step is included in the sample code for this section.

The beauty of this solution is that arrays are expandable and process very quickly. Arrays of thousands of values are both common and reasonable.

When processing arrays, as was done here, it is often necessary to clear the array when crossing boundary conditions ❷. In this example a loop was used to set each value to missing and an alternate technique would be to avoid the DO loop by taking advantage of the CALL MISSING routine.

```
do i = 1 to 16;
   vdate{i}=.;
end;
```

```
call missing(of vdate{*});
```

3.1.3 Using the LAG Function

The LAG function can be used to track values of a variable from previous observations. This is known as a look-back read. Effectively the LAG function retains values from one observation to the next. The function itself is executable and values are loaded into memory when the function is executed. This has caused users some confusion. In the following example the statement `lagvisit= lag(visit);` ❷ loads the current value of VISIT into memory where it is held, along with the value from the previous observation. Whenever the variable LAGVISIT is used in an expression, the value of VISIT from the previous observation is returned. Because the current value must be loaded for each observation, the LAG function must be executed for each observation. When the LAG function is conditionally executed with an IF statement or inside of a conditionally executed DO block, the LAG function may not return what you expect.

The following example uses the LAG function to determine the number of days since the previous visit. The data are sorted and the BY statement is used ❶ to establish the FIRST.SUBJECT temporary variable. The LAG function is used to save the value of the VISIT and LABDT ❷ variables. The first observation for each subject is used to establish a base visit date and the remaining observations ❸ are used to calculate interval length from the previous visit ❹. For the first observation of each subject LAGVISIT and LAGDATE will contain the last values from the previous subject. These meaningless values are not used because they are excluded by the IF statement ❸.

```
data labvisits(keep=subject visit lagvisit
               interval lagdate labdt);
   set labdates;
   by subject; ❶

   lagvisit= lag(visit); ❷
   lagdate = lag(labdt); ❷

   if not first.subject then do; ❸
      interval = labdt - lagdate; ❹
      if interval ne . then output;
   end;
   format lagdate mmddyy10.;
   run;
```

This PROC PRINT listing of the resultant data table shows the relationship between the current and lagged values.

```
3.1.3 Using the LAG Function

SUBJECT     lagvisit    VISIT       lagdate         LABDT       interval

  200          1          2      07/06/2006     07/13/2006         7
               2          5      07/13/2006     07/21/2006         8
               5          6      07/21/2006     07/29/2006         8
               6          7      07/29/2006     08/04/2006         6
               7          8      08/04/2006     08/11/2006         7
               8          9      08/11/2006     09/12/2006        32
               9          9      09/12/2006     09/13/2006         1
               9         10      09/13/2006     10/13/2006        30

  201          1          2      07/07/2006     07/14/2006         7
               2          5      07/14/2006     07/21/2006         7
               5          4      07/21/2006     07/26/2006         5

          . . . .Portions of the table are not shown . . . .
```

The DIF function is designed to calculate the difference between a value and its lag value, as we have done here. In the previous example the INTERVAL could have been calculated using the DIF function.

```
interval= dif(labdt);
```

The full code for this solution is shown in example program E3_1_3b.sas.

SEE ALSO
Schreier (2007) discusses in detail the issues associated with conditionally executing the LAG function and shows how to do it appropriately.

3.1.4 Look-Ahead Using a MERGE Statement

While the LAG function can be used to remember or look-back to previous observations, it is more problematic to anticipate information on an observation that has not yet been read. The MERGE statement can be used to read two observations at once, the one of current interest and a portion of the next one.

In this example we need to calculate the number of days until the next laboratory date (LABDT), which will be on the next observation. The visits have been sorted by date within SUBJECT.

```
options mergenoby=nowarn ; ❶
data nextvisit(keep=subject visit labdt days2nextvisit);
   merge labdates(keep=subject visit labdt) ❷
         labdates(firstobs=2 ❸
                  keep=subject labdt ❹
                  rename=(subject=nextsubj labdt=nextdt)); ❺
   Days2NextVisit = ifn(subject=nextsubj,nextdt-labdt, ., .); ❻
   run;
```

❶ Since the MERGE statement is purposefully being used without a BY statement, the warning is turned off by using the SAS system option MERGENOBY= set to the value of NOWARN.

❷ The current observation is being read and only the variables of interest are kept.

❸ The FIRSTOBS= data set option causes this read of the LABDATES to be one observation ahead of the current observation ❷. The value of this option could be adjusted to allow a look-ahead of any number of observations.

❹ Only those variables specifically needed for the look-ahead are read.

❺ The look-ahead variables are renamed so that they can coexist on the Program Data Vector.

❻ The look-ahead calculations are performed. Here the number of days until the patient's next visit is calculated.

When the last observation is read from the primary ❷ data set, there will be no corresponding observation in the look-ahead data set ❸ and its associated variables will be missing.

For large data sets this technique has the disadvantage or requiring two passes of the data. It does not, however, require sorting but it does assume that the data are correctly arranged in the look-ahead order.

MORE INFORMATION
The complete code for this example shows the use of the GETOPTION function to collect the current setting of the MERGENOBY option and then reset it after the program's execution. The MERGENOBY option is discussed in Section 14.1.2.

SEE ALSO
Mike Rhodes was one of the first SAS programmers to propose a look-ahead technique similar to the one described in this section during a SAS-L conversation. It is likely that this "look-ahead" or "simulating a LEAD function" was first published in the original *Combining and Modifying SAS Data Sets: Examples, Version 6, First Edition,*" in example 5.6.

3.1.5 Look-Ahead Using a Double SET Statement

Using two SET statements within the same DATA step can have a similar effect as the MERGE statement. While this technique can offer you some additional control, there may also be some additional overhead in terms of processing.

Like in the example in Section 3.1.4, the following example calculates the number of days to the next visit. An observation is read ❶ and then the look-ahead observation is conditionally read using a second SET statement ❹.

```
data nextvisit(keep=subject visit labdt days2nextvisit);
   set labdates(keep=subject visit labdt) ❶
       end=lastlab; ❷
   if not lastlab then do; ❸
      set labdates(firstobs=2 ❹
                   keep=subject labdt
                   rename=(subject=nextsubj labdt=nextdt));
      Days2NextVisit = ifn(subject=nextsubj,nextdt-labdt, ., .); ❺
   end;
   run;
```

❶ The primary or current observation is read with the first SET statement.

❷ The END= option on the SET statement creates the temporary variable LASTLAB that will take on the value of 1 only when the last observation is being read.

❸ When the current observation is not the last, there will be at least one more look-ahead observation. Prepare to read that look-ahead observation. This is a minor additional overhead that the example in Section 3.1.4 does not have.

❹ The look-ahead observation is read by using the FIRSTOBS= data set option to provide an initial off-set from the current observation. This value could be changed to look-ahead more than one observation.

❺ The look-ahead calculations are performed.

A solution similar to the one shown here has been proposed by Jack Hamilton.

MORE INFORMATION
A double SET statement is used with the POINT= option to look both forward and backward in the second example in Section 3.8.1.

3.1.6 Look-Back Using a Double SET Statement

A look-back for an unknown number of observations is not easily accomplished using the LAG function. Arrays can be used (see Section 3.1.2), but coding can be tricky. Two SET statements can be applied to the problem without resorting to loading and manipulating an array.

In this example we would like to find all lab visits that fall between the first and second POTASSIUM reading that meets or exceeds 4.2 inclusively. Patients with fewer than two such readings are not to be included, nor are any readings that are not between these two (first and second) peaks. Clearly we are going to have to find the second occurrence for a patient, if it exists, and then look-back and collect all the observations between the two observations of interest. This can be done using two SET statements. The first SET statement steps through the observations and notes the locations of the peak values. When it is needed the second SET statement is used to read the observations between the peaks.

```
data BetweenPeaks(keep=subject visit labdt potassium);
   set labdates(keep=subject labdt potassium);
   by subject labdt;
   retain firstloc .  ❶
          found ' ';
   obscnt+1;  ❷
   if first.subject then do;  ❸
      found=' ';  ❹
      firstloc=.;
   end;
   if found=' ' and potassium ge 4.2 then do;  ❺
      if firstloc=. then firstloc=obscnt;  ❻
      else do;
         * This is the second find, write list;
         found='x';  ❼
         do point=firstloc to obscnt;  ❽
            set labdates(keep= subject visit labdt potassium)
                point=point;  ❾
            output betweenpeaks;  ❿
         end;
      end;
   end;
   run;
```

❶ The variables that are used to remember information across observations are retained.

❷ The observation is counted. In this case _N_ could have been used instead of OBSCNT; however, since _N_ counts passes of the DATA step, it is not as robust when data are read from within a DO loop, such as is done here.

❸ The retained variables must be initialized for each subject.

❹ Initialize the flag variable FOUND. This variable notes whether or not a second peak value has been found.

❺ When true, either the first or second peak (value of POTASSIUM >= 4.2) has been found.

❻ This must be the first peak and the current observation number is stored. If a second peak is found, this will become the starting point for reading the data between peaks.

❼ The flag variable FOUND notes that the second peak has been found and that we no longer need to search for additional observations for this subject.

❽ The DO loop index variable POINT cycles through the observation numbers between the two peaks.

❾ The POINT= option is used to indicate the temporary variable (POINT) that holds the observation number that is to be read.

❿ The observation is written to the new data set.

The program only collects the observations between the first two peaks. It could be modified to collect information between additional peaks by reinitializing the flag FOUND and by resetting FIRSTLOC to OBSCNT. This step also continues to process a subject even if a second peak has been found.

MORE INFORMATION
A double SET statement is used with the POINT= option to look both forward and backward in the second example in Section 3.8.1. A look-back is performed using an array in Section 3.10.2.

SEE ALSO
SAS Forum discussions of similar problems include both DATA step and SQL step solutions http://communities.sas.com/message/46165#46165.

3.1.7 Building a FIFO Stack

When processing across a series of observations for the calculation of statistics, such as running averages, a stack can be helpful. A stack is a collection of values that have automatic entrance and exit rules. Within the DATA step, implementation of stack techniques is through the use of arrays. In Section 3.1.2 an array was used to process across a series of values; however, the values themselves were not rotated through the array as they are in a stack.

Stacks come in two basic flavors: First-In-First-Out (FIFO) and Last-In-First-Out (LIFO). For moving averages the FIFO stack is most useful. In a FIFO stack the oldest value in the stack is removed to make room for the newest value.

In the following example a three-day moving average of potassium levels is to be calculated for each subject. The stack is implemented through the use of an array with the same dimension as the number of elements in the moving average.

```
data Average(keep=subject visit labdt
                  potassium Avg3day);
   set labdates;
   by subject;

   * dimension of array is number of
   * items to be averaged;
   retain temp0-temp2 ❶
          visitcnt .; ❷
   array stack {0:2} temp0-temp2; ❸
   if first.subject then do; ❹
      do i = 0 to 2 by 1; ❺
         stack{i}=.;
      end;
      visitcnt=0;
   end;
   visitcnt+1; ❻
   index = mod(visitcnt,3); ❼
   stack{index} = potassium; ❽
   avg3day = mean(of temp:); ❾
   run;
```

❶ The array elements are retained.

❷ The visits within subject are to be counted.

❸ The array has the same dimension as the number of elements to be used in the moving average. Notice that the array is indexed to start at 0, because the index is calculated with the MOD function ❼.

❹ For each subject it is necessary to clear the stack (array) and the counter (VISITCNT).

❺ The loop index steps through the elements of the array so that the individual values can be cleared. This DO loop could have been replaced with a `call missing(of stack{*});`

(see example program E3_1_7b.sas).

❻ The visit within subject is counted.

❼ The array index is calculated using the MOD function. This function is the key to rotating the values in and out of the stack. The newest value will always replace, and therefore remove, the oldest element in the stack. This is what makes this a FIFO stack.

❽ The value of POTASSIUM is loaded into the correct element in the stack.

❾ The average of the array elements (the values in the stack) is calculated.

Some coding alternatives can be found in example program E3_1_7b.sas.

MORE INFORMATION
A multi-label format is used to calculate moving averages without building a stack in Section 12.3.2.

3.1.8 A Bit on the SUM Statement

As we have seen in the other subsections of Section 3.1, in the DATA step it is necessary to take deliberate steps if we intend to work across observations. In this DATA step we want to keep an

```
data totalage;
   set sashelp.class;
   retain totage 0;  ❶
   totage = totage+age;  ❷
   run;
```

accumulator on AGE. ❶The variable must first be retained and initialized to 0. ❷ Then for each observation the AGE is added to the total (TOTAGE).

```
data totalage;
   set sashelp.class;
   totage+age;  ❸
   run;
```

The coding can be simplified by using the SUM statement. Since the SUM statement has an implied RETAIN statement and automatically initializes to 0, the RETAIN statement is not needed.

Some programmers assume that these two methods of accumulation are equivalent; however, that is not the

case, and the difference is non-trivial. Effectively the SUM statement calls the SUM function, which ignores missing values. If AGE is missing, the accumulated total value for either ❸ or ❹ will not be affected; however, the total at ❷ will be set to missing and will be unable to do further accumulations.

```
data totalage;
   set sashelp.class;
   retain totage 0;
   totage = sum(totage,age);  ❹
   run;
```

MORE INFORMATION
The sasCommunity tip
http://www.sascommunity.org/wiki/Tips:SUM_Statement_and_the_Implied_SUM_Function
mentions the use of the implied SUM function.

3.2 Calculating a Person's Age

The calculation of an individual's age can be problematic. Dates are generally measured in terms of days (or seconds if a datetime value is used), so we have to convert the days to years. To some extent, how we calculate age will depend on how we intend to use the value. The society's concept of age is different than the mathematical concept. *Age* in years is further complicated by the very definition of a year as one rotation of the earth around the sun. This period does not convert to an integer number of days, and it is therefore further complicated by leap years. Since approximately every fourth year contains an extra day, society's concept of a year as a unit does not have a constant length.

In our society we get credit for a year of life on our birthday. Age, therefore, is always an integer that is incremented only on our birthday (this creates what is essentially a step function). When we want to use age as a continuous variable, say as a covariate in a statistical analysis, we would lose potentially valuable information using society's definition. Instead we want a value that has at least a couple of decimal places of accuracy, and that takes on the characteristics of a continuous variable rather than those of a step function.

The following examples calculate a patient's age, at the date of their death (this is all made up data—no one was actually harmed in the writing of this book), using seven different formulas.

```
3.2 Calculating Age

    DOB     DEATH    age1      age2     age3   age4   age5    age6     age6a    age7

21NOV31   13APR86   54.3929   54.4301    55     54     55    54.3918   54.3918    54
03JAN37   13APR88   51.2745   51.3096    51     51     51    51.2759   51.2740    51
19JUN42   03AUG85   43.1239   43.1534    43     43     43    43.1233   43.1233    43
19JAN42   03AUG85   43.5373   43.5671    43     43     43    43.5370   43.5370    43
23JAN37   13JUN88   51.3867   51.4219    51     51     51    51.3878   51.3863    51
18OCT33   21JUL87   53.7550   53.7918    54     53     54    53.7562   53.7562    53
17MAY42   03SEP87   45.2977   45.3288    45     45     45    45.2986   45.2986    45
07APR42   03AUG87   45.3224   45.3534    45     45     45    45.3233   45.3233    45
01NOV31   13APR86   54.4476   54.4849    55     54     55    54.4466   54.4466    54
18APR33   21MAY87   54.0890   54.1260    54     54     54    54.0904   54.0904    54
18APR43   21MAY87   44.0903   44.1205    44     44     44    44.0904   44.0904    44
```

As an aside, if you are going to use the age in years as a continuous variable in an analysis such as a regression or analysis of covariance, there is no real advantage (other than a change in units) in converting from days to years. Consider using age in days to avoid the issues associated with the conversion to years.

SEE ALSO
A well-written explanation of the calculation of age and the issues associated with those calculations can be found in Sample Tip 24808 by William Kreuter (2004). Cassidy (2005) also discusses a number of integer age calculations.

3.2.1 Simple Formula

When you need to determine age in years and you want a fractional age (continuous values), a fairly well accepted industry standard approximates leap years with a quarter day each year.

```
age1 = (death - dob) / 365.25;
```

Depending on how leap years fall relative to the date of death and birth, the approximation could be off by as much as what is essentially two days over the interval. Over a person's lifetime, or even over a period of just a few years, two days will cause an error in at most the third decimal place.

There are several other, somewhat less accurate, variations on this formula for age in years.

Group	Operators
`age2 = (death - dob) / 365;`	Ignores leap years. Error is approximately 1 day in four years.
`age3 = year(death) - year(dob);`	Treats all days within the year of birth and the year of death as equal.
`age4 = year(death-dob) - 1960;`	Similar inaccuracy as AGE3. If this formula makes intuitive sense, then you probably have deeper issues and you may need to deal with them professionally.

3.2.2 Using Functions

The INTCK function counts the number of intervals between two dates (see Section 3.4.3 for more on the INTCK function). When the selected interval is `'year'`, it returns an integer number of years. Since by default this function always measures from the start of the interval, the resulting calculation would be the same as if the two dates were both first shifted to January 1. This means that the result will ignore dates of birth and death and could be incorrect by as much as a full year. AGE3 and AGE5 give the same result, as they both ignore date within year.

```
age5 = intck('year',dob,death);
```

Unlike the INTCK function the YRDIF function does not automatically shift to the start of the interval and it partially accounts for leap years. This function was designed for the securities industry to calculate interest for fixed income securities based on industry rules, and returns a fractional age. Note the use of the third argument (*basis*), since there is more than one possible entry that starts with the letters 'act', 'act' *is **not*** an acceptable abbreviation for 'actual'.

```
age6 = yrdif(dob,death,'actual');
```

With a *basis* of ACTUAL the YRDIF function does not handle leap days in the way that we would hope for when calculating age. Year 2000 was a leap year and year 2001 was not. In terms of a calculated value for age, we would expect both TEST2000 and TEST2001 to have a value of 1.0. Like the formula for

```
data year;
test2000 = yrdif('07JAN2000'd,'07JAN2001'd,"ACTual");
test2001 = yrdif('07JAN2001'd,'07JAN2002'd,"ACTual");
put test2000=;
put test2001=;
run;
```
```
test2000=1.0000449135
test2001=1
```

```
test2004 = yrdif('07JAN2000'd,'07JAN2004'd,"ACTual");
put test2004=;
```

AGE1 shown above, the leap day is being averaged across four years. If we were to examine a full four-year period (with exactly one leap day), the YRDIF function returns the correct age in years (age=4.0).

When dealing with longer periods, such as the lifetime of an individual, the averaging of leap days would introduce an error of at most ¾ of a day over the period. As such this function is very comparable to the simple formula (AGE1 in Section 3.2.1), which could only be off by at most 2 days over the same period. Both of these formulas tend to vary only in the third or fourth decimal place.

Caveat for YRDIF with a Basis of ACTUAL

As is generally appropriate, the YRDIF function does not include the last day of the interval (the date of the second argument) when counting the total number of days. SAS Institute strongly recommends that YRDIF, with a basis of ACTUAL, not be used to calculate age. http://support.sas.com/kb/3/036.html and http://support.sas.com/kb/36/977.html .

Starting with SAS 9.3

The YRDIF function supports a *basis* of AGE. This is now the most accurate method for calculating a continuous age in years, as it appropriately handles leap years.

```
age6a = yrdif(dob,death,'age');
```

SEE ALSO

If you need even more accuracy consult Adams (2009) for more precise continuous formulas.

3.2.3 The Way Society Measures Age

Society thinks of age as whole years, with credit withheld until the date of the anniversary of the birth. The following equation measures age in whole years. It counts the months between the two dates, subtracts one month if the day boundary has not been crossed for the last month, and then converts months to years.

```
age7 = floor(( intck( 'month', dob, death)
           - ( day(death) < day(dob)))/ 12);
```

CAVEAT

This formula, and indeed how we measure age in general, has issues with birthdays that fall on February 29.

MORE INFORMATION

This formula is used in a macro function in Section 13.7.

SEE ALSO

Chung and Whitlock (2006) discuss this formula as well as a version written as a macro function. Sample code #36788 applies this formula using the FCMP procedure http://support.sas.com/kb/36/788.html.

And Sample Code # 24567 applies it in a DATA step http://support.sas.com/kb/24/567.html.

3.3 Using DATA Step Component Objects

DATA step component objects are unlike anything else in the DATA step. They are a part of the DATA Step Component Interface, DSCI, which was added to the DATA step in SAS®9. The objects are compiled within the DATA step and task-specific methods are applied to the object.

Because of their performance advantages, knowing how to work with DATA step component objects is especially important to programmers working with large data sets. Aside from the performance advantages, these objects can accomplish some tasks that are otherwise difficult if not impossible.

The two primary objects (there were only two in SAS 9.1) are HASH and HITER. Both are used to form memory resident hash tables, which can be used to provide efficient storage and retrieval of data using keys. The term *hash* has been broadly applied to techniques used to perform direct addressing of data using the values of key variables.

The hash object allows us to store a data table in memory in such a way as to allow very efficient read and write access based on the values of key variables. The processing time benefits can be huge especially when working with large data sets. These benefits are realized not only because all the processing is done in memory, but also because of the way that the key variables are used to access the data. The hash iterator object, HITER, works with the HASH object to step through rows of the object one at a time.

Additional objects have been added in SAS 9.2 and the list of available objects is expected to continue to grow. Others objects include:

- Java object
- Logger and Appender objects

Once you have started to understand how DATA step component objects are used, the benefits become abundantly clear. The examples that follow are included to give you some idea of the breadth of possibilities.

MORE INFORMATION
In other sections of this book, DATA step component objects are also used to:

- eliminate duplicate observations in Section 2.9.5
- conduct many-to-many merges in Section 3.7.6
- perform table look-ups in Section 6.8

SEE ALSO
An index of information sources on the overall topic of hashing can be found at http://www.sascommunity.org/wiki/Hash_object_resources.

Getting started with hashing text can be found at http://support.sas.com/rnd/base/datastep/dot/hash-getting-started.pdf.

Detailed introductions to the topic of hashing can be found in Dorfman and Snell (2002 and 2003); Dorfman and Vyverman (2004b); and Ray and Secosky (2008). Additionally, Jack Hamilton (2007), Eberhardt (2010), as well as Secosky and Bloom (2007) each also provide a good introduction to DATA step objects. Richard DeVenezia has posted a number of hash examples on his Web site http://www.devenezia.com/downloads/sas/samples/ .

One of the more prolific authors on hashing in general and the HASH object is Paul Dorfman. His very understandable papers on the subject should be considered required reading. Start with Dorfman and Vyverman (2005) or the slightly less recent Dorfman and Shajenko (2004a), both papers contain a number of examples and references for additional reading.

SAS 9.2 documentation can be found at http://support.sas.com/kb/34/757.html, and with a description of DATA step component objects at http://support.sas.com/documentation/cdl/en/lrcon/61722/HTML/default/a002586295.htm.

A brief summary of syntax and a couple of simple examples can be found in the SAS®9 HASH OBJECT Tip Sheet at http://support.sas.com/documentation/cdl/en/lrcon/61722/HTML/default/a002586295.htm.

The construction of stacks and other methods are discussed by Edney (2009).

3.3.1 Declaring (Instantiating) the Object

The component object is established, instantiated, by the DECLARE statement in the DATA step. Each object is named and this name is also established on the DECLARE statement, which can be abbreviated as DCL.

```
declare hash hashname();
```

The name of the object is followed by parentheses which may or may not contain constructor methods. The name of the object, in this example HASHNAME, is actually a variable on the DATA step's PDV. The variable contains the hash object and as such it cannot be used as a variable in the traditional ways.

```
dcl hash hashname;
hashname = _new_ hash();
```

You can also instantiate the object in two statements. Here it is more apparent that the name of the object is actually a special kind of DATA step variable. Although not a variable in the traditional sense, it can contain information about the object that can on occasion be used to our advantage.

When the object is created (declared), you will often want to control some of its attributes. This is done through the use of arguments known as constructors. These appear in the parentheses, are followed by a colon, and include the following:

- DATASET: name of the SAS data set to load into the hash object
- HASHEXP: exponent that determines the number of key locations (slots)
- ORDERED: determines how the key variables are to be ordered in the hash table

The HASH object is used to create a hash table, which is accessed using the values of the key variables. When the table needs to be accessed sequentially, the HITER object is used in conjunction with the hash table to allow sequential reads of the hash table in either direction

The determination of an efficient value for HASHEXP: is not straightforward. This is an exponent so a value of 4 yields $2^4 = 16$ locations or slots. Each slot can hold an infinite number of items; however, to maximize efficiency, there needs to be a balance between the number of items in a slot and the number of slots. The default size is 8 (2^8=256 slots). The documentation suggests that for a million items 512 to 1024 slots (HASHEXP = 9 or 10) should offer good performance.

3.3.2 Using Methods with an Object

The DECLARE statement is used to create and name the object. Although a few attributes of the object can be specified using constructor arguments when the object is created, additional methods are available not only to help refine the definition of the object, but how it is used as well. There are quite a few of these methods, several of which will be discussed in the examples that follow.

Methods are similar to functions in how they are called. The method name is followed by parentheses that may or may not contain arguments. When called, each method returns a value indicating success or failure of its operation. For each method success is 0. Like with DATA step routines, you might choose to utilize or ignore this return code value.

Since there may be more than one object defined within the DATA step, it is necessary to tie a given method call to the appropriate object. This is accomplished using a dot notation. The method name is preceded with the name of the object to which it is to be associated, and the two names are separated with a dot.

```
hashname.definekey('subject', 'visit');
hashname.definedata('subject','visit','labdt') ;
hashname.definedone() ;
```

Methods are used both to refine the definition of the object, as well as to operate against it. Methods that are used to define the object follow the DECLARE statement and include:

- DEFINEKEY   list of variables forming the primary key
- DEFINEDATA   list of data set variables
- DEFINEDONE   closes the object definition portion of the code

During the execution of the DATA step, methods are also used to read and write to the object. A few of these methods include:

- ADD   adds the specified data on the PDV to the object
- FIND   retrieves information from the object based on the values of the key variables
- MISSING   initializes a list of variables on the PDV to missing
- OUTPUT   writes the object's contents to a SAS data set
- REPLACE   writes data from the PDV to an object; matching key variables are replaced

3.3.3 Simple Sort Using the HASH Object

Because the hash object can be ordered by keys, it can be used to sort a table. In the following example we would like to order the data set ADVRPT.DEMOG by subject within clinic number. This sort can be easily accomplished using PROC SORT; however, as a demonstration a hash object can also be used.

```
proc sort data=advrpt.demog(keep=clinnum subject lname fname dob)
          out=list nodupkey;
   by clinnum subject;
   run;
```

A DATA _NULL_ step is used to define and load the hash object. After the data have been loaded into the hash table, it is written out to the new sorted data set. Only one pass of the DATA step is made during the execution phase and no data are read using the SET statement.

```
data _null_; ❶
   if 0 then set advrpt.demog(keep=clinnum subject lname fname dob); ❷
   declare hash clin (dataset:'advrpt.demog', ordered:'Y'); ❸
      clin.definekey ('clinnum','subject'); ❹
      clin.definedata ('clinnum','subject','lname','fname','dob'); ❺
      clin.definedone (); ❻
   clin.output(dataset:'clinlist'); ❼
   stop; ❽
   run;
```

❶ A DATA _NULL_ step is used to build and take advantage of the hash object.

❷ The SET statement is used only during the DATA step's compilation phase to add the variables, their attributes, and initial value (missing) to the PDV. The IF statement can never be true; consequently, no observations can be read during the execution of the DATA step.

❸ The hash object is instantiated with the name CLIN. The object is further defined to contain the data set ADVRPT.DEMOG and will be ordered using the key variables in ascending order. The use of the DATASET: constructor method is sufficient to cause the entire data set to be loaded into the hash object, which is held in memory. The ORDERED: constructor can be used to specify either an ascending ('a', 'ascending', 'yes', or 'y') or descending ('descending', 'd', 'n', or 'no') ordering.

❹ The sort key variables are included in the same order as they would be in the BY statement of PROC SORT. The resulting data set will necessarily contain no duplicate key values (the NODUPKEY option on the PROC SORT statement).

❺ List the variables that will be on the data set that is being created. Key variables listed on the DEFINEKEY method ❹ are not automatically included in this list.

❻ The hash object has been defined. Close the definition section initiated by the DECLARE statement ❸.

❼ The contents of the CLIN hash object are written to the data set WORK.CLINLIST using the OUTPUT method and the DATASET: constructor.

❽ The STOP statement closes the implied loop created by the SET statement ❷.

When a method is called as was the OUTPUT method above ❼, a return code is generated. If you want to write more robust steps, you should capture and potentially query this return code. The statement at ❼ becomes: `rc=clin.output(dataset:'clinlist');` Although not a problem in this step, when a method is not successful and is unable to pass back a return code value, as would be the case shown in the example ❼, an error is generated which results in an abnormal end to the DATA step. While this seems to be less of an issue for the methods used in the declaration of the object, it definitely is an issue for those methods that read and write to and from the hash object, e.g., FIND, OUTPUT, ADD. It has been my experience that you should always capture the return code from one of these methods, even if you are not going to test or otherwise use the return code.

CAVEAT
Although the HASH object can be used to sort a data table, as was shown above, using the HASH object will not necessarily be more efficient than using PROC SORT. Remember that the DATA step itself has a fair amount of overhead, and that the entire table must be placed into memory before the hash keys can be constructed. While the TAGSORT option can be used with PROC SORT for very large data sets, it may not even be possible to fit a very large data set into memory. As with most tools within SAS you must select the one appropriate for the task at hand.

3.3.4 Stepping through a Hash Table
Unlike in the previous example where the data set was loaded and then immediately dumped from the hash table, very often we will need to process the contents of the hash table item by item. The advantage of the hash table is the ability to access its items using an order based on the values of the key variables.

There are a couple of different approaches to stepping through the items of a hash table. When you know the values of the key variables they can be used to *find* and retrieve the item of interest. This is a form of a table look-up, and additional examples of table look-ups using a hash object can be found in Section 6.8. A second approach takes advantage of the hash iterator object, HITER, which is designed to work with a series of methods that read successive items (forwards or backwards) from the hash table.

The examples used in this section perform what is essentially a many-to-many fuzzy merge. Some of the patients (SUBJECT) in our study have experienced various adverse events which have been recorded in ADVRPT.AE. We want to know what if any drugs the patient *started taking* within the 5 days prior to the event. Since a given drug can be associated with multiple events and a given event can be associated with multiple drugs, we need to create a data set containing all the combinations for each patient that matches the date criteria.

Using the FIND Method with Successive Key Values

The FIND method uses the values of the key variables in the PDV to search for and retrieve a record from the hash table. When we know all possible values of the keys, we can use this method to *find* all the associated items in the hash table.

```
data drugEvents(keep=subject medstdt drug aestdt aedesc sev);
   declare hash meds(ordered:'Y') ; ❶
      meds.definekey ('subject', 'counter');
      meds.definedata('subject', 'medstdt','drug') ;
      meds.definedone () ;

   * Load the medication data into the hash object;
   do until(allmed); ❷
      set advrpt.conmed(keep=subject medstdt drug) end=allmed;
      by subject; ❸
      if first.subject then counter=0;❹
      counter+1; ❺
      rc=meds.add();
   end;

   do until(allae); ❻
      set advrpt.ae(keep=subject aedesc aestdt sev) end=allae;
      counter=1; ❼
      rc=meds.find(); ❽
      do while(rc=0);
         * Was this drug started within 5 days of the AE?;
         if (0 le aestdt - medstdt lt 5) then output drugevents; ❾
         counter+1; ❿
         rc=meds.find();❽
      end;
   end;
   stop;
   run;
```

❶ The MEDS hash table is declared and its keys and data variables are defined.

❷ A DO UNTIL loop is used to read the medication observations into the MEDS hash table. We have not used the DATASET: constructor as was done in Section 3.3.3, because we are performing IF-THEN processing and creating an additional variable (COUNTER) within this loop.

❸ Because we are using FIRST.SUBJECT to initialize the counter variable, the BY statement is needed. This adds a restriction to the incoming data set—it must be in sorted order. The next example shows a way around this restriction.

❹ A unique numeric variable, COUNTER, is established as a key variable. Not only will this variable guarantee that each observation from ADVRPT.CONMED is stored in the hash table (each row has a unique key), but we will be able to use this counter to step through the rows of the table ❽. For a given patient the item (row) counter is initialized to 0.

❺ The row counter is incremented for each incoming row for this patient and the observation is then written to the MEDS object using the ADD method.

❻ Establish a loop to read each observation from the ADVRPT.AE data set. For each of these observations, which gives us the SUBJECT number and the event date (AESTDT), we want to find and check all the entries for this subject that reside in the hash table.

❼ Initialize the COUNTER so that we will start retrieving the first item for this specific patient.

❽The FIND method will be used to retrieve the item that matches the current values of the key variables (SUBJECT and COUNTER). Since COUNTER=1 this will be the first item for this subject. This and each successive item for this subject is checked against the 5-day criteria inside the DO WHILE loop.

❾ If the onset of the adverse event (AESTDT) is within 5 days of the medication start date (MEDSTDT), the observation is saved to WORK.DRUGEVENTS.

❿ The key value is incremented and the next item is retrieved using the FIND method. The DO WHILE executes until the FIND method is no longer able to retrieve anymore items for this subject and the return code for FIND (RC) is no longer 0.

In the previous example we loaded the MEDS hash table using a unique counter for each medication used by each subject. This counter became the second key variable. In that example the process of initializing and incrementing the counter depended on the data having already been grouped by patient. For very large data sets, it may not be practical to either sort or group the data.

If we had not used FIRST. processing, we would not have needed the BY statement, and as a result we would not have needed the data to be grouped by patient. We can eliminate these requirements by storing the subject number and the count in a separate hash table. Since we really only need to store one value for each patient—the number of medications encountered so far, we could do this as an array of values. In the following example this is what we do by creating a hash table that matches patient number with the number of medications encountered for that patient.

```
data drugEvents(keep=subject medstdt drug aestdt aedesc sev);
   * define a hash table to hold the subject counter;
   declare hash subjcnt(ordered:'y'); ❶
      subjcnt.definekey('subject');
      subjcnt.definedata('counter');
      subjcnt.definedone();

   declare hash meds(ordered:'Y') ;
      meds.definekey ('subject', 'counter');
      meds.definedata('subject', 'medstdt','drug','counter') ;
      meds.definedone () ;

   * Load the medication data into the hash object;
   do until(allmed);
      set advrpt.conmed(keep=subject medstdt drug) end=allmed; ❷
      * Check subject counter: initialize if not found,
                              otherwise increment;
      if subjcnt.find() then counter=1; ❸
      else counter+1; ❹
      * update the subject counter hash table;
      rc=subjcnt.replace(); ❺
      * Use the counter to add this row to the meds hash table;
      rc=meds.add(); ❻
   end;

   do until(allae);
      . . . . . the remainder of the DATA step is unchanged from the previous example . . . . .
```

The hash table SUBJCNT contains the number of medications that have been read for each patient at any given time. As additional medications are encountered, the COUNTER variable is incremented.

❶ A hash table to hold the counter for each patient (SUBJECT) is declared. The key variable is SUBJECT and the only data variable is COUNTER.

❷ An observation is read from the medications data set. This loads a value of SUBJECT into the PDV.

❸ The COUNTER is initialized to 1 when this value of SUBJECT has not yet been loaded into the SUBJCNT hash table. Remember that a successful FIND will return a 0 value. This means that the expression SUBJCNT.FIND() will be true when the current value of SUBJECT is the first time that subject has been encountered. When SUBJCNT.FIND() successfully returns a value (this is not the first time this SUBJECT has been read from the medications data set), the expression evaluates to false and the COUNTER is incremented ❹. **Either way the SUBJCNT.FIND() has been executed.**

❹ When this value of SUBJECT is found in the hash table, the COUNTER is returned and loaded into the PDV. This SUM statement then causes COUNTER to be incremented.

❺ The updated COUNTER value is written to the hash table. The REPLACE method causes this value of COUNTER to overwrite a previous value.

❻ The medication information along with the updated value of COUNTER is saved.

Essentially we have used the SUBJCNT hash table to create a dynamic single dimension array with SUBJECT as the index and the value of COUNTER as the value stored. For any given subject we can dynamically determine the number of medications that have been encountered so far and use that value when writing to the MEDS hash table.

Using the Hash Iterator Object

In the previous examples we stepped through a hash object by controlling the values of its key variables. You can also use the hash iterator object and its associated methods to step through a hash object.

Like the previous examples in this section we again use a unique key variable (COUNTER) to form the key for each patient medication combination. The solution shown here again assumes that the medication data are grouped by subject, but we have already seen how we can overcome this limitation. The difference in the solution presented below is the use of the hash iterator object, HITER. Declaring this object allows us to call a number of methods that will only work with this object.

```
data drugEvents(keep=subject medstdt drug aestdt aedesc sev);
   declare hash meds(ordered:'Y') ;
      declare hiter medsiter('meds'); ❶
      meds.definekey ('subj', 'counter');
      meds.definedata('subj', 'medstdt','drug','counter') ;
      meds.definedone () ;

   * Load the medication data into the hash object;
   do until(allmed); ❷
      set advrpt.conmed(keep=subject medstdt drug) end=allmed;
      by subject;
      if first.subject then do;
         counter=0;
         subj=subject;
      end;
      counter+1;
      rc=meds.add();
   end;

   do until(allae);
      set advrpt.ae(keep=subject aedesc aestdt sev) end=allae;
      rc = medsiter.first(); ❸
      do until(rc); ❹
         * Was this drug started within 5 days of the AE?;
         if subj=subject❺ & 0<=aestdt-medstdt<5❻ then output drugevents;
         if subj gt subject then leave; ❼
         rc=medsiter.next(); ❽
      end;
   end;
   stop;
   run;
```

❶ The hash iterator object is declared and named MEDSITER. Notice that its one argument is the name of the hash object (MEDS) with which it is to be associated.

❷ The MEDS hash object is loaded as it was in the previous examples.

❸ The FIRST method returns the very first item in the MEDSITER hash object. Notice that the name of the method is preceded by the name of the iterator object, MEDSITER. Since FIRST does not take the values of the key variables into consideration, except for the first patient, we are forced to cycle through earlier patients until we get to the patient of interest. Because the MEDSITER object is linked to the MEDS object we are actually retrieving from the MEDS object via the MEDSITER object.

❹ The DO UNTIL reads successive items from the hash object until the NEXT method ❽ is unable to return another item or until all the items for the current subject have been exhausted ❼. Remember that the return code (RC) for methods is 0 for success and non-zero for failure.

❺ We are only interested in those medications that are associated with the current patient (SUBJECT), ❻ and that meet the date criterion.

❼ The DO UNTIL loop started at ❹ steps through all the medications stored in the hash object. Since they are ordered by subject, once we have finished with the current patient (SUBJECT) we can leave the loop.

❽ The NEXT method is used to read the next item from the hash object. The next item is determined by the key variables and the way the hash object was ordered. MEDS was specified to be in ascending order by the ORDERED: constructor method. Although not used in this example there is also a PREV method to retrieve the previous item.

The order of the observations in the ADVRPT.AE data set in the preceding examples does not matter. If the data were in known SUBJECT order we could have saved on memory usage by loading the MEDS hash table one subject (BY group) at a time. To remove the values for the previous SUBJECT the CLEAR method could be used to clear the hash table values and would be executed for each FIRST.SUBJECT. The example in Section 3.3.5 has a hash object that stores data for only a single clinic at a time; however, in that example the object is deleted and re-instantiated for each clinic.

3.3.5 Breaking Up a Data Set into Multiple Data Sets

We have been given a data set that is to be broken up into a series of subsets, each subset being based on some aspect of the data. In the example that follows we want to create a data set for each clinic. That means a data set for each unique value of the variable CLINNUM. The brute force approach would require knowing, and then hard coding, the individual clinic codes, using a DATA step such as the one to the left.

```
data clin011234 clin014321 clin023910 clin024477;
   set advrpt.demog;
   if clinnum= '011234' then output clin011234;
   else if clinnum= '014321' then output clin014321;
   else if clinnum= '023910' then output clin023910;
   else if clinnum= '024477' then output clin024477;
   run;
```

Actually there are many more clinic codes than shown here, but I find hard coding to be very tiring so I only did enough to show the intent of the step. Clearly this is neither a practical, nor a smart, solution.

There have been any number of papers offering macro language solutions to this type of problem (Fehd and Carpenter, 2007); however, all of those solutions require two passes of the data. One pass of the data to determine the list of values, and a second pass to utilize that list. By using a hash table we can accomplish the task in a single pass of the data.

A DATA _NULL_ step is used to create the data sets. Since we are not specifying the names of the data sets that are to be created, they will have to be declared using the OUTPUT hash method.

```
data _null_;
   if 0 then set advrpt.demog(keep=clinnum subject lname fname dob); ❶
   * Hash ALL object to hold all the data;
   declare hash all (dataset: 'advrpt.demog', ordered:'Y');
      all.definekey ('clinnum','subject');
      all.definedata ('clinnum','subject','lname','fname','dob');
      all.definedone ();
   declare hiter hall('all');

   * CLIN object holds one clinic at a time;
   declare hash clin; ❷

   * define the hash for the first clinic; ❸
   clin = _new_ hash(ordered:'Y');
      clin.definekey('clinnum','subject');
      clin.definedata('clinnum','subject','lname','fname','dob');
      clin.definedone();

   * Read the first item from the full list;
   done=hall.first(); ❹
   lastclin = clinnum;

   do until(done);  *loop across all clinics;
      clin.add(); ❺
      done = hall.next(); ❻
      if clinnum ne lastclin or done then do; ❼
         * This is the first obs for this clinic or the very last obs;
         * write out the data for the previous clinic;
         clin.output(dataset:'clin'||lastclin); ❽
         * Delete the CLIN hash object;
         clin.delete();❾
         clin = _new_ hash(ordered:'Y');
            clin.definekey('clinnum','subject');
            clin.definedata('clinnum','subject','lname','fname','dob');
            clin.definedone();
         lastclin=clinnum; ❿
      end;
   end;
   stop;
   run;
```

❶ During the compilation phase of the DATA step, the SET statement is used to establish the attributes for the variables on the PDV. These attributes will be used to build the data sets that are written by the OUTPUT method ❽.

❷ Declare the CLIN hash object, which will hold the data for one clinic at a time.

❸ Instantiate the hash object, CLIN, which was declared earlier ❷. This object will hold the data for each individual clinic. Each clinic will be loaded into the CLIN object one at a time ❺.

❹ Using the iterator object for the ALL object, HALL, retrieve the very first item (set of values) from the hash object. Save this clinic number in the LASTCLIN variable for comparison with later values. This value is used to detect when we cross clinic boundaries ❼.

❺ The current values on the PDV are written to the CLIN object. For the very first item retrieved from the HALL object these values were read by the FIRST method ❹; otherwise, the values were read by the NEXT method ❻ in the previous pass of this DO UNTIL (DONE) loop.

❻ Unlike the END= option on the SET statement, which is assigned a true value when processing the last observation, the NEXT method returns a 0 until it attempts to read *past* the last item. This means that DONE is not true on the last observation that will be read out of the HALL object. One more pass of this loop will be required. We take advantage of this behavior to write out the contents of CLIN for the last clinic number.

❼ When the clinic number that was just retrieved is different from the previous one (or when we are done reading items), we know that it is time to write out the contents of the CLIN object to a data set. The values in the PDV which were just loaded by the NEXT method contain items from a different clinic, but they have yet to be loaded into the CLIN object.

❽ The OUTPUT method is used to write the contents of the CLIN object, which contains the rows for only one clinic, to the named SAS data set. The data set name contains the value of the clinic number that is stored in the data set, and the name of the data set is determined during the execution of the statement.

❾ We have written out the contents of the clinic specific hash object (CLIN) ❽ so we are finished using it. In this example we will delete CLIN and then reestablish it for the next clinic. Since we have already declared CLIN ❷, it can be re-created using the _NEW_ keyword on the assignment statement. Rather than deleting and then reestablishing the CLIN object, we could have cleared it by using the CLEAR method. `rc=clin.clear();` (see the sample programs associated with this section for the full DATA step).

❿ The current clinic number is saved for comparison ❼ against the next retrieved item ❻. If this is the very last item (DONE=1), this is unnecessary, but costs us little.

MORE INFORMATION
The example in Section 3.3.6 also creates multiple data subsets using nested hash objects.

SEE ALSO
Hamilton (2007) discusses this topic in very nice detail, including background and alternate approaches.

3.3.6 Hash Tables That Reference Hash Tables

The value of the variable that names a hash table holds information that is unique to that table.

```
hashnum = _new_ hash(ordered:'Y');
```

The assignment statement shown to the left, instantiates the hash table HASHNUM, and when it is executed a unique value associated with this object is stored in the variable HASHNUM. While this variable exists on the PDV, it is not a variable in the traditional DATA step numeric/character sense—in a real sense the value held by this variable is the whole hash object. This implies that we can instantiate multiple hash tables using the same name as long as the value of the hash table variable changes.

The example in Section 3.3.5 used two independent hash tables to break up one data set into multiple data-dependent data tables—one table for each clinic. That solution loads the data for a specific clinic into a hash table from a master hash table. Once loaded the data subset is written to a data set and the associated hash table is deleted or cleared. This process requires that each observation has three I/O passes.

1. It is read from the incoming data set and loaded into the master hash table.
2. The data for a given clinic is read from the master and loaded into a hash table containing data only for that clinic.
3. The data are written to the new clinic-specific data set using the OUTPUT method.

In the following example the data for individual clinics are loaded directly into the hash object designated for that hash object. Although a hash object is used to organize and track the hash objects used for the individual clinics, a master hash object containing all the data is not required.

```
data _null_;
    * Hash object to hold just the HASHNUM pointers;
    declare hash eachclin(ordered:'Y'; ❶
        eachclin.definekey('clinnum');
        eachclin.definedata('clinnum','hashnum');
        eachclin.definedone ();
    declare hiter heach('eachclin');

    * Declare the HASHNUM object;
    declare hash hashnum; ❷

    do until(done);
        set advrpt.demog(keep=clinnum subject lname fname dob) end=done; ❸
        * Determine if this clinic number has been seen before;
        if eachclin.check() then do; ❹
            * This is the first instance of this clinic number;
            * create a hash table for this clinic number;
            hashnum = _new_ hash(ordered:'Y'); ❺
                hashnum.definekey ('clinnum','subject');
                hashnum.definedata ('clinnum','subject','lname','fname','dob');
                hashnum.definedone ();
            * Add to the overall list;
            rc=eachclin.replace(); ❻
        end;
        * Retrieve this clinic number and its hash number;
        rc=eachclin.find(); ❼
        * Add this observation to the hash table for this clinic.;
        rc=hashnum.replace(); ❽
    end;

    * Write the individual data sets;
    * There will be one data set for each clinic;
    do while(heach.next()=0); ❾
        * Write the observations associated with this clinic;
        rc=hashnum.output(dataset:'clinic'||clinnum); ❿
    end;
    stop;
    run;
```

❶ The EACHCLIN hash object and its associated iterator, HEACH, are declared and instantiated. EACHCLIN is ordered by its key variable CLINNUM, and although it is a key variable, it is also included as a data element. The other data variable is HASHNUM, which will contain a value that will allow us to access the information associated with the hash table that holds the clinic-specific information. CLINNUM is only needed for the process that follows because it is used in the name of the data table that is created using the OUTPUT method ❿.

❷ The hash object HASHNUM is declared but not instantiated until ❺. This object will be used to hold the data for the individual clinics. The DECLARE statement creates the variable HASHNUM that will contain a distinct identifying value for each of the individual hash objects.

❸ An observation is read from the incoming data set and the values of the variables that will be added to the hash objects are stored in the PDV.

❹ The CHECK method is used to ascertain whether or not this is the first occurrence of this current clinic number, CLINNUM. This method returns a 0 if it has already been encountered and the DO block is not executed.

❺ When a hash object is Instantiated for this clinic number, a value is assigned to the variable HASHNUM for this hash object. HASHNUM is also a data variable in the EACHCLIN object.

❻ Since this is the first time that this clinic number has been encountered, it is added to the EACHCLIN hash object. EACHCLIN is ordered by CLINNUM, but also stores the associated value of HASHNUM, which is the variable holding the value of the hash object for the current clinic.

❼ The FIND method is used to retrieve the HASHNUM value associated with the clinic number (CLINNUM) that was added to the PDV by the SET statement ❸.

❽ This observation is added to the hash object that is associated with this clinic. The value of the variable HASHNUM has been retrieved from the hash object EACHCLIN ❼ that contains the list of clinicspecific hash objects.

❾ The NEXT method is used to successively retrieve the HASHNUM values from the HEACH iterator object, which contains one item for each clinic number.

❿ The data stored in the hash object that are identified by the variable HASHNUM are written to the data set using the OUTPUT method. Notice that the name of the data set is constructed during the execution of the DO loop. The value stored in the variable HASHNUM was retrieved from the HEACH iterator object using the NEXT method ❾, and this value, when used with the OUTPUT object, identifies which of the clinic-specific hash tables is to be written. The NEXT method also returns a value of CLINNUM which is used in the construction of the name of the data set.

SEE ALSO
An early paper by Dorfman and Vyerman (2005) contains a number of examples including one that is very similar to this one. Some of the earliest published examples of hash objects that point to hash objects were presented by Richard DeVenzia on SAS-L (DeVenezia, 2004).

3.3.7 Using a Hash Table to Update a Master Data Set

When you want to update a SAS data set using a transaction data set, the UPDATE and MODIFY statements can be used. UPDATE requires sorted data sets, while the MODIFY statement's efficiency can be greatly improved with sorting or indexes. A similar result can be achieved using a hash table.

In this example a transaction data set (TRANS) has been created using FNAME and LNAME as the key variables, and the value of SEX is to be updated. To illustrate what happens when values of the key variables are incorrect, the last name of Peter Antler has been misspelled (this name will not exist in the master).

```
* Build a transaction file;
data trans;
   length lname $10 fname $6 sex $1;
   fname='Mary'; lname='Adams';  sex='N'; output;
   fname='Joan'; lname='Adamson';sex='x'; output;
   * The last name is misspelled;
   fname='Peter';lname='Anla';   sex='A'; output;
   run;
data newdemog(drop=rc);
   declare hash upd(hashexp:10);
      upd.definekey('lname', 'fname');  ❶
      upd.definedata('sex');  ❷
      upd.definedone();
   do until(lasttrans);
      set trans end=lasttrans;
      rc=upd.add();  ❸
   end;
   do until(lastdemog);
      set advrpt.demog end=lastdemog;  ❹
      rc=upd.find();  ❺
      output newdemog;  ❻
   end;
   stop;  ❼
   run;
```

❶ The key variables are defined as LNAME and FNAME.

❷ The variable that we want to update is added to the hash table.

❸ Each of the transaction observations are added to the UPD hash table.

❹ An observation is read from the master data set.

This loads the values of the key variables into the PDV.

❺ Using the key values for this master record, a transaction record is recovered from the UPD hash table. If there is no update record, the PDV is not altered and the observation is unchanged.

```
3.3.7  Update a Master

Obs    fname    lname        sex

 1     Mary     Adams         N
 2     Joan     Adamson       x
 3     Mark     Alexander     M
 4     Peter    Antler        M  ❽
```

❻ The updated record from the master file is written. More typically the master data set would be replaced; here a temporary copy (WORK.NEWDEMOG) is created.

❼ Because all the processing takes place within loops, the STOP statement is needed to terminate the DATA step.

❽ The misspelled transaction (Antla) is NOT added to the master data set, nor is the value of SEX changed for Peter Antler.

SEE ALSO

A similar solution was suggested by user @KSharp in a SAS Forum discussion on HASH objects http://communities.sas.com/message/53968.

3.4 Doing More with the INTNX and INTCK Functions

The INTNX and INTCK functions are used to work with date, time, and datetime intervals. Both can work with a fairly extensive list of interval types; however, you can add even more flexibility to these two functions by using interval multipliers, shift operators, and alignment options.

Using these two functions is not always straightforward; however, you need to be aware of how they make their interval determinations. Of primary importance is that by default they both make their calculations based on the start of the current interval. For instance when using a YEAR interval type for any date in 2009, the current interval will start on January 1, 2009. As a result of the interval start, the two function calls shown here will both return an interval length of one year.

```
twoday = intck('year','31dec2008'd,'01jan2009'd);
twoyr  = intck('year','01jan2008'd,'31dec2009'd);
```

SEE ALSO

Interval multipliers and shift operators are complex topics. Fortunately the documentation for the INTNX and INTCK functions is well written and should be consulted for additional important details.

These two functions are carefully described by Cody (2010), and this is a good source for further information on the topics in this section.

3.4.1 Interval Multipliers

Interval multipliers allow you to alter the definition of the interval length. Interval multipliers are simply implemented as integers that are appended to the standard interval. The interval 'WEEK' has a length of 7 days while the same interval with a multiplier of 2 (WEEK2) will have an interval length of 14 days.

In the following rather silly example we would like to schedule a follow-up exam in two weeks (14 days). EXAMDT_2 is calculated to be one two-week interval in the future using an interval multiplier of two ❶.

```
data ExamSchedule;
   do visdt = '25may2009'd to '14jun2009'd;
      examdt_2 = intnx('week2'❶,visdt,1);
      examdtx2 = intnx('week',visdt,2 ❷);
      output;
   end;
   format visdt examdt_2 examdtx2 date9.;
   run;
```

EXAMDTX2, on the other hand, is determined by requesting two one-week intervals ❷. Nominally we would expect that the two future dates would be the same; however, because the two INTNX functions measure intervals from the start of their respective interval, the resulting dates are not always the same. The point of this example is to understand why.

June 7, 2009 was a Sunday and since a week interval starts on a Sunday, each of these uses of the INTNX function advances the date to a Sunday. Clearly interval multipliers change the way that the function views the start of the interval.

```
3.4.1 Interval Multipliers

Obs        visdt      examdt_2     examdtx2
  1      25MAY2009    07JUN2009    07JUN2009
  2      26MAY2009    07JUN2009    07JUN2009
  3      27MAY2009    07JUN2009    07JUN2009
  4      28MAY2009    07JUN2009    07JUN2009
  5      29MAY2009    07JUN2009    07JUN2009
  6      30MAY2009    07JUN2009    07JUN2009
  7      31MAY2009    07JUN2009    14JUN2009
  8      01JUN2009    07JUN2009    14JUN2009
  9      02JUN2009    07JUN2009    14JUN2009
 10      03JUN2009    07JUN2009    14JUN2009
 11      04JUN2009    07JUN2009    14JUN2009
 12      05JUN2009    07JUN2009    14JUN2009
 13      06JUN2009    07JUN2009    14JUN2009
 14      07JUN2009    21JUN2009    21JUN2009
 15      08JUN2009    21JUN2009    21JUN2009
 16      09JUN2009    21JUN2009    21JUN2009
 17      10JUN2009    21JUN2009    21JUN2009
 18      11JUN2009    21JUN2009    21JUN2009
 19      12JUN2009    21JUN2009    21JUN2009
 20      13JUN2009    21JUN2009    21JUN2009
 21      14JUN2009    21JUN2009    28JUN2009
```

When an interval is expanded the new interval start date will relate back to the beginning of time (January 1, 1960). May 24th was a Sunday and it started both the WEEK and WEEK2 interval. May 25, therefore, was advanced to June 7 for both interval types. May 31st (also a Sunday), however, did *NOT* start a WEEK2 interval, but it did start a WEEK interval. Consequently these two INTNX functions give different results when based on dates in the range of May 31 to June 6, 2009 (Obs=7-13).

If we use an interval multiplier to create a three-year interval (YEAR3), the interval start date would be determined based on the first three-year interval, which would start on January 1, 1960.

MORE INFORMATION
Alignment options are available for the INTNX function that can be helpful when the start of the interval that you are measuring from is not what you want. See Section 3.4.3.

3.4.2 Shift Operators

Both the INTNX and INTCK functions by default measure from the start of the base interval. Weeks start on Sunday; years start on January 1st, and so on. Shift operators can be used to change the way that the function determines the start of the interval. A week could start on Monday, or a fiscal year could start on July 1st.

The shift operator is designated by a number following a decimal point at the end of the interval name. The units of the *shift* depend on how the interval is defined. Weeks contain seven days and start on Sunday, which has the value of 1. The interval WEEK.2, therefore, would indicate a seven day week that starts on a Monday. The following example shows a series of shifts on a week interval (June 7, 2009 was a Sunday).

```
data ExamSchedule;
   do visdt = '01jun2009'd to '15jun2009'd;
      day  = intnx('week',visdt,1);
      day1 = intnx('week.1',visdt,1);
      day2 = intnx('week.2',visdt,1);
      day3 = intnx('week.3',visdt,1);
      day4 = intnx('week.4',visdt,1);
      day5 = intnx('week.5',visdt,1);
      day6 = intnx('week.6',visdt,1);
      day7 = intnx('week.7',visdt,1);
      output;
   end;
   format visdt day: date7.;
   run;
```

The WEEK interval starts on a Sunday and WEEK.1 does not change the interval start. WEEK.2, however, will change the start to a Monday.

Using a PROC PRINT on the resulting data set shows how the dates progress. More importantly, it shows us that the date is reset to the start of the adjusted interval first and then advanced 7 days. ❶ In the LISTING output below, notice the values based on the VISDT of Wednesday, June 3, 2009. DAY4 is advanced to June 10th (DAY4 was defined using WEEK.4, which is an interval starting on a Wednesday), so advancing 7 days yields June 10th. For DAY5 (WEEK.5 defines a week starting on Thursday), on the other hand, Wednesday is at the end of the interval and the measurement is taken from the previous Thursday (May 29).

```
3.4.2 Shift Operators

   visdt      day       day1      day2      day3      day4      day5      day6      day7

01JUN09    07JUN09   07JUN09   08JUN09   02JUN09   03JUN09   04JUN09   05JUN09   06JUN09
02JUN09    07JUN09   07JUN09   08JUN09   09JUN09   03JUN09   04JUN09   05JUN09   06JUN09
03JUN09    07JUN09   07JUN09   08JUN09   09JUN09   10JUN09   04JUN09   05JUN09   06JUN09 ❶
04JUN09    07JUN09   07JUN09   08JUN09   09JUN09   10JUN09   11JUN09   05JUN09   06JUN09
05JUN09    07JUN09   07JUN09   08JUN09   09JUN09   10JUN09   11JUN09   12JUN09   06JUN09
06JUN09    07JUN09   07JUN09   08JUN09   09JUN09   10JUN09   11JUN09   12JUN09   13JUN09
               . . . . portions of the listing are not shown . . . .
```

A typical use of a shift operator is to create a fiscal year with the interval start on July 1. Since years are made up of months, the interval 'YEAR.7' would shift the start of the year by seven months. Interval multipliers and shift operators can be used together. A five-year interval starting on July 1st could be specified as YEAR5.7.

3.4.3 Alignment Options

Although alignment options are now available for both INTNX and INTCK, they are not the same for the two functions.

Alignment with the INTNX Function

Since it is not always convenient to advance values based on the start of the interval, as was done in Sections 3.4.1 and 3.4.2, the INTNX function has the ability to change this behavior through alignment options. These options may be specified as an optional fourth argument, which can change how the function offsets from the interval start point. Without using the alignment options all displacements are

```
new = intnx('year','03jun2000'd,1);
```

measured from the start of the interval; consequently, if we advance a date by one year from June 3, 2000 the resulting date is January 1, 2001. Alignment options allow us to measure the displacement other than from the start of the interval.

The alignment option positions the result of the function relative to the original interval. It can take on the values of:

- beginning   b   interval start (default)
- middle   m   interval center
- end   e   interval end
- same   s   same relative position as the initial interval

Each of these options is demonstrated in the DATA step that follows. A date in June is advanced one month into the future (July) using each of the alignment options. The result is predicable and,

```
data ExamSchedule;
   do visdt = '01jun2007'd to '10jun2007'd;
      next_d  = intnx('month',visdt,1);
      next_b  = intnx('month',visdt,1,'beginning');
      next_m  = intnx('month',visdt,1,'middle');
      next_e  = intnx('month',visdt,1,'end');
      next_s  = intnx('month',visdt,1,'same');
      output;
   end;
   format visdt next: date7.;
   run;
```

as we might anticipate, the 'END' alignment option correctly advances to July 31st even though June has 30 days. For months with 31 days the 'MIDDLE' option will give a different result than it will for months with fewer days.

```
3.4.3 Alignment Options

Obs      visdt     next_d     next_b     next_m     next_e     next_s

 1      01JUN07    01JUL07    01JUL07    16JUL07    31JUL07    01JUL07
 2      02JUN07    01JUL07    01JUL07    16JUL07    31JUL07    02JUL07
 3      03JUN07    01JUL07    01JUL07    16JUL07    31JUL07    03JUL07
 4      04JUN07    01JUL07    01JUL07    16JUL07    31JUL07    04JUL07
 5      05JUN07    01JUL07    01JUL07    16JUL07    31JUL07    05JUL07
 6      06JUN07    01JUL07    01JUL07    16JUL07    31JUL07    06JUL07
 7      07JUN07    01JUL07    01JUL07    16JUL07    31JUL07    07JUL07
 8      08JUN07    01JUL07    01JUL07    16JUL07    31JUL07    08JUL07
 9      09JUN07    01JUL07    01JUL07    16JUL07    31JUL07    09JUL07
10      10JUN07    01JUL07    01JUL07    16JUL07    31JUL07    10JUL07
```

If you ask the INTNX function to advance a date to an illegal value, you will not receive an error message. Each of these two statements use the 'SAMEDAY' alignment option to advance a date

```
leap = intnx('year', '29feb2008'd, 1, 's');
short= intnx('month','31may2008'd, 1, 's');
```

```
leap=28FEB2009 short=30JUN2008
```

to a value that does not exist. The LOG shows that the INTNX function returns a reasonable alternative, in this case the actual last day of the month.

Alignment with the INTCK Function

By default the INTCK function counts intervals by counting the number of interval starts. Thus if your start and end dates span a single Sunday they are considered to be one week apart. As was demonstrated in the example in Section 3.4, this can result in the counting of partial intervals equally with full intervals.

The alignment option on the INTCK function has two settings:

- C   continuous
- D   discrete (this is the default)

The difference between these two option values can be demonstrated by counting the intervals between two dates. In this example the number of intervals (weeks) between the base date (START), which is fixed at Wednesday, September 14, 2011, and END which is a date that advances up to a month beyond START.

```
data check;
   start = '14sep2011'd; * the 14th was a Wednesday;
   do end = start to intnx('month',start,1,'s');
      weeks = intck('weeks',start,end);
      weeksc= intck('weeks',start,end,'c');
      weeksd= intck('weeks',start,end,'d');
      output check;
   end;
format start end date9.;
run;
```

The resulting data set contains the number of elapsed weeks as calculated by the INTCK function using the alignment option.

The variables WEEKS and WEEKSD are both incremented each time the interval boundary is crossed (Sunday – 18 and 25 September). However, the continuous alignment option causes WEEKSC to be incremented only when a full interval has elapsed—the interval boundary has effectively been adjusted to start at the date that starts the interval.

Obs	start	end	weeks	weeksc	weeksd
1	14SEP2011	14SEP2011	0	0	0
2	14SEP2011	15SEP2011	0	0	0
3	14SEP2011	16SEP2011	0	0	0
4	14SEP2011	17SEP2011	0	0	0
5	14SEP2011	18SEP2011	1	0	1
6	14SEP2011	19SEP2011	1	0	1
7	14SEP2011	20SEP2011	1	0	1
8	14SEP2011	21SEP2011	1	1	1
9	14SEP2011	22SEP2011	1	1	1
10	14SEP2011	23SEP2011	1	1	1
11	14SEP2011	24SEP2011	1	1	1
12	14SEP2011	25SEP2011	2	1	2

. . . . portions of the listing are not shown . . . .

3.4.4 Automatic Dates

Although the INTNX function is designed to advance a date or time value, it can used in a number of other situations where its immediate application is not as obvious.

Collapsing Dates

The INTNX function can be used to collapse a series of dates into a single date, thus allowing the new date to be used as a classification variable. When a format is available, most procedures can use the formatted value to form groups (ORDER=FORMATTED; see Section 2.6.2). However, when a format is not available the INTNX function can be used as an alternative.

To collapse dates we take advantage of the characteristic of the function that adjusts dates to the start of the interval (or the middle or end using the alignment option). If we then *advance* each date by 0 intervals the dates are collapsed into a single date. In the manufacturing data (ADVRPT.MFGDATA) items are being built continuously with the manufacturing

```
hourgrp = intnx('hour',datetime,0);
```

```
twohr   = intnx('hour2',datetime,0);
```

time stored as a datetime value. We would like to group the items into a one-hour periods. Using the first INTNX function call shown here, all items manufactured within the same hour will have the same value of HOURGRP. For instance this will group all times between 06:00 and 06:59 into the same group (06:00). If we had needed to create two-hour interval groups we could have used an interval multiplier (TWOHR).

Expanding Dates

The INTNX function can also be used to expand a single date or datetime value into a series of equally spaced values. The expansion is as simple as a DO loop. This DATA step creates 12 observations with DATE taking on the value of the first day of each month in 2007.

```
data monthly(drop=i);
   do i = 0 to 11;
      date = intnx('month','01jan2007'd,i);
      output monthly;
   end;
   format date date9.;
   run;
```

This usage of the INTNX function is written specifically so that the resulting dates always fall on the first of the month. Sometimes we need the date to be centered on the interval. This is problematic for months, because they do not have equal length. The midpoint alignment option for the INTNX function (shown here to generate MIDMON) only works to a point. The resulting dates will fall on the 14th, 15th, or 16th depending on the length of the month. Consistency is usually more important than technical accuracy (relative to the midpoint which does not really even exist for most months). The variable MON15 will always contain a date that falls on the 15th of each month. This consistency is achieved by adding 14 days to the beginning of the month so variable MON15 will always contain a date that falls on the 15th of each month.

```
midmon = intnx('month','01jan2007'd,i,'m');
```

```
mon15 = intnx('month','01jan2007'd,i) + 14;
```

Date Intervals or Ranges

In the following example the macro variable &DATE contains a date (in DATE9. form), and we need to subset the data for all dates that fall in the same month. The goal is to specify the start and end points of the correct interval, in this case the correct month of the correct year.

```
%let date=12jun2007;
data june07;
set advrpt.lab_chemistry;
      if intnx('month',labdt,0) ❶
         le "&date"d
            le intnx('month',labdt,0,'end'); ❷
run;
```

❶Advancing to the start of an interval is the default. Here the date is advanced 0 months—effectively the start of the current month.

❷The last day of the month is obtained by specifying the 'end' alignment option. Another common way to find the last day of the month is to find the first day of the following month and subtract one day.

```
(intnx('month',labdt,1)-1)
```

Previous Month by Name

The INTNX and INTCK functions can also be utilized by the macro language. We will be given a three-letter month abbreviation and our task is to return the abbreviation of the previous month. To do this we need to use the INTNX function to advance the month one month into the past. The macro function %SYSFUNC will be used to allow us to access the INTNX function outside of the DATA step.

```
%let mo=Mar;  ❶
%* Create a date for this month (01mar2010);
%let dtval❸  = %sysfunc(inputn(&mo.2010,monyy7.));  ❷

%* Previous month;
%let last    = %sysfunc(intnx(month,&dtval,-1));  ❹

%* Determine the abbreviation of the previous month;
%let molast = %sysfunc(putn(&last,monname3.));  ❺
%put mo=&mo dtval=&dtval molast=&molast;
```

❶ A three-letter month abbreviation is created. One month prior to 'Mar' is 'Feb'.

❷ The INPUTN function converts the three-letter month abbreviation into a SAS date. The year number used here is unimportant. Although the PUT and INPUT functions cannot be used with %SYSFUNC, their execution time analogues can be used with %SYSFUNC. These analogues are type-specific. Here the INPUTN function, which writes a numeric value ❸ (a SAS date in this case), is used instead of the INPUT function.

❹ The date contained in &DTVAL is advanced one month into the past. Notice that the interval name is constant text and is not quoted when using the INTNX function within a %SYSFUNC.

❺ The PUTN function converts the numeric SAS date contained in &LAST to a three-letter month abbreviation. &MOLAST correctly now contains 'Feb'.

```
140   %put mo=&mo dtval=&dtval molast=&molast;
mo=Mar dtval=18322 molast=Feb
```

The intermediate macro variables are not really needed, but for illustration purposes they do simplify the code. The more complex statement without these macro variables is shown in the sample code for this section.

MORE INFORMATION
A SAS date is created from a macro variable using the PUTN function in Section 3.5.1. A related example to the one shown here is also shown in Section 3.5.2.

SEE ALSO
A more complex version of this code example was used in a SAS Forum thread http://communities.sas.com/message/47615.

3.5 Variable Conversions

When we use the term *variable conversions*, we most often are referring to the conversion of the variable's type from numeric to character or character to numeric. We could also be referring to the conversion of the units associated with the values of the variable.

3.5.1 Using the PUT and INPUT Functions

When a numeric variable is used in a character expression or when a character variable is used in a numeric expression, the variable's type has to be converted before the expression can be evaluated. By default these conversions are handled automatically by SAS. However, whenever a variable's type is converted, SAS writes a note in the LOG. Although this note is fairly innocuous, in some situations or even industries the note itself is sufficient to cast doubt on your program.

In the DATA step shown here, the variable SUBJECT is character, and we need to create a numeric analog. Since subject number is just an identification string, one could argue that it is more appropriately character. However, for this example I would like to convert the character value to numeric.

```
data ae(drop=subjc);
   set advrpt.ae(rename=(subject=subjc));
   length subject 8; ❶
   subject=subjc; ❷
   run;
```

❶ The variable SUBJECT is added to the Program Data Vector, PDV, as a numeric variable.

❷ The conversion takes place when the character value of SUBJC is forced into the numeric variable SUBJECT. The LOG shows:

```
NOTE: Character values have been converted to numeric values at the
places given by:
      (Line):(Column).
      114:15
```

There is nothing wrong with allowing SAS to perform these automatic conversions. In fact there is evidence (Virgle, 1998) to suggest that these are the most efficient conversions. However, since as was mentioned above, there are some programming situations where even this rather benign note in the LOG is unacceptable, we need alternatives that do not produce this note. The PUT and INPUT families of functions provide this alternative.

When SAS performs an automatic conversion of a numeric value to a character, the result is right justified (behind the scenes a PUT function is used with a BEST. format). Usually you will want the character value to be left justified and this is most easily accomplished using the LEFT function, which operates on character strings. When converting from character to numeric, as was done above, this is not an issue.

The PUT and INPUT functions can be used directly to convert from numeric to character and character to numeric. Added power is provided through the use of a format. The PUT function is used to convert from numeric to character and the INPUT function is used to convert from character to numeric.

- PUT   always results in a character string. The format matches the type of the incoming variable.
- INPUT   results in a variable with the same type as the informat.

MORE INFORMATION
The PUTN and INPUTN functions are used with %SYSFUNC in a macro language example in Section 3.4.4.

Character to Numeric
In the AE data the subject is coded as character and we would like to have it converted to a numeric variable. Converting the value by forcing the character variable into numeric variable, as was done above, will get the job done; however, the conversion message will appear in the LOG.

```
data ae(drop=subjc);
   set advrpt.ae(rename=(subject=subjc));
   subject = input(subjc,3.);
   run;
```

When the INPUT function is used with a numeric informat, the incoming value (SUBJC) is converted to numeric without the note appearing in the LOG.

```
data conmed;
   set advrpt.conmed;
   startdt = input(medstdt_,mmddyy10.);
   run;
```

Character dates are converted to SAS dates in the same manner. Again the key is that a numeric infomat causes the INPUT function to return a numeric value. The selection of the informat depends on the form of the character date.

SEE ALSO

The SAS Forum thread http://communities.sas.com/message/29331 discusses character to numeric conversion when special characters are involved.

Numeric to Character

The PUT function is generally used to convert a numeric value to a character string. Because a numeric format is used, the resulting string is right justified. Very often a LEFT function is then applied to left justify the string. The LEFT function can be avoided by using the format justification modifier. Here

```
worddt1 = put(medstdt,worddate18.);
worddt2 = left(put(medstdt,worddate18.));
worddt3 = put(medstdt,worddate18.-l);
```

WORDDT1 will be a right justified string. WORDDT2 and WORDDT3 will be left justified. When WORDDT3 is formed the -L causes the format to left justify the string without using the LEFT function.

Using User-Defined INFORMATS

In a SAS Forum thread the following question (and I paraphrase) was posted. "How can I convert the name of a color to a numeric code?" One of the suggested solutions highlights a common misunderstanding of the relationship of formats and informats.

```
proc format;
value $ctonum  ❸
  'yellow' = 1
  'blue'   = 2
  'red'    = 3;
  run;
data colors;
  color='yellow'; output colors;
  color='blue';   output colors;
  color='red';    output colors;
  run;
data codes;
  set colors;
  x = put(color,$ctonum.);  ❹
  z = input(x,3.);  ❺
  run;
```

The data set COLORS has the variable COLOR which takes on the values of 'yellow', 'blue', and so on.

❸ We define a format ($ctonum.) that converts the colors to numbers. The format attempts to make the resultant value numeric by not quoting the values on the right side of the assignments in the VALUE statement.

❹ The format $CTONUM cannot be used with the INPUT function, so the PUT is used to generate the numeric value as a character variable.

❺ The character variable X is then converted to the numeric code (Z) through the use of the INPUT function.

The reason that this format will not work with the PUT function is actually simple. There is a distinct difference between formats and informats. The INPUT function expects an informat. The previous example can be simplified by creating CTONUM. as a numeric informat using the INVALUE statement.

When the INPUT function is used with a numeric informat the result will be a numeric value. Consequently, we need to create a numeric informat that will convert color to a numeric code.

```
proc format;
invalue ctonum ❻
  'yellow' = 1 ❼
  'blue'   = 2
  'red'    = 3;
  run;
data colors;
  color='yellow'; output colors;
  color='blue';   output colors;
  color='red';    output colors;
  run;
data codes;
  set colors;
  x = input(color,ctonum.); ❽
  run;
```

❻ Since we want to create a numeric informat, the format name CTONUM. is not preceded by a $.

❼ Notice that the value to be assigned (the right side of the equal sign) is not quoted. It was not quoted in the previous example ❸; however, there we were creating a character format (as evidenced by the $ in the name), and the quotes were assumed.

❽ The numeric informat CTONUM. is used to convert the color string to a numeric code. The variable X will be numeric.

Execution or Run-Time Versions

Generally when we use the PUT or INPUT functions we know what format is to be used, and we can specify it like in the previous examples. When specified this way, these formats are applied when the statement is compiled. Sometimes the format that is to be applied is unknown until the DATA step actually executes. Usually this means that the format itself is not constant for all the observations and is either supplied on the data itself or it is dependent on the data.

The PUT and INPUT functions each come with an execution time analogue for both numeric and character values (PUTN, PUTC, INPUTN, and INPUTC). For each of these four functions, the format/informat used by the function is determined during the execution of the function.

In the following example, the incoming dates are supplied in a variety of forms and each has a format that is to be used in its conversion to a SAS date. The date is read as a character value, as is the format that will be used in the conversion. The variable FMT, which contains the informat that is to be applied in the conversion, becomes the second argument of the INPUTN function.

```
data dates;
   input @4 cdate $10. @15 fmt $9.;
   ndate = inputn(cdate,fmt);
   format ndate date9.;
   datalines;
   01/13/2003 mmddyy10.
   13/01/2003 ddmmyy10.
   13jan2003  date9.
   13jan03    date7.
   13/01/03   ddmmyy8.
   01/02/03   mmddyy8.
   03/02/01   yymmdd8.
   run;
```

```
3.5.1 PUT and INPUT Functions
Using INPUTN

Obs    cdate       fmt           ndate
 1    01/13/2003  mmddyy10.    13JAN2003
 2    13/01/2003  ddmmyy10.    13JAN2003
 3    13jan2003   date9.       13JAN2003
 4    13jan03     date7.       13JAN2003
 5    13/01/03    ddmmyy8.     13JAN2003
 6    01/02/03    mmddyy8.     02JAN2003
 7    03/02/01    yymmdd8.     01FEB2003
```

Examination of the LISTING output of the data set WORK.DATES shows that the incoming character strings have been correctly translated into SAS dates using the informats supplied in the data.

MORE INFORMATION
In many cases when dealing with inconsistent date forms, one of the "anydate" informats can also be successfully applied (see Section 12.6).

SEE ALSO
The INPUTN function was a solution to a question posed in the SAS Forums http://communities.sas.com/thread/30362?tstart=0.

Using the PUTN Function with the %SYSFUNC Macro Function
As was done in the last example of Section 3.4.4, when you need to perform a numeric/character conversion using the INPUT or PUT functions in the macro language you will need to use one of the execution time versions. The function itself is accessed using the %SYSFUNC macro function.

The following example writes the date value stored in the automatic macro variable &SYSDATE9 to the LOG using the WORDDATE18. format. Without the PUTN

```
%put %sysfunc(putn("&sysdate9."d,worddate18.));
```

function, the date constant would not be recognized as such by the macro language, and the macro variable &SYSDATE9 would not be converted to a SAS date value.

The PUTN function can be applied to other date formats. The following macro function will return the name of the previous month and its year. The PUTN function converts a date value to the name of the month by using the MONNAME. format.

```
%macro lastmy;
%local prevdt tmon tyr;  ❶
%let prevdt =
        %sysfunc(intnx(month,%sysfunc(today()),-1));  ❷
%let tmon = %sysfunc(putn(&prevdt,monname9.));  ❸
%let tyr  = %sysfunc(year(&prevdt));  ❹
&tmon/&tyr
%mend lastmy;

* Write last month's month and year into a title;
TITLE2 "Counts for the Previous Month/Year (%lastmy)";  ❺
```

❶ All macro variables are forced onto the local symbol table.

❷ The INTNX function is used to advance the date one month into the past. &PREVDT is now a SAS date value representing the first day of the previous month.

❸ The PUTN function is used to write the name of the month of the date held in &PREVDT.

❹ The four-digit year associated with the previous month is retrieved using the YEAR function.

❺ The macro %LASTMY is called from within the TITLE2 statement. When the macro executes the macro call is replaced by the month name and its associated year. The TITLE shown here would result if the macro was called for any date in October 2011.

```
3.5.1 PUT and INPUT Functions
Counts for the Previous Month/Year (September/2011)
```

3.5.2 Decimal, Hexadecimal, and Binary Number Conversions

The conversion of decimal values to hex, octal, and binary is often accomplished through the use of formats and informats in conjunction with the PUT and INPUT functions.

When converting from a decimal number to binary, hex, or octal, you can use the PUT function along with the respective formats (BINARY., HEX., or OCTAL.). In this example the decimal number 456 is being converted. The PUT statement writes these values to the LOG, which shows that the conversion was successful. Because these are integer conversions, decimal fractions are lost (through truncation).

```
data convert;
length bin $20;
* Converting from Decimal;
dec = 456;
bin = put(dec,binary9.);
oct = put(dec,octal.);
hex = put(dec,hex.);
put dec= bin= oct= hex=;

* Converting to Decimal;
bdec = input(bin,binary9.);
odec = input(oct,octal.);
hdec = input(hex,hex.);
put bdec= odec= hdec=;
run;
```

When you need to convert from one of these number systems to decimal, the INPUT function is used. Informats of the same name as the formats that were used in the PUT function are used in this conversion as well. In this example the PUT function is again used to confirm the conversion by writing the values to the LOG.

```
dec=456 bin=111001000 oct=710 hex=000001C8
bdec=456 odec=456 hdec=456
```

The LOG shows that the original decimal number of 456 has been converted to other number systems and back to decimal.

In SAS/GRAPH both the RGB color scale and the gray scale use hex numbers to specify specific color values. The 256 (16^2) possible shades of gray are coded in a hex number. The codes for a gray scale number will range from GRAY00 to GRAYFF. For RGB colors there are 256 shades of each of the three primary colors of red, green, blue. Some color wheels use decimal values rather than hex values and a specific color value might require conversion. As was shown above, the HEX. format would be used with the PUT function to provide the converted value. A macro (%PATTERN) that performs a series of these conversions for a gray scale example can be found in Carpenter (2004, Section 7.4.2).

The functions ANYXDIGIT and NOTXDIGIT can be used to parse a character string for hex numbers (see Section 3.6.1). ANYXDIGIT returns the position of the first number or any of the letters A through F. The NOTXDIGIT returns the functional opposite and returns the position of the first character that cannot be a part of a hex number.

3.6 DATA Step Functions

It is simply not possible to enumerate all of the useful and important DATA step functions in a single section of a book such as this one. In fact the topic fills a complete book (Cody, 2010), which should be required reading for every SAS programmer. This section only covers a few of the functions that seem to be underutilized, either because they are newer to the language, have newer functionality, or because they just have trouble making friends.

It should be noted that many of the newer functions, as well as some of the old standbys have additional modifiers that greatly expand the utility and flexibility of the functions. A classic example would be the COMPRESS function, which has been available for a very long time. While its default behavior remains unchanged, it can now do much more. It is important for even advanced SAS programmers to reread and refamiliarize themselves with these functions.

SEE ALSO

Cody (2010) is an excellent source of information on the syntax and use of functions.

3.6.1 The ANY and NOT Families of Functions

The ANY family of functions is group of character search functions with names that start with ANY, and like the INDEX function they search for the first occurrence of the stated target and return the position.

- ANYALNUM   first alpha or numeric value
- ANYALPHA   first alpha character
- ANYDIGIT   first digit (number)
- ANYPUNCT   first punctuation (special character—not alpha numeric)
- ANYSPACE   first space (although the definition of a space is broader than just a blank)
- ANYUPPER   first uppercase letter
- ANYXDIGIT   first character that could be a part of a hexadecimal number

In the example below the variables SODIUM, POTASSIUM, and CLORIDE are to be converted from character to numeric. Before the conversion takes place we would like to verify that the conversion will be successful; that is, that there are no non-numeric values. Using the INPUT function directly will perform the conversion and will correctly produce missing values; however, values that cannot be converted (because they contain non-numeric characters) will also produce errors in the log. These errors can be eliminated by first checking the value with the ANYALPHA function.

```
data lab_chem_n(keep=subject visit labdt
                    sodium_n potassium_n chloride_n)
    valcheck(keep=subject visit variable value note); ❶
  set lab_chem;
  length variable $15 value $5 note $25;
  array val {3} $6 sodium potassium chloride;
  array nval {3} sodium_n potassium_n chloride_n;
  do i = 1 to 3;
    if anyalpha(left(val{i})) then do; ❷
      variable = vname(val{i});
      value=val{i};
      note = 'Value is non-numeric';
      output valcheck; ❶
    end;
    else do;
      * Convert value;
      nval{i} = input(val{i},best.); ❸
    end;
  end;
  output lab_chem_n; ❹
run;
```

❶ The values which will not properly convert are saved in a data set.

❷ The ANYALPHA function is used to determine if the conversion will be successful, and any values that will cause a problem are written out to the data set VALCHECK for further evaluation.

❸ The remaining values are converted using the BEST. informat. The ?? format modifier could

```
nval{i} = input(val{i},?? best.);  ❸
```

also be used to suppress the error messages in the LOG; however, in this example we want notification of the invalid values.

❹ The observation with the converted values is written to the new data set.

This solution will not work for character values containing scientific notation. While the BEST. informat will successfully convert values containing scientific notation, the ANYALPHA function will also flag the 'E'. Additional logic such as used in the autocall macro function %DATATYP would be required.

The NOT family of functions, which also contains over a dozen functions, are nominally the functional opposite of the ANY functions. These functions are used to detect text that is *not* present. For instance the NOTALPHA function returns the position of the first non-alpha character. The NOTDIGIT is very similar to the ANYALPHA function; however, NOTDIGIT could not be substituted for ANYALPHA at ❷ above. NOTDIGIT detects trailing blanks, plus signs, minus signs, and decimal points even though they could be part of a number. A nearly equivalent use of NOTDIGIT to the ANYALPHA shown above could be coded as:

```
if notdigit(trim(left(compress(val{i},'+-.')))) then do;  ❷
```

All of the functions in these two families have an optional second argument, which adds a great

```
text = '1234x6yz9';
pos = anyalpha(text,-6);
```

deal of flexibility to what these functions can accomplish. This argument, which is the start position, can be either a positive or negative integer. When negative, the search is right to left rather than left to right as it is when positive. In either case the value returned is the position counting left to right. In the example shown here, the ANYALPHA will find the letter 'x' and will return a 5.

MORE INFORMATION
The ANYDIGIT function is used in one of the examples in Section 3.6.5 and one of the examples in Section 3.6.6 uses the ANYALPHA function. The ?? format modifier is introduced in Section 1.3.1 and used with the INPUT function in Section 2.3.1.

3.6.2 Comparison Functions

Performing inexact comparisons has always been, well, inexact, not to mention tedious and difficult. Traditional comparison functions have included SOUNDEX and SPEDIST. This family of comparison functions has been expanded and now provides several ways to look at the similarities or differences between two strings. These additional functions include:

- COMPARE   compares two strings
- COMPLEV   computes a distance between two strings based on similarities
- COMPGED   computes a generalized distance between two strings
- COMPCOST   this routine adjusts the comparison criteria for COMPGED

Each of these functions supports a number of arguments that allow a variety of types of comparisons.

In the following example the data contains names of various metals that have possibly been misspelled. We need to determine the correct spelling. Similar problems arise when trying to match names or drugs which also often include abbreviations. For this example we have a data set that contains all the correctly spelled metals (WORK.METALS).

```
data perfect (keep=datname value)  ❶
     potential(keep=datname name value mincost);  ❷
  array metals {10} $9 _temporary_ ;  ❸
  do until(done);  ❹
    set metals end=done;
    cnt+1;
    metals{cnt} = name;
  end;

  do until(datdone);
    set namelist(keep=datname value)
         end=datdone;  ❺
    mincost=9999999;
    do i = 1 to 10;  ❻
      cost = compged(datname,metals{i},'il');  ❼
      if cost=0 then do;  ❽
        output perfect;
        goto gotit;
      end;
      else if cost lt mincost then do;  ❾
        mincost=cost;
        name = metals{i};
      end;
    end;
    output potential;  ❿
    gotit:
  end;
  stop;
  run;
```

In an attempt to simplify the code a bit for this example, the number of metals (10) and the maximum length of the metal's name ($9) have been hardcoded.

❶ The observations with perfect matches are saved in WORK.PERFECT, while the best guess for the mismatched values is saved in WORK.POTENTIAL ❷.

❸ A temporary array is defined to hold the known good spellings.

❹ The table that contains the correct spellings is loaded into the array ❸.

❺ The data (with the potentially misspelled metal names in the variable DATNAME) is read.

❻ The loop, which steps through the list of correctly spelled metal names, is entered.

❼ The current potentially misspelled name is compared to one of the correctly spelled names and a measure of their similarity is stored in COST. The more similar they are the lower the cost. Exact matches will have a cost of 0. The third argument of the COMPGED function is used to specify one or more comparison modifiers. These include:

- I   (the letter i) ignore case
- L   (the letter L) remove leading blanks

❽ When there is a perfect match, the COMPGED function returns a 0. This observation is saved in WORK.PERFECT and there is no need to check any other spellings.

❾ The non-zero cost is checked against those already found for this observation. If it is less than the lowest found so far, we have a better match and its NAME and COST are saved.

❿ After checking all 10 possibilities, the closest match is written to WORK.POTENTIAL. This data set can be further examined in the process of building a mapping dictionary.

The benefits of using the comparison functions over direct comparison include the ability to make the comparison case insensitive, ignore leading blanks, compare strings of unequal length, and to remove quotes from the comparison.

Although the functions COMPBL, COMPRESS, and COMPOUND also start with the letters COMP, they are not a part of this family of comparison functions.

MORE INFORMATION
A further discussion of DATA steps with two SET statements can be found in Section 3.8.3 and the use of the DOW loop in Section 3.9.1.

3.6.3 Concatenation Functions

Functions are now available that allow us to perform concatenation operations without resorting to the concatenation operator (| |). These include:

- CAT   same as | |, it preserves leading and trailing blanks
- CATQ   adds a delimiter and quotes individual items
- CATS   removes leading and trailing blanks
- CATT   removes only trailing blanks
- CATX   removes leading and trailing blanks, but also adds a separator between strings (you get to choose the separator)

The following statement, which was used in a PROC REPORT compute block, places a text string containing both the mean and standard deviation in a single report item.

```
_c5_ = cats(put(_c3_,6.2),' (',put(_c4_,7.3),')');
```

The resulting value might appear as something like: `15.23 (4.567)`.

As a general rule the CAT functions are considered to be preferred to the concatenation operator.

SEE ALSO
The CATS function is used in a CALL EXECUTE example in Fehd and Carpenter (2007).

3.6.4 Finding Maximum and Minimum Values

When finding the maximum or minimum values from within a list of variables and/or values, the MAX and MIN functions are no longer the only functions from which to choose. Functions that can return the maximum and minimum values include:

- LARGEST   Returns the nth largest value from a list of values
- MAX   Returns the largest value from a list of values
- SMALLEST   Returns the nth smallest value from a list of values (ignores missing values)
- MIN   Returns the smallest value from a list of values (ignores missing values)
- ORDINAL   Returns the nth smallest value from a list of values (includes missing values)

The MAX and MIN functions can only return the single largest or smallest value. When you use the LARGEST and SMALLEST functions; however, you can choose something other than that single extreme. In addition the ORDINAL function allows the consideration of missing values as a minimum value.

In this example we would like to determine the dates of the first two and last two visits for each subject. Since we need more than just the maximum and minimum date for each subject, the MAX and MIN functions cannot be used. In this data set we cannot depend on the visit number as subjects sometimes complete visits out of order.

```
data Visitdates(keep=subject firstdate seconddate
                      lastdate next2last);
   set advrpt.lab_chemistry;
   by subject;
   array dates {16} _temporary_; ❶
   if first.subject then call missing(of dates{*}); ❷
   * Save dates;
   dates{visit} = labdt; ❸
   if last.subject then do; ❹
      firstdate  = smallest(1,of dates{*});
      seconddate = smallest(2,of dates{*}); ❺
      next2last  = largest(2,of dates{*});
      lastdate   = largest(1,of dates{*});
      output visitdates;
   end;
   format firstdate seconddate
          lastdate next2last date9.;
   run;
```

❶ Create an array to contain the (up to 16) visit dates.

❷ The CALL MISSING routine is used to clear the array for each new subject.

❸ Load the dates into the array using the visit number as the array index.

❹ The SMALLEST and LARGEST functions are applied to the array of values.

❺ The first argument of the SMALLEST (and LARGEST) function determines which extreme value is to be selected. When this argument is a 1, these functions mimic the MIN and MAX functions. In this function call, the two (2) selects the next to smallest value.

MORE INFORMATION
A comparison of the MAX and MIN functions to the MAX and MIN operators (and why the operators should *never* be used), can be found in Section 2.2.5.

3.6.5 Variable Information Functions

Variable information functions can be used to provide information about the characteristics of the variables in a data set during DATA step execution. Usually you already know these characteristics while you are programming; however, this is not always the case. Generalized macro applications often are designed to work against data sets whose characteristics are unknown during macro development. Much of this information can be retrieved using these functions.

There are over two dozen functions in the Variable Information category. The following list is adapted from the SAS documentation.

- VNEXT   Returns the name, type, and length of a variable that is used in a DATA step.

- VARRAY   Returns a value that indicates whether the specified name is an array.

- VARRAYX   Returns a value that indicates whether the value of the specified argument is an array.

- VFORMAT   Returns the format that is associated with the specified variable.

- VFORMATD   Returns the decimal value of the format that is associated with the specified variable.

- VFORMATDX   Returns the decimal value of the format that is associated with the value of the specified argument.

- VFORMATN   Returns the format name that is associated with the specified variable.

- VFORMATNX   Returns the format name that is associated with the value of the specified argument.

- VFORMATW   Returns the format width that is associated with the specified variable.

- VFORMATWX   Returns the format width that is associated with the value of the specified argument.

- VFORMATX   Returns the format that is associated with the value of the specified argument.

- VINARRAY   Returns a value that indicates whether the specified variable is a member of an array.

- VINARRAYX   Returns a value that indicates whether the value of the specified argument is a member of an array.

- VINFORMAT   Returns the informat that is associated with the specified variable.

- VINFORMATD   Returns the decimal value of the informat that is associated with the specified variable.

- VINFORMATDX Returns the decimal value of the informat that is associated with the value of the specified variable.

- VINFORMATN   Returns the informat name that is associated with the specified variable.

- VINFORMATNX Returns the informat name that is associated with the value of the specified argument.

- VINFORMATW   Returns the informat width that is associated with the specified variable.

- VINFORMATWX Returns the informat width that is associated with the value of the specified argument.

- VINFORMATX   Returns the informat that is associated with the value of the specified argument.

- VLABEL   Returns the label that is associated with the specified variable.

- VLABELX   Returns the label that is associated with the value of the specified argument.

- VLENGTH   Returns the compile-time (allocated) size of the specified variable.

- VLENGTHX   Returns the compile-time (allocated) size for the variable that has a name that is the same as the value of the argument.
- VNAME   Returns the name of the specified variable.
- VNAMEX   Validates the value of the specified argument as a variable name.
- VTYPE   Returns the type (character or numeric) of the specified variable.
- VTYPEX   Returns the type (character or numeric) for the value of the specified argument.
- VVALUE   Returns the formatted value that is associated with the variable that you specify.
- VVALUEX   Returns the formatted value that is associated with the argument that you specify.

You may not see an immediate need for many of these functions and routines, but when you start building programs that need to dynamically gather information about a data set, they can be indispensible. I believe that you should at least understand them well enough to know to look them up when you do need to use them.

The VNEXT routine can be especially helpful as it can return not only the variable's name, but its type (numeric/character) and length as well. In addition it can be used to step through, one-at-a-time, all the variables (including temporary variables) in a data set.

```
data labdat;
   set advrpt.lab_chemistry;
   retain p_type ' ' p_len .;  ❶
   if _n_=1 then do;  ❷
      call vnext(potassium,p_type,p_len);  ❸
   end;
   run;
```

❶ P_TYPE will be used to store the type of the variable POTASSIUM (N or C). The variable P_LEN will hold the length of the selected variable.

❷ These values will be constant for the entire data set (the attributes of the variable POTASSIUM can't change), so we only need to call the VNEXT routine once.

❸ Notice that the arguments are variable names—not character strings.

```
3.6.5 Using Variable Information Functions
VNEXT and a Specific Variable

Obs   SUBJECT   VISIT      LABDT   potassium   sodium   chloride   p_type   p_len

  1     200       1    07/06/2006     3.7        14.0      103        C        3
  2     200       2    07/13/2006     4.9        14.4      106        C        3
  3     200       1    07/06/2006     3.7        14.0      103        C        3
  4     200       4    07/13/2006     4.1        14.0      103        C        3
               .... portions of the table are not shown ....
```

Although this same information can also be gathered a number of ways (e.g., PROC CONTENTS, SASHELP.VCOLUMNS, DICTIONARY.COLUMNS), a practical use of the VNEXT routine is to build elements of a data dictionary. The following example is a first attempt at using VNEXT to cycle through all the variables of a data set.

```
%let dsn = advrpt.lab_chemistry;
data listallvar(keep=dsn name type len);
   if 0 then set &dsn;
   length name $15;
   retain dsn "&dsn" type ' ' len .;
   name= ' ';  ❹
   do until (name=' ');  ❺
      call vnext(name,type,len);
      output listallvar;  ❻
   end;
   stop;  ❼
run;
```

❹ The variable NAME will be used to hold the name of the variable to be retrieved by VNEXT. Since this variable is blank, the VNEXT routine will retrieve the name, type, and length of the next variable on the Data Set Data Vector.

❺ Loop through all the variables on the data set. VNEXT will return a blank when it is unable to retrieve another variable name. Although the variable NAME is initialized to blank ❹, the DO UNTIL loop will still execute at least once as it is evaluated at the bottom of the loop.

❻ The name, type, and length of each variable are written to the data set LISTALLVAR.

❼ Since we are only interested in the variable attributes (metadata), we do not actually need to read any data, so the step is stopped.

Obs	name	dsn	type	len
1	SUBJECT	advrpt.lab_chemistry	C	3
2	VISIT	advrpt.lab_chemistry	N	8
3	LABDT	advrpt.lab_chemistry	N	8
4	potassium	advrpt.lab_chemistry	N	8
5	sodium	advrpt.lab_chemistry	N	8
6	chloride	advrpt.lab_chemistry	N	8
7	name	advrpt.lab_chemistry	C	15 ❽
8	dsn	advrpt.lab_chemistry	C	20
9	type	advrpt.lab_chemistry	C	1
10	len	advrpt.lab_chemistry	N	8
11	_ERROR_	advrpt.lab_chemistry	N	8
12	_N_	advrpt.lab_chemistry	N	8
13		advrpt.lab_chemistry		0 ❾

❽ The variables used by the VNEXT routine are a part of the Program Data Vector and consequently they are processed by VNEXT as well.

❾ On the last iteration, the VNEXT routine fails to return a value (name=' '); however, the OUTPUT statement ❻ is not conditionally executed so the observation is written to the data set.

Notice that all variables on the PDV, including temporary variables such as _ERROR_ and _N_ are also retrieved by VNEXT. If a BY statement had been present, FIRST. and LAST. variables would have also appeared in the list.

```
%let dsn = advrpt.lab_chemistry;
data listvar(keep=dsn name type len);
   if 0 then set &dsn;
   length name $15;
   retain dsn "&dsn" type ' ' len .;
   name= ' ';
   do until (name='name');  ❿
      call vnext(name,type,len);
      if name ne 'name' then output listvar;
   end;
   stop;
run;
```

You can limit the variable list to only those on the incoming data set by a simple modification to the loop logic ❺. In this code we search until we find the first variable that we have defined for use with VNEXT, the variable NAME ❿. Since the OUTPUT statement is conditionally executed, the observation with name='name' is not written to the data set.

Since the VNEXT routine can be used to retrieve variable attributes, these attributes can then be used during DATA step execution to retrieve the data itself. In the following example, character variables have been stored as codes and their associated formats have been added to the data set's metadata. We need to create a new data set with the same variable names, but we want the values to be formatted rather than stored as codes. In the motivating problem we do not know how many variables are to be converted or even their names. A similar problem was resolved using a macro language solution by Rosenbloom (2011a).

The variables SEX, SYMP, and RACE in the demographics data set (ADVRPT.DEMOG) are to be recoded using their associated formats, which are defined and added to the metadata in the sample program partially shown here. In the program shown below, we are simulating that we do not know the names of the coded variables (SEX, SYMP, and RACE).

```
3.6.5 Using Variable Information Functions
Retrieving and Using Formats

Obs    lname        fname      sex    symp    race

  1    Adams        Mary       F      02      2
  2    Adamson      Joan       F      10      2
  3    Alexander    Mark       M              1
  4    Antler       Peter      M      10      2
       .... portions of the listing are not shown ....
```

Each of these variables has a format assigned to it using the following FORMAT statement.

```
format sex $gender. symp $symptom. race $race.;
```

Notice that the name of the format is not necessarily the same as its variable.

Remember that for the purposes of this example we are assuming that we do not know either the names of the variables or the names of their formats.

```
proc sort data=advrpt.demog
               (keep=lname fname sex symp race)
          out=codedat;
   by lname fname; ❶
   format sex $gender. symp $symptom. race $race.;
   run;
data namelist(keep=lname fname varname varvalue); ❷
   set codedat;
   length varname name type $15 varvalue $30; ❸
   array vlist{25} $15 _temporary_ ; ❹
   if _n_=1 then do until (name=' '); ❺
      call vnext(name,type); ❻
      if upcase(name) not in:('LNAME' 'FNAME' ❼
                              'NAME' 'TYPE'
                              'VARNAME' 'VARVALUE')
                    and type='C' then do;
         cnt+1;
         vlist{cnt}=vnamex(name); ❽
      end;
   end;
   do i = 1 to cnt;
      varname = vlist{i}; ❾
      varvalue = vvaluex(varname);
      output namelist;
   end;
   run;
```

❶ The data are sorted using the key variables whose names we do know. The sort order becomes important when we transpose the data back into its original form (see ❿ below).

❷ Only the key variables and the two derived variables are to be written to the new data set. This data set will have one observation for each of the unknown variables for each of the original observations.

❸ The length of VARVALUE must be large enough to store the longest possible formatted value.

❹ The temporary array holds the names of the variables of interest. The dimension has to be sufficiently large. In this example a dimension of three would have been sufficient.

❺ A DO UNTIL loop is used to step through the unknown number of variables and to store their names in the array. Since we are dealing with the metadata at this point we only need to process this information one time.

❻ The VNEXT routine retrieves successive variable names and stores the name and type in the variables NAME and TYPE.

❼ All character variables (TYPE='C'), excluding those used in the DATA step, are selected for loading into the array of variable names.

❽ The VNAMEX function is used to store the value contained in the variable NAME in the name list array. This function allows the resolved value of NAME to be recovered. The VNAME function would have stored the unresolved value (NAME).

❾ The variable name is recovered from the array and stored in VARNAME. VARNAME now contains the name of the variable whose code we also need to recover. The VVALUEX function not only retrieves the value of that variable, but also the formatted value of that variable. Since we want the formatted value, this is perfect.

VIEWTABLE: Work.Namelist			
	lname	fname	varname	varvalue
1	Adams	Mary	sex	Female
2	Adams	Mary	race	Black
3	Adams	Mary	symp	Coughing
4	Adamson	Joan	sex	Female
5	Adamson	Joan	race	Black
6	Adamson	Joan	symp	Shortness of Breath
7	Alexander	Mark	sex	Male
8	Alexander	Mark	race	Caucasian
9	Alexander	Mark	symp	
10	Antler	Peter	sex	Male

The NAMELIST data set will have one observation for each of the unknown variables. VARNAME contains the original variable name and VARVALUE contains its formatted value.

The general form of the original data is reconstructed by transforming these rows into columns using PROC TRANSPOSE.

```
proc transpose data=namelist
    out=original(drop=_name_);
  by lname fname;
  id varname; ❿
  var varvalue;
  run;
```

❿ The variable holding the name of the variable of interest is used as the ID variable in the PROC TRANSPOSE step.

After the transpose step the data reflects the original form of the data; however, the coded values have been converted to the formatted values.

```
3.6.5 Using Variable Information Functions
Retrieving and Using Formats

Obs   lname       fname      sex      race       symp

  1   Adams       Mary       Female   Black      Coughing
  2   Adamson     Joan       Female   Black      Shortness of Breath
  3   Alexander   Mark       Male     Caucasian
  4   Antler      Peter      Male     Black      Shortness of Breath
  5   Atwood      Teddy      Male                Nausea
          . . . . portions of the listing are not shown . . . .
```

Rather than use the original variable name we may want to use the format name as the name of the new variable. Even this is only a slight alteration of the previous code. Here the VFORMATX

```
do i = 1 to cnt;
    varvalue = vvaluex(vlist{i});
    varname = substr(vformatx(vlist{i}),
                  2,
                  (anydigit(vformatx(vlist{i}))-2));
    output namelistc;
end;
```

function recovers the format name from which we grab just the name portion using the SUBSTR function

(excluding the leading $ sign and the trailing numbers).

MORE INFORMATION

The VTYPE function is used to retrieve a variable's type in the second example in Section 11.2.2. Metadata information can be retrieved through a variety of techniques. Additional approaches are discussed in Section 13.8.

SEE ALSO

The VNEXT routine documentation contains a simplified version of the first example and can be found in the SAS documentation http://support.sas.com/documentation/cdl/en/lrdict/64316/HTML/default/viewer.htm#a002295699 .htm.

3.6.6 New Alternatives and Functions That Do More

While most DATA step functions are fairly straightforward, some have uses that one might not at first anticipate. Others have seldom used optional arguments that give the function added utility. As a general rule using and understanding these functions is not the difficulty—knowing that they exist and remembering to use them is the issue.

To make matters even more interesting, a number of new functions were introduced with SAS®9. Many of these have similar utility to existing functions, but have been augmented so as to provide more flexibility and power.

The ARCOS Function

When you need a value to approximate the constant pi, avoid hard coding a less accurate value. The ARCOS(-1) returns the value of pi to as many significant digits as should be needed for most applications. The value of pi is also one of the constant values that can be returned by the CONSTANT function

```
pi1 = arcos(-1);
pi2 = constant('pi');
```

The COALESCE Function

The COALESCE function is used to find the first non-missing value in a list of values (variables). This function does not have any modifiers that allow it to search other than from left to right. However, it is possible to control the order of the values/variables listed in the call to the function. This allows one to return either the first, or by reversing the order, the last non-missing value.

SEE ALSO

Mike Zdeb provided a tip on sasCommunity.org that uses the COALESCE function to take the difference between the first and last non-missing values in a list of values http://www.sascommunity.org/wiki/Tips:Find_the_LAST_Non-Missing_Value_in_a_List_of_Variables.

The Counting Functions

The functions in the COUNT family return the number of items in a string. These can be strings of characters or words. Each function supports three arguments. The first is always the string that is to be searched. The usage of the second argument varies for each of these functions, and the real power and flexibility of these functions is achieved through the use of the third argument, which can take on a number of different values.

- COUNT counts appearances of a specific string of characters
- COUNTC counts the characters that either do or do not appear in a string of characters
- COUNTW counts the number of words in a string

The COUNTC and COUNTW functions both support well over a dozen modifiers for the third argument. These modifiers allow you to add characters or digits to the list, count from the left or right, and add a number of different types of special characters.

These functions can also be used in the macro language. Here the COUNTW function is used with %SYSFUNC to count the number of names in a list.

```
proc sql noprint;
select lname
    into :namelist separated by '/'
        from advrpt.demog(where=(lname=:'S'));
quit;
%put &namelist;
%put the number of names is %sysfunc(countw(&namelist,/));
```

```
28   %put &namelist;
Saunders/Simpson/Smith/Stubs
29   %put the number of names is %sysfunc(countw(&namelist,/));
the number of names is 4
```

SEE ALSO
The COUNTW function was used in the SAS Forum thread
http://communities.sas.com/thread/14720.

The DIM Function
The DIM function was designed to return the dimension of an array. This implies that it counts variables and in the past, prior to the advent of the COUNTW function – shown above, it has been used to count the number of words in a list.

```
%macro wcount(list);
%* Count the number of words in &LIST;
%global count;
   data _null_;
       array nlist{*} &list;
       call symputx('count', dim(nlist));
       run;
%mend wcount;
%wcount(&namelist)
%put The total number of words in &namelist is: &count;
```

Here we fool the DIM function by using a list of words as variable names. The DIM function then counts these words. This approach for counting is more restrictive than the COUNTW function shown above because the words must conform to SAS variable naming conventions: the list must be space separated, the &COUNT macro variable is placed on the global symbol table, and the DATA step is required.

SEE ALSO

This example was adapted from one shown by Cheng (2011). The technique itself was first proposed by Michael Friendly (1991).

The GEOMEAN Function

There are several different types of means. Usually when we refer to a mean we are actually referring to the arithmetic average. Another type of mean is the Geometric mean, which is calculated by the GEOMEAN function. An artifact of the geometric mean's formula is that it can also be used to calculate the nth root of a number.

```
title2 'GEOMEAN';
data roots;
do x = 0 to 30 by 5;
  *Square root;
  root2 = sqrt(x);
  nth2  = x**(1/2);
  g2    = geomean(x,1);
  *Cube root;
  nth3  = x**(1/3);
  g3    = geomean(x,1,1);
  *4th root;
  nth4  = x**(1/4);
  g4    = geomean(x,1,1,1);
  output;
end;
run;
```

The 5th root of X would be coded as GEOMEAN(X,1,1,1,1) the value and a series of ones (1 less than the root).

The IFC and IFN Functions

The IFC and IFN functions give us the ability to consolidate a set of certain types of IF-THEN/ELSE statements with a single function call. Generally these functions are used for a single comparison that results in TRUE/FALSE/MISSING, which in turn is used to determine a variable assignment.

The IFN function is used to return a numeric result, while the IFC function returns a character string. For both functions the arguments are:

- 1st expression
- 2nd result returned when the expression is true
- 3rd result returned when the expression is false
- 4th result returned when the expression is missing (optional)

In the following example the patients are being divided into GENERATION according to birth year.

```
data generation;
   set advrpt.demog(keep=lname fname dob);
   length generation $10;
   if year(dob) = . then generation='Unknown';
   else if year(dob) lt 1945 then generation= 'Greatest';
   else if year(dob) ge 1945 then generation = 'Boomer';
   run;
```

The three IF/ELSE/IF statements can be replaced by a single assignment statement that takes advantage of the IFC function.

```
data generation2;
   set advrpt.demog(keep=lname fname dob);
   length generation $10;
   generation =ifc(year(dob) ge 1945,'Boomer','Greatest','Unknown');
   run;
```

The two solutions are *not* identical. The IFC function shown above will *never* return a missing value. The expression can only resolve to 0 or 1. A missing DOB will result in a missing year which will necessarily be less than a constant. When using IFC or IFN with the intent that it can select the 4th (missing) argument you must make sure that it is possible that the expression can indeed resolve to a missing value. That is not the case here.

The solution is simple; as programmers the issue is for us to remember to be careful. One solution is to multiply the stated expression by the variable that could be missing. Here the expression is multiplied by the year of the DOB. The sense of the expression is not changed, but it can now take on a missing value.

```
ifc(year(dob)*(year(dob) ge 1945),'Boomer','Greatest','Unknown');
```

MORE INFORMATION
A similar IFC function is discussed in terms of the FINDC function in the next subsection.

SEE ALSO
When used with the %SYSFUNC macro function, Fehd (2009) shows how the IFN and IFC functions can be used to conditionally execute global statements.

The INDEX and FIND Families
While the three functions in the INDEX family (INDEX, INDEXC, and INDEXW) remain unchanged, the newer FIND functions (FIND, FINDC, and FINDW) provide the same basic functionality with a great deal of additional flexibility. The FIND functions support both the ability to state a start position, as well as, modifiers that can be used to fine tune the search.

In the following example a string of comma separated words (LIST) is to be subsetted by removing the last word in the list (unless it is the only word). The FINDC function is used to find the location of the last word delimiter, in this case a comma. Like the INDEX function FINDC returns the position of the first occurrence of the second argument ❶.

```
data lists;
input list $;
datalines;
A
A,B
A,B,C
A,B,C,D
A,B,C,D,
run;

data shorter;
   set lists;                    ❶    ❷    ❸
   commaloc=findc(list, ',','b',-length(list));
   if commaloc=0 then newlist=list;  ❹
   else newlist=substr(list,1,commaloc-1);  ❺
run;
```

The third ❷ and fourth ❸ arguments can be in either order (one must be a character code and the other an integer. In this example the third argument ❷ is a 'b'. This instructs SAS to search from right to left, rather than the usual left to right. A right to left search can also be requested by using a negative 4th argument. In this example the 4th argument ❸ requests

that the search be right to left and that it should start at the last position in the string. When the 4th argument is negative the 'b' modifier should not be needed. However in my experience, a positive integer in the 4th argument will not necessarily search right to left in the presence of the 'b' modifier.

```
3.6.6 Using Other Functions
FINDC

Obs     list           commaloc      newlist

1       A                 0           A
2       A,B               2           A
3       A,B,C             4           A,B
4       A,B,C,D           6           A,B,C
5       A,B,C,D,          8           A,B,C,D
```

❹ If a comma is not found there is only one word and nothing is eliminated.

❺ When a comma is found COMMALOC will contain the position of the rightmost comma. SUBSTR is used to keep everything to the left of that comma.

NEWLIST could have been assigned without the IF-THEN/ELSE through the use of the IFC function. This function would yield the same values, but it could cause an error to be written to

```
newlist = ifc(commaloc,substr(list,1,commaloc-1),list);
```

the LOG. When there is only one word in the list (there are no commas), the value of COMMALOC will be 0. When COMMALOC=0 (FALSE) the value of LIST is assigned, although the second argument (TRUE) will not be executed it is still evaluated. The result of the expression COMMALOC-1 will be minus 1 (-1), and that is an illegal argument for the SUBSTR function, hence the ERROR in the LOG—even though the SUBSTR function is not executed.

SEE ALSO

A variation of this problem was posed on a SAS Forum post and this solution was proposed by @Patrick http://communities.sas.com/message/100071.

Another common problem is to find or detect all locations of a character within a larger string. The INDEX function will only detect the first location. Unlike the INDEX function, the FINDC function has the ability to start the search in a position other than the leftmost position.

In this example we want to enumerate the location of each delimiter in a string.

```
data listloc (keep=id cnt position);
   informat id $30.;
   input id;
   delimiter='!'; ❻
   cnt=0; ❼
   position=0; ❽
   do until(position=0); ❾
     position=findc(id,delimiter,position+1); ❿
     cnt+ ^^position; ❼
     if cnt=0 or position ne 0 then output listloc;
   end;
cards;
1!2!3445!!!
!!!
123
run;
```

❻ The delimiter is declared.

❼ A counter is added just to count the occurrences of the delimiter. When POSITION=0 nothing is added to the counter, otherwise 1 is added.

❽ The position is initialized to zero. This allows us to increment by 1 in the third argument of the FINDC function ❿.

❾ A DO UNTIL loop is used to step through the string. The loop will terminate when FINDC fails to find another occurrence of the delimiter. Since the DO UNTIL always executes at least once, it is ok that position was initialized to zero.

❿ The FINDC function returns the next location of the delimiter starting at POSITION+1. If none is found a zero is returned.

```
3.6.6 Using Other Functions
FINDC

Obs     id            cnt     position

 1      1!2!3445!!!    1         2
 2      1!2!3445!!!    2         4
 3      1!2!3445!!!    3         9
 4      1!2!3445!!!    4        10
 5      1!2!3445!!!    5        11
 6      !!!            1         1
 7      !!!            2         2
 8      !!!            3         3
 9      123            0         0
```

SEE ALSO
Variations on this solution were posted by @ArtT and @Ksharp in response to a question on a SAS Forums thread http://communities.sas.com/thread/30629?tstart=0.

The ROUND Function
The ROUND function is most typically used to round a number to the nearest integer; however, it also has a less commonly used second argument that allows us to round to any value. Here the weights of the individuals in our study (the weights are measured to the nearest pound) are being grouped by rounding to the nearest 50 pounds.

```
data wtgroup;
   set advrpt.demog;
   wtgroup = round(wt,50);
   run;
```

```
3.6.6 Using Other Functions
ROUND

Obs    lname       fname      wt    wtgroup

 1     Adams       Mary       155     150
 2     Adamson     Joan       158     150
 3     Alexander   Mark       175     200
 4     Antler      Peter      240     250
 5     Atwood      Teddy      105     100
 6     Banner      John       175     200
          . . . . portions of the report not shown . . . .
```

The midpoints of the intervals are centered on the even 50 pound increments. This technique is often used to form consolidated age intervals such as decades (round to the nearest 10 years).

The SCAN Function

The SCAN function is used to retrieve a word from a string. The word extracted by this function is determined by the numeric second argument of the scan function. When the word number is positive the words are counted from the left end of the string and when it is negative the words are counted from the right.

In SAS 9.2 the SCAN function has a number of enhancements. Like a number of the newer SAS®9 functions, SCAN now supports an optional fourth argument which can be used to modify the way that the SCAN function operates. There are over 20 modifiers available for the function, and they add a great deal of flexibility to the word selection process.

```
data locations;
autoloc = " 'c:\my documents' 'c:\temp' sasautos";
do i = 1 to 3;
   woq = scan(autoloc,i,' ');      ❶
   wq  = scan(autoloc,i,' ','q');   ❷
   wqr = scan(autoloc,i,' ','qr');  ❸
   output ;
end;
run;
```

In this example the character variable AUTOLOC contains the three locations used for the autocall macro library. Two are quoted physical paths and one of these contains an embedded blank (the word delimiter).

❶ Using the SCAN function without a modifier does not separate the words correctly because of the embedded blank (see the value in the variable WOQ).

❷ Adding the 'Q' modifier as the fourth argument to the function causes the SCAN function to ignore word delimiters within quoted strings, and correctly separates the three words (WQ).

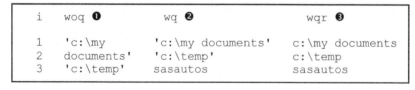

i	woq ❶	wq ❷	wqr ❸
1	'c:\my	'c:\my documents'	c:\my documents
2	documents'	'c:\temp'	c:\temp
3	'c:\temp'	sasautos	sasautos

❸ Including the 'R' modifier along with the 'Q' modifier correctly separates the words and removes the quotes from the two quoted words.

MORE INFORMATION

A macro that uses the SCAN function to separate the autocall macro locations can be found in Section 13.8.2.

The SUBSTR Function

The SUBSTR function has the capability of not only extracting one or more characters from a string, it can also be used to insert characters into an existing string. This is accomplished by placing the SUBSTR function on the left side of the equal sign. In the following example the variable MEDSTDT_ is a character date in the form of mm/dd/yyyy. Unknown values, such as month, have been recorded using XX. The IF statement checks the values of the fourth and fifth characters (day of month), and it replaces any values of 'XX' with '15'.

```
if substr(medstdt_,4,2)='XX' then substr(medstdt_,4,2)='15';
```

The real power of this substitution can be seen when it is coupled with the use of a format. The text date MEDSTDT_ may contain a month code in the first two positions. We would like to substitute a month number for the code, but we of course would rather not write a series of IF-THEN/ELSE statements.

```
proc format ; ❹
   value $moconv
   'XX', 'xx' = '06'
   'LL', 'll' = '01'
   'ZZ', 'zz' = '12';
   run;

data conmed204(keep=subject medstdt_);
   set advrpt.conmed(where=(subject='204'));
   if ❺ anyalpha(substr(medstdt_,1,2)) then
       substr(medstdt_,1,2)=put(substr(medstdt_,1,2),$moconv2.);
   run;       ❻                              ❼
```

❹ A user defined format, $MOCONV., is created with the codes to map to month number.

❺ The ANYALPHA function is used to detect a non-numeric value in the month field.

❻ The SUBSTR function on the left indicates that the value of the PUT will be inserted into the first two columns.

❼ The SUBSTR function is also used to obtain the code that is to be applied to the format. The code allows us to substitute the beginning, middle, or final month depending on the coded value.

MORE INFORMATION

An example in Section 2.3.1 also substitutes date values using the SUBSTR on the left side of the = sign. That example also takes advantage of the ?? format modifier.

The TRANWRD Function

The TRANWRD function is used to replace words within a text string with other text. The function is straightforward in how it is used; however, there is a potential problem. By default, unless otherwise specified, the length of the returned string is $200. This means that you should be sure to specify a length for the variable that is receiving the translated text.

```
data _null_;
   length newstatement1 $34;
   statement = "I enjoy going to SUGI conferences.";
   newstatement1 = tranwrd(statement,"SUGI", "SGF");❽
   newstatement2 = tranwrd(statement,"SUGI", "SGF");❾
   length_newstatement1 = lengthc(newstatement1);
   length_newstatement2 = lengthc(newstatement2);
   put length_newstatement1 = ;
   put length_newstatement2 = ;
   run;
```

❽ The variable

NEWSTATEMENT1 will have a length of $34, because its length was set using a LENGTH statement.

❾ NEWSTATEMENT2, on the other hand, will have a length of $200, because its length was not otherwise specified.

SEE ALSO
A further discussion of the hidden gotcha of the TRANWRD function can be found in the sasCommunity.org article titled "Caution with the TRANWRD Function!" http://www.sascommunity.org/wiki/Caution_with_the_TRANWRD_Function!

The WHICHN Function
The WHICHN function searches a list of values for a specific value and returns the position of the result. In this example the last three visits (by latest date) are selected and the visit numbers are checked to see if the visits have been taken in order.

```
data Visitdates(keep=subject date visit note);
   set advrpt.lab_chemistry;
   by subject;
   array dates {16} _temporary_ ; ❶
   array maxvis {3} _temporary_ ; ❷
   if first.subject then call missing(of dates{*});
   * Save dates;
   dates{visit} = labdt; ❸
   if last.subject then do i = 1 to 3;
      date = largest(i,of dates{*}); ❹
      visit = whichn(date,of dates{*}); ❺
      if i=1 then do;
         call missing(of maxvis{*});
         note=' '; ❻
      end;
      else if visit>=min(of maxvis{*}) then note='*'; ❼
      else note=' ';
      output visitdates;
      maxvis{i}=visit; ❽
   end;
   format date date9.;
   run;
```

❶ Define the array to hold the visit dates.

❷ We are interested in only the last three visits.

❸ Load the visit dates into the DATES array using the visit number as the array index. There are at most 16 visits.

❹ Determine the three latest dates. These *should* be the largest visit numbers.

❺ The WHICHN function is used to detect the visit number associated with this date. For ties the first value detected is returned, consequently for subjects 206 and 208 the incorrect visit number is returned for the second of the tied dates.

```
3.5.1 PUT and INPUT Functions
Using WHICHN to check Visit Order

Obs   SUBJECT   VISIT        date   note
    .... portions of the table are not shown ....
 19     206       7      07FEB2007
 20     206       7      07FEB2007     *
 21     206       8      05JAN2007     *
 22     207      10      09MAR2007
 23     207       9      31JAN2007
 24     207       8      03JAN2007
 25     208       7      30MAR2007
 26     208       7      30MAR2007     *
 27     208      10      09MAR2007     *
 28     209      16      27JUN2007
 29     209      13      07JUN2007
 30     209      15      23MAY2007     *
    .... portions of the table not shown ....
```

❻ The last (largest date) visit cannot be out of order.

❼ If this visit number is larger than any of the previous visits then at least one visit was taken out of order, and this visit should be flagged. Notice that the current visit has not yet been added to MAXVIS array.

❽ Place this visit number in the list of visits associated with the three latest (largest) dates.

SEE ALSO
A related SAS Forum thread uses the WHICHN and VNAME functions to retrieve variable names associated with the largest values http://communities.sas.com/thread/30487?tstart=0.

3.6.7 Functions That Put the Squeeze on Values

A number of character functions are available that can be used to remove characters from a text string. These include, but are not limited to:

- COMPRESS   Removes characters from a text string.
- COMPBL   Removes multiple blanks by translating them into single blanks.
- %COMPRES   Like COMPBL this macro function removes multiple blanks.
- DEQUOTE   Removes matching quotes from a string that starts with a quote.
- STRIP   Removes leading and trailing blanks.
- TRANSLATE   Replaces characters in a text string at the character level.
- TRANSWRD   Replaces character groups.
- TRANSTRN   Replaces character groups.

Functions that trim and left justify a list of characters also remove blanks. These include: LEFT, %LEFT, %QLEFT, TRIM, TRIMN, %TRIM, and %QTRIM. The CATS, CATT, and CATX functions can also be used to remove leading and/or trailing blanks.

The COMPRESS Function
The COMPRESS function can remove much more than just blanks from a string. The first argument of this function is the string that is to be compressed and the second argument can be used to specify one or more characters that are to either be removed or not to be removed. The third argument can specify a modifier, and there are over a dozen that can be used to specify groups or classes of characters to either remove or retain. Taken together the second and third arguments provide an extremely flexible tool.

```
string1 = 'ABCDEABCDE';
string2 = compress(string1,'CAE','k');
```

```
string1=ABCDEABCDE
string2=ACEACE
```

In this example, the second argument usually specifies the characters to remove; however, because the third argument is specified as 'k' they are instead the characters that are kept.

The following example uses the COMPRESS function to count the number of lines of code in a SAS program by counting the semicolons. The 'k' is used to remove every character except semicolons in the COMPRESS function. The LENGTH function is then used to count the semicolons for that physical line. Since some physical lines of a program may not have a semicolon the INDEX function is used to determine if a semicolon is present. The count is then written to a macro variable. To speed up the processing no IF-THEN/ELSE statements are used (executing the SYMPUTX on every incoming row can sometimes be more efficient than executing an IF statement to check for the last observation).

```
filename code "c:\sascode\ABC.sas";
data _null_;
  infile code truncover;
  input ;
  justsemi = compress(_infile_,';','k');
  cnt+index(justsemi,';')*length(justsemi);
  call symputx('cnt',cnt);
  run;
  %put line count is: &cnt;
```

MORE INFORMATION

The COMPRESS function is used with the NOTDIGIT function in an example in Section 3.6.1.

SEE ALSO

A number of examples that expound on the use of the COMPRESS function's third argument can be found in Murphy and Proskin (2006).

The STRIP Function

The STRIP function removes both leading and trailing blanks from a character string. Unlike the TRIM and TRIMN functions, the STRIP function can result in a string with a length of zero. The STRIP function was originally intended to work with the concatenation operator. The statement shown here detects all values of PRODUCT that are exactly '0' after removing any leading and trailing spaces.

```
if strip(product) eq "0" then output dontwant;
```

This example is taken from an answer provided by @ArtT in the SAS Forum thread http://communities.sas.com/thread/30382?tstart=0.

The TRANSLATE Function

The TRANSLATE function is designed to replace characters and the replacement character

```
data test;
string1 = 'ABCDE';
string2 = translate(string1,' ','A');
string3 = translate(string1,' ','E');
string4 = string3||'x';
put string1=;
put string2=;
put string3=;
put string4=;
len2 = length(string2);
len3 = length(string3);
put len2= len3=;
run;
```

cannot have a null length, consequently this function generally does not result in a shorter string. There are a couple of situations, however, where this is not true. By default the new variable created by TRANSLATE will have the same length as the original variable (STRING1 in this example). When the first letter in the string is replaced with a blank, the blank is not preserved (STRING2), but the length is not changed so we have effectively moved the blank to the end (the string has essentially been left justified). Converting the last character to a blank, STRING3, is more complex. As is shown by STRING4 the trailing blank is preserved on the PDV. However the trailing blank is truncated and the variable's length is adjusted when the variable is written to the new data set.

```
string1=ABCDE
string2=BCDE
string3=ABCD
string4=ABCD x
len2=5 len3=4
```

For this function remember that the order of the to/from arguments is different than from the other functions in the translation family.

Removing Quotes—The DEQUOTE, COMPRESS, and TRANSTRN Functions

It is occasionally helpful to be able to remove quotes from a string. The DEQUOTE, COMPRESS, and TRANSTRN functions can each be used to remove quotes, but they do not necessary yield the same result. DEQUOTE only removes pairs of quotes, but it will also truncate the remainder of the string. COMPRESS and TRANSTRN can replace all occurrences without looking for quote pairs.

```
data quoteless;
string1 = "'CA', ""OR"", 'WA'";   ❶
string2 = "Tom's Truck";   ❷
dq1 = dequote(string1);   ❸
dq2 = dequote(string2);   ❹
cprs1=compress(string1,"%bquote(')");   ❺
cprs2=compress(string2,"%bquote(')");   ❻
cprs3=compress(string2,,'p');   ❼
trns1=transtrn(string1,"%bquote(")",trimn(''));   ❽
trns2=transtrn(string2,"'",trimn(''));   ❾
put string1=; ❶
put string2=; ❷
put dq1=;   ❸
put dq2=;   ❹
put cprs1=;   ❺
put cprs2=;   ❻
put cprs3=;   ❼
put trns1=;   ❽
put trns2=;   ❾
run;
```

❶ STRING1 contains both single and double quote pairs.

❷ STRING2 contains an unmatched single quote (apostrophe).

❸ DEQUOTE removes the quotes from CA and truncates the remainder of the string.

❹ The apostrophe is unmatched and therefore untouched by DEQUOTE.

❺ ❻ COMPRESS can be given a list of characters, here both single and double quotes. All are removed.

```
string1='CA', "OR", 'WA'  ❶
string2=Tom's Truck  ❷
dq1=CA  ❸
dq2=Tom's Truck❹
cprs1=CA, OR, WA❺
cprs2=Toms Truck❻
cprs3=Toms Truck❼
trns1='CA', OR, 'WA'❽
trns2=Toms Truck❾
```

❼ The 'p' 3rd argument modifier on the COMPRESS function (replace punctuation) also removes single and double quotes.

❽ The TRANSTRN function can replace a character with a null string (specified here with the TRIMN function), but it cannot replace a series of individual characters as can the COMPRESS function. Only double quotes have been removed.

❾ The single quote (apostrophe) is replaced with a null string.

3.7 Joins and Merges

Although merges and joins are both commonly used and generally used successfully, you should be aware that there are some caveats, as well as things to keep in mind when doing this type of processing.

3.7.1 BY Variable Attribute Consistency

Merges and joins are very susceptible to inconsistencies in the joining criteria. The variable(s) that are used in the BY statement must have the same attributes or unfortunate things can happen.

Inconsistent BY Variable Type

In the following example we would like to add the patient's first and last names to the lab data. The common variable is the SUBJECT number, and we use SUBJECT as a BY variable in a data set MERGE. The incoming data sets have been sorted, but the step fails to execute.

```
data labnames;
   merge advrpt.demog(keep=subject lname fname)
         advrpt.lab_chemistry(keep=subject visit labdt
                                    in=inlab);
   by subject;
   if inlab;
   run;
```

Fortunately the error message in the LOG is very helpful.

```
ERROR: Variable subject has been defined as
both character and numeric.
```

Typically when we misuse a variable's type, such as when we use a character variable in an arithmetic statement, SAS will attempt to convert the variable's type. When the variable is in the BY statement, a conversion is not possible and the step fails.

Converted Type

In the previous example we were unable to perform the merge because the BY variable SUBJECT was character in one data set and numeric in the other. In the DATA step below, the numeric SUBJECT (which has three digits) in DEMOG is converted to character prior to its use as a BY variable.

```
data demog_c;
   set advrpt.demog(keep=subject lname fname
                    rename=(subject=ptid));
   subject = put(ptid,4.);
   run;

data labnames;
   merge demog_c(keep=subject lname fname)
         advrpt.lab_chemistry(keep=subject visit labdt
                                    in=inlab);
   by subject;
   if inlab;
   run;
```

Unfortunately only part of the problem has been solved. Looking at the resulting data set we see that we were unable to retrieve any names.

```
Inconsistent Joining Criteria
3.7.1b Converted Type

Obs     lname     fname     subject     VISIT          LABDT

  1                            200         1        07/06/2006
  2                            200         2        07/13/2006
  3                            200         1        07/06/2006
  4                            200         4        07/13/2006
  5                            200         4        07/13/2006
            . . . . portions of the table not shown . . . .
```

The problem is in the way that we have used the PUT function. When we converted the numeric
value to character, we used the numeric format 4. Numeric formats create right justified character
strings, consequently the resulting value starts with a blank. Adding a LEFT function would have

```
subject = left(put(ptid,4.));
```

solved this problem, but would have
introduced a more subtle one.

In the data set ADVRPT.LAB_CHEMISTRY, the variable SUBJECT has a length of $3;
however, in the previous statement the resulting variable will have a length of $4. In this
particular example the inconsistent length will not cause a problem, but as is shown next, it can
under some circumstances cause a problem that can be harder to detect.

Inconsistent Length

Remember that a variable and its attributes are added to the PDV when the variable is first
encountered as the DATA step is processed during the compilation phase. Once the attributes are
established they will not be changed even if additional or contradictory information is found while
compiling the remainder of the DATA step. The following rather silly example illustrates the
problem.

I would like to use the data set WORK.PETS to add the family pet to the demographic
information in ADVRPT.DEMOG.

The pet information contains the owner's first and last name.

VIEWTABLE: Work.Pets

	lname	FNAME	pet
1	Adams	Joan	Dog
2	Adams	Mary	Cat
3	Alexa	Mark	Cat
4	Antle	Peter	Dog

VIEWTABLE: Work.Demogsymp

	subject	lname	fname	symp
1	200	Adams	Mary	02
2	201	Adamson	Joan	10
3	202	Alexander	Mark	
4	203	Antler	Peter	10

```
proc sort data=pets;
   by lname fname;
   run;
proc sort data=advrpt.demog(keep=subject lname fname symp)
         out=demogsymp;
   by lname fname;
   run;

data petsymptoms;
   merge pets(keep=lname fname pet)
         demogsymp(keep=subject lname fname symp);
   by lname fname;
   run;
```

Before performing the merge both data sets are sorted; however, the DATA step fails and the errors include a "not properly sorted" message.

```
WARNING: Multiple lengths were specified for the BY variable lname by
input data sets. This may cause unexpected results.
ERROR: BY variables are not properly sorted on data set WORK.DEMOGSYMP.
lname=Adams FNAME=Mary pet=Cat subject=101 SYMP=02 FIRST.lname=1
LAST.lname=0 FIRST.FNAME=1
LAST.FNAME=1 _ERROR_=1 _N_=2
NOTE: The SAS System stopped processing this step because of errors.
```

In this example the truncation occurs because the length of the variable LNAME in the data set PETS ($5) determines the length for LNAME on the PDV. The result is truncation when values

```
data petsymptoms;
   length lname $10;
   merge pets(keep=lname fname pet)
         demogsymp(keep=subject lname fname symp);
         . . . . code not shown . . . .
```

from DEMOGSYMP are read. In fact, because of the truncation of LNAME in the data set DEMOGSYMP, Joan Adamson becomes Joan Adams, and since Joan Adams now follows Mary Adams the rows are no longer physically sorted. It could have been worse. If Joan Adamson had a first name alphabetically after Mary, say Tricia, Tricia Adams would have followed Mary alphabetically and no sort error would have been reported.

The truncation problem could have been avoided with the use of a LENGTH statement prior to the MERGE statement (the length of LNAME on DEMOGSYMP is $10). This problem would have also been solved by simply reordering the two data sets on the MERGE statement.

Numeric BY Variables

Extreme care must be taken if you ever need to use numeric BY variables, especially variables with non-integer values. Because of the way that numbers are stored within the computer, even numbers that appear to be integers may not actually be integers. This can be simply demonstrated by creating a value that is slightly different from 1.

```
data similar;
x = 1;
y =  3.000000000000001/3;
if x=y then put 'the same';
put x= best32.;
put y= best32.;
put x= hex16.;
put y= hex16.;
run;
```

Examining the LOG shows that even the BEST32.

```
x=1
y=1
x=3FF0000000000000
y=3FF0000000000001
```

format displays this value (Y=) as 1. The HEX16. format does show these two numbers differently, but really how often would you use the HEX format to double-check the integers?

```
data other;
  y=1; a='a';
  run;
data both;
  merge similar
        other;
  by y;
  run;
```

Worse if we were to use this variable as a BY variable as is done next, the difference is sufficient to sabotage the merge. The LOG shows that the data set BOTH has two observations—one for each value of Y, where there would only have been one observation if the values were seen as equal.

```
NOTE: There were 1 observations read from the data set
WORK.SIMILAR.
NOTE: There were 1 observations read from the data set WORK.OTHER.
NOTE: The data set WORK.BOTH has 2 observations and 3 variables.
```

At some point the fuzz rules come into play and the difference is so small that SAS considers them to be equal. In this example adding one more zero to the number of decimal places in the first definition of Y would have been sufficient for the merge to have been successful.

The take-away point is, be very careful when using numeric BY variables in a merge.

SEE ALSO

Ron Cody wrote Sample Note 33-407 on issues associated with variable attribute inconsistencies, and suggests an automated solution http://support.sas.com/kb/33/407.html .

3.7.2 Variables in Common That Are Not in the BY List

After a merge or join, variables common to more than one data set will appear only once in the new data set. This means that there can be variables that overwrite each other.

In the following example we merge two data sets by SUBJECT. Each also contains the variable DATE; however, DATE is not included on the BY statement. In order to make the example a bit easier to follow the SORT steps have used the NODUPKEY option so that each SUBJECT appears only once.

```
proc sort data=advrpt.lab_chemistry(keep=subject labdt
                                    rename=(labdt=date)) ❶
         out=labchem nodupkey; ❷
   by subject;
   run;
proc sort data=advrpt.ae(keep=subject aestdt
                         rename=(aestdt=date)) ❶
         out=ae nodupkey; ❷
   by subject;
   run;

data aelab;
   merge labchem(where=(date<'01sep2006'd)) ❸
         ae;
   by subject;
   run;
```

```
3.7.2 Variables in Common

Obs    SUBJECT         date

 1       200       07/28/2006
 2       201       07/06/2006
 3       202              .    ❹
 4       203       09/13/2006
 5       204       09/27/2006
    . . . . portions of the table not shown . . . .
```

❶ For this example the date variables have both been renamed to DATE, and then sorted by subject using the ❷ NODUPKEY option.

❸ A WHERE clause restricting the DATE has been placed on the LABCHEM data; however, the clause does not change the resultant table.

Inspection of the data set AELAB shows that although we have restricted lab dates to those before '01sep2006'd, we seem to have dates that do not meet the criteria. In fact these are actually the AE start dates that have overwritten the dates from the LABCHEM data set. ❹ SUBJECT 202 in the LABCHEM data set has a date of '07jul2006'd, but that value has been replaced by the one in AE, which is missing.

Because the PDV is constructed from left to right, the LABCHEM date label is used in the new data set. It is also because the data are read from the rightmost data set last that the AE date overwrites the LABCHEM date.

3.7.3 Repeating BY Variables

When merging, the BY variables should identify down to the row level in all, but at most one of the data sets named on the MERGE statement. This means that at most only one of the incoming data sets will not have a sufficient key (BY variables do not identify down to the row level). When the BY variables do not form a primary key (identify down to the row level) for more than one data set, a NOTE is issued to the LOG, and more importantly, within the BY

```
NOTE: MERGE statement has more than one data set
with repeats of BY values.
```

group the merge takes place as a one-to-one merge and this is rarely desirable.

Here the LABCHEM and AE data sets are merged BY SUBJECT. For SUBJECT 200 there are 14 LABCHEM observations, but only 4 AE observations. The fourth AE observation is repeated for the remaining LABCHEM observations. Clearly this will be unacceptable in virtually all

```
data aelab;
  merge labchem
        ae;
  by subject;
run;
```

Obs	SUBJECT	VISIT	LABDT	AEDESC
1	200	1	07/06/2006	DIARRHEA (X1)
2	200	2	07/13/2006	PAIN-NECK
3	200	1	07/06/2006	PAIN-MUSCULAR CHEST
4	200	4	07/13/2006	INCREASED EOS (6)
5	200	4	07/13/2006	INCREASED EOS (6)
6	200	5	07/21/2006	INCREASED EOS (6)
7	200	6	07/29/2006	INCREASED EOS (6)

. . . . portions of the table are not shown . . . .

situations. It is essential that you understand the data and whether or not the BY variables form a sufficient key.

3.7.4 Merging without a Clear Key (Fuzzy Merge)

When a clear set of BY variables are not available (as was the case in the example in Section 3.7.3) logic will be needed to create the appropriate assignments. For this reason these types of merges are collectively known as fuzzy merges.

As a general rule these types of merges are best handled with an SQL step rather than the DATA step. The SQL join holds all combinations of the rows from both tables in memory (Cartesian product). This allows the programmer to apply logic to select the appropriate rows.

In this example we would like to identify all the adverse events for each patient that occurred

```
proc sql noprint;
create table aelab as
   select a.subject,labdt, visit, aestdt, aedesc
      from labchem as L, ae as a
         where (l.subject=a.subject)
            & (labdt le aestdt le labdt+5)
            ;
quit;
```

within 5 days of a laboratory visit date. The subject numbers are equated in the WHERE clause as is the logic needed to evaluate the proximity of the two dates.

The DATA step can also be used to perform a fuzzy merge. In Section 6.4 a DATA step with two SET statements performs a merge. A similar technique can be applied to a fuzzy merge through logic; however, the coding can become quite tricky.

SEE ALSO

Heaton (2008) discusses the use of hash objects to perform many-to-many merges, and has a good set of references to other papers having to do with the use of hash objects.

3.8 More on the SET Statement

Although a majority of DATA steps use the SET statement, few programmers take advantage of its full potential. The SET statement has options that can be used to control how the data are to be read.

- END=   used to detect the last observation from the incoming data set(s) (see Section 3.9.1).

- KEY=   specifies an index to be used when reading (see Section 6.6.2).

- INDSNAME=   used to identify the current data source (see Section 3.8.2).

- NOBS=   number of observations (see Section 3.8.1).

- OPEN=   determines when to open a data set.

- POINT=   designates the next observation to read (see Section 3.8.1).

- UNIQUE   used with KEY= to read from the top of the index (see Section 6.6.2).

MORE INFORMATION

Several of these options are also used in the examples in Section 3.9.

3.8.1 Using the NOBS= and POINT= Options

The SET statement by default performs a sequential read; that is, one observation after another; first observation to last. It is also possible to perform a non-sequential read using the POINT= option to tell the SET statement which observation to read next. Very often the POINT= option is used in conjunction with the NOBS= option, which returns the number of observations in the data set at DATA step compilation.

The POINT= option identifies a temporary variable that indicates the number of the next observation to read. The NOBS= option also identifies a temporary variable, which after DATA step compilation, will hold the number of observations on the incoming data set.

This short example reads the last 10 observations from the incoming data set. The temporary variable OBS (defined by the NOBS= option) will hold the number of observations available to read. A DO loop with PT (defined by the POINT= option) as the index variable is then used to cycle through the last few observations.

```
data lastfew;
   if obs ge 10 then do pt =obs-9 to obs by 1;
      set sashelp.class point=pt nobs=obs;
      output lastfew;
   end;
   else put 'NOTE: Only ' obs ' observations.';
   stop;
   run;
```

Note the use of the STOP statement to terminate the DATA step after reading the 10 observations. Normally, when the last observation is read from the incoming data set, the DATA step is automatically terminated. The use of the POINT= option disables the DATA step's ability to detect that it has finished reading from the incoming data set.

The POINT= option allows us to read observations in a non-sequential manner (in any order). When the value of the next observation to read is determined randomly, it is possible to draw a random subsample.

```
%macro rand_wo(dsn=,pcnt=0);
   * Randomly select observations from &DSN;
   data rand_wo(drop=cnt totl);
      * Calculate the number of obs to read;
      totl = ceil(&pcnt*obscnt); ❶
      array obsno {10000} _temporary_; ❷

      do until(cnt = totl);
         point = ceil(ranuni(0)*obscnt); ❸
         if obsno{point} ne 1 then do; ❹
            * This obs has not been selected before;
            set &dsn point=point nobs=obscnt; ❺
            output;
            obsno{point}=1; ❻
            cnt+1;
         end;
      end;
      stop; ❼
   run;
%mend rand_wo;
%rand_wo(dsn=advrpt.demog,pcnt=.3)
```

The %RAND_WO macro shown here uses these two options to randomly read (without replacement) a subset of the observations from the incoming data set.

Because the user only specifies the fraction of the total number of observations, the macro must know the total number of available observations so that the subset size can be calculated. This value is stored in the temporary variable OBSCNT, which is defined on the SET statement ❺ through the use of the NOBS= option.

❶ The total number of observations to be selected is calculated as a fraction of the total number of observations (OBSCNT). Although it may seem that the OBSCNT variable is being used before it is defined ❺, in fact OBSCNT is established and assigned a value during DATA step compilation.

❷ An array is used to track whether or not a given observation has already been selected. The array dimension must exceed the number of observations on the incoming data set. This version of the macro will accommodate up to 10,000 observations; however, arrays can easily handle much larger dimensions.

❸ The variable POINT is randomly generated with an integer value that ranges from 1 to the number of observations in the data set (OBSCNT). This variable will be used to determine the next observation to be read.

❹ A check is made against the flag in the array to determine if the selected observation has already been read. If it has not already been selected, it is then read. Using an array to store the flag is the fastest form of a look-up (see Chapter 6).

❺ The SET statement uses the POINT= and NOBS= options to name the temporary variables. Tradition, although certainly not a necessity, often uses the variable names to be the same as the options (POINT=POINT and NOBS=NOBS).

❻ When an observation has been selected, a flag is set in the array. This prevents the observation from being read again. Here the flag is a numeric 1 which takes 8 bytes of storage. If a character $1 flag had been used the array could have been defined as a character array and 70,000 bytes of memory could have been saved.

❼ Whenever you use a SET statement inside of a loop, especially when using the POINT= option, the automatic detection of the last observation is disabled. Be sure to include a STOP to prevent an infinite loop.

The POINT= and NOBS= options can also be helpful when performing look-ahead or look-back reads of the data. In the following example we need to detect observations with certain thresholds and then determine if the value is aberrant by reporting the previous observation and the following two observations as well as the extreme value. Each observation is counted and the counter is used to establish the value used by POINT.

```
data surrounded(keep=subject visit sodium);
   set advrpt.lab_chemistry(keep=subject sodium
                            rename=(subject=sub1));
   cnt+1;  ❽

   if sodium ge 14.4 then do point=(cnt-1) to (cnt+2);  ❾
      if 1 le point le nobs then do;
         set advrpt.lab_chemistry point=point nobs=nobs;  ❿
         if sub1=subject then output surrounded;
      end;
   end;
run;
```

❽ The observation is counted. CNT will determine the range of values (observation numbers) taken on by the temporary variable POINT.

❾ If a given observation has a sodium value of 14.4 or greater we need to print the previous observation and the next two observations (up to 4 observations—within a subject).

❿ The temporary variables POINT and NOBS are associated with the SET statement options of the same name.

This solution does not take into consideration whether or not a given observation has already been written to the data set. An array can be used to flag an observation once it has been used without adding much additional overhead. The sample program E3_8_1b.SAS contains a program that utilizes an array to allow a given observation to be printed only once.

SEE ALSO
Hamilton (2001) includes limitations and alternatives to the NOBS= option. A more sophisticated version of the %RAND_WO macro can be found in Carpenter (2004, Section 11.2.3).

3.8.2 Using the INDSNAME= Option

The INDSNAME= option was added to the SET statement in SAS 9.2. This option stores the name of the data set from which the current observation was read. Prior to its introduction, the IN= data set option was used to make this determination.

In this example we want to concatenate the two data sets (BOOMER and OTHERS) and we want to create a variable (GROUP) to identify the data source. Two solutions, one using IN= and the other using INDSNAME= are shown and contrasted.

❶ The IN= data set option (see Section 2.1 for more on data set options) names a temporary numeric variable that takes on the values of 0 or 1 depending on whether or not a given observation is from this data set.

```
data grouped1;
   set boomer(in=inboom)  ❶
       others(in=inoth);
   if inboom then group='BOOMER';  ❷
   else if inoth then group='OTHERS';
   run;
```

❷ IF-THEN/ELSE processing is used to determine the data source and to assign a value to the variable GROUP.

For large data sets the IF-THEN/ELSE can be time consuming and can be avoided altogether by using the INDSNAME= SET statement option.

```
data grouped2;
   set boomer
       others indsname=dsn;  ❸
   length group $6;  ❺
   group=scan(dsn,2,'.');  ❹
   run;
```

❸ The INDSNAME= option identifies a temporary character variable (DSN) that holds the name of the data set from which the current observation has been read.

❹ Since the variable DSN will contain a two-level name ('WORK.BOOMER'), the *libref* portion is removed using the SCAN function, and the name portion (the second word) is stored in the variable GROUP.

❺ The length of the GROUP variable is declared; otherwise, the SCAN function would return a length of $200.

INDSNAME= has a default length of $41. This may not be long enough if you are using a physical path (which is generally not recommended by this author).

3.8.3 A Comment on the END= Option

The END= option can be used to create a numeric (0/1) temporary variable that indicates that the last record has been read. In the following example the EOF variable ❷, which has been defined using the END=option, is used to control when a PUT statement is to be executed.

```
data a;
  if eof then put total=;  ❶
  set sashelp.class end=eof;  ❷
  end=eof;
  total+age;
  put 'last ' age= total= eof=;
  run;
```

The IF statement ❶ is true only once, and its action (the PUT statement) is executed only on the last pass of the DATA step.

However notice that the IF statement ❷ is before the SET statement. This reminds us that by default the DATA step is not fully terminated until the attempt is made to execute the SET statement after the last observation has been read.

```
last Age=12 total=212 eof=0
last Age=15 total=227 eof=0
last Age=11 total=238 eof=0
last Age=15 total=253 eof=1
total=253
```

3.8.4 DATA Steps with Two SET Statements

As can be seen in numerous examples throughout this chapter, the DATA step may contain multiple SET statements. Multiple SET statements can give you a great deal of power and flexibility over the process of reading the data. However, as you take control of the read process, exercise caution and be sure that you understand what you are requesting the DATA step to execute.

```
data new;
   set a;
   set b;
   run;
```

This simplest case of a double SET statement is essentially a one-to-one merge with restrictions. And the restrictions (conditions if you will) are very important.

Without other controls (usually supplied by the programmer), the number of observations in the new data set is determined by the number of observations in the smallest original data set. **As soon as SAS reads the last observation from**

either data set the full DATA step is not fully executed again. You will notice that in all of the other examples with two SET statements, that there are some restrictions or controls on how the SET statements are executed. Generally we want the step to terminate on our conditions, and not necessarily just because a last observation is read from one of the data sets.

Like in a MERGE, if there are variables in common, the values that are read in from the last data set replace those read in from earlier ones. Also like in a MERGE the PDV will contain all variables from either of the incoming data sets and each variable will be assigned attributes based on its first encounter during the compilation of the DATA step. As always any variable that is read from an incoming data set is automatically retained.

As was seen in Sections 3.1.5, 3.1.6, and 3.8.1, it is possible and sometimes even very advantageous to be able to use multiple SET statements. Just be sure that you understand what is happening when you do so, and be sure that you exercise caution as you take control of the read process.

MORE INFORMATION

Two SET statements are used in the second example of Section 3.8.1. The example in Section 3.6.2 uses DOW loops to read two data sets using two SET statements.

SEE ALSO

A solution to a SAS Forum question utilized a DATA step with two SET statements http://communities.sas.com/message/42266.

3.9 Doing More with DO Loops

The four principle forms of the DO statement are well known and commonly applied to great advantage. However, there is so much more that we can do with this statement and sometimes in surprising ways. This section discusses a few of these techniques.

SEE ALSO

Paul Dorfman (2002) gives a very nice overview of the DO loop and demonstrates many of its behaviors. Fehd (2007) discusses the differences between the DO UNTIL and DO WHILE loops. An extensive list of references and links can be found on sasCommunity.org at http://www.sascommunity.org/wiki/Do_until_last.var.

3.9.1 Using the DOW Loop

While it may have been first proposed by Don Henderson, the DOW loop, which is also known as the DO-W loop, was named for Ian Whitlock who popularized the technique and was one of the first to demonstrate its efficiencies. The DOW loop can often be used to improve DATA step performance, and in its simplest form the DOW loop takes control of the DATA step's implied loop.

Consider the DATA step's implied loop. During the execution phase each executable statement in the DATA step will execute once for each observation in the incoming data set (WORK.BIG).

```
data implied; ❶
   set big;
   output implied;
   run;
```

This includes a fair amount of behind the scenes processing. When the DATA statement ❶ is executed, values of derived variables are cleared and the value of the temporary variable _N_ is incremented. For the step shown here, we do not care about these things. By using the DOW loop to circumvent the implied loop, these operations, and others, do not take place.

```
data dowloop;
   do until(eof);  ❷
      set big end=eof;  ❸
      output dowloop;
   end;
   stop;  ❹
   run;
```

To create a DOW loop place the SET statement within the control of a DO loop ❷. Then take control of the reading process. Here the END= option ❸ is used to detect the end of file; this is used to terminate the DO UNTIL loop. When the loop terminates, we have read all the data and we are ready to terminate the DATA step. The STOP ❹ statement prevents the execution of another iteration of the implied loop ❶.

Another typical use of a DOW loop is seen when using multiple SET statements to merge data sets. Here the mean weight of the individuals in the study is calculated and then used to determine the percent difference from the mean. Since the mean weight is calculated in a separate step, the means must be merged back onto the original data.

A common solution is to use an IF statement to conditionally execute the first SET statement ❺. Since _n_=1 will only be true once, the single observation from WORK.MEANS will only be read once. The implied loop of the

```
proc summary data=advrpt.demog;
   var wt;
   output out=means mean=/autoname;
   run;
data Diff1;
   if _n_=1 then set means(keep=wt_mean);  ❺
   set advrpt.demog(keep=lname fname wt);  ❻
   diff = (wt-wt_mean)/wt_mean;
   run;
```

DATA step will then be used to read all the observations from the analysis data set ❻ . This solution requires that the IF statement ❺ be checked for every incoming observation of the analysis data set ❻. This is unnecessary and could be very time consuming. A DOW loop can be employed to

remove the IF statement and to improve the processing efficiency of the step.

Since only one pass is made through the DATA step, the IF, which was used to control the read of the summary data set, is not needed ❼.

```
data Diff2;
   set means(keep=wt_mean);  ❼
   do until(eof);❽
      set advrpt.demog(keep=lname fname wt)
         end=eof;❾
      diff = (wt-wt_mean)/wt_mean;
      output diff2;
   end;
   stop;  ❿
   run;
```

❽ A DOW loop, which will execute for each observation on the analysis data set, is initiated using a DO UNTIL loop.

❾ The END= SET statement option is used to create an end of file flag that will terminate the DO UNTIL loop ❽.

❿ The STOP statement terminates the DATA step with only one pass of the implied loop.

MORE INFORMATION
A DOW loop is used in Section 2.9.5 to load a hash object.

SEE ALSO
Dorfman (2009) details the DOW loop and its history.

3.9.2 Compound Loop Specifications

The iterative DO loop is commonly used to step through a list of values. What is less commonly known is that we are not restricted to a single list. Here the variable COUNT takes on the values

```
do count=1 to 3, 5 to 20 by 5, 26, 33;
```

of 1, 2, 3, 5, 10, 15, 20, 26, and 33. This DO statement actually has four distinct loop specifications. The first (1 to 3) has an implied BY and the last two consist of a single value. In fact the TO and the BY are not required as is demonstrated by the last two specifications. The numbers themselves do not need to be numeric constants, but can also be stated as expressions that resolve to a number.

To illustrate the use of expressions this example includes an expression; however, the iterative DO *is* limited to a single index variable. In the DO statement shown here, the writer would like to

```
do count=1 to 3, cnt=4 to 8 by 2;
```

iterate across COUNT (1, 2, 3) and then across CNT (4, 6, 8). However this is not what happens.

The CNT=4 is interpreted as a logical expression which will resolve to 0 or 1. If CNT is not equal to 4 the second loop specification will cause COUNT to take on the values of 0, 2, 4, 6, 8;

```
do count=1 to 3, (cnt=4) to 8 by 2;
```

otherwise, the specification results in the values 1, 3, 5, 7. Effectively the DO statement is coded as if parentheses surrounded the expression.

```
do month = 'Jan', 'Feb', 'Mar';
```

Since the individual values are expressions, you may also use expressions that resolve to character values.

3.9.3 Special Forms of Loop Specifications

Iterative DO loops are evaluated at the bottom of the loop. After each pass, at the END statement, the loop counter is incremented and then evaluated. This is shown in the following simple loop.

```
data _null_;
   do count=1 to 3;
      put 'In loop ' count=;
   end;
   put 'Out of loop ' count=;
   run;
```

The LOG shows that the variable has been incremented to 4 before it exits the loop.

```
In loop count=1
In loop count=2
In loop count=3
Out of loop count=4
```

Usually this behavior is acceptable; however, we may want to control whether or not the counter will be incremented the final time. We can add an UNTIL to the DO statement to provide additional control over how the loop is exited. The LOG shows that the UNTIL clause is executed before the counter (COUNT) is incremented.

```
data _null_;
   do count=1 to 3 until(count=3);
      put 'In loop ' count=;
   end;
   put 'Out of loop ' count=;
   run;
```

```
In loop count=1
In loop count=2
In loop count=3
Out of loop count=3
```

A variation on the use of the UNTIL can also be seen in the following example which counts the number of visits within clinics (CLINNUM). PROC FREQ could also have been used and would have probably been more efficient, but that is quite beside the point.

```
data frq;
   set demog;
   by clinnum;
   if first.clinnum then cnt=0;
   cnt+1;
   if last.clinnum then output frq;
   run;
```

A common approach to this type of counting problem is to use FIRST. and LAST. processing to detect the group (clinic number) boundaries. This solution requires us to track and maintain the counter (CNT) and to control the process with two IF statements. We can simplify the code and increase efficiencies by taking advantage of DO loops.

The DO loop surrounds the SET statement (see more about DOW loops in Section 3.9.1), and the UNTIL is used to terminate the loop. Since we do not know the upper bound of the loop, notice

```
data frq;
   do cnt = 1 by 1 until(last.clinnum);
      set demog;
      by clinnum;
   end;
   run;
```

that the iterative portion of the loop specification (cnt=1 by 1) does not contain a TO keyword, which effectively creates an infinite loop. The loop is terminated with the UNTIL. A side benefit of this approach is that the counter variable, CNT, is automatically taken care of for us. By using the DOW loop and by eliminating the IF statements, this DATA step will execute more quickly than the first approach.

```
3.9.3 Special Loop Specifications

Obs    cnt    clinnum

 1      2     011234
 2      2     014321
 3      3     023910
 4      4     024477
 5      2     026789
 6      4     031234
.... portions of the table are not shown ....
```

SEE ALSO
The SAS Forums thread
http://communities.sas.com/message/57412
has a similar counting example with alternate solutions.

In this example we need to assign a value (of the variable I) from the last observation in the data set to a macro variable using

```
data _null_;
   set big;
   call symputx('bigx',i);
   run;

data _null_;
   set big end=eof;
   if eof then call symputx('bigx',i);
   run;
```

SYMPUTX, what is the best approach? Two typical solutions are shown here. Which will be more efficient—the step that executes SYMPUTX for each observation, or the one that executes the IF for each observation, but the SYMPUTX only once?

It turns out that the SYMPUTX has more overhead than even the IF, so the second approach is faster. However, while discussing this issue with John King, he suggested the following even more efficient approach. It is presented here mostly as an aid in understanding DATA step execution.

```
data _null_;
   if eof then stop; ❶
   do _n_ = nobs to 1 by -1 until(_error_ eq 0); ❷
      _error_ = 0; ❸
      set BIG point=_n_ nobs=nobs; ❹
      end;
   if _error_ eq 0 then call symputx('bigx',i); ❺
   stop; ❻
   set BIG(drop=_all_) end=eof; ❼
   run;
```

❶ This will be true only for zero observation data sets. The EOF variable is created using the END= option at ❼.

❷ The loop reads from the last observation first. This is the key that makes this approach the faster of the three shown here. The UNTIL forces the exit of the loop after a single pass.

❸ The _ERROR_ flag is set to 0. This flag will be reset if there is a problem when the SET statement attempts to read the next observation.

❹ The POINT= and NOBS= options are specified. The END= option cannot be declared here as this SET statement will not be executed for zero observation data sets.

❺ The assignment of the variable I is made using SYMPUTX.

❻ Once the value has been determined the DATA step is stopped. This prevents the execution of the second SET statement ❼.

❼ A second SET statement protects us from data sets with zero observations. The END= option is declared here. Because of the DROP= option this step will fail if the incoming data set has no variables.

3.10 More on Arrays

Arrays have been included in examples in a number of sections in this book. While their use generally seems fairly straightforward, there are a number of aspects of their definition and application that are not as generally well known.

SEE ALSO
Stroupe (2007) discusses array basics as does Waller (2010) who also includes the use of implicit arrays.

3.10.1 Array Syntax

The ARRAY statement gives us a way to address a list of values using a numeric index. The most common array syntax uses a list of variables. However, there are a number of alternative forms, some of which can have surprising consequences.

```
array chem {3} potassium sodium chloride;
```

ARRAY Statement Syntax	Comments About This Syntax
array list {3} aa bb cc;	Array dimension of 3, indexed from 1 to 3, LIST{2} addresses BB
array list {1:3} aa bb cc;	Array dimension of 3, indexed from 1 to 3, LIST{2} addresses BB
array list {0:2} aa bb cc;	Array dimension of 3, indexed from 0 to 2, LIST{1} addresses BB (see Section 3.1.7)
array vis {16} visit1-visit16;	Undefined variables within the list will be added to the PDV
array vis {*} visit1-visit16;	SAS determines the dimension of the array by counting the elements. Variables are created as needed before the array dimension is determined.
array visit {16} ;	Will create variables VISIT1-VISIT16
array nvar {*} _numeric_;	Array includes all numeric variables in PDV order
array nvar {*} _character_;	Array includes all character variables in PDV order
array clist {3} $2 aa bb cc;	Array elements are character with a length of 2
array clist {3} $1 ('a', 'b','c'); array clist {4:6} $1 ('a', 'b','c');	The variables CLIST1-CLIST3 will be created and loaded with the values of 'a', 'b', 'c' respectively

SEE ALSO

Additional syntax options and examples for the ARRAY statement can be found at http://www.cpc.unc.edu/research/tools/data_analysis/sastopics/arrays.

3.10.2 Temporary Arrays

Each of the examples of ARRAY statements in Section 3.10.1 worked with a list of variables. If the variables did not already exist the ARRAY statement would create them. Sometimes, however, you want to be able to have access to the power of an array without creating variables. Temporary arrays create unnamed, temporary, but addressable, variables that will be retained during the processing of the DATA step. Because these variables are temporary they will not be written to the new data set.

Temporary arrays are defined using the keyword _TEMPORARY_ instead of the list of variables. When using _TEMPORARY_ you must provide the array dimension.

ARRAY Statement Syntax	Comments About This Syntax
array visdate {16} _temporary_;	Values are initialized to numeric missing
array list {5} _temporary_ (11,12,13,14,15);	LIST{3} is initialized to 13
array list {5} _temporary_ (11:15);	LIST{3} is initialized to 13
array list {6} _temporary_ (6*3);	All array values are initialized to 3
array list {6} _temporary_ (2*1:3);	LIST{3} is initialized to 3, LIST{4} is initialized to 1

MORE INFORMATION

A temporary array is used in Section 3.1.2.

SEE ALSO

Keelan (2002) has examples of several forms of temporary arrays.

3.10.3 Functions Used with Arrays

Most functions will accept array values as arguments; however, some functions are designed to work with arrays, and others have particular use with arrays. Some of these functions have been shown in other sections of the book as well as here.

The DIM Function

The DIM function (introduced in Section 3.6.6) returns the dimension of an array. It is especially useful when the programmer does not know the dimension of the array when writing the program.

```
data newchem(drop=i);
   set advrpt.lab_chemistry
         (drop=visit labdt);
   array chem {*} _numeric_;   ❶
   do i=1 to dim(chem);   ❷
      chem{i} = chem{i}/100;
   end;
   run;
```

In this example we want to divide each of the chemistry values by 100. ❶We select all numeric variables by using the _NUMERIC_ shortcut, but we do not necessarily know how many numeric variables there are in the list.

❷ The upper bound of the iterative DO loop is specified using the DIM function. The dimension is established during the compilation of the DATA step and is available to the DIM function during DATA step execution.

The LBOUND and HBOUND Functions

The LBOUND and HBOUND functions can be especially helpful when you want to step through the elements of an array whose index does not start at one. This type of indexing is often done when the index value itself has meaning or is stored as a part of the data.

In this example we would like to find for any given subject all the other subjects that are within one inch of having the same height. This particular solution uses two passes of the data and DOW loops.

```
data CloseHT;
array heights {200:276} _temporary_;   ❸
do until(done);
   set advrpt.demog(keep=subject ht) end=done;
   heights(subject)=ht;   ❹
end;
done=0;
do until(done);
   set advrpt.demog(keep=subject ht) end=done;
   do Hsubj = lbound(heights) to hbound(heights);   ❺
      closeHT = heights{hsubj};   ❻
      if (ht-1 le closeht le ht+1)   ❼
            & (subject ne hsubj) then output closeHT;
   end;
end;
stop;   ❽
run;
```

❸ The array is specified using the lowest and highest subject numbers.

❹ The height for this subject is loaded into the array. Parentheses are used here; however, I suggest that curly braces should always be used for array calls.

❺ The iterative DO loop steps through the subject numbers based on the range definition in the ARRAY statement ❸.

❻ The height for the other subject (HSUBJ) is recovered from the array.

❼ The two height values are compared for proximity.

❽ The STOP is not really needed here, but is included as a visual reminder to the programmer that we are controlling the data read using DOW loops.

Normally a code such as SUBJECT would be stored as a character field; however, storing it as a numeric field, as is done in ADVRPT.DEMOG, allows for its use as an array index.

Other Handy Functions

A number of functions that were not necessarily designed to be used with arrays also have utility when processing across arrays. The WHICHN (see Section 3.6.6) and VNAME (see Section 3.6.5) functions, and the CALL MISSING (see Sections 2.9.5 and 2.10.4) routine are particularly helpful.

These three functions are used together in this example, which compares a given visit date with all the previous visit dates with the aim of detecting duplicate visit dates. The name of the duplicate visit is returned.

```
data dupdates(keep=subject visit labdt dupvisit);
   array vdates {16} visit1-visit16; ❶
   set advrpt.lab_chemistry;
   by subject;
   retain visit1-visit16 .; ❷
   length dupvisit $7;
   if first.subject then
                call missing(of vdates{*}); ❸
   dup = whichn(labdt, of vdates{*}); ❹
   if dup then do;
      dupvisit = vname(vdates{dup}); ❺
      if dup ne visit then output dupdates;
   end;
   vdates{visit}=labdt; ❻
   run;
```

❶ The array to hold the visit dates is established. A temporary array could have been used, except we want to retrieve the variable name through the use of the VNAME function ❺.

❷ The values of the array variables are retained. Since this is not a temporary array the values are not automatically retained across observations.

❸ The array is cleared (all values set to missing) through the use of the CALL MISSING routine.

❹ WHICHN returns the number of the *first* duplicate date stored in the array (the date of the current visit has not yet been added to the array).

❺ The name of the DUP[th] array element is returned. In this example the array index starts at one; consequently, the visit number and the index number are the same. VNAME would be especially needed when this was not the case.

❻ The current visit date is added to the array.

SEE ALSO
The WHICHN and DIM functions are used in the SAS Forum thread
http://communities.sas.com/thread/30377?tstart=0.

3.10.4 Implicit Arrays

Implicit arrays (sometimes incorrectly referred to as non-indexed arrays) have been in the SAS language longer than the more recent explicitly indexed arrays. The implicit arrays utilize an implicit index – one that is not generally specified by the user. Array calls do not include an index, and consequently, the array calls can be easily confused with variable names. Most SAS programmers, including this author, try to avoid the use of implicit arrays.

This type of array was only documented through SAS 6, and then only for backward compatibility. They were completely deprecated starting with SAS 7 and are no longer supported.

SEE ALSO

SAS Usage Note #1780 (http://support.sas.com/kb/1/780.html) discusses the removal of implicit arrays. The use of implicit arrays is discussed by Waller (2010).

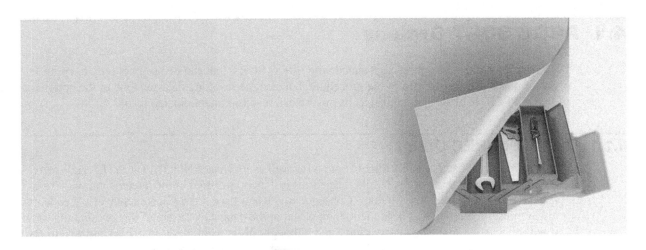

Chapter 4

Sorting the Data

Sorting data is always a resource-intensive operation; therefore, using PROC SORT wisely can save you both time and computing effort. Fortunately, there are both options and strategies to assist you in selecting more efficient, if not optimal methodologies.

MORE INFORMATION

Indexes can be an alternative to sorting the data. Section 5.3 discusses the creation and maintenance of indexes.

4.1 PROC SORT Options

There are a number of options associated with PROC SORT that can be used not only to control performance and capabilities of the procedure, but also the resulting data set. One of the options, NODUPREC, is of special interest as its misuse can result in unanticipated results.

4.1.1 The NODUPREC Option

It is my opinion that the NODUPREC option (as well as its aliases NODUPLICATES and NODUP) is used far too often. While it performs just as is specified in the documentation, it does *not* do what many users think it does. Consequently, when I see it used in someone else's code, it always raises a red flag that begs me to look closer at the data. Most users of this option think that it will remove all duplicate observations, and although this is what it nominally is supposed to do, it does *not* necessarily cause PROC SORT to remove *all* duplicate observations. In fact it *only* removes duplicate observations *that are adjacent* after sorting. When the sorting process results in a data set in which duplicate observations are not next to each other (they do not come one after the other sequentially), they will not be detected and the duplicate observation(s) will not be removed.

The following PROC SORT uses the NODUPREC option with the intent that it will remove any duplicate observations.

```
title1 '4.1.1a NODUPLICATES in PROC SORT';
proc sort data=advrpt.lab_chemistry
          out=lab_chem
          noduprec;
   by subject;
   run;
```

In a listing of the resulting data set (lab_chem) we can see that observations 1 and 3 are duplicates that have not been detected by PROC SORT.

```
4.1.1a NODUPLICATES in PROC SORT

Obs    SUBJECT    VISIT         LABDT    SODIUM    POTASSIUM    CHLORIDE

 1       200        1      07/06/2006     140        3.6         103✔
 2       200        2      07/13/2006     144        4.81        106
 3       200        1      07/06/2006     140        3.6         103✔
 4       200        4      07/13/2006     140        4.02        103
 5       200        4      07/13/2006     140         4          103
 6       200        5      07/21/2006     142        4.57        104
              . . . portions of the listing are not shown . . . .
```

When key fields in the BY statement *are* sufficient to form a primary key, the observations will necessarily be reordered sufficiently to cause the NODUPREC option to work as we would hope that it would. In the previous example if the BY statement had included VISIT and LABDT as well as SUBJECT, the duplicate record would have been removed.

```
NOTE: There were 169 observations read from the data set
ADVRPT.LAB_CHEMISTRY.
NOTE: 3 duplicate observations were deleted.
NOTE: The data set WORK.LAB_CHEM has 166 observations and 6
variables
```

In fact the LOG shows that although three observations were indeed removed in this example (4.1.1a), the two rows that are checked (obs 1 and 3) are also duplicates and neither was removed.

When you do not have a primary key, or if you do not know which variables will form a primary key, the only way to guarantee that duplicate observations are removed is to list all the data set's variables in the BY statement. The list of all variables could be abbreviated by using the _ALL_ list abbreviation.

```
by _all_;
```

In practice this tends to be a very inefficient solution to the problem of duplicate observations. Although inefficient, if your data set size is such that the extra sorting resources do not impact you to a very great degree, then using _ALL_ at least works and does what we need it to do when using NODUPREC. Generally, although not foolproof, the inclusion of a derived variable (such as a variance or standard deviation) along with the probable primary key variables is often sufficient to successfully reorder the observations so that duplicates will be in adjacent rows.

The NODUPKEY option does not have this problem as only the key fields are evaluated during the check for duplicate values.

MORE INFORMATION
The Hash object is used to eliminate duplicate observations in Section 2.9.5. The NODUPREC option is also discussed in the context of the NOEQUALS option in Section 4.1.6.

4.1.2 The DUPOUT= Option

When the NODUPREC or the NODUPKEY options are used, the LOG will note when observations are removed; however, which observations were removed is not apparent. If you want to be able to see these observations, the DUPOUT= option can be used to save the eliminated observations into a separate data table.

In the example that follows, the BY statement now includes a sufficient key to reorder the problem observations noted in the previous section. The removed observations are written to a separate data set (WORK.REMOVEDOBS).

```
title1 '4.1.2 NODUPLICATES and DUPOUT= in PROC SORT';
proc sort data=advrpt.lab_chemistry
          out=lab_chem
          dupout=RemovedObs
          noduprec;
   by subject visit labdt;
   run;
proc print data=removedobs(obs=10);
   run;
```

Because we are using a sufficient key to reorder the problem observations noted in the previous section, that duplicate is now also being deleted. ❶

Obs	SUBJECT	VISIT	LABDT	SODIUM	POTASSIUM	CHLORIDE
1	200	1	07/06/2006	140	3.6	103 ❶
2	200	9	09/13/2006	139	4.06	103
3	201	2	07/14/2006	140	4.15	101
4	202	6	07/29/2006	139	5.68	96

4.1.2 NODUPLICATES and DUPOUT= in PROC SORT

Although we have detected these duplicate observations, without using what we *know* to be a sufficient key, we cannot guarantee that these are *all* of the duplicate observations.

Removal of duplicate observations can also be accomplished using Hash objects (see Sections 2.9.5 and 3.3); however, costs can be similar to those experienced when sorting with _ALL_.

4.1.3 The TAGSORT Option

As the data set to be sorted increases in size (number of rows, number of key variables, or number of variables in the table), more and more resources are required to complete the sorting process. The process itself can result in a number of temporary copies of all or part of the data set that is being sorted, and if the data set is large enough, these temporary tables can exceed the amount of available storage in the WORK directory.

When successful sorting is hampered by a lack of intermediate storage, the TAGSORT option can be used. This option causes PROC SORT to separate the key fields from the rest of the data. The key fields are then sorted, and after the sorting is complete, the data set is reconstructed using the new sort order.

While the TAGSORT option decreases storage requirements during the sort process, the overall time to complete the sort will usually increase.

4.1.4 Using the SORTSEQ Option

PROC SORT uses what is known as the 'collating sequence' to determine the sorted order of values. Traditionally there have been two collating sequences, EBCDIC (for mainframe computers) and ASCII (for most other machines running Operating Systems like Windows and UNIX). You have long been able to select one or the other of these two different collating sequences by specifying the EBCDIC or ASCII options on the PROC SORT statement.

With the introduction of National Language Support, NLS, additional collating sequences have been made available to support languages other than English. Like ASCII and EBCDIC these other collating sequences are also selected through options, which include: DANISH, POLISH, SWEDISH, and NATIONAL. The NATIONAL sequence is selected when your site has specified a customized sequence.

The SORTSEQ option allows you to further refine the way the selected collating sequence is used. This includes subsets or locals within a national collating sequence. Even without changing the base collating sequence the SORTSEQ option can be beneficial.

Reordering Numeric Strings

When character strings that contain numbers are sorted the values are sorted alphabetically. This can be visually unappealing in that the values *seem* out of order, e.g., '10' < '2'. This is shown when we sort on REGION, a $2 character string containing numeric values that range from '1' to '10'. The result has region '10' sorted between regions '1' and '2'.

```
proc sort data=advrpt.clinicnames
                    (keep=region)
           out= regions1 nodupkey;
   by region;
   run;
```

Obs	region
1	1
2	10
3	2
4	3
5	4
6	5
7	6
8	7
9	8
10	9

The SORTSEQ option can be used to change this default behavior. When the SORTSEQ option is assigned the keyword LINGUISTIC ❶, a number of additional keyword qualifiers can also be specified. Turning on the NUMERIC_COLLATION ❷ causes the regions to be ordered as if they were numeric. Region '10' will now be sorted last.

```
proc sort data=advrpt.clinicnames(keep=region)
          out= regions2
          sortseq=linguistic ❶ (numeric_collation=on) ❷
          nodupkey;
  by region;
  run;
```

Case-Sensitive Reordering

Depending on the collating sequence the lowercase letters will all sort either before or after the uppercase letters. The examples in this section have taken the first five names from the ADVRPT.DEMOG data set and copied them into all uppercase and all lowercase, as well as the original mixed case.

```
proc sort data=anames
          out=anamesE
          sortseq=ebcdic;
  by lname;
  run;
title3 'EBCDIC Sequence';
proc print data=anamesE;
  run;
```

EBCDIC Sequence		ASCII Sequence	
Obs	lname	Obs	lname
1	adams	1	ADAMS
2	adamson	2	ADAMSON
3	alexander	3	ALEXANDER
4	antler	4	ANTLER
5	atwood	5	ATWOOD
6	Adams	6	Adams
7	Adamson	7	Adamson
8	Alexander	8	Alexander
9	Antler	9	Antler
10	Atwood	10	Atwood
11	ADAMS	11	adams
12	ADAMSON	12	adamson
13	ALEXANDER	13	alexander
14	ANTLER	14	antler
15	ATWOOD	15	atwood

The side-by-side comparison shows the difference between these two primary collating sequences (the code for generating the ASCII sequence is in the sample programs).

```
proc sort data=anames
          out=anamesc
          sortseq=linguistic (case_first=upper); ❸
  by lname;
  run;
```

We can further refine the sequencing by using keyword qualifiers.

The CASE_FIRST=UPPER ❸ qualifier causes uppercase to take priority over lowercase. Notice, however, that the order is still not the same as ASCII which also gives priority to uppercase letters. Here the sensitivity is within the word not across the list of words. The CASE_FIRST keyword qualifier can also take on the value of LOWER.

Case_First=Upper	
Obs	lname
1	ADAMS
2	Adams
3	adams
4	ADAMSON
5	Adamson
6	adamson
7	ALEXANDER
8	Alexander
9	alexander
10	ANTLER
11	Antler
12	antler
13·	ATWOOD
14	Atwood
15	atwood

4.1.5 The FORCE Option

When the OUT= option is not used on the PROC SORT statement, the incoming data set is replaced with its sorted analogue. When the data set is indexed or if the metadata sort indicators (such as is created by the SORTEDBY= data set option – see Section 4.4) indicate that the data set is already sorted, the sorting does not take place. For indexed data sets this protects the index, and for data sets that are already sorted this conserves resources. When this default behavior is not what you want, the FORCE option can be used.

4.1.6 The EQUALS or NOEQUALS Options

Typically when PROC SORT reorders observations based on the levels of the BY variables, the block of observations within a given level or BY group do not change their order. Generally we do not care about the order of the rows within a BY group; if we did we would add another variable to the BY list. While this default behavior can be controlled at the operating system option level using the SORTEQUALS or the NOSORTEQUALS option, it can also be controlled at the PROC SORT step level using the EQUALS or NOEQUALS options.

Under earlier versions of SAS this order preservation made sense from an operational point of view. Fewer resources were expended by handling the rows as a block. Under the current versions of SAS this default behavior may no longer be our best choice. With multi-threading available to the SORT procedure, portions of these blocks may be divided up across processors. When the rows are returned from the different threads, additional resources may actually be expended just to preserve the order within a block, an order that we probably do not care about.

The NOEQUALS option can be used on the PROC SORT statement to allow SAS to not worry about maintaining the original order within groups. Allowing the within block order to change (by not forcing it to be preserved) through the use of the NOEQUALS option can save resources; however, it can also have other impacts. The order returned, especially when multiple processors are involved, can change from one sort to the next. Since the NODUPREC option (see Section 4.1.1) relies on observation order, its results may also vary from one run to the next when the NOEQUALS option is used.

MORE INFORMATION
Caveats associated with the use of the NODUPREC option are discussed in Section 4.1.1.

4.2 Using Data Set Options with PROC SORT

One of the primary efficiency techniques used to speed up our programs is to eliminate variables and/or observations as soon as possible in the data handling process. Just as we will seldom carry a parka in our luggage when visiting Miami, we should not carry the extra baggage of variables or observations that are not needed. Trimming up the data can have a major impact in the time needed to complete a PROC SORT.

Fortunately for us the process of culling unneeded variables and observations can be handled within the PROC SORT step itself through the use of data set options.

If you associate the KEEP= or WHERE= data set options with the data set that is being generated

```
proc sort data=realbig
          out=onoutgoing(keep=sodium2);
   by sodium2;
   run;
```

(on the OUT= data set), variables and observations are removed after the sort has been completed. Although this will help with efficiency in subsequent steps, it will do little to help with the current PROC SORT.

When data set options are associated with the incoming data set, they are applied before the

```
proc sort data=realbig(keep=sodium2)
          out=onincoming;
   by sodium2;
   run;
```

PROC SORT is executed. This can substantially reduce the processing requirements of the PROC SORT.

Additional efficiency gains can be achieved by eliminating observations.

Like in the previous example, which eliminated columns, eliminating observations before they are read (on the DATA= data set) as opposed to as they are being written to the final data set (on the OUT= data set), can make a substantial difference.

SEE ALSO
The use of the WHERE= data set option on the incoming data set is discussed by Benjamin (2007).

4.3 Taking Advantage of Known or Knowable Sort Order

While there can be a negative impact associated with the use of PROC SORT, we obviously still need to be able to use it to reorder the data. Or do we? Often thinking about your program, its flow, and how it is organized, can help you make sure that you only use PROC SORT when it is actually needed. Some strategies to help minimize the number of SORT steps could include the following:

Plan Your Sorts
Since sorts can be costly, plan your program and data flow around your sorts rather than programming sorts as they are needed in your program. If several different steps use a specific sort order, sort the data once for all the steps rather than placing the steps so that the data must be sorted, resorted, and then sorted back to the first order a second time.

Use CLASS Statements
Unless you are going to explicitly use a BY statement, most procedures do not require the data in a specific order. Obviously there are exceptions; however, the point is that you often do not necessarily need to sort your data. This is especially true of procedures that use implicit or explicit classification variables. CLASS statements do NOT require sorted or even ordered data.

When using the MEANS or SUMMARY procedures for instance, the procedure will probably execute faster when a BY statement is used instead of a CLASS statement (of course the results may not contain exactly the same information). However, the BY statement requires sorted data and the sorting itself may increase the overall processing time such that using the CLASS statement would have ultimately been more efficient. The CLASS statement will avoid sorting, but will generally require more memory.

MORE INFORMATION
The use of threads to improve efficiency is discussed in Section 4.5. Differences between the BY and CLASS statements for the MEANS or SUMMARY procedures are discussed in Section 7.12.

Anticipate Procedure Output Order
For procedures that create output data sets, the order of the data is generally known or at least knowable, and knowing the order of the generated data, or planning the procedure so that it generates data in the desired order can eliminate the necessity of a subsequent PROC SORT. To control the possible orderings of the output data set, be sure to take advantage of the ORDER= option. Generally speaking, the order of the classification variables on the incoming data does not affect the output order unless the ORDER= option is set to DATA (see Section 2.6.2).

Even procedures that do not support the CLASS statement may have implied classification variables (*e.g.,* PROC FREQ), and the values of these variables, along with the ORDER= option, help to determine the order of any generated data sets.

The following table is a listing of a data set that was created by a PROC SUMMARY step. By inspection you can see the sort order and you could even infer the CLASS statement. You can also infer the ORDER= option associated with each variable on the CLASS statement.

```
4.3  Predicting Sort Order

                                         mean
Obs   race   edu   symp   _TYPE_   _FREQ_   HT

  1                         0        8     66.25
  2            .     01     1        2     64.00
  3            .     02     1        4     66.50
  4            .     03     1        2     68.00
  5           12            2        4     67.50
  6           14            2        2     64.00
  7           15            2        2     66.00
  8           12    02      3        2     67.00
  9           12    03      3        2     68.00
 10           14    01      3        2     64.00
 11           15    02      3        2     66.00
 12     1     .             4        6     67.00
 13     4     .             4        2     64.00
 14     1     .     02      5        4     66.50
 15     1     .     03      5        2     68.00
 16     4     .     01      5        2     64.00
 17     1    12            6        4     67.50
 18     1    15            6        2     66.00
 19     4    14            6        2     64.00
 20     1    12    02      7        2     67.00
 21     1    12    03      7        2     68.00
 22     1    15    02      7        2     66.00
 23     4    14    01      7        2     64.00
```

Assuming that the classification variables are not formatted, inspection of this table suggests the following CLASS statement (for this procedure INTERNAL is the default value for the ORDER= option).

```
class race edu symp /
      order=internal;
```

If this data set was to be used in a subsequent step, each of these BY statements could be used without first using a PROC SORT.

```
by _type_;
```

```
by race edu symp;
```

```
by _type_ race edu symp;
```

Avoid Sorting by Using Indexes

Indexes provide you with a way to establish one or more virtual sort orders against a data set. While an index must be created and maintained, for stable data sets this cost may be minimal relative to the cost of sorting and then re-sorting the data.

When an index is created it is stored in a separate file from the data set itself. Whenever the data set is modified, even if the modification does not alter the order of the rows, the index must be recreated. The index file itself will take storage space. The amount of required space will depend on several factors, including the number of rows in the table, the number of indexes that have been established, and the number of variables that make up each index. Additional benefits of indexes include optimized searches with WHERE clauses and the ability to perform double SET statement merges and table look-ups without sorting (see Sections 6.4 and 6.6).

MORE INFORMATION
Indexes are discussed in more detail in Section 5.3.

Using PROC SQL to Avoid Sorts
When PROC SQL operates on a data table, the entire table is loaded into memory. While this means that PROC SQL can be limited in what it can do with larger tables (limited by available memory), it also means that the sort order of incoming data is rarely an issue within the SQL step.

A match merge in the DATA step requires a BY statement; however, this is not the case with the equivalent JOIN in an SQL step. Also, the GROUP clause, which is analogous to the CLASS statement, is also available in the SQL step. Before sorting the data, consider whether or not the use of an SQL step might yield the same result, while avoiding a SORT, as well as, an additional pass of the data.

If a data set is too large to sort, especially if you have tried the TAGSORT option, then it is likely to be too large to be effectively handled by an SQL step.

4.4 Metadata Sort Information

When data are sorted or indexed, information about the sort is stored as a part of the table's metadata. In this example a simple PROC SORT is executed, and the data set's metadata is then displayed using PROC CONTENTS.

```
title1 '4.4a Showing SORT Meta-data';
proc sort data=advrpt.lab_chemistry
          out=lab_chem
          noduplicates; |
   by subject visit labdt; {
   run;
proc contents data=lab_chem;
   run;
```

Among other things, information about how the data is sorted is contained in two different sections of the PROC CONTENTS output.

The upper-most section shows the internal sorted flag ❶, which takes on the values of either YES or NO.

```
4.4a Showing SORT Metadata

The CONTENTS Procedure

Data Set Name        WORK.LAB_CHEM              Observations          165
Member Type          DATA                       Variables             6
Engine               V9                         Indexes               0
Created              Thu, Nov 05, 2009 02:26:48 PM  Observation Length  56
Last Modified        Thu, Nov 05, 2009 02:26:48 PM  Deleted Observations  0
Protection                                      Compressed            NO
Data Set Type                                   Sorted                YES ❶
```

When the SORTED flag ❶ contains YES, indicating that the data are sorted, an additional section is added to the PROC CONTENTS output. This section lets us know more about the conditions of the sort, and these include the BY variables ❷ and sort options ❹.

```
                Sort Information

Sortedby        SUBJECT VISIT LABDT ❷
Validated       YES ❸
Character Set   ANSI
Sort Option     NODUPREC ❹
```

❸ When SAS does the sorting the VALIDATED flag is set to YES.

The SORTEDBY Data Set Option

When the data are already sorted, but not by SAS, the SORTED metadata flag will not be changed to YES, and we can miss out on performance enhancements that take advantage of known sort order. We can let SAS know that the data are actually sorted by setting the SORTED flag set to YES. You can set the SORTED metadata indicator flag directly by using the SORTEDBY data set option.

```
title1 '4.4b Using the SORTEDBY Option';
data lab2(sortedby=subject visit);
   set lab_chem;
   run;
proc contents data=lab2;
   run;
```

```
                Sort Information

Sortedby        SUBJECT VISIT
Validated       NO ❺
Character Set   ANSI
```

Not only is the SORTED flag set to YES, but the sort information section is also completed. Notice, however, that the VALIDATED indicator is still set to NO ❺. We are trusted, but only trusted so far.

4.5 Using Threads

Some operations within a computer are computationally intensive. This is especially true for sorting operations and also for the calculation of large numbers of summary statistics. When multiple CPUs are available, some procedures will follow the principle of 'divide and conquer,' and they can split up computationally challenging tasks by spreading the work out among the available CPUs. This distributed work load can offer substantial improvements in the elapsed time to complete tasks.

The system option THREADS is used to allow SAS to take advantage of multiple CPUs. Although some of the documentation refers to the option in the singular, it needs to be THREADS. When multi-threads are used, a note is added to the LOG.

A number of Base SAS procedures, as well as several in SAS/STAT®, support multi-threaded operations. Additionally, support can also be found in some SAS Enterprise Miner® procedures and in some SAS/ACCESS® engines. Base procedures which have multi-threaded capabilities include: SORT, MEANS, SUMMARY, SQL, TABULATE, and REPORT.

Although the use of multiple threads will generally improve processing thru-put, this does not necessarily have to be the case. Since resources are used not only to create and maintain the threads, but also to coordinate the information flow between threads, it is possible that the use of these resources can outweigh the advantage of the multiple threads. It is important for you to test your environment with and without threads to determine which has the better performance. You should test with both THREADS and NOTHREADS and, depending on your data and code, determine which is more efficient.

When more than two CPUs are present on the system, you may not want all of them to be available for use by SAS. The system option, CPUCOUNT, is used to control how many of the available CPUs can be used by SAS.

SEE ALSO

Additional information on multi-threading can be found on the SAS R&D site http://support.sas.com/rnd/scalability/procs/index.html.

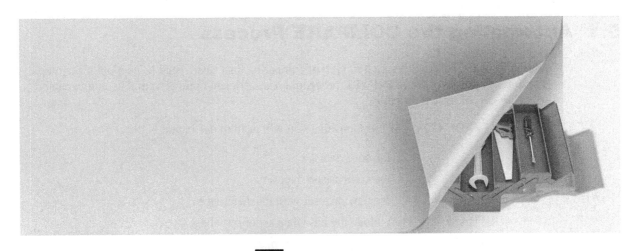

Chapter **5**

Working with Data Sets

While we are usually most interested in the data values and the analysis results that they generate, there are a number of operations that take place at the data table level. Sometimes we need to be able to work with the entire table and not just at the observation level. Fortunately SAS provides us with a number of tools for this type of processing.

5.1 Automating the COMPARE Process

While PROC COMPARE does a good job of comparing data sets, visual inspection of the output is tedious. Fortunately the COMPARE procedure has sufficient options so that its output can be utilized in an automated process.

The primary PROC COMPARE options that you will need to use include:

- DATA   the base data set
- COMPARE   the comparison data set
- OUT   the new data set with the differences
- OUTBASE   include the base data set observation
- OUTCOMP   include the compare data set observation
- OUTNOEQUAL   only write differences

It is of primary importance that the combination of the BY and ID statements include sufficient variables to form a unique key. If a unique key is not formed the COMPARE procedure will be unable to keep the two data sets synchronized. This implies that you may need to do a pre-analysis to check for duplicates in the key variables.

The first step in the automated comparison process is to create a data set containing only those rows that have differences. These rows are written to a data set using the OUTNOEQUAL option ❺.

```
proc compare
      data=lab_chem    ❶
      compare=lab_chem2 ❶
      out=cmpr  ❷
      outbase outcomp  ❸
      noprint  ❹
      outnoequal  ❺;
   id subject visit labdt;
   run;
```

❶ The DATA= and COMPARE= options determine which two data sets are to be compared.

❷ A data set is created using the OUT= option, which will contain the observations with the differences.

❸ The data set of the differences should contain the original observations from both of the data sets that are being compared.

❹ Since the process is being automated, there is no need to create any printed output.

❺ Only those observations that contain differences need to be written to the data set.

For this example changes have been artificially inserted in the values of SODIUM for two observations in the data set LAB_CHEM2 (see the sample code for the full program).

Because the OUTNOEQUAL ❺ option has been specified, when a difference is detected, the OUTBASE OUTCOMP ❸ options cause the entire observation from each of the two incoming data sets to be written to the data set named in the OUT= option. This data set has the additional variables _TYPE_ and _OBS_ to help identify the original observations.

```
5.1 Automated Comparisons
Obs with differences

Obs  _TYPE_   _OBS_  SUBJECT  VISIT      LABDT   SODIUM  POTASSIUM  CHLORIDE

 1   BASE       2      200      2    07/13/2006   144      4.9        106
 2   COMPARE    2      200      2    07/13/2006   1A4      4.9        106
 3   BASE      20      202      1    07/07/2006   139      4.8         96
 4   COMPARE   20      202      1    07/07/2006   1A9      4.8         96
```

TYPE refers to the two data sets being compared: ❶

- BASE   the data set identified with the DATA= option
- COMPARE   the data set identified in the COMPARE= option

Each pair of rows in the data set WORK.CMPR ❷ has at least one difference; however, searching these pairs of rows for the differences is only incrementally easier than searching the original output. We need to further isolate the individual values that are different. Since we know that there is at least one difference within each BASE/COMPARE pair of observations, we need to examine each pair of values for each variable individually within this observation pair. This can be more easily accomplished if we first transpose the data so that each value pair can be isolated.

PROC TRANSPOSE is used to create a data set with one observation per pair of variables within each of the two observations. Again the BY and ID statements are used to isolate the key variables. The variable _OBS_, which reflects the original observation number, is added to the BY statement to help make the identification process easier.

```
proc sort data=cmpr;  ❷
   by subject visit labdt _obs_;
   run;

proc transpose data=cmpr
               out=tdiff(drop=_label_
                         rename=(_name_=variable));
   by subject visit labdt _obs_;  ❻
   var _numeric_ _character_;  ❼
   id _type_;  ❽
   run;
```

❻ The BY statement is used to identify down to the two rows with differences. The ID statement ❽ further identifies down to the specific row.

❼ In the VAR statement we need to list all the variables that need to be compared. Since we do not necessarily know the names of the variables we can use variable list abbreviations (see Section 2.6.1). _ALL_ could also have been used here. When the variables to be compared are known (as is technically the case in this example) they can be named explicitly.

❽ The ID variable _TYPE_, which contains 'BASE' and 'COMPARE', is used to name the two new columns formed in the transformation process. These will be character variables when the variables in the VAR statement are either all character or a mixture of numeric and character. When there is a mixture of variable types, as is the case in this example, a conversion note is written to the LOG.

The data set TDIFF will now have one row for each original variable, and the values from the two original data sets are stored in the variables BASE and COMPARE. By selecting for unequal BASE and COMPARE values, we can determine the differences that have been detected in the comparison process.

```
title3 'After Transpose';
proc print data=tdiff(where=(variable ne '_TYPE_'
                             & base ne compare❾));
   run;
```

❾ The variables BASE and COMPARE now contain the original values of the variable named VARIABLE, which was renamed from _NAME_.

```
5.1 Automated Comparisons
Obs with differences
After Transpose

Obs   SUBJECT   VISIT       LABDT   _OBS_   variable   BASE   COMPARE

  6    200        2     07/13/2006      2   SODIUM     144    1A4
 14    202        1     07/07/2006     20   SODIUM     139    1A9
```

In this example the differences are only printed; however, they could have been easily stored in a data set for further processing.

MORE INFORMATION

A macro that further generalizes the comparison process can be found in program E5_1b.sas, which is located in the sample programs accompanying this book.

5.2 Reordering Variables on the PDV

The order of the variables on the Program Data Vector, PDV, is generally of no concern to the SAS programmer. Indeed there are no tools in the language that are specifically designed to help us to change the order once it has been established. Although they come up fairly infrequently, there are legitimate occasions that force us to either know or change the order of the variables on the PDV.

Sometimes when we create an EXCEL spreadsheet, the resulting columns need to be in a specific order. PROC EXPORT (see Section 1.2) will use the PDV order of the variables to determine the order of the EXCEL columns. To change this order we need to change the order of the variables going into PROC EXPORT.

The current order of the variables can be seen in a number of ways. Some of the most straightforward of these include:

- When PROC PRINT is used without an ID, BY, or VAR statement, the variables are printed in the order of their position.
- VIEWTABLE in the Display Manager displays the variables in position order.
- PROC CONTENTS displays the position of each variable, and with the VARNUM option (formerly the POSITION option) it will also list the variables in position order.
- The COLUMNS window in the Display Manager displays the columns in position order by default.

Within a DATA step the order of the variables on the PDV is determined as the step is compiled. If the step reads a data set, as with a SET statement, the incoming data set is examined and its variables are added to the PDV using the same order. Once the order is determined on the PDV it is fixed and cannot be altered. If we want to control the order we must do so before it is fixed.

Using the VARNUM option in a PROC CONTENTS step reveals, among other attributes, the order of the variables in ADVRPT.DEMOG. The left-most column is the variable or position number.

```
5.2 Reordering Variables on the PDV

The CONTENTS Procedure

                    Variables in Creation Order

#     Variable    Type    Len    Format    Label

1     subject     Num      8
2     clinnum     Char     6                clinic number
3     lname       Char    10                last name
4     fname       Char     6                first name
5     ssn         Char     9                social security number
6     sex         Char     1                patient sex
7     dob         Num      8     DATE7.     date of birth
8     death       Num      8     DATE7.     date of death
9     race        Char     1                race
10    edu         Num      8                years of education
11    wt          Num      8                weight in pounds
12    ht          Num      8                height in inches
13    symp        Char     2                symptom code
14    death2      Num      8     DATE9.
```

In the example that follows, only a few of the variables from ADVRPT.DEMOG flow through to the new data set, and the order of these variables is changed by the use of the LENGTH statement.

```
data demog2(keep=subject lname fname sex ht wt dob symp); ❶
   length lname $10 fname $6 sex $1 symp $2; ❷
   set advrpt.demog(keep=subject lname fname sex edu ❶
                        death ht wt dob symp);
   where death and edu>15; ❸
   run;
```

❶ The KEEP and DROP statements/options do not change or affect the order of variables on the PDV.

❷ The LENGTH statement is used not only to assign the length attribute to these four variables, but it also adds them to the PDV. Since the LENGTH statement appears before the SET statement these variables and their associated attributes are added to the PDV before any variables or attributes are contributed by the SET statement. Attributes for these variables that have not been specified, for example formats or labels, will be picked up from the metadata of the incoming data set named on the SET statement.

❸ Although the variables DEATH and EDU are not to be included on the new data set, they must be available on the PDV for use by the WHERE statement.

```
5.2 Reordering Variables on the PDV

Obs    lname    fname    sex    symp    subject        dob    wt    ht

 1     James    Debra     F      05      232      19JUN42    163    63
 2     Manley   Debra     F      05      241      19JAN42    163    63
```

Notice that while the order has changed for the variables in the LENGTH statement, the order for the remaining variables is the same as it was on the incoming data.

The ARRAY, FORMAT, INFORMAT, RETAIN, and ATTRIB statements can also be used in a similar manner to reorder variables on the PDV. The KEEP, DROP, and RENAME statements cannot be used to change variable order. Generally the recommended choice of statements for reordering the variables is the RETAIN statement as it does not require any other attributes and does not otherwise change behavior of variables that are being brought into the DATA step via the SET or MERGE statements.

In the previous example the LENGTH statement ❷ could have been replaced with a RETAIN statement, which does not require additional knowledge of the attributes of the variables that are to be reordered.

```
retain lname fname sex symp;
```

It is also possible to reorder variables using an SQL step. Here the SELECT statement ❹ is used to specify the new variable order. Notice that unlike the DATA step a full list of variables must be specified unless you use the asterisk (*) ❺ to specify all variables, in which case you will see them in PDV order.

```
proc sql ;
create table demog4 as
  ❹ select lname, fname, sex, symp, subject, dob, wt, ht
      from advrpt.demog(keep=subject lname fname sex
                            edu death ht wt dob symp)
        where death and edu>15;
    select * ❺
      from demog4;
quit;
```

When you read more than one data set in a step, the order of variables is determined to some extent by the order that the data sets are read in. Variables are added to the PDV in the order that they are encountered by the compiler. The order of the variables taken from the first data set seen by the compiler will be written to the PDV first. Variables not already on the PDV will be added to the PDV in the order that they are encountered on subsequent data sets. This means that you may have some control of variable order by controlling the order in which your incoming data sets are first seen.

MORE INFORMATION
Although I do not necessarily recommend their use, some of the variable list abbreviations require you to have specific knowledge of the order of the variables (see Section 2.6.1).

SEE ALSO
SAS Usage Note 8395 discusses the reordering of variables at http://support.sas.com/kb/8/395.html.

5.3 Building and Maintaining Indexes

It is a bit surprising how few programmers actually take advantage of indexing. True the topic can be a bit complex; after all, a book has been written on the subject (Raithel, 2006). Complex or not you do not need a lot of knowledge to take advantage of them.

Indexes give us the ability to virtually sort a data table without physically sorting it. More than one index can be created for any give data table and with multiple indexes it is possible to effectively sort the data set multiple ways at the same time. Once created the index allows the use of the BY statement and other processing techniques as if the data had been sorted using PROC SORT. The index can also be very helpful in subsetting and merging situations.

MORE INFORMATION

Indexes are used with the KEY= option in a table lookup example in Section 6.6, and discussed relative to the use of the WHERE statement in Section 2.7.

SEE ALSO

The SAS Press book written by Michael Raithel (2006) is the definitive work on SAS indexes. A shorter introduction to the practical aspects of indexes can be found in Raithel (2004). Clifford (2005) addresses a number of frequently asked questions about indexes. Andrews and Kress (year unknown) compare the DATASETS and SQL procedures for the building of indexes.

5.3.1 Introduction to Indexing

Indexes provide a search tool that allows the detection and extraction of a data subset. Well-defined indexes can be especially useful in increasing the efficiency of the subsetting process. The highest efficiency gains can be had as the data subset becomes smaller relative to the size of the overall data set. An index that segments the data into subsets that are no more than 10 or 15% of the total data set will tend to provide the most efficiency benefits.

Data set variables are used to define indexes. The selected variables should be chosen so that they maximize the ability to discriminate or break up the data into smaller subsets. Variables that take on only a few levels, such as GENDER or RACE, would probably make poor candidates, while variables such as SUBJECTID or NAME, which take on many more levels relative to the size of the overall data set, would tend to make better index variables. You can also use two or more variables in combination in order to increase the ability of the index to discriminate among data subsets. When the data are also sorted, the variables used to sort the data are also good index variable candidates.

Indexes can be simple (a single variable) or composite (two or more variables), and they can be created in a DATA step, an SQL step, or through PROC DATASETS. Once created, the user can take advantage of indexes in several different ways. Sometimes SAS will even take advantage of available indexes without the user's knowledge. The system option MSGLEVEL=I will cause index usage notes to be written to the LOG.

There are options available in some statements that will specifically invoke indexes (see Section 6.6). While the user can use indexes simply by including the appropriate BY statement, this is not necessarily the best use of indexes. BY-group processing with the index can be inefficient, especially when the full data set is being processed.

Indexes are named and for a simple index, which consists of a single variable, the name is the same as that variable. For composite indexes, which use two or more variables, a name must be provided (the name must be different from any variables on the data set). The index name is used to identify the index file, but is not used by the user to retrieve the index. Indexes are used by specifying the names of the variables that make up the index (simple or composite).

```
proc contents data=advrpt.demog;
   run;
```

The metadata shown by the CONTENTS procedure shows that the INDEXES flag ❶ is now set and shows the number of indexes associated with this table.

```
5.3.1 Metadata for an Indexed Table

The CONTENTS Procedure

Data Set Name    ADVRPT.DEMOG                      Observations        77
Member Type      DATA                              Variables           14
Engine           V9                                Indexes              3   ❶
Created          Tue, Sep 22, 2009 10:57:37 AM     Observation Length  96
```

```
Alphabetic List of Indexes and Attributes

                      # of
                     Unique
   #    Index        Values      Variables
        ❷                        ❸
   1    group          23        sex race edu
   2    ssn            76
   3    subject        77
```

The index definitions are also included in the data set's metadata as can be seen by looking at the listing generated by PROC CONTENTS. The name of each index is shown under the column labeled Index ❷. For composite indexes the variable list is also shown under the column labeled Variables ❸.

When an index has been selected, its use will be mentioned in the LOG.

```
66    proc print data=advrpt.demog;
67       by sex race edu;
NOTE: An index was selected to execute the BY statement.
      The observations will be returned in index order rather than
      in physical order.  The selected index is for the variable(s):
  sex
  race
  edu
```

The indexes are not actually stored in the data set itself. Instead they reside in a separate file. Under Windows the index file has the same name as the data set with which it is associated; however, the extension is different.

5.3.2 Creating Simple Indexes

Indexes can be created using a DATASETS step, an SQL step, or a DATA step. Each of the next three steps creates a simple index. Later in this section these indexes are used in several PROC PRINT steps.

```
proc datasets lib=advrpt;
   modify demog; ❶
   index create clinnum; ❷ ❸
   quit;
```

❶ The data set receiving the index is named. The data set may be either permanent or temporary.

❷ The index to be created is named. For simple indexes the name is the same as the single variable used to form the index.

```
proc sql noprint;
   create index clinnum ❷
      on advrpt.demog ❶ (clinnum)❸;
   quit;
```

❸ The variable used to define the index is named (separately in the SQL step).

```
data demog2❶(index=(clinnum ❷❸));
   set advrpt.demog;
   run;
```

The three methods for creating the index will tend to have different efficiencies based on your particular data. Each method of creating the index will read and handle the data differently.

Of these three methods of creating indexes, the DATA step is the only one that reads the entire data set. PROC SQL and PROC DATASETS don't read the data in the same way; primarily they just add the index file. This means that the generation of the index can be more costly when done with a DATA step, and its cost will grow as the volume of data grows.

When the MSGLEVEL system option is set to I ❹, a note will be written to the LOG when an index is utilized. In this PROC PRINT step the data are subsetted using the SSN variable. The LOG shows that the index for SSN was selected for use with the WHERE clause.

```
* Create index on ssn;
proc sql noprint;
   create index ssn
      on advrpt.demog (ssn);
   quit;

options msglevel=i;  ❹

* Use the ssn index;
proc print data=advrpt.demog;
   var lname fname;
   where ssn < '3';
   id ssn;
   run;
```

```
139   proc print data=advrpt.demog;
140      var lname fname;
141      where ssn < '3';
INFO: Index ssn selected for WHERE
clause optimization.  ❹
```

Once the indexes have been created, a BY statement using the indexed variable(s) will cause the index to be used. In the two PROC PRINT steps, two different BY ❺ statements are used with the same incoming data set. Since both are indexes, that data may not be sorted by either of the two BY variables; however, both steps will execute successfully.

```
proc print data=advrpt.demog;
   by clinnum;  ❺
   id clinnum;
   run;

proc print data=advrpt.demog;
   by ssn;  ❺
   id ssn;
   run;
```

An index can be removed from a data set through the use of PROC DATASETS. The index to be removed is identified by its name (simple or composite) on the INDEX statement along with the DELETE option. In this step the simple index CLINNUM is being removed from the ADVRPT.DEMOG data set.

```
proc datasets lib=advrpt;
   modify demog;
   index delete clinnum;
   quit;
```

5.3.3 Creating Composite Indexes

Since composite indexes are made up of two or more variables, the index name must necessarily be different from the individual variables and different from any other variable on the data set. In the first three steps below, composite indexes are created using three different methods. The PROC PRINT steps that follow then make use of two of these indexes.

```
proc datasets lib=advrpt;
   modify conmed; ❶
   index create drgstart ❷=(drug medstdt); ❸
   quit;
```

❶ The data set receiving the index is named. A given data set may contain multiple indexes at any given time.

```
proc sql noprint;
   create index drgstart ❷
      on advrpt.conmed ❶ (drug medstdt)❸;
   quit;
```

❷ The index to be created is named and the name must be different from other variables in the data set.

```
data cmed2❶(index=(drgstart❷=(drug medstdt)❸));
   set advrpt.conmed;
   run;
```

❸ The list of variables making up the index is written inside the parentheses.

The name of a composite index is not used in the BY statement; however, the variables used to define the index can be used in the BY statement. When using a composite index you can specify

```
proc means data=advrpt.conmed noprint;
   by drug; ❹
   var mednumber;
   output out=sumry max= n=/autoname;
   run;
```

any inclusive subset of the variables starting from the left.❹ Consequently just as a data set that is sorted by DRUG MEDSTDT, must necessarily also be sorted DRUG, this sorted hierarchy will be true for indexes as well.

5.3.4 Using the IDXWHERE and IDXNAME Options

In the absence of an index SAS will satisfy the conditions of a WHERE expression by reading the data sequentially. When an index is present, SAS determines whether or not the utilization of the index will be optimal. Without the index the data is read sequentially; however, the use of the index can produce a non-sequential processing of the data. You can force the use of the index through the use of the IDXWHERE data set option. The IDXNAME data set option can also be used to specify a specific index when more than one exists.

Both of the examples below print portions of the same data set. The composite index DRGSTART was established for this data set in Section 5.3.3. Without the IDXWHERE option, the DRGSTART index has not been applied. And, as is shown by the consecutive numbers in the OBS column, a sequential read of the data took place. In the second PROC PRINT (to the right) the use of the index is requested and a non-sequential read takes place.

```
title2 'Without IDXWHERE';
proc print data=advrpt.conmed;
   where drug < 'C';
   var drug medspdt;
   run;
```

```
title2 'With IDXWHERE';
proc print
      data=advrpt.conmed
            (idxwhere=yes
             idxname=drgstart);
   where drug < 'C';
   var drug medspdt;
   run;
```

```
5.3.4 Using IDXWHERE
Without IDXWHERE

Obs     drug              medspdt

 21     B1-VIT            01/01/2010
 22     ACCUPRIL/HCT2     01/01/2010
 44     B1-VIT            01/01/2010
 45     ACCUPRIL/HCT2     09/09/2006
 67     B1-VIT            10/24/2006
 68     ACCUPRIL/HCT2     12/10/2006
 90     B1-VIT            01/01/2010
 91     ACCUPRIL/HCT2     01/01/2010
113     B1-VIT            01/01/2010
114     ACCUPRIL/HCT2     01/01/2010
136     B1-VIT            05/13/2007
137     ACCUPRIL/HCT2     03/05/2007
159     B1-VIT                     .
160     ACCUPRIL/HCT2     04/30/2007
182     B1-VIT            08/25/2007
183     ACCUPRIL/HCT2     09/24/2007
```

```
5.3.4 Using IDXWHERE
With IDXWHERE

Obs     drug              medspdt

114     ACCUPRIL/HCT2     01/01/2010
 22     ACCUPRIL/HCT2     01/01/2010
 45     ACCUPRIL/HCT2     09/09/2006
 68     ACCUPRIL/HCT2     12/10/2006
 91     ACCUPRIL/HCT2     01/01/2010
137     ACCUPRIL/HCT2     03/05/2007
160     ACCUPRIL/HCT2     04/30/2007
183     ACCUPRIL/HCT2     09/24/2007
113     B1-VIT            01/01/2010
136     B1-VIT            05/13/2007
 21     B1-VIT            01/01/2010
 44     B1-VIT            01/01/2010
 67     B1-VIT            10/24/2006
 90     B1-VIT            01/01/2010
159     B1-VIT                     .
182     B1-VIT            08/25/2007
```

SEE ALSO

During SQL joins SAS will determine whether or not it is optimal to utilize an index. The _METHOD option can be used to determine when an index has been utilized (Lavery, 2005).

5.3.5 Index Caveats and Considerations

While the use of indexes can provide a number of efficiency gains, their use is not without a price. The user should have sufficient information to make an informed decision as to when to build and use indexes.

Remember that the indexes are stored in a separate file from the data set. The indexes must therefore be deliberately maintained. If you update a data set without updating its indexes, the indexing will be lost. If you copy a data set, the index file must also be copied. PROC DATASETS and PROC COPY know to look for and copy index files, but data set copies made at the OS level require that the index file be explicitly copied.

The index file will take up space. How much space depends on a number of factors, but the volume can be non-trivial. Indexes also take time to build and are therefore most appropriate for fairly stable data sets. Here stability is measured against the cost and effort of building and maintaining the index.

5.4 Protecting Passwords

When using SQL pass-through statements to access remote databases we often have to pass user identification and password information to the remote database. If we use those passwords in our program, our user ID and password will be surfaced for all to see. Very often we need to protect our passwords and to a lesser extent our user ID. The level of protection will vary from industry to industry, but the basics are the same.

Fortunately there are a number of ways to protect our passwords.

5.4.1 Using PROC PWENCODE

The PWENCODE procedure can be used to encode or encrypt passwords. *Encoding* is a text substitution technique that disguises your password through a series of text substitutions. *Encryption*, which is a more secure method than encoding (available starting in SAS 9.2), uses mathematical operations in the transformation of the text.

I have had very limited success at using this procedure to protect passwords. First, encoding is not a very strong protection, but more importantly the encoded or encrypted text can often be used instead of the password. It does not particularly matter if the bad guys cannot 'see' the actual password if the encoded text string, which they can see, will work in its place.

Encoding or encrypting your password through PROC PWENCODE creates and then allows you to use a text string instead of the actual password. It protects your password from being viewed directly, since only the encoded/encrypted string is visible. However, remember that the visible string, while not the password, can still be used as the password. This means that we will also need to protect this encoded/encrypted string.

```
filename pwfile "&path\results\pwfile.txt";
proc pwencode in='pharmer' out=pwfile;
   run;
```

In this PWENCODE step our password 'pharmer' is to be encrypted (under SAS 9.1 the default was encoding). As the procedure executes, the encrypted value of 'pharmer' is written to the text file (PWFILE.TXT) as: {sas002}81F6943F251507393B969C0753B2D73B and is not otherwise surfaced for viewing.

Once the value has been stored in the text file, it can be recovered and used at some point in the future. The SAS documentation for PROC PWENCODE shows how this value can be stored in a macro variable or written to the LOG. Both approaches are not practical, because if the user (or someone else) can see the encoded/encrypted value, they can then use it instead of the real password. They may not know your real password, but that does not matter; they still have access to your data.

We need to be able to use the password without ever surfacing its value, either in a macro variable or to the LOG. In the following DATA step the LIBNAME function is used to create a *libref* which establishes an ODBC connection to an SQL server. The password is recovered from the text file and the value is inserted into the PASSWORD= option.

```
data _null_;
length tmp $1024 opt $1200;
infile pwfile truncover;
input tmp;
opt='dsn=SQLServer user=myid password="'||left(trim(tmp))||'"';
rc=libname('sqlsrv',,'odbc', opt);
txt=sysmsg();
put rc= txt=;
run;
```

The encoded/encrypted password cannot be utilized in all coding situations that require the use of a password. For example, the RENAME function (see Section 5.6.1) allows the use of passwords, but restricts them to 8 characters (this limitation should be fixed in a future release of SAS).

MORE INFORMATION
The discussion in Section 5.4.2 takes an alternate approach to the protection of passwords.

SEE ALSO
Although slightly dated by subsequent releases of SAS, Steven (2007) describes the use of the PWENCODE procedure.

5.4.2 Protecting Database Passwords

The following method places the sensitive information, such as passwords and user identification codes, in a protected data set. And then we write our program (the macro %SECRETSQL) to extract the password without surfacing it.

The data set ADVRPT.PASSTAB includes the user ID and password for several different databases. The data set itself is encrypted and password protected using data set options.

```
data advrpt.passtab(encrypt=yes   pwreq=yes
                    read=readpwd write=writepwd
                    alter=chngpwd );
  format dsn uid pwd $8.;
  dsn='dbprod'; uid='mary'; pwd='wish2pharm'; output;
  dsn='dbprod'; uid='john'; pwd='data4you';   output;
  dsn='dbdev';  uid='mary'; pwd='hope2pharm'; output;
  run;
%let syslast=;
```

Notice that the macro variable &SYSLAST has been cleared to remove the name of this data set from the global symbol table.

The macro %SECRETSQL retrieves the passwords from the password data set and then uses the information in the SQL pass-through in such a way that the password is never surfaced to the LOG.

```
%macro secretsql(dbname, username);
  %local dd uu pp;  ❶
  Proc SQL noprint nofeedback;
    (
      SELECT dsn, uid, pwd into :dd, :uu, :pp
      FROM advrpt.passtab(read=readpwd)  ❷
        WHERE dsn=trim(symget('dbname'))  ❸
            AND uid=trim(symget('username'))
    );

    connect to odbc(dsn=%superq(dd) uid=%superq(uu) pwd=%superq(pp));  ❹

    create table mylib.mytable as select * from connection to odbc(
      %passthru  ❺  /* contains your pass-thru SQL statement(s) */
    );

    disconnect from odbc;
    quit;
%mend secretsql;

%secretsql(dbname=dbprod, username=John) ❻
```

❶ The macro variables that will contain the sensitive information are placed in the temporary local symbol table.

❷ We will read the password from the secret password data set. Notice that the user of this macro does not even need to know of the existence of the password data set.

❸ Macro variables are retrieved by using the SYMGET function rather than the more common macro variable reference with an ampersand (&UU or &DBNAME). This is done because the SYMGET function does not write macro variable values in the LOG, even when macro debugging options, such as SYMBOLGEN, are turned on.

❹ The SYMGET function does not always execute in the CONNECT statement. The macro quoting function %SUPERQ will also resolve the macro variable without surfacing its value to the LOG.

❺ The user creates a macro called %PASSTHRU that contains only those SQL statements that are to be processed by the remote database. The user does not need to see anything inside the %SECRETSQL macro.

❻ The macro call only contains information that is not sensitive.

When the %SECRETSQL macro is kept in a stored compiled macro library, the source statements will not be available to the person using it (see Section 13.9 and Sun and Carpenter (2011) for information on protecting the macro code itself).

MORE INFORMATION
Data set options that provide data set protections are described in more detail in Section 2.1.2. SQL pass-through is introduced in Section 1.5. The PWENCODE procedure can potentially also be used to provide password security (see Section 5.4.1). Issues dealing with macro source code security are also discussed in Section 13.9.

SEE ALSO
The %SECRETSQL macro and a number of related techniques are described in more detail by Sherman and Carpenter (2007).

5.5 Deleting Data Sets

There are a number of ways of deleting data sets, both from within SAS and from the operating system. Although generally we do not need to delete our data sets during the execution of a program, sometimes when processing especially large data sets, it can be necessary to clear data sets from the WORK library in order to free up disk space.

PROC DATASETS is the tool most often used from within a SAS program. When using PROC DATASETS, there are two basic ways of carrying out the deletions.

```
proc datasets library=work ❶
              memtype=data ❷
              nolist ❸
              kill ❹;
   quit;
```

```
proc datasets library=work ❶
              memtype=data ❷
              nolist ❸;
   delete male female; ❺
   quit;
```

❶ The library from which the items are to be deleted is specified.

❷ Select the type of item to be deleted. Data sets have the MEMTYPE=DATA. When deleting catalogs the CATALOG procedure can also be used.

❸ The NOLIST option suppresses the list of members prior to the deletion from being written to the LOG.

❹ The KILL option deletes all items of the specified type ❷ to be deleted.

❺ The DELETE statement lists one or more items to delete from the specified library ❶.

Although no longer documented, the DELETE procedure is still available. This procedure ❻ is one of the very few, if not only, procedures to allow more than one data set name to be associated with the DATA= option.

```
proc delete data=male allgender ❻;
   run;
```

Data sets can also be deleted from within an SQL step by using the DROP TABLE statement ❼. Notice that more than one table can be listed on the DROP TABLE statement by separating the names with a comma.

```
proc sql;
   drop table allgender, male; ❼
   quit;
```

SEE ALSO
Rosenbloom and Lafler (2011d) discuss the use of PROC DATASETS to delete data sets.

5.6 Renaming Data Sets

Data sets can be renamed using a variety of methods including a number of ways through the use of the OS tools. Renaming data sets from within a program is also possible, and for some situations even preferable as the process can be automated using the macro language.

5.6.1 Using the RENAME Function

The DATA step function RENAME can be used to rename data sets, catalogs, and even directories. Like most DATA step functions it can also be utilized by the macro language. Here the data set WORK.MALE is being renamed to WORK.MALES.

```
data  male
      female;
   set sashelp.class;
   if sex='M' then output male;
   else output female;
   run;

%let rc=%sysfunc(rename(work.male,Males,data));
%put &RC;
```

Notice that the *libref* for the new name (second argument) is implied and is *not* explicitly included. The function returns a 0 for a successful rename operation.

5.6.2 Using PROC DATASETS

Within the DATASETS procedure there are two primary methods for renaming data sets and catalogs:

- CHANGE   changes or renames a data set
- AGE   renames a group of data sets to form a series of previous versions

```
data current;
   created = datetime();
   format created datetime18.;
   run;

proc datasets library=work nolist;
   change current=now;
   quit;
```

The CHANGE statement is designed to rename one or more data sets. The data sets are listed on the CHANGE statement in OLDNAME=NEWNAME pairs. In this example the data set WORK.CURRENT is renamed to WORK.NOW. Because the data sets are being renamed, not copied (which uses the COPY statement), the library for the old and new name will always be the same.

If a data set with the NEW name already exists the rename will not take place.

When you need to retain one or more snapshots (backup) copies of a data set, the AGE statement can be used to perform the operation. The oldest data set is deleted and then in order of age the data sets are renamed one at a time. In this example CURRENTV7 would be deleted, CURRENTV6 is renamed to CURRENTV7, and so on until the most recent version of

```
proc datasets library=mydata nolist;
   age current currentV1 - currentV7;
   quit;
```

CURRENT is renamed to CURRENTV1. If this AGE statement was executed every morning, there would be a backup or 'aged' copy of CURRENT for each day of the week.

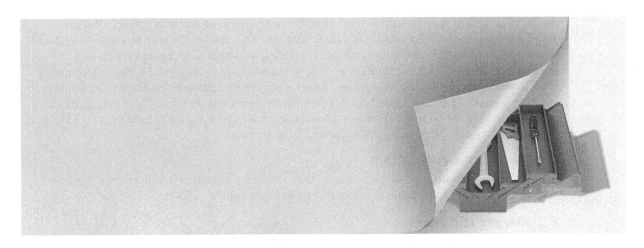

C h a p t e r 6

Table Lookup Techniques

A table lookup is performed when you use the value of a variable (e.g., a clinic number) to determine the value of another variable (e.g., clinic name). Often this second piece of information must be 'looked up' in some other secondary table or location. The process of finding the appropriate piece of information is generally fast; however, as the number of items and/or observations increases, the efficiency of the process becomes increasingly important. Fortunately there are a number of techniques for performing these table lookups.

These techniques can be radically different both in terms of programming complexity and in performance. As the programmer, you will be faced with not only a variety of techniques, but also performance issues. Which technique will perform best, or even adequately, will vary from situation to situation. This means that there is no easy answer to the question, 'Which technique should I use?' It is possible, however, to give you some general guidelines.

In this chapter data set size is often used as a decision point for determining which of these techniques to use. But what exactly is a large data set? This answer too will be situational. It will depend a lot on your OS and your hardware (including available memory and processing capabilities). You will know a data set is large when you have to be careful with the techniques used to process it. For table lookups this chapter will give you alternatives. Each is a compromise between processing efficiency and coding complexity. Each section should give you sufficient information to help you make an informed decision. In all cases you will probably want to test your candidate techniques using your data and your hardware.

6.1 and 6.2 IF – THEN/ELSE
This is the slowest and least sophisticated of the techniques discussed in this chapter; however, these techniques may be adequate for small data sets and simple lookups.

6.3 Merges and Joins
These techniques tend to be slower than the use of formats (6.5); however, they do allow the retrieval of multiple items.

6.4 Double SET Statement DATA Steps
Replacing the MERGE statement with a double SET statement generally provides a performance boost, but the technique is more programming intensive.

6.5 Format-Driven Lookup
These techniques tend to be substantially faster than the use of merges and joins. These techniques are probably the best first choice for most situations with small to somewhat large data sets.

6.6 Using Indexes
Indexes can improve the performance of merge and join techniques, especially when the indexes have already been created. Depending on the situation these techniques might outperform the use of formats.

6.7 Key Indexing (Array Processing)
The use of DATA step arrays and memory eliminates the need for sorting. Although there are potential memory constraints, key indexing typically out performs merges and joins, and for larger data sets out performs the use of formats. When it is possible to use them, array (direct addressing) techniques, such as these, will also generally outperform hash table techniques.

6.8 Hash Tables
Hash tables (hashing) is more flexible than array processing techniques, and except for array processing techniques, these techniques are typically the fastest of the lookup techniques. Coding requires an understanding of the DATA step component hash objects and methods. For very large data sets, the use of hash tables may be the only viable choice.

In each of the examples in this chapter we will be using lookup techniques to determine the clinic name based on the clinic number.

SEE ALSO
An overview of lookup techniques, which includes several nice examples with explanations, can be found in Liu (2008). Aker (2002) and Carpenter (2001b) each discuss differences and programming techniques for lookups including those for match merging, format lookups, and the use of arrays. An overview of lookup techniques with emphasis on hash tables and hash objects can be found in Dorfman and Vyverman (2004b). Comparison papers on the efficiency of table lookup techniques include Stroupe and Jolley (2008) as well as Jolley and Stroupe (2007).

6.1 A Series of IF Statements—The Logical Lookup

The simplest form of a table lookup makes use of the IF-THEN statement. Although easy to code, this is one of the slowest table lookup methods. Essentially this technique creates the new variable with its associated value through the use of IF-THEN processing. Effectively we are 'hard coding' the clinic name within the program. This technique is demonstrated in the following DATA step.

```
data demognames;
   set advrpt.demog(keep=clinnum lname fname);
   length clinname $35;
   if clinnum='011234' then clinname = 'Boston National Medical';
   if clinnum='014321' then clinname = 'Vermont Treatment Center';
   if clinnum='107211' then clinname = 'Portland General';
   if clinnum='108531' then clinname = 'Seattle Medical Complex';
   if clinnum='023910' then clinname = 'New York Metro Medical Ctr';
   if clinnum='024477' then clinname = 'New York General Hospital';
   run;
```

The problem with this approach is that it is not practical if there are more than a very few codes to lookup; besides this is VERY inefficient. SAS must execute each IF statement even if an earlier IF statement was found to be true. To make matters worse, IF statements require a fair bit of processing time.

This is an example of a 100% lookup. It is a sequential search and one where every code is checked regardless of whether or not the answer has already been found.

6.2 IF -THEN/ELSE Lookup Statements

A substantially faster method than the IF-THEN, is to use the IF-THEN / ELSE statement combination. The following DATA step executes more quickly than the previous one because as soon as one IF statement is found to be true, its associated ELSE is not executed. Consequently as soon as an expression is found to be true, none of the remaining IF-THEN / ELSE statements are executed. This technique can be made somewhat faster if the more likely outcomes are placed earlier in the list.

```
data demognames;
   set advrpt.demog(keep=clinnum lname fname);
   length clinname $35;
   if clinnum='011234' then clinname = 'Boston National Medical';
   else if clinnum='014321' then clinname = 'Vermont Treatment Center';
   else if clinnum='107211' then clinname = 'Portland General';
   else if clinnum='108531' then clinname = 'Seattle Medical Complex';
   else if clinnum='023910' then clinname = 'New York Metro Medical Ctr';
   else if clinnum='024477' then clinname = 'New York General Hospital';
   run;
```

In terms of performance efficiency this technique is similar to that of the DATA step's SELECT statement and the CASE statement in SQL.

The SELECT statement is on par with the IF-THEN / ELSE combination when performing table lookups. It can even be a bit faster (Virgle, 1998). Again processing time is minimized when the most likely match is located early in the list.

```
data demognames;
   set advrpt.demog(keep=clinnum lname fname);
   select(clinnum);
    when( '011234') clinname='Boston National Medical';
    when( '014321') clinname='Vermont Treatment Center';
    when( '107211') clinname='Portland General';
    when( '108531') clinname='Seattle Medical Complex';
    when( '023910') clinname='New York Metro Medical Ctr';
    when( '024477') clinname='New York General Hospital';
    otherwise;
   end;
   run;
```

Interestingly Virgle (1998) found that the efficiency of the SELECT statement can sometimes be enhanced by placing the entire expression on the WHEN statement.

```
data demognames;
   set advrpt.demog(keep=clinnum lname fname);
   select;
    when(clinnum='011234') clinname='Boston National Medical';
    when(clinnum='014321') clinname='Vermont Treatment Center';
    when(clinnum='107211') clinname='Portland General';
    when(clinnum='108531') clinname='Seattle Medical Complex';
    when(clinnum='023910') clinname='New York Metro Medical Ctr';
    when(clinnum='024477') clinname='New York General Hospital';
    otherwise;
   end;
   run;
```

There are two overriding issues with these techniques. The primary problem is that the search is sequential. When the list is long the average number of comparisons goes up quickly, even when you carefully order the list. The second, but no less important, issue is that these techniques hard code the values in the program. This is just not smart programming.

Fortunately, the other lookup techniques in this chapter not only avoid hard coding altogether, but also minimize the need for sequential searches.

6.3 DATA Step Merges and SQL Joins

The use of joins in an SQL step and the MERGE in the DATA step is another very common way to perform table lookups by matching values between two data sets. The MERGE statement (when used with the BY statement as it usually is) requires sorted or indexed data sets, while the SQL step does not. There are advantages and disadvantages to both processes.

MERGE Statement
The MERGE statement is used to combine two or more data sets. For the purpose of this discussion, one of these data sets will contain the information that is to be looked up. The BY statement is used to make sure that the observations are correctly aligned. The BY statement

should include sufficient variables to form a unique key in all but at most one of the data sets. For the example below ADVRPT.CLINICNAMES has exactly one observation for each value of CLINNUM.

Because the BY statement is used, the data must be sorted. Sorting can be time consuming, or even on occasion impossible for very large data sets or for data sets on tape. In the following steps PROC SORT is used to reorder the data into temporary (WORK) data sets. These are then merged using the MERGE statement.

```
proc sort data=advrpt.demog
          out=demog;
   by clinnum;
   run;

proc sort data=advrpt.clinicnames
          out=clinicnames;
   by clinnum;
   run;

data demognames(keep=clinnum clinname
                     lname fname);
   merge demog(in=indemog)
         clinicnames(in=innames);
   by clinnum;
   if indemog; ❶
   run;
```

For a successful lookup using the MERGE statement, both of the incoming data sets must be indexed or in sorted order and as was discussed in Chapter 4 sorting can be operationally expensive.

The following PROC PRINT listing of the first 10 observations of the merged data shows that the CLINICNAME has been successfully acquired.

As anticipated the observations are sorted by CLINNUM. Notice also that the variable CLINNAME, which came from the right-most data set in the MERGE statement, is last on the PDV.

```
6.3 Lookup By Joining or Merging Two Tables
10 Observations of the merged data

Obs    clinnum    lname     fname         clinname

  1    011234     Nabers    David     Boston National Medical
  2    011234     Taber     Lee       Boston National Medical
  3    014321     Lawless   Henry     Vermont Treatment Center
  4    014321     Mercy     Ronald    Vermont Treatment Center
  5    023910     Atwood    Teddy     New York Metro Medical Ctr
  6    023910     Harbor    Samuel    New York Metro Medical Ctr
  7    023910     Leader    Zac       New York Metro Medical Ctr
  8    024477     Haddock   Linda     New York General Hospital
  9    024477     Little    Sandra    New York General Hospital
 10    024477     Maxwell   Linda     New York General Hospital
         .... portions of the table are not shown ....
```

The IF statement ❶ has been used to eliminate any clinic numbers in CLINICNAMES that do not appear in DEMOG. This logic will not eliminate cases where there is a clinic number in DEMOG that does not appear in CLINICNAMES (the clinic name will be missing). If we want to restrict the lookup to only those clinic numbers with matches in CLINICNAMES, the IF statement ❶ could be replaced with `if indemog and innames;`. This result is achieved directly in the SQL join discussed next ❷.

SQL Join

When using SQL, the merging process is called a *join*. The SQL join operations do not require sorting and can be more efficient than the DATA step MERGE, unless the tables are so large that they do not fit well into memory.

Just because an SQL join does not require the incoming data to be sorted, does not mean that no resources are going to be expended in preparation for the join (lookup). Hidden from view, but within the processing of the SQL step, lookup techniques are being applied. These behind the scenes operations are very similar to the hash techniques that we can explicitly apply and control in the DATA step (see Sections 6.7 and 6.8).

There are a number of different types of joins within SQL, and one that closely matches the previous step is shown below.

In this example we have added the requirement (through the use of a WHERE clause ❷) that the clinic number be in both data tables. The WHERE clause is used to select rows that have matching values of CLINNUM on both incoming tables. This is a more stringent criteria than was used in the DATA step shown above ❶.

```
proc sql noprint;
   create table demognames2 as
     select a.clinnum, b.clinname, lname, fname
       from advrpt.demog a, advrpt.clinicnames b
         where a.clinnum=b.clinnum;   ❷
   quit;
```

SQL does not require either of the two data sets to be sorted prior to the join, and unless we specifically request that the resulting data table be sorted (ORDER BY clause) it will reflect the order of the incoming data. This can be seen in the order of the clinic number in the PROC PRINT results shown below.

```
6.3 Lookup By Joining or Merging Two Tables
10 Observations of the Joined data

Obs    clinnum    clinname                      lname        fname

 1     049060     Atlanta General Hospital      Adams        Mary
 2     082287     Denver Security Hospital      Adamson      Joan
 3     066789     Austin Medical Hospital       Alexander    Mark
 4     063742     Houston General               Antler       Peter
 5     023910     New York Metro Medical Ctr    Atwood       Teddy
 6     066789     Austin Medical Hospital       Banner       John
 7     046789     Tampa Treatment Complex       Baron        Roger
 8     049060     Atlanta General Hospital      Batell       Mary
 9     095277     San Francisco Bay General     Block        Will
10     031234     Bethesda Pioneer Hospital     Candle       Sid
            .... portions of the table are not shown ....
```

6.4 Merge Using Double SET Statements

There are a number of schemes that have been published that utilize two SET statements in the same DATA step. These SET statements replace the single MERGE statement and the programmer takes charge of the joining process (keeping the two data sets in sync). These techniques can be faster but more complicated than a MERGE. However they do still require that both incoming data sets be sorted.

In this example, the two incoming data sets have already been sorted (by CLINNUM). The

```
data withnames(keep=subject clinnum clinname);
  set demog(rename=(clinnum=code)); ❶
  * The following expression is true only when
  * the current CODE is a duplicate.;
  if code=clinnum then output; ❷
  do while(code>clinnum); ❸
    * lookup the clinic name using the code (clinnum)
    * from the primary data set;
    set clinicnames(keep=clinnum clinname); ❹
    if code=clinnum then output; ❺
  end;
run;
```

primary data set contains the observations for which we need the clinic name. The secondary data set (CLINICNAMES) contains just the names that are to be retrieved. It is both possible and even likely that the lookup data set will contain values that have no match in the first or primary data set. This is fine as long as we plan for the possibility ❸.

❶ An observation is read from the primary data set. Because DEMOG and CLINICNAMES both use the variable CLINNUM to hold the clinic number, when it is read from the DEMOG data set it is renamed CODE. This allows us to access and compare the clinic numbers from both data sets at the same time ❷❸❺.

❷ The value of CODE (clinic number from DEMOG) is compared to the value of CLINNUM, which comes from the second data set (CLINICNAMES). On the very first pass, no observation will have been read from CLINICNAMES, and CLINNUM will be missing. Codes that do not have matching names will not be written out ❺.

❸ The DO WHILE is used to read successive rows from the second data set.

❹ It is possible that there are codes and names in the CLINICNAMES data set that are not in the primary data set (DEMOG). These observations will necessarily have a CLINNUM that is less than CODE. This loop ❸ cycles through any extra names until the second data set 'catches up' to the first (CODE=CLINNUM).

❺ This code matches the current value of CLINNUM and the observation is written out.

As in the MERGE example shown earlier, the data still have to be sorted before the above DATA step can be used. Although the sorting restrictions are the same as when you use the MERGE statement, the advantage of the double SET can be a substantial reduction in processing time.

MORE INFORMATION
The use of two SET statements in one DATA step is introduced in Section 3.8.4 and used in examples in Sections 3.1.5 and 3.1.6.

6.5 Using Formats

The use of FORMATS allows us to step away from the logical processing of assignment statements, and to take advantage of the search techniques that are an inherent part of the use of FORMATS. When a value is retrieved from a format, a binary search is used and this means that we can search 2^N items in N or fewer tries. With 10 guesses we can search over 1000 items. Since binary searches operate by iteratively splitting a list in half until the target is found, these searches tend to be faster than sequential searches—especially as the number of items increases.

For many users, especially those with smaller data sets and lookup tables, the efficiency gains realized here may be sufficient for most if not all tasks. Formats with several thousands of items have been used successfully as lookup tables.

Formats can be built and added to a library (permanent or temporary) through the use of PROC FORMAT (see Chapter 12). The process of creating a format is both fast and straightforward. The following format ($CNAME.) contains an association between the clinic code and its name.

```
proc format;
   value $cname
   '011234'='Boston National Medical'
   '014321'='Vermont Treatment Center'
   '107211'='Portland General'
   '108531'='Seattle Medical Complex'
   '023910'='New York Metro Medical Ctr'
   . . . . some code not shown . . . .
   '024477'='New York General Hospital';
   run;
```

Of course typing in a few values is not a 'big deal'; however, as the number of entries increases the process tends to become tedious and error prone. Fortunately it is possible to build a format directly from a SAS data set. The CNTLIN= option identifies a data set that contains specific variables. These variables store the information needed to build the format, and as a minimum must include the name of the format (FMTNAME), the incoming value (START), and the value that the incoming value will be translated to (LABEL). The following DATA step builds the data set CONTROL, which is used by PROC FORMAT. Notice the use of the RENAME= option and the RETAIN statement. One advantage of this technique is that the control data set does not need to be sorted.

Since we already have a data set with the matched value pairs (ADVRPT.CLINICNAMES), it is a perfect candidate for building a format automatically.

```
data control; ❶
   set advrpt.clinicnames(keep=clinname clinnum
                    rename=(clinnum=start ❷
                            clinname=label)); ❸
   retain fmtname '$cname'; ❹
   run;

proc format cntlin=control; ❺
   run;
```

❶ The control data set containing the variables (FMTNAME, START, and LABEL) is created based on the data set ADVRPT.CLINICNAMES.

❷ The START variable (left side of the = sign in the value statement) is created by renaming CLINNUM.

❸ The LABEL variable (right side of the = sign in the value statement) is created by renaming CLINNAME.

❹ The format name is a constant and is created using the RETAIN statement.

❺ The format is created by PROC FORMAT through the use of the CNTLIN= option which points to the control data set.

Once the format has been defined, the PUT function ❻ can be used to assign a value to the variable CLINICNAME by using the $CNAME. format.

```
data fmtnames(keep=subject clinnum clinname dob);
   set demog(keep = subject dob clinnum);
   clinname = left(put(clinnum,$cname.));  ❻
   run;
```

Remember the PUT function always returns a character string; when a numeric value is required, the INPUT function can be used. The length of the new variable is determined by the format used in the PUT function. If no length is specified, as in this example, the variable's length will be based on the longest value in the format ($27. in this example). A shorter format width, say $CNAME20., would cause the variable to have the shorter length (truncation would be a possibility). Values longer than the longest formatted value will not increase the variable's length past the longest formatted value.

The previous DATA step will be substantially faster than the IF-THEN/ELSE or SELECT processing steps shown above. The difference becomes even more dramatic as the number of items in the lookup list increases. The lookup itself will use the format $CNAME., and hence will employ a binary search. As a rule of thumb, format searches should be very efficient up until the number of items to look up exceeds 20,000 or so items.

MORE INFORMATION
An assignment statement technique that outperforms a PUT function lookup when creating a numeric result is discussed in Section 2.2.3. Data are used to create user-defined formats in Section 12.7.

6.6 Using Indexes

Indexes are a way to logically sort your data without physically sorting it. If you find that you are sorting and then re-sorting data to accomplish your various merges, you may find that indexes will be helpful.

Indexes must be created, stored, and maintained. They are usually created through either PROC DATASETS (shown below) or through PROC SQL; however, they can also be created in a DATA step. The index stores the order of the data as if it had been physically sorted. Once an index exists, SAS will be able to access it, and you will be able to use the data set with the appropriate BY statement even though the data has never been physically sorted.

Resources are required to create an index, and these resources should be taken into consideration. Indexes are stored in a separate file, and the size of this file can be substantial especially as the number of indexes, observations, and variables used to form the indexes increases.

Indexes can substantially speed up processes. They can also SLOW things down (Virgle, 1998). Not all data sets are good candidates to be indexed and not all variables will form good indexes. Be sure to read about indexes (see Section 5.3 for more on indexes), and then experiment carefully before investing a lot in the use of indexes.

The following example shows the creation of indexes for the two data sets of interest. The library containing the data sets, ADVRPT, is identified ❶. The NOLIST option prevents PROC DATASETS from writing a list of all the objects in this library to the LOG.

```
proc datasets library=advrpt nolist; ❶
   modify clinicnames; ❷
     index create clinnum ❸/ unique; ❹
   modify demog; ❷
     index create clinnum; ❸
   quit;
```

❷ The MODIFY statement is then used to name the data sets that are to receive the indexes. And the INDEX statement ❸ defines the index for each data set.

❹ The unique option forces unique values for CLINNUM.

MORE INFORMATION
The building, maintenance, and use of indexes are discussed further in Section 5.3.

6.6.1 Using the BY Statement

Making use of an index can be as simple as using a BY statement. When the BY variable is an index, the index is automatically used, and the data does not need to be sorted. However relying on an index to perform a merge is not necessarily as fast as a merge on sorted data. The advantage is that we do not have to sort the data prior to the merge, and the time required to perform the sort should be taken into consideration. Assuming the indexes have already been created, one of the following techniques should generally give you better performance over an indexed merge.

```
data mrgnames;
   merge demog(keep=subject clinnum edu)
         clinicnames(keep=clinnum clinname);
   by clinnum;
   run;
```

6.6.2 Using the KEY= Option

You can also look up a value when an index exists on only the data set that contains the values to be looked up. The KEY= option on the SET statement identifies an index that is to be used.

```
data keynames;
   set advrpt.demog
         (keep=subject clinnum lname fname); ❺
   set advrpt.clinicnames key=clinnum/unique; ❻
   if _iorc_ ne 0 then clinname=' '; ❼
run;
```

❺ An observation is read from the primary, potentially unsorted data set. This loads a value for the index variable (CLINNUM) into the PDV.

❻ An observation is read from the lookup data set. Because the KEY= option has been specified, the observation corresponding to the current value of CLINNUM is returned. Since this is an indexed read, the observations read from ADVRPT.DEMOG can be in any order, and values of CLINNUM can be repeated.

❼ The temporary variable _IORC_ will be 0 when an indexed value is successfully read. If the value of CLINNUM is not found, _IORC_ will not be equal to 0 and we will need to supply a missing value for the clinic name. Otherwise the value of CLINNAME will have been retained from the previous observation.

The values returned to _IORC_ may change in future releases of SAS. Rather than depend directly on the value of _IORC_, the SAS supplied autocall library macro %SYSRC can be used

```
data rckeylookup;
    set advrpt.demog(keep=subject clinnum lname fname);
    set advrpt.clinicnames key=clinnum/unique;
    select (_iorc_); ❽
        when (%sysrc(_sok)) do; ❾
            * lookup was successful;
            output;
        end;
        when (%sysrc(_dsenom)) do; ❿
            * No matching clinic number found;
            clinname='Unknown';
            output;
        end;
        otherwise do;
            put  'Problem with lookup ' clinnum=;
            stop;
        end;
    end;
run;
```

to 'decode' the values contained in _IORC_. The following example is the same as the previous one, but it takes advantage of two of the over two dozen values accepted by %SYSRC. ❽ The SELECT statement is used to compare the returned _IORC_ value with codes of interest. _IORC_ will be an

integer: 0 for success and >0 for various types of failure.

❾ For a given error mnemonic, the %SYSRC macro returns the number associated with the associated error. _SOK is the mnemonic for success and %SYSRC returns a 0, which matches the value of _IORC_.

❿ When a specific value of CLINNUM is not on the index, the error mnemonic is _DSENOM.

SEE ALSO
Additional examples and discussion on the use of the KEY= option can be found in Aker (2000).

6.7 Key Indexing (Direct Addressing)—Using Arrays to Form a Simple Hash

Sometimes when sorting is not an option or when you just want to speed up a search, the use of arrays can be just what you need. Also known as *direct addressing*, variations on this form of lookup tend to be the fastest of the lookup techniques discussed in this chapter; however, there are some restrictions on their use that can limit their flexibility. These techniques require you to create arrays, sometimes very large arrays. Fortunately under current versions of SAS you can build arrays that can contain millions of values (Dorfman, 2000a, 2000b).

6.7.1 Building a List of Unique Values

To introduce this topic consider the problem of creating a list of unique values from a data set. In terms of the data sets being used in this set of examples, we would like to establish a list of unique

```
proc sort data=advrpt.demog
          out=uniquenums
          nodupkey;
    by clinnum;
    run;
```

clinic numbers within the data set ADVRPT.DEMOGS. One of several ways that you could use to solve this problem is shown to the left. Here PROC SORT uses a NODUPKEY option to build a data set with unique values of CLINNUM.

This works, assuming that the data set can be sorted and that the cost of the resources expended in the sorting process is reasonable. An alternate method appropriate for data sets of all sizes makes use of DATA step arrays.

To avoid sorting, we somehow have to "remember" which clinic codes we have already seen. The way to do this is to use the ARRAY statement. The beauty of this technique is that the search is very quick because it has to check only one item. We accomplish this by using the clinic code itself as the index to the array.

```
data uniquekey;
   array check {999999} _temporary_ ;   ❶
   set advrpt.demog;   ❷
   if check{input(clinnum,6.)}=. then do;   ❸
      output;   ❹
      check{input(clinnum,6.)}=1;   ❺
   end;
run;
```

❶ Establish an array with sufficient dimension to handle all the clinic numbers.

❷ Read a clinic number.

❸ When the array element is missing, this is the first occurrence of this clinic number. Write it out and then mark it ❺ so it will not be written again.

❹ Write out this clinic number.

❺ Mark this number as having been seen.

As an observation is read from the incoming data set, the character clinic code is converted to a number using the INPUT function and then is used as the index for the ARRAY CHECK. If the array value is missing, this is the first (unique) occurrence of this clinic number. It is then marked as found (the value is set to 1). Notice that this step will allow a range of clinic codes from 1 to 999,999. Larger ranges, into the 10s of millions, are easily accommodated.

The array used in the previous example is numeric; however, we could have stored a single byte character flag and reduced the memory requirements by a factor of 8.

```
data uniquekey;
   array check {999999} $1 _temporary_ ;   ❻
   set advrpt.demog;
   if check{input(clinnum,6.)}=' '   ❼ then do;
      output;
      check{input(clinnum,6.)}='x';❽
   end;
run;
```

❻ The array is declared to be a character array of $1 elements.

❼ The check is made for a missing (blank) array value.

❽ A non-blank character is stored to indicate that this clinic number has been found.

6.7.2 Performing a Key Index Lookup

In the previous example an array was used to look up whether or not an item had been found before. This process of looking up a value is exactly what we do when we merge two data sets. In the following DATA step the list of codes are read sequentially, *once*, into an array that stores the clinic name (instead of just the number 1) again using the clinic code as the array subscript. The second DO UNTIL then reads the data set of interest. In this loop the clinic name is recovered from the array and assigned to the variable CLINNAME.

In addition to its speed of execution, a major advantage of this technique is that neither of the incoming data sets needs to be sorted.

This technique is known as *key indexing* because the index of the array is the value of the variable that we want to use as the lookup value.

The array itself may be numeric or character depending on whether a numeric or character value is to be retrieved. The index, however, must be numeric (or convertible to numeric as in this example). Large arrays are common. In this example there are almost a million array elements, when the example needs a mere 27. Memory is fast, cheap, and generally available on most modern machines, thus making this *overkill* a small price.

```
data clinnames(keep=subject lname fname clinnum clinname);
  array chkname {999999} $35 _temporary_; ❶
  do until(allnames); ❷
     set advrpt.clinicnames end=allnames;
     chkname{input(clinnum,6.)}=clinname; ❸
  end;
  do until(alldemog);
     set advrpt.demog(keep=subject lname fname clinnum) ❹
              end=alldemog;
     clinname = chkname{input(clinnum,6.)}; ❺
     output clinnames;
  end;
  stop; ❻
  run;
```

❶ A character array of temporary values is established. This array will hold the values to be retrieved (clinic names), and will be indexed by the clinic number (CLINNUM). The length of the array elements must be sufficient to hold each value being inserted into the array.

❷ A loop is used to read all of the observations from the data set that contains the values to be looked up. Because the temporary variable ALLNAMES is defined using the END= option on the SET statement, it will be 0 for all observations except the last one, and then it will be assigned the value of 1 (true).

❸ The value of the clinic name, CLINNAME, is stored in the array element identified by the clinic number. The INPUT function is used to convert the character variable CLINNUM into an integer.

❹ An observation is read from the primary data set. This loads the value of the clinic number into the Program Data Vector, PDV, where it can be used to retrieve the clinic name from the CHKNAME array.

❺ The clinic name is retrieved from the CHKNAME array using the value of the clinic number just retrieved from the primary data set.

❻ Because we have placed the SET statements inside of DO loops, it is necessary to terminate the DATA step directly.

```
data crnames(keep=subject lname fname clinnum clinname region);
  array chkname {999999} $35 _temporary_;
  array chkregn {999999} $2 _temporary_;
  do until(allnames);
     set advrpt.clinicnames end=allnames;
     chkname{input(clinnum,6.)}=clinname;
     chkregn{input(clinnum,6.)}=region;  ❼
  end;
  do until(alldemog);
     set advrpt.demog(keep=subject lname fname clinnum)
                end=alldemog;
     clinname = chkname{input(clinnum,6.)};
     region   = chkregn{input(clinnum,6.)};  ❼
     output crnames;
  end;
  stop;
run;
```

Because we are working with arrays this technique is not limited to the retrieval of a single value. In this example we want to retrieve both the clinic name and the region associated with the clinic number. ❼ The only real difference is the addition of another array.

As was noted above memory is fast and usually readily available. We should still be at least conscious of our memory usage. In this example the maximum length of a clinic name is 27 characters. Since the ARRAY definition will not affect the length of the new variable (array elements were defined as $35 and 35>27), this array statement has wasted almost 8 megabytes (999,999 * 8 bytes) of memory. Although not a lot of memory for a small array such as this one, you should at least be aware of the overall cost of your array. This technique will not work in all situations. As the number of array elements increases, the amount of memory used also increases. Paul Dorfman (2000a) discusses memory limitations. Certainly most modern computers should accommodate arrays with the number of elements in the millions.

For situations where this technique requires unreasonable amounts of memory, other techniques such as bitmapping and hashing are available. Again Paul Dorfman is the acknowledged expert in this area and his cited papers should be consulted for more details.

MORE INFORMATION
In the sample programs associated with this section there is a key indexing example that stores and retrieves multiple values of multiple variables.

6.7.3 Using a Non-Numeric Index

One of the limitations of the key indexing techniques is that the index to the array must be numeric. This limitation is overcome completely by the use of hash objects (see Section 6.8). In the examples in Section 6.7.2 the index is a character string that contains a number; therefore, the INPUT function can be used to create the numeric index value. What if the character string does not readily convert to a number?

When the number of items to be looked up is fairly small, for example fewer than 20 or 30 thousand, a format can be used to convert the character key to a number. In this example let's assume that CLINNUM could not be converted directly to a number. Instead we create a numeric informat to create a unique artificial index number.

```
data control(keep=fmtname start label type);  ❶
   set advrpt.clinicnames(keep=clinnum
                          rename=(clinnum=start))
      end=eof;
   retain fmtname 'nam2num' type 'I';  ❷
   label=_n_;  ❸
   output control;
   if eof then call symputx('levels',_n_);  ❹
   run;

proc format cntlin=control;  ❺
   run;

data clinnames(keep=subject lname fname
                    clinnum clinname);
   array chkname {&levels❹} $35 _temporary_;
   do until(allnames);
      set advrpt.clinicnames end=allnames;
      chkname{input(clinnum,nam2num.❻)}=clinname;
   end;
   do until(alldemog);
      set advrpt.demog(keep=subject lname fname
                            clinnum)
             end=alldemog;
      clinname = chkname{input(clinnum,nam2num.❻)};
      output clinnames;
   end;
   stop;
   run;
```

❶ A control file is created that will be used to build the conversion format NAM2NUM. (see Section 12.7).

❷ The TYPE variable declares this to be a numeric informat.

❸ The label is the numeric counter.

❹ Since we are reading the whole data set (one observation per unique clinic number), we can save the number of possible values. This value can be used to provide a dimension to the array.

❺ The CONTROL data set is used by PROC FORMAT to create the NAM2NUM. format.

❻ The format is used to convert the character value into a usable array index.

6.8 Using the HASH Object

Users with very large data sets are often limited by constraints that are put on them by memory or processor speed. Often, for instance, it is not practical or perhaps even possible to sort a very large data set. Unsorted data sets cannot be merged using a BY statement unless the data set is indexed, and this type of merge is generally not feasible (see Section 6.6.1). Joins in SQL may be possible by using the BUFFERSIZE option, but this still may not be a useful solution. Fortunately there are a number of techniques for handling these situations as well.

In Section 6.7.2 an array was used to hold and retrieve values. This is a form of a simple hash table. In SAS®9 the DATA step has a HASH object that will hold and access the array portion of this hash array. This hash object is a DATA step component object (DATA step component objects are introduced and discussed in Section 3.3).

While key indexing is fast and works well, it does have limitations that the hash object can overcome.

- Key indexing requires a numeric value as the array index. While techniques have been developed to work around this limitation, hash objects are designed to work with character or numeric keys.

- Unless using a multi-dimensional array, key indexing can use only a single key while hash objects can use composite keys.

- Multiple fields can be returned with a hash object.

Essentially the hash object defines an array in memory, initializes its values with data from a table, and sets up an indexing variable or variables that can be either numeric or character.

```
data hashnames(keep=subject clinnum clinname lname fname);

   * Define the attributes for variables on lookup table;
   if 0 then set advrpt.clinicnames; ❶

   * Create and load the hash object;
   declare hash lookup(dataset: 'advrpt.clinicnames', ❷
                       hashexp: 8); ❸
   lookup.defineKey('clinnum'); ❹
   lookup.defineData('clinname'); ❺
   lookup.defineDone();

   * Read the primary data;
   do until(done); ❻
      set advrpt.demog(keep=subject clinnum lname fname) ❼
          end=done; ❽
      if lookup.find() = 0 then output hashnames; ❾
   end;
   stop; ❿
   run;
```

❶ The attributes for the variables that are to be retrieved from the hash object need to be established on the PDV. This SET statement is used only during DATA step compilation to determine variable attributes.

❷ The HASH object itself is defined, named, and loaded using the DECLARE statement. The attributes of the object are then defined using the DEFINEKEY, DEFINEDATA, and DEFINEDONE methods. This hash object has been named LOOKUP, and has been loaded with the data that contains the values (CLINNAME) that we want to be able to look up.

❸ The number of bins (2^8=256) used by the hash table is specified.

❹ The DEFINEKEY method is used to list one or more key variables whose values are used to index the LOOKUP hash table.

❺ The DEFINEDATA method lists those variables to be added to the LOOKUP hash table. The values of these variables can be retrieved using the FIND method. Although not needed here, you may want to include the key variables here as well if they are also to be retrieved.

❻ A DO UNTIL loop is used to cycle through the observations in the primary data set.

❼ An observation is read and its value for CLINNUM is loaded into the PDV. Since CLINNUM is a key variable for the hash object ❹, its value will automatically be used by the FIND method ❾ when retrieving the value of the clinic name.

❽ The temporary variable DONE will be set to 1 when the last observation is read from ADVRPT.DEMOG.

❾ The clinic name, which is being held in the hash table, is retrieved through the use of the FIND method. This method returns its success (0) or failure. When the retrieval is successful, we write out the resulting observation (with the clinic name defined).

❿ Since the SET statement is inside a loop, we should always stop the implied DATA step loop manually.

```
data hashnames(keep=subject clinnum clinname lname fname);
   if _n_ = 1 then do;
      * Define the attributes for variables on lookup table;
      if 0 then set advrpt.clinicnames; ❶

      * Create and load the hash object;
      declare hash lookup(dataset: 'advrpt.clinicnames', ❷
                          hashexp: 8); ❸
      lookup.defineKey('clinnum'); ❹
      lookup.defineData('clinname'); ❺
      lookup.defineDone();
   end;

   * Read the primary data;
   set advrpt.demog(keep=subject clinnum lname fname); ❼
   if lookup.find() = 0 then output hashnames; ❾
run;
```

The DATA step shown above is commonly coded using something similar to the following simpler step that is shown here. The definition and loading of the hash object is done inside a DO block that is executed only once, and the SET statement that reads the ADVRPT.DEMOG data set ❼ is not within a DO UNTIL loop (thus eliminating the need to include a STOP statement ❿). From a performance perspective, it is valuable to understand the difference between these two DATA steps. While the code used in the former step is more complex, it will probably process faster than the code shown here. This performance advantage will be more apparent as the size of the data set ADVRPT.DEMOG increases.

Part 2

Data Summary, Analysis, and Reporting

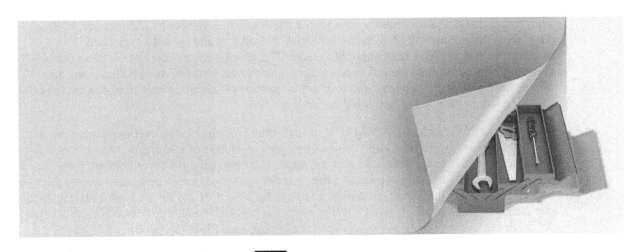

Chapter 7

MEANS and SUMMARY Procedures

While the MEANS and SUMMARY procedures have been a part of Base SAS for a long time (MEANS is an original procedure), and while these procedures are used extensively, many users of these procedures actually take advantage of only a fraction of their capabilities. Primarily this is true because a great deal can be accomplished with fairly simple procedure steps.

With recent enhancements (especially in SAS 8 and SAS®9), a number of additional capabilities have been added to the MEANS and SUMMARY procedures, and even the 'seasoned' programmer may not have been exposed to them. This chapter covers some of the more useful of these capabilities. Because these two procedures have the same capabilities and have very few differences, most of the examples and text in this chapter will highlight only one of them. In each case either of the two procedures could be used.

Prior to SAS 6, MEANS and SUMMARY were distinct procedures with overlapping capabilities. Currently the same software is used behind the scenes regardless of which procedure the user calls; therefore, their capabilities are now the same. The only real differences between these procedures are seen in their defaults, and then primarily in the way each procedure creates printed tables. By default MEANS *always* creates a table to be printed. If you do not want a printed table you must explicitly turn it off (NOPRINT option). On the other hand, the SUMMARY procedure *never* creates a printed table unless it is specifically requested (PRINT option).

SEE ALSO

Carpenter (2008) discusses these two procedures in more detail, including an introduction as well as additional options not covered in this book.

7.1 Using Multiple CLASS Statements and CLASS Statement Options

Although the following discussion concerning the use of multiple CLASS statements and CLASS statement options is within the context of the MEANS and SUMMARY procedures, it can be generalized to most procedures that use the CLASS statement.

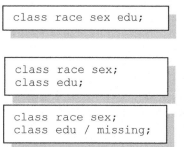

The CLASS statement can be specified as a single statement or it can be broken up into a series of CLASS statements. The order of the CLASS statements determines the overall order of the classification variables.

The CLASS statement now accepts options and for most procedures that accept the CLASS statement, a single class statement can be replaced by a series of CLASS statements. This allows us to control the application of CLASS statement options to specific classification variables. One or more options are specified on a CLASS statement by preceding the option with a slash. While it is not necessary to have multiple CLASS statements just to apply CLASS statement options, multiple CLASS statements allow you to apply these options differentially. For instance when you use the MISSING option on the PROC statement, it is applied to *all* of the classification variables. By using multiple CLASS statements along with the MISSING option on the CLASS statement, you can choose which classification variables are to utilize the MISSING option.

CLASS statement options include:

ASCENDING/DESCENDING (Section 7.1.1)

Analogous to the DESCENDING option in PROC SORT and other procedures, these options allow you to reverse the order of the displayed values.

GROUPINTERNAL and EXCLUSIVE (Section 7.1.2)
You can use these two options to control how formats associated with CLASS variables are to be used when forming groups (see Section 12.1 for the related topic on preloaded formats).

MISSING (Section 7.1.1)
Observations with missing levels of the classification variables are normally excluded from the analysis. This option allows missing values to represent valid levels of the classification variable.

MLF (Section 12.3)
Multilevel formats allow overlapping formatted levels.

ORDER= (Section 7.1.3)
This option allows you to control the order of the classification variables levels. The ORDER= option is discussed in more detail in Section 2.6.1.

PRELOADFMT and EXCLUSIVE (Section 12.1.3)
When formats are preloaded they can be used to establish data filters when forming groups.

The following example performs a simple SUMMARY step and generates the data set STATS. In this step the two classification variables (RACE and EDU) are used to summarize the data for the two analysis variables (HT and WT).

```
title1 '7.1 Single Class Statement';
proc summary data=advrpt.demog;
   class race edu;
   var ht wt;
   output out=stats
          mean= htmean wtmean
          stderr=htse wtse
          ;
   run;
proc print data=stats;
   run;
```

Examination of the partial listing below shows that only 75 (of the potential 77) observations were used in the calculation of the summary statistics. At this point it is not clear why two of the observations were excluded from the analysis.

```
7.1 Single Class Statement

Obs   race   edu   _TYPE_   _FREQ_   htmean   wtmean    htse      wtse

 1             .       0       75    67.6000  161.200  0.40670   3.9272
 2            10       1       11    71.3636  194.091  0.96552   5.7532
 3            12       1       19    67.0526  168.105  0.65102   6.3628
 4            13       1        4    70.0000  197.000  1.15470  10.3923
 5            14       1       10    64.2000  108.400  0.13333   1.4236
 6            15       1        7    65.2857  155.571  0.86504  11.1160
 7            16       1       10    70.4000  165.200  0.54160   6.1946
 8            17       1       10    65.2000  145.200  0.74237   7.9342
 9            18       1        4    69.0000  174.000  2.30940  15.5885
10     1       .       2       42    68.4762  176.143  0.58756   4.0053
11     2       .       2       17    67.6471  162.000  0.76668   8.1633
            .... portions of the output are not shown ....
```

In the examples that follow, this analysis will be repeated using various CLASS statement options.

MORE INFORMATION
The CLASS statement is not the only statement that can be split. Both the VAR statement and the CLASS statement are commonly split to allow the assignment of options in PROC TABULATE, and there is a PROC PRINT example in Section 8.5.2 with multiple VAR statements.

7.1.1 MISSING and DESCENDING Options

It is *very* important to understand that all SAS procedures eliminate an entire observation from the analysis if any one of the classification variables has a missing value. This is true for both explicitly declared classification variables (through the use of the CLASS statement), or implicitly declared classification variables such as those on the TABLES statement in PROC FREQ (which does not use a CLASS statement). Since the entire observation is eliminated, this can affect data summaries that do not even include the offending classification variable. This is a problem that can result in incorrect analyses.

The data table ADVRPT.DEMOG has 77 rows; however, because of missing values in one or both of the classification variables, only 75 observations have been used in the previous summary (Section 7.1). From the LISTING above, or even by inspection of the LOG, it is unclear which classification variable has the missing values.

In the next example, the DESCENDING option is applied to RACE and the MISSING option is applied to the classification variable EDU.

```
proc summary data=advrpt.demog;
    class race/descending; ❶
    class edu/missing; ❷
    var ht wt;
    output out=stats
        mean= htmean wtmean
        stderr=htse wtse
        ;
    run;
```

❶ The groups formed by RACE are now shown in decreasing (DESCENDING) order.

❷ A missing value for the variable EDU will now be considered to be a valid level and will be included in the report. Any observation with a missing value for RACE will still be excluded.

```
7.1.1 Multiple Class Statements
MISSING and DESCENDING Options
```

Obs	race	edu	_TYPE_	_FREQ_	htmean	wtmean	htse	wtse
1		.	0	76	67.6053	162.237	0.40135	4.0115
2		. ❷	1	1	68.0000	240.000	.	.
3		10	1	11	71.3636	194.091	0.96552	5.7532
4		12	1	19	67.0526	168.105	0.65102	6.3628
5		13	1	4	70.0000	197.000	1.15470	10.3923
6		14	1	10	64.2000	108.400	0.13333	1.4236
7		15	1	7	65.2857	155.571	0.86504	11.1160
8		16	1	10	70.4000	165.200	0.54160	6.1946
9		17	1	10	65.2000	145.200	0.74237	7.9342
10	❶	18	1	4	69.0000	174.000	2.30940	15.5885
11	5	.	2	4	66.5000	147.000	0.86603	0.0000
12	4	.	2	4	64.5000	113.500	0.28868	0.8660
13	3	.	2	8	65.0000	112.000	0.65465	4.5826
14	2	.	2	18	67.6667	166.333	0.72310	8.8325
15	1	.	2	42	68.4762	176.143	0.58756	4.0053

. . . . portions of the table are not shown . . . .

The overall number of observations is now 76, and we can see that there is one observation with a missing value of EDU (OBS=2 in the listing). Since there are 77 observations in the data set, there must be an observation with a missing value for RACE as well.

7.1.2 GROUPINTERNAL Option

When a classification variable is associated with a format, that format is used in the formation of the groups. In the next example, the EDULEVEL. format maps the years of education into levels of education.

```
title1 '7.1.2 CLASS Statement Options';
proc format;
   value edulevel ❶
      0-12 = 'High School'
      13-16= 'College'
      17-high='Post Graduate';
   run;

title2 'GROUPINTERNAL not used';
proc summary data=advrpt.demog;
   class edu; ❸
   var ht wt;
   output out=stats
         mean= MeanHT MeanWT
      ;
   format edu edulevel.; ❷
   run;
proc print data=stats;
   run;
```

❶ The EDULEVEL. format maps years of education into three ranges.

❷ In the SUMMARY step the FORMAT statement has been used to create the association between EDU and the EDULEVEL. format.

❸ The MISSING option has not been applied; consequently missing values of EDU will not be included in the summary.

A PROC PRINT LISTING of the resulting data table shows that the SUMMARY procedure has used the format to collapse the individual levels of EDU into the three levels of the formatted classification variable.

```
7.1.2 CLASS Statement Options
GROUPINTERNAL not used

Obs         edu        _TYPE_  _FREQ_   MeanHT    MeanWT

 1              .          0      76    67.5526   160.461
 2      High School        1      30    68.6333   177.633
 3      College            1      32    67.0938   147.438
 4      Post Graduate      1      14    66.2857   153.429
```

To use the original data values (internal values) to form the groups, rather than the formatted

```
class edu/groupinternal;
```

values, the GROUPINTERNAL option is added to the CLASS statement.

```
7.1.2 CLASS Statement Options
Using GROUPINTERNAL

Obs          edu        _TYPE_  _FREQ_   MeanHT    MeanWT

 1               .          0      76    67.5526   160.461
 2      High School         1      11    71.3636   194.091
 3      High School         1      19    67.0526   168.105
 4      College             1       4    70.0000   197.000
 5      College             1      11    64.1818   108.091
 6      College             1       7    65.2857   155.571
 7      College             1      10    70.4000   165.200
 8      Post Graduate       1      10    65.2000   145.200
 9      Post Graduate       1       4    69.0000   174.000
```

Notice that although the original values of EDU are used to form the groups, the formatted values are still displayed. In this example we could have achieved similar results by using the

ORDER=INTERNAL option shown in Section 7.1.3.

7.1.3 Order= Option

When procedures create ordered output, often based on the classification variables, there are several different criteria that can be used to determine the order. The ORDER= option is used to establish the scheme, which establishes the ordering criteria. The ORDER= option can generally appear on the PROC statement where it applies to all the classification variables (implicit or explicit), or as an option on the CLASS statement where it can be applied to selected classification variables.

These schemes include:

- DATA   order is based on the order of the incoming data
- FORMATTED   values are formatted first and then ordered
- FREQ   the order is based on the frequency of the class level
- INTERNAL   same as UNFORMATTED or GROUPINTERNAL

The default ordering is always INTERNAL (whether or not the variable is formatted) except for PROC REPORT. In PROC REPORT, formatted variables have a default order of FORMATTED.

Using the ORDER=FREQ option on the CLASS statement causes the table to be ordered according to the most common levels of education.

```
class edu/order=freq;
```

In this table EDU has been left unformatted. Notice that the order of the rows for EDU is based on the value of _FREQ_.

```
7.1.3 CLASS Statement Options
Using ORDER=FREQ

Obs    edu    _TYPE_    _FREQ_    MeanHT    MeanWT

 1      .       0         76      67.5526   160.461
 2     12       1         19      67.0526   168.105
 3     14       1         11      64.1818   108.091
 4     10       1         11      71.3636   194.091
 5     17       1         10      65.2000   145.200
 6     16       1         10      70.4000   165.200
 7     15       1          7      65.2857   155.571
 8     18       1          4      69.0000   174.000
 9     13       1          4      70.0000   197.000
```

7.2 Letting SAS Name the Output Variables

In each of the examples in Section 7.1, the statistics that are to be calculated and written to the output data set are explicitly specified. However, you do not necessarily have to specify the statistics or provide names for the variables that are to hold the calculated values.

OUTPUT Statement without Specified Statistics

When no statistics are specified on the OUTPUT statement, the resulting data set will contain a specific set of statistics and will be in a different form. Rather than one column per statistic, the statistics will be in a transposed form—one row per statistic. The type of statistic is named in the _STAT_ column.

This SUMMARY step uses the OUTPUT statement with only the OUT= option. No statistics have been requested; consequently, a standard suite of statistics (the same list for the default printed statistics) are calculated and included in the data set. One column (named after the analysis variable) holds the value for each of the statistics noted under the variable _STAT_.

```
title1 '7.2 No Statistics Specified';
proc summary data=advrpt.demog;
   class race;
   var ht;
   output out=stats;
   run;
```

While this form of data set can have its uses, you need to be careful when using it, as the variable HT contains different types of information for each row.

```
7.2 No Statistics Specified

Obs     race    _TYPE_    _FREQ_    _STAT_        ht

 1                 0        76       N         76.0000
 2                 0        76       MIN       62.0000
 3                 0        76       MAX       74.0000
 4                 0        76       MEAN      67.6053
 5                 0        76       STD        3.4989
 6        1        1        42       N         42.0000
 7        1        1        42       MIN       62.0000
 8        1        1        42       MAX       74.0000
 9        1        1        42       MEAN      68.4762
10        1        1        42       STD        3.8078
        . . . . portions of the table are not shown . . . .
```

AUTONAME and AUTOLABEL Options

The OUTPUT statement has always used options to name the summary data set (OUT=), and usually the summary statistics of interest (*e.g.,* MEAN=, N=, MAX=). A second type of option can be placed on the OUTPUT statement. These options follow a slash (/) on the OUTPUT statement and include:

- AUTONAME   allows MEANS and SUMMARY to determine names for the generated variables

- AUTOLABEL   allows MEANS and SUMMARY to supply a label for each generated variable

- LEVELS   adds the _LEVELS_ column to the summary data set (see Section 7.11)

- WAYS   adds the _WAYS_ column to the summary data set (see Section 7.11)

When a statistic is requested on the OUTPUT statement, the variable that it generates has the default name of the corresponding analysis variable in the VAR statement. Since only one statistic can be generated using the default name, there will be a naming conflict if default naming is used when two or more statistics are requested. For this reason the following PROC SUMMARY will fail because of naming conflicts in the new data set STATS. Actually it only partially fails, which is probably worse. An error is produced in the LOG, but a partial table is still produced.

```
proc summary
  data=advrpt.demog;
  class race;
  var ht;
  output out=stats
         n=
         mean=
         stderr=
         ;
  run;
```

The AUTONAME option allows you to select multiple statistics without picking a name for the resulting variables in the OUTPUT table. The generated names are unique and therefore naming conflicts are eliminated. Similarly the

```
output out=stats
       n=
       mean=
       stderr=/autoname
       ;
```

AUTOLABEL option creates a label, which is based on the analysis variable's existing label, for variables added to the OUT= data set.

Conveniently the names of the generated variables are both reasonable and predictable, and are in the form of *variable_statistic*.

```
7.2 Using AUTONAME


                                                      ht_Std
Obs      race      _TYPE_      _FREQ_     ht_N     ht_Mean      Err

 1                    0          76        76      67.6053     0.40135
 2        1           1          42        42      68.4762     0.58756
 3        2           1          18        18      67.6667     0.72310
 4        3           1           8         8      65.0000     0.65465
 5        4           1           4         4      64.5000     0.28868
 6        5           1           4         4      66.5000     0.86603
```

7.3 Statistic Specification on the OUTPUT Statement

When creating variables based on statistics specified on the OUTPUT statement, there are several ways to name the variables and to associate the resultant variable with the original analysis variable.

The most traditional way of specifying the statistics and naming the generated variables is shown

```
var ht wt;
output out=stats
       n    = n_ht n_wt
       mean= mean_HT Mean_WT
       ;
```

to the left. The option specifying the statistic (N= and MEAN= are shown here) is followed by a variable list. This form requires the programmer to make sure that the list of analysis variables and the list of new variables that hold the values of the selected statistics are in the same position and order. The disadvantages of this form include:

- The order of the statistics is tied to the order of the analysis variables.

- A statistic must be generated for each of the analysis variables; that is, in order to calculate a mean for WT you must also calculate a mean for HT (since HT is first on the VAR statement).

```
var ht wt;
output out=stats
       n(wt) = n_wt ❶
       mean(wt ht) = mean_WT Mean_HT ❷
       ;
```

This is not the only, or even necessarily the most practical, way of specifying the statistics and their associated variables. A list of analysis variables can be included in parentheses as a part of the statistic option. This allows you to specify a subset of the analysis variables ❶, as was done here with the N statistic. You can also use this technique to control the order of the usage of the analysis variables ❷.

It is also possible to split up the specification of the statistics of interest. A given statistic can be specified multiple times, each with a different analysis variable. This form of option specification gives you quite a bit of flexibility, not only over which statistics will be calculated for which analysis variables, but also over the order of the generated variables on the resultant data set.

```
var ht wt;
output out=stats
        n(wt)    = n_wt
        mean(wt) = mean_WT
        n(ht)    = n_ht
        mean(ht) = Mean_HT
        ;
```

A PROC PRINT of the data set WORK.STATS generated by the OUTPUT statement to the left shows the variable order. You might also notice that SAS remembers the case of the name of the variable as it is first defined: Mean_HT as opposed to mean_WT.

```
7.3 Splitting the Stat(varlist)=

Obs     race   _TYPE_    _FREQ_   n_wt    mean_WT    n_ht    Mean_HT

 1               0        76       76     162.237     76     67.6053
 2        1      1        42       42     176.143     42     68.4762
 3        2      1        18       18     166.333     18     67.6667
 4        3      1         8        8     112.000      8     65.0000
 5        4      1         4        4     113.500      4     64.5000
 6        5      1         4        4     147.000      4     66.5000
```

It is also possible to specify more than one OUTPUT statement within a given PROC step. Each OUTPUT statement could have a different combination of statistics.

7.4 Identifying the Extremes

When working with data, it is not at all unusual to want to be able to identify the observations that contain the highest or lowest values of the analysis variables. These extreme values are automatically displayed in PROC UNIVARIATE output, but must be requested in the MEANS and SUMMARY procedures.

While the MIN and MAX statistics show the extreme value, they do not identify the observation that contains the extreme. Fortunately there are a couple of ways to identify the observation that contains the MAX or MIN.

7.4.1 Using the MAXID and MINID Options

The MAXID and MINID options in the OUTPUT statement can be used to identify the observations with the maximum and minimum values (the examples in this section are for MAXID; however, MINID has the same syntax). The general form of the option is:

```
MAXID(analysis_var_list(ID_var_list))=new_var_list
```

The MAXID option is used in the following example to identify which subjects had the maximums for each value of any classification variables. This option allows us to add a new variable to the OUTPUT data set, which takes on the value of the ID variable for the maximum observation.

```
title1 '7.4.1a Using MAXID';
title2 'One Analysis Variable';
proc summary data=advrpt.demog;
   class race;
   var ht;
   output out=stats
           mean= meanHT
           max=maxHt  ❶
           maxid(ht(subject))=maxHtSubject  ❷
           ;
   run;
proc print data=stats;
   run;
```

The maximum for the analysis variable HT is requested ❶ for each RACE. We would also like to know which SUBJECT had the maximum HT ❷ (the subject number is to be stored in the variable MAXHTSUBJECT).

Using the same generalized option syntax as was discussed in the previous section, there are several variations of the syntax for the MAXID option shown in this example. In this case there is a single analysis variable and a single ID variable.

When there is more than one analysis variable, the MAXID statement can be expanded following the same syntax rules as were discussed in Section 7.3.

```
7.4.1a Using MAXID
One Analysis Variable


                                      max      maxHt
Obs    race   _TYPE_   _FREQ_  meanHT    Ht    Subject
                                      ❶        ❷
 1              0        76    67.6053   74      209
 2      1       1        42    68.4762   74      209
 3      2       1        18    67.6667   72      201
 4      3       1         8    65.0000   68      215
 5      4       1         4    64.5000   65      244
 6      5       1         4    66.5000   68      212
```

In this example the subject number of the tallest and of the heaviest subjects in the study are to be displayed.

```
var ht wt;
output out=stats
       mean= meanHT MeanWT
       max=maxHt maxWT
       maxid(ht(subject) wt(subject))=maxHtSubject MaxWtSubject
       ;
```

```
7.4.1b Using MAXID
Two Analysis Variables


                                               max   max   maxHt    MaxWt
Obs    race   _TYPE_   _FREQ_  meanHT   MeanWT   Ht    WT  Subject  Subject


 1              0        76    67.6053  162.237  74   240    209      203
 2      1       1        42    68.4762  176.143  74   215    209      208
 3      2       1        18    67.6667  166.333  72   240    201      203
 4      3       1         8    65.0000  112.000  68   133    215      215
 5      4       1         4    64.5000  113.500  65   115    244      230
 6      5       1         4    66.5000  147.000  68   147    212      211
```

Because of the flexibility for structuring options in the OUTPUT statement, the previous MAXID option could also have been written as:

```
var ht wt;
output out=stats
       mean= meanHT MeanWT
       max=maxHt maxWT
       maxid(ht(subject))=maxHtSubject
       maxid(wt(subject))=maxWtSubject
       ;
```

When more than one variable is needed to identify the observation with the extreme value, the MAXID supports a list. As before when specifying lists, there is a one-to-one correspondence between the two lists (the list of ID variables and the list of generated variables). In this OUTPUT statement both the SUBJECT and SSN are used in the list of identification variables. Consequently a new variable is created for each in the summary data set.

```
var ht wt;
output out=stats
       mean= meanHT MeanWT
       max=maxHt maxWT
       maxid(ht(subject ssn))= MaxHtSubject MaxHtSSN
       maxid(wt(subject ssn))= MaxWtSubject MaxWtSSN
       ;
```

7.4.2 Using the IDGROUP Option

The MAXID and MINID options allow you to capture only a single extreme. It is also possible to display a group of the extreme values using the IDGROUP option.

Like the MAXID and MINID options, this option allows you to capture the maximum or minimum value and associated ID variable(s). More importantly, however, you may select more than just the single extreme value.

```
proc summary data=advrpt.demog;
   class race;
   var wt;
   output out=stats
          mean= MeanWT
          max(wt)=maxWT  ❶
          idgroup (max(wt)out[2](subject sex)=maxsubj)
          ;         ❷        ❸          ❹        ❺
   run;
```

In this example the maximum WT has been requested using the MAX statistic ❶. In addition the IDGROUP option has been requested to identify the two ❸ individuals (identified by using SUBJECT and SEX ❹) with the largest values of WT ❷.

❺ The prefix for the variable name that will hold SUBJECT number with the maximum weight is MAXSUBJ. Since there is no corresponding prefix for SEX, the original variable will be used as the prefix. Because we have only one analysis variable, there will not be a naming conflict; however, not specifying the new variable's name is generally to be avoided. Even with the /AUTONAME option in force there can be naming conflicts with only moderately complex IDGROUP options that do not name the new variables.

Since we have requested the top two ❸ values, the values are written to MAXSUBJ_1, MAXSUBJ_2, SEX_1, and SEX_2. Notice that a number indicating the relative position is appended to the variable name. In this example we can see that the second heaviest subject in the study had a subject number of 236 and a SEX of M.

```
7.4.2a Using IDGROUP

                                   max
Obs    race   _TYPE_   _FREQ_   MeanWT   WT   maxsubj_1 maxsubj_2   sex_1 sex_2

 1             0        76      162.237  240     203       236        M     M
 2      1      1        42      176.143  215     208       216        M     M
 3      2      1        18      166.333  240     203       236        M     M
 4      3      1         8      112.000  133     215       256        M     M
 5      4      1         4      113.500  115     230       240        F     F
 6      5      1         4      147.000  147     211       212        M     M
```

The request for the MAX in IDGROUP is actually independent of the MAX= request issued at ❶.

In the previous example we are able to see who the second heaviest subject was, but because we used the MAX option, which shows only one value—the heaviest, we cannot see the weight of the second heaviest individual. This problem disappears with a slight modification of the IDGROUP option.

In the following example we want to identify the two oldest individuals within each group (minimum value of DOB). Since we want to see the date of birth for each of the oldest two individuals, DOB ❻ has been included in the list of ID variables. Notice that the MIN statistic, which would show only one DOB, is not being used at all.

```
proc summary data=advrpt.demog;
   class race;
   var dob;
   output out = stats
           idgroup (min(dob)out[2](dob ❻ subject sex)=
                   MinDOB OldestSubj OldestGender)
        ;
   run;
```

```
7.4.2b Using IDGROUP with the Analysis Variable

                         MinDOB_  MinDOB_  Oldest  Oldest  Oldest    Oldest
Obs   race   _TYPE_  _FREQ_   1        2     Subj_1  Subj_2  Gender_1  Gender_2

 1             0      76    03NOV21  05NOV24   252     269      M         M
 2     1       1      42    03NOV21  05NOV24   252     269      M         M
 3     2       1      18    15JAN34  15AUG34   203     236      M         M
 4     3       1       8    02JUL46  11JUN47   234     268      F         M
 5     4       1       4    13FEB48  28FEB49   230     240      F         F
 6     5       1       4    18FEB51  18JUN51   212     214      M         M
```

SEE ALSO

The IDGROUP option is used to transpose data in King and Zdeb (2010). It is also used in a subsetting question in the SAS Forum thread http://communities.sas.com/message/102002102002.

7.4.3 Using Percentiles to Create Subsets

The percentile statistics can be used to create search bounds for potential outlier boundaries. Several percentile statistics are available including the 1% and 5% bounds. In this example we would like to know if any observations fall outside of the 1% percentiles.

```
title1 '7.4.3 Using Percentiles';
proc summary data=advrpt.lab_chemistry;
   var potassium;
   output out=stats
          p1=   ❶
          p99= /autoname;
   run;

data chkoutlier;
   set stats(keep=potassium_p1 potassium_p99); ❷
   do until (done);
      set advrpt.lab_chemistry ❷
             (keep=subject visit potassium)
         end=done; ❸
      if  potassium_p1 ge potassium
        or potassium ge potassium_p99 ❹
                      then output chkoutlier;
   end; ❺
   run;
options nobyline; ❻
proc print data=chkoutlier;
 by potassium_p1 potassium_p99; ❼
 title2 'Potassium 1% Bounds are #byval1, #byval2'; ❽
 run;
```

❶ The 1st and 99th percentiles are calculated and saved in the data set STATS.

❷ The single observation of WORK.STATS is added to the Program Data Vector (PDV).

❸ The analysis data are read one row at a time in a DO UNTIL loop. The END= option on the SET statement creates the numeric 0/1 variable DONE, which is used to end the loop.

❹ Check to see if the current POTASSIUM reading is above or below the 1st and 99th percentiles.

❺ A STOP statement has not been used. Although the SET statement is inside the DO UNTIL loop, the STOP is not necessary because all observations have been read from the STATS data set.

❻ The NOBYLINE system option removes the BY variable values from the table created by PRINT.

❼ A BY statement is used so that the values can be loaded into the #BYVAL options on the TITLE statement.

❽ Since the bounds are constants, the #BYVAL option is used to place them in the title. Generally TITLE statements are placed outside of the PROC step; however, for better clarity, when I use the #BYVAR and #BYVAL options I like to move the TITLE statement so that it follows the BY statement.

MORE INFORMATION
The TITLE statement option #BYVAL is introduced in Section 15.1.2.

7.5 Understanding the _TYPE_ Variable

One of the variables automatically included in the summary data set is _TYPE_. By default this is a numeric variable, which can be used to help us track the level of summarization, and to distinguish the groups of statistics. It is not, however, intuitively obvious how to predict its value.

```
proc summary
      data=advrpt.demog
         (where=(race in('1','4')
               & 12 le edu le 15
               & symp in('01','02','03')))
      ;
   class race edu symp;
   var ht;
   output out=stats
         mean= meanHT
      ;
   run;
```

In this SUMMARY step there are three variables in the CLASS statement (RACE, EDU, and SYMP).

Examination of a listing of the data set STATS shows that _TYPE_ varies from 0 to 7 (8 distinct values). With the _TYPE_=0 associated with the single row that summarizes across the entire data set (all three classification variables are ignored), and with _TYPE_=7 summarizing the interaction of all three classification variables (all three classification variables are used). The remaining values of _TYPE_ represent other combinations of classification variables and vary according to which are used and which are ignored.

```
7.5 Understanding _TYPE_
```

Obs	race	edu	symp	_TYPE_	_FREQ_	mean HT
1			.	0	8	66.25
2			01	1	2	64.00
3			02	1	4	66.50
4			03	1	2	68.00
5		12		2	4	67.50
6		14		2	2	64.00
7		15		2	2	66.00
8		12	02	3	2	67.00
9		12	03	3	2	68.00
10		14	01	3	2	64.00
11		15	02	3	2	66.00
12	1	.		4	6	67.00
13	4	.		4	2	64.00
14	1	.	02	5	4	66.50
15	1	.	03	5	2	68.00
16	4	.	01	5	2	64.00
17	1	12		6	4	67.50
18	1	15		6	2	66.00
19	4	14		6	2	64.00
20	1	12	02	7	2	67.00
21	1	12	03	7	2	68.00
22	1	15	02	7	2	66.00
23	4	14	01	7	2	64.00

The following table summarizes the eight possible combinations of these three classification variables for the LISTING shown above. Under the classification variables, a 0 indicates that levels of the classification variable are being ignored when calculating the summary statistics, while a 1 indicates that the classification variables is being used. When considered together, these three zeros and ones (representing each classification variable) form a 3-digit binary number (one digit for each of the three classification variables. When this binary value is converted to decimal, the result yields _TYPE_.

	CLASS VARIABLES				
Observation Number	**RACE**	**EDU**	**SYMP**	**Binary Value**	**_TYPE_**
1	0	0	0	000	0
2 - 4	0	0	1	001	1
5 - 7	0	1	0	010	2
8 - 11	0	1	1	011	3
12 - 13	1	0	0	100	4
14 - 16	1	0	1	101	5
17 - 19	1	1	0	110	6
20 - 23	1	1	1	111	7
	$2^2=4$	$2^1=2$	$2^0=1$		

The conversion of a binary number to decimal involves the use of powers of 2. A binary value of $110 = 1*2^2 + 1*2^1 + 0*2^0 = 1*4 + 1*2 + 0*1 = 6 = $ _TYPE_.

The NWAY option limits the output data set to the highest order interaction and consequently only the highest value of _TYPE_ would be displayed.

MORE INFORMATION
Interestingly enough, some SAS programmers find converting a binary number to a decimal number to be inconvenient. The CHARTYPE option (see Section 7.6) makes that conversion unnecessary.

7.6 Using the CHARTYPE Option

The CHARTYPE option displays _TYPE_ as a character binary value rather than the decimal value. The following example repeats the example shown in Section 7.5, while adding the CHARTYPE option on the PROC statement.

```
proc summary
     data=advrpt.demog
        (where=(race in('1','4')
              & 12 le edu le 15
              & symp in('01','02','03')))
     chartype;
  class race edu symp;
  var ht;
  output out=stats
        mean= meanHT
           ;
  run;
```

Instead of being numeric, _TYPE_ is now created as a character variable with a length corresponding to the number of classification variables.

```
7.6 Using the CHARTYPE Option
```

Obs	race	edu	symp	_TYPE_	_FREQ_	mean HT
1		.		000	8	66.25
2		.	01	001	2	64.00
3		.	02	001	4	66.50
4		.	03	001	2	68.00
5		12		010	4	67.50
6		14		010	2	64.00
7		15		010	2	66.00
8		12	02	011	2	67.00
9		12	03	011	2	68.00
10		14	01	011	2	64.00
11		15	02	011	2	66.00
12	1	.		100	6	67.00
13	4	.		100	2	64.00
14	1	.	02	101	4	66.50
15	1	.	03	101	2	68.00
16	4	.	01	101	2	64.00
17	1	12		110	4	67.50
18	1	15		110	2	66.00
19	4	14		110	2	64.00
20	1	12	02	111	2	67.00
21	1	12	03	111	2	68.00
22	1	15	02	111	2	66.00
23	4	14	01	111	2	64.00

7.7 Controlling Summary Subsets Using the WAYS Statement

When you do not need to calculate all possible combinations of the classification variables, you can save not only the resources used in calculating the unneeded values, but the effort of eliminating them later as well. There are several ways that you can specify which combinations are of interest. The WAYS statement can be used to specify the number of classification variables to utilize.

Combinations of the WAYS statement for three classification variables include the following summarizations:

- `ways 0;`   across all classification variables
- `ways 1;`   each classification variable individually (no cross products)
- `ways 2;`   each two-way combination of the classification variables (two-way interactions)
- `ways 3;`   three-way interaction. For three classification variables, this is the same as using the NWAY option
- `ways 0,3;`   lists of numbers are acceptable

When the number of classification variables becomes large the WAYS statement can utilize an incremental list much like an iterative DO.

```
ways 0 to 9 by 3;
```

In the following example main effect summaries and the three-way interaction are eliminated; as a matter of fact, they are not even calculated.

```
proc summary data=advrpt.demog
        (where=(race in('1','4')
              & 12 le edu le 15
              & symp in('01','02','03')));
   class race edu symp;
   var ht;
   ways 0,2;
   output out=stats
          mean= meanHT
        ;
   run;
```

The WAYS statement has been used to request calculation of only the overall summary and the two-way interactions.

Notice in the listing shown below that _TYPE_ does not take on the values of 1 or 2. These would be the main effect summaries for SYMP and EDU, respectively. A full examination of the table shows that _TYPE_ appropriately only takes on the values of 0, 3, 5, and 6.

```
7.7 Using the WAYS Statement

                                                   mean
Obs     race     edu     symp     _TYPE_    _FREQ_   HT

 1        .                          0         8    66.25
 2       12       02                 3         2    67.00
 3       12       03                 3         2    68.00
 4       14       01                 3         2    64.00
 5       15       02                 3         2    66.00
 6        1        .       02        5         4    66.50
 7        1        .       03        5         2    68.00
 8        4        .       01        5         2    64.00
 9        1       12                 6         4    67.50
10        1       15                 6         2    66.00
11        4       14                 6         2    64.00
```

7.8 Controlling Summary Subsets Using the TYPES Statement

Like the WAYS statement, the TYPES statement can be used to select and limit the data roll-up summaries. As an added bonus, the TYPES statement eliminates much of the need to understand and to be able to use the _TYPE_ automatic variable. While the WAYS statement (see Section 7.7) lists which levels of summarization are desired, TYPES designates specific summarization levels (effects and interactions).

```
proc summary data=advrpt.demog
        (where=(race in('1','4')
            & 12 le edu le 15
            & symp in('01','02','03')));
class race edu symp;
var ht;
types edu race*symp;
output out=stats
        mean= meanHT
        ;
run;
```

The TYPES statement used here explicitly requests that statistics be calculated only for the main effect for EDU, and the interaction between RACE and SYMP. None of the other effects or summarizations will even be calculated.

```
7.8 Using the TYPES Statement

                                                   mean
Obs     race     edu     symp     _TYPE_    _FREQ_   HT

 1               12                 2         4    67.5
 2               14                 2         2    64.0
 3               15                 2         2    66.0
 4        1        .       02        5         4    66.5
 5        1        .       03        5         2    68.0
 6        4        .       01        5         2    64.0
```

For the following CLASS statement:

```
class race edu symp;
```

Variations of the TYPES statement could also include:

- `types ();`   overall summary
- `types race*edu edu*symp;` two two-way interactions
- `types race*(edu symp);`   two two-way interactions
- `types race*edu*symp;`   three-way interaction—same as NWAY

7.9 Controlling Subsets Using the CLASSDATA= and EXCLUSIVE Options

While the WAYS and TYPES statements control the combinations of classification variables that are to be summarized, you can also specify which levels of the classification variables are to appear in the report or output data set by creating a data set that contains the combinations and levels of interest. The data set can even include levels of classification variables that do not exist in the data itself, but that nonetheless are to appear in the data set or report.

```
data selectlevels(keep=race edu symp);
   set advrpt.demog
       (where=(race in('1','4')
              & 12 le edu le 15
              & symp in('01','02','03')));
   output;
* For fun add some nonexistent levels;
if _n_=1 then do;
   edu=0;
   race='0';
   symp='00';
   output;
   end;
run;
```

This DATA step builds a data set that will be used with the CLASSDATA= option. As an illustration, it also adds a level for each classification variable that does not exist in the data.

```
Show the SELECTLEVELS Data

Obs     race      edu      symp

 1        0         0       00
 2        1        12       02
 3        1        12       03
 4        1        15       02
 5        4        14       01
```

The data set specified with the CLASSDATA option becomes a sophisticated filter for the data entering into the analysis.

The CLASSDATA option can be paired with the EXCLUSIVE option to radically change the observations that are available to the procedure. When the EXCLUSIVE option is not used, all

```
proc summary data=advrpt.demog
            classdata=selectlevels;
   class race edu symp;
   var ht;
   output out=stats mean= meanHT;
   run;
```

levels of the classification variables that exist either in the analysis data or in the CLASSDATA= data set are included in the summary data set. Since we specifically included a level for each of the classification variables that are not in the data, we should expect to see them

summarized in the summary data set.

```
CLASSDATA without EXCLUSIVE

Obs    race    edu    symp    _TYPE_    _FREQ_    meanHT

 1       .               0         63       67.2381
 2       .      00        1          0          .
 3       .      01        1          4       67.5000
 4       .      02        1         10       66.8000
          . . . . portions of the table are not shown . . . .
```

In the data that is being summarized the variable SYMP never takes on the value of '00', but since it is a value of SYMP in the CLASSDATA= data set it appears in the summary data.

When the EXCLUSIVE option is paired with the CLASSDATA= option the makeup of the summary data set can be altered dramatically. The EXCLUSIVE option forces only those levels that are in the CLASSDATA= data set to appear in the summary report. This includes the levels of the classification variables that do not appear in the data set.

```
proc summary data=advrpt.demog
            classdata=selectlevels
            exclusive;
   class race edu symp;
   var ht;
   output out=stats mean= meanHT;
   run;
```

```
7.9 Using the CLASSDATA and EXCLUSIVE Options

                                                 mean
Obs    race    edu    symp    _TYPE_    _FREQ_    HT

 1       .               0          8       66.25
 2       .      00        1          0          .
 3       .      01        1          2       64.00
 4       .      02        1          4       66.50
 5       .      03        1          2       68.00
 6       0               2          0          .
 7      12               2          4       67.50
 8      14               2          2       64.00
 9      15               2          2       66.00
          . . . . portions of the table are not shown . . . .
```

The summary lines for observations 2 and 6 represent levels of the classification variables that do not appear in the data. They were generated through a combination of the CLASSDATA= data set and the EXCLUSIVE option.

Through the use of these two options we have the capability of creating a sophisticated filter for the classification variables. This combination not only gives us the ability to remove levels, but to add them as well.

The ability to add levels at run time without altering the analysis data set has some potentially huge advantages. First, we can modify the filter by changing the CLASSDATA= data set without changing the program that utilizes the data set. Second, we do not need to 'sparse' the data (see Section 2.5 for other sparsing techniques) prior to the analysis, thus increasing the program's efficiency.

MORE INFORMATION
The CLASSDATA= and EXCLUSIVE options are also available in the TABULATE procedure (see Section 8.1.4).

7.10 Using the COMPLETETYPES Option

All combinations of the classification variables may not exist in the data and therefore those combinations will not appear in the summary table. If all possible combinations are desired, regardless as to whether or not they exist in the data, you can use the COMPLETETYPES option on the PROC statement.

```
proc summary data=advrpt.demog
              (where=(race in('1','4')
                    & 12 le edu le 15
                    & symp in('01','02','03')))
            completetypes;
   class race edu symp;
   var ht;
   output out=stats mean= meanHT;
   run;
```

In the data (ADVRPT.DEMOG) there are no observations with both EDU=12 and SYMP='01'; however, since both levels exist somewhere in the data (individually or in combination with another classification variable), the COMPLETETYPES option causes the combination to appear in the summary data set (obs=8).

```
7.10 Using the COMPLETETYPES Option

                                                  mean
 Obs    race    edu    symp    _TYPE_    _FREQ_    HT

   1            .                  0        8      66.25
   2            .      01          1        2      64.00
   3            .      02          1        4      66.50
   4            .      03          1        2      68.00
   5            12                 2        4      67.50
   6            14                 2        2      64.00
   7            15                 2        2      66.00
   8            12     01          3        0       .
   9            12     02          3        2      67.00
  10            12     03          3        2      68.00
  11            14     01          3        2      64.00
  12            14     02          3        0       .
  13            14     03          3        0       .
  14            15     01          3        0       .
  15            15     02          3        2      66.00
         . . . . portions of the table are not shown . . . .
```

MORE INFORMATION
COMPLETETYPES is also used to create sparsed data in Section 2.5.3.

The procedures REPORT and TABULATE also have the ability to display non-existent combinations. See Section 8.1.4 for a TABULATE example. Preloaded formats can also be used to similar advantage, see Section 12.1 for examples with the MEANS, SUMMARY, TABULATE, and REPORT procedures.

7.11 Identifying Summary Subsets Using the LEVELS and WAYS Options

LEVELS and WAYS are options that can be used on the OUTPUT statement. They add the variables _LEVEL_ and _WAY_, respectively to the generated data table. Together or individually these variables can be used to help navigate the summary data set.

```
proc summary data=advrpt.demog;
   class race edu;
   var ht;
   output out=stats
      mean= meanHT /levels ❶
                    ways; ❷
   run;
```

The LEVELS option ❶ adds the variable _LEVEL_ to the OUT= data table. This numeric variable contains a sequential counter of rows within a given value of _TYPE_. This can be useful when working with rows within _TYPE_. Not only does the combination of _TYPE_ and _LEVEL_ form a unique sorted key for the new data set but, for further subsetting and subsequent summarization, when FIRST._TYPE_ is true, _LEVEL_ will necessarily equal 1.

The WAYS option ❷ adds the variable _WAY_ to the OUT= data table. This numeric variable equals the number of classification variables that were used to calculate each observation. A two-way interaction between two classification variables will have _WAY_=2.

```
7.11 Using the LEVELS and WAYS Options

Obs     race     edu     _WAY_     _TYPE_     _LEVEL_     _FREQ_     meanHT
                           ❷                     ❶

 1                 .        0          0          1          75       67.6000
 2                10        1          1          1          11       71.3636
 3                12        1          1          2          19       67.0526
 4                13        1          1          3           4       70.0000
 5                14        1          1          4          10       64.2000
 6                15        1          1          5           7       65.2857
 7                16        1          1          6          10       70.4000
 8                17        1          1          7          10       65.2000
 9                18        1          1          8           4       69.0000
10       1         .        1          2          1          42       68.4762
11       2         .        1          2          2          17       67.6471
12       3         .        1          2          3           8       65.0000
13       4         .        1          2          4           4       64.5000
14       5         .        1          2          5           4       66.5000
15       1        10        2          3          1          11       71.3636
16       1        12        2          3          2          16       67.2500
17       1        13        2          3          3           4       70.0000
18       1        15        2          3          4           5       64.2000
19       1        16        2          3          5           2       71.0000
20       1        17        2          3          6           2       63.0000
           .... portions of the table are not shown ....
```

7.12 CLASS Statement vs. BY Statement

Although the CLASS and BY statements will often produce similar results, the user should be aware of the differences, not only in performance, but in function as well, for these two statements.

In terms of general operation the BY statement requires the incoming data to be sorted. Given that the data is sorted, the data is processed in BY groups—one group at a time. This requires less memory and processing resources than when the CLASS statement is used. However, when the data is not already sorted, the sorting of the data itself will generally outweigh the performance advantages of the BY statement.

When the CLASS statement is used, it is possible to calculate any of the possible interactions among the classification variables. This is not possible when using BY group processing. We can examine statistics within each unique combination of BY variables, but not across BY variables.

When a classification variable takes on a missing value, the entire observation is removed from the analysis (see Section 7.1.1 for the use of the MISSING option to change this behavior). Missing levels of the BY variables are considered valid levels and are not eliminated.

Since the MEANS and SUMMARY procedures allow for multi-threaded processing, if you execute SAS on a server or a machine with multiple CPUs you may see a performance difference in the use of BY vs. CLASS statements. The procedure will take advantage of multi-threading for both types of summarizations; however, the internals are not necessarily the same. You may want to experiment a bit on your system.

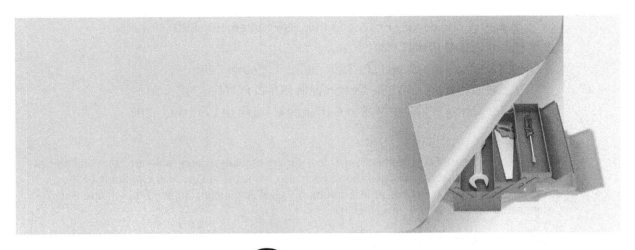

Chapter 8

Other Reporting and Analysis Procedures

A number of Base SAS procedures provide a variety of analysis and summarization techniques. Although some have similar capabilities, each also has some unique features. Some of these features rely on newer options or less commonly used statements. Some of these options and statements are discussed in this chapter.

MORE INFORMATION

The MEANS and SUMMARY procedures are discussed in Chapter 7.

SEE ALSO

Cynthia Zender (2008) discusses a number of techniques for the generation of complex reports.

8.1 Expanding PROC TABULATE

PROC TABULATE has been confounding new users of the procedure for a number of years. Actually it is not just new users, but any user who is new to TABULATE. For the most part, this is because the TABLE statement, which is the procedure's primary statement, is constructed differently than any other procedure statement. Understanding the structure of the TABLE statement is the key to successfully writing a TABULATE step. Fortunately the building blocks that form the primary syntax structure of the TABLE statement are not that difficult to master. Once the fundamentals are understood, the more complex topics can be tackled more successfully.

SEE ALSO

The definitive go-to reference for this procedure is Lauren Haworth's 1999 book *PROC TABULATE by Example.* Also Dianne Rhodes (2005) provides a very crisp explanation of the origins of TABULATE and the relationships among the various elements of the TABLE statement. Carpenter (2010a) introduces not only the beginning elements of TABULATE, but also discusses a number of advanced techniques that are not covered in this book.

8.1.1 What You Need to Know to Get Started

Like most procedures, PROC TABULATE has a number of statements that define how the procedure is to summarize the data. Of these statements, virtually every TABULATE step will have the following three:

- CLASS   variables used to form groups within either rows or columns
- VAR   numeric variables that are to be summarized
- TABLE   table definition

The TABLE statement is the heart of the TABULATE step. It is the complexity of the TABLE statement that tends to thwart the user who is new to the procedure. The key to its use is to remember that it has parts (dimensions) and definitions within those parts. Break it down a piece at a time and it should make more sense.

The first and primary building blocks of the TABLE statement are the table *dimensions*. The table(s) generated by TABULATE can have up to three comma-separated dimensions to their definition: *page*, *row*, and *column*. These dimensions always appear in page, row, column order:

- page   defines how the individual pages are formed (used less often)
- row   defines the rows of the table within each page (almost always present)
- column   defines the columns within rows and pages (always present)

You will always have at least a column dimension and you cannot have a page dimension without also having both row and column dimensions. The general makeup of the TABLE statement therefore looks something like the following. It is *very* important to notice that the three dimensions are comma separated. This is the only time that commas are used in the TABLE statement; the commas separate these three dimensions (definition parts).

```
table page, row, column;
```

Generally you will want your entire table on one page; it's easier to read, so there will not be a page dimension and your TABLE statement looks like:

```
table row, column;
```

To build the individual page, row, and column dimensions, you will use a combination of option and element phrasing. The three types of phrases are:

- singular   used when a single element is needed
- concatenated   multiple elements are joined using a space
- nested   one element is nested within another using an asterisk

There are several symbols or operators that are commonly used to work with these various elements. These include the following:

Operator	What It Does
space	Forms concatenations
*	Nests elements—forms hierarchies
( )	Forms groups of elements
'*text*'	Adds text
F=	Assigns a format

Singular Elements

A singular element has, as the name implies, a single variable. In the following table statement there is a single classification variable (RACE) in the row dimension and a single analysis variable (WT) in the column dimension.

```
ods pdf file="&path\results\E8_1_1a.pdf"
        style=journal;
title1 '8.1.1a Proc Tabulate Introduction';
title2 'Singular Table';

proc tabulate data=advrpt.demog;
   class race;
   var wt;
   table race,wt;
   run;
ods pdf close;
```

Since RACE is a classification variable, the resulting table will have a single row for each unique value of RACE.

8.1.1a Proc Tabulate Introduction Singular Table

	weight in pounds
	Sum
race	
1	7398.00
2	2994.00
3	896.00
4	454.00
5	588.00

The analysis variable, WT, is specified in the VAR statement, and a single column, with a heading showing the variable's label, will be generated for the statistic based on WT.

Since no statistic was specifically requested, the default statistic (SUM) is displayed.

Concatenated Elements

Concatenated tables allow us to easily combine multiple elements within columns and/or rows. A concatenated definition is formed when two or more space separated elements are included in the same dimension.

```
proc tabulate data=advrpt.demog;
   class race sex;
   var ht wt;
   table sex race,wt ht;
   run;
```

This example augments the table from the previous example (8.1.1a) by adding a second classification variable and a second analysis variable. The label associated with each analysis variable is by default used in the column header.

The analysis and classification variables can be used in page, row, or column dimensions.

8.1.1b Proc Tabulate Introduction Concatenated Table

	weight in pounds	height in inches
	Sum	Sum
patient sex		
F	4481.00	2017.00
M	7849.00	3121.00
race		
1	7398.00	2876.00
2	2994.00	1218.00
3	896.00	520.00
4	454.00	258.00
5	588.00	266.00

Nested Elements

Nested definitions allow us to create tables within tables.
The nested elements can be classification variables, analysis variables, statistics, options, and modifiers; and are designated as nested elements through the use of the asterisk.

In this TABLE statement, the row dimension is singular (RACE), while the column dimension has the analysis variable (WT) nested within a classification variable (SEX).

```
proc tabulate data=advrpt.demog;
   class race sex;
   var wt;
   table race,sex*wt*(n mean);
   run;
```

8.1.1c Proc Tabulate Introduction Nested Table

	patient sex			
	F		M	
	weight in pounds		weight in pounds	
	N	Mean	N	Mean
race				
1	14	159.00	28	184.71
2	10	148.60	8	188.50
3	3	105.00	5	116.20
4	4	113.50	.	.
5	.	.	4	147.00

Notice also that two space-separated statistics are concatenated into a group with parentheses, and then the group is nested under the analysis variable WT, which, as was mentioned, is nested within SEX.

Combinations of Elements

In most practical uses of TABULATE, the TABLE statement will contain a combination of nested and concatenated elements. These will include not only variables and statistics, but options as well. The TABULATE procedure is rich in options, and once you have started to build simple tables such as those shown above, you would be well advised to seek out more complete references to the procedure.

The following example contains additional options, and demonstrates a few of the more complex techniques that are commonly used with many of the tables generated by TABULATE.

```
proc tabulate data=advrpt.demog format=8.3 ; ❶
   class race sex ;
   var wt;
   table sex ❷ all='Across Gender' race all,
        wt*(n*f=2.0 mean*f=7.1 var median*f=6.0) ❸
        / box='Syngen Protocol'; ❹
   keylabel mean = 'Average' ❺
           var  = 'Variance';
   run;
```

❶ An overall format is designated for the analysis cells in the table. This default format can be overwritten by associating individual formats with each statistic ❸. Because the other statistics have specific formats, this default format is applied only to the VARIANCE in this table.

❷ The table definition has two concatenated elements in the row dimension. The ALL keyword summarizes across the associated element. Here it creates a row that summarizes across all values of SEX. A text label can be assigned to an element by using the equal sign. Without the text label the word 'All' appears in the table (as it does for RACE).

❸ Formats can be associated with specific variables and statistics by nesting the F= option under the desired element.

❹ There are a number of options that can be applied on the TABLE statement (following the /) for the table as a whole. The BOX= option adds text in the upper left corner of the table.

8.1.1d Proc Tabulate Introduction Demonstration of a Few Other Options

Syngen Protocol	weight in pounds			
	N	Average	Variance	Median
patient sex				
F	31	144.5	1027.589	155
M	45	174.4	1011.749	177
Across Gender	76	162.2	1223.010	161
race				
1	42	176.1	673.784	177
2	18	166.3	1404.235	158
3	8	112.0	168.000	105
4	4	113.5	3.000	114
5	4	147.0	0.000	147
All	76	162.2	1223.010	161

❺ The KEYLABEL statement allows you to assign a text label to statistics and to the keyword ALL.

In example 8.1.1c there were no males for RACE 4 nor were there any females for RACE 5. This is reflected in missing values for the N and MEAN. Missing values can be replaced with the MISSTEXT= option ❻.

```
proc tabulate data=advrpt.demog;
   class race sex;
   var wt;
   table race,
         sex*wt='Pounds'*(n mean)
         / misstext='0'  ❻;
   run;
```

Notice that each of the missing values has been replaced by a zero (we could have selected other text, such as an asterisk). In this example a zero for the N is appropriate; however, the mean really is not zero. We need a way to indicate that it is not a calculable value.

8.1.1e Proc Tabulate Introduction
MISSTEXT - Replace Missing with 0

	patient sex			
	F		M	
	Pounds		Pounds	
	N	Mean	N	Mean
race				
1	14	159.00	28	184.71
2	10	148.60	8	188.50
3	3	105.00	5	116.20
4	4	113.50	0	❻ 0
5	0	❻ 0	4	147.00

Fortunately a user-defined format can be used to provide the reader with the necessary cues.

```
proc format;
   value mzero
      .='----'  ❼
      other=[6.2];
   run;

proc tabulate data=advrpt.demog;
   class race sex;
   var wt;
   table race,                        ❽
         sex='Gender'*wt=' '*(n mean*f=mzero.)
         /box='Weight in Pounds'
          misstext='0';  ❾
   run;
```

8.1.1f Proc Tabulate Introduction
Replace Missing with a Format

Weight in Pounds	Gender			
	F		M	
	N	Mean	N	Mean
race				
1	14	159.00	28	184.71
2	10	148.60	8	188.50
3	3	105.00	5	116.20
4	4	113.50	0	----
5	0	----	4	147.00

❼ The MZERO. format will translate a missing value into four dashes.

❽ The MZERO. format is associated with the mean.

❾ Since the format is applied before the MISSTEXT option, we can still use MISSTEXT=0 to replace the missing value for N.

8.1.2 Calculating Percentages Using PROC TABULATE

Because of the need to determine the denominator, the calculation of percentages in the TABULATE procedure can be problematic. Although there are situations where the determination of the denominator has to be done outside of the TABULATE step, the procedure does offer a number of tools that make this necessity less common.

PCTN and PCTSUM Options

The PCTN and PCTSUM options request the calculation of percentages based on the denominator specified using angle brackets. PCTN bases the percentages on counts (N), while PCTSUM bases the percentages on the total of an analysis variable.

The following example requests percentages based on counts. An analysis variable (VAR statement) is not needed in this step since the percentages are based on counts and no other statistics are requested.

```
proc tabulate data=advrpt.demog;
   class race edu;
   table (race all)*pctn<edu>='%'  ❶,
         edu;  ❷
   run;
```

❶ Within each value of RACE, calculate the percentage of observations for each value of EDU. Since PCTN is nested within RACE, the denominator <EDU> is the total count for that value of RACE.

❷ The column dimension is based on the classification variable EDU. There is no analysis variable; therefore, the count is converted to a percent.

Although the determination of the denominator is straightforward in this example, it is often more complex. The procedure's documentation and Haworth (1999) show more complex examples.

8.1.2a Proc Tabulate Percentages Using Angle Braces

| | | \multicolumn{8}{c|}{years of education} | | | | | | | |
race		10	12	13	14	15	16	17	18
1	%	26.19	38.10	9.52	.	11.90	4.76	4.76	4.76
2	%	.	17.65	.	.	.	35.29	47.06	.
3	%	.	.	.	75.00	25.00	.	.	.
4	%	.	.	.	100.00	.	.	.	.
5	%	.	.	.	.	.	50.00	.	50.00
All	%	14.67	25.33	5.33	13.33	9.33	13.33	13.33	5.33

Percentage Generation Statistics

Sometimes it can be difficult to obtain the correct denominator by using the angle brackets. Fortunately there are also several percentage generation statistics. For each of these statistics, the denominator (which can be based on the report, the page, the row, or the column) is predetermined.

Percentage applies to:	Percent Frequency (N)	Percent Total (SUM)
Report	reppctn	reppctsum
Page	pagepctn	pagepctsum
Column	colpctn	colpctsum
Row	rowpctn	rowpctsum

In the following example the percentages are for the columns rather than rows. The displayed

```
proc tabulate data=advrpt.demog;
   class race;
   var wt;
   table race all,
        wt*(n colpctn mean colpctsum);
   run;
```

percentages are calculated using both the N (COLPCTN) and the total WT (COLPCTSUM).

8.1.2b Proc Tabulate Percentages Column Percents

		weight in pounds		
	N	ColPctN	Mean	ColPctSum
race				
1	42	55.26	176.14	60.00
2	18	23.68	166.33	24.28
3	8	10.53	112.00	7.27
4	4	5.26	113.50	3.68
5	4	5.26	147.00	4.77
All	76	100.00	162.24	100.00

The following example summarizes survey data. Here the response variable (RESP) takes on the values of 0 or 1 (no or yes).

```
proc tabulate data=survey;
   class question;
   var resp;
   table question,
        resp='responses'*(n='total responders' *f= comma7. ❸
                          sum='total yes' *f= comma7. ❸
                          pctsum='response rate for this question'*f=5.1
                          pctn='rate of Yes over whole survey' *f= 5.
                          mean='mean Q resp' * f=percent7.1
                          ) /rts=40;
   run;
```

8.1.2c Proc Tabulate Percentages Using PCTSUM and PCTN

	responses				
	total responders	total yes	response rate for this question	rate of Yes over whole survey	mean Q resp
question					
1	10	7	26.9	22	70.0%
2	8	4	15.4	18	50.0%
3	9	6	23.1	20	66.7%
4	7	2	7.7	16	28.6%
5	11	7	26.9	24	63.6%

Notice that unlike the first example the denominator for PCTSUM and PCTN has not been specified. In this TABULATE step, the assumed denominator will be across the whole report.

❸ The COMMA7. format has been applied to these two statistics. For the LISTING destination, the width of the format will be taken into consideration when forming the width of the column. For other destinations, such as PDF (style=minimal) which is shown here, the format width is used only in the display of the number itself and will have no affect on the column width.

SEE ALSO

The survey example is discussed with alternative coding structures in the SAS Forum thread at http://communities.sas.com/message/42094.

8.1.3 Using the STYLE= Option with PROC TABULATE

The TABULATE procedure is one of three procedures that accept the STYLE override option. Its use in TABULATE is similar, but not the same as its use in the PRINT (see Section 8.5.2) and REPORT (see Section 8.4.6) procedures. This option allows the user to control how various aspects of the table are to appear by overriding the ODS style attributes.

Styles can be applied to a number of areas within the table from general overall attributes, down to the attributes of a specific cell. These areas include:

	Table Area	STYLE= Used on
❶	Box Cell	BOX= option
❷	Class Heading	CLASS statement
❸	Class Levels	CLASSLEV statement
❹	Analysis Variable Headings	VAR statement
❺	Statistics Headings (keywords)	KEYWORD statement
❻	Value Cells	PROC and TABLE statements
❻	Individual Cells	PROC and TABLE statements

8.1.3a TABULATE Using the Journal Style

	height in inches ❹				weight in pounds			
❶	N	Min.❺ Median	Max	N	Min	Median	Max	
race ❷								
1	42	62	69.0	74	42	109	177.0	215
2	18	62	67.0	72	18	98	158.0	240
3 ❸	8	64	64.0	68 ❻	8	105	105.0	133
4	4	64	64.5	65	4	112	113.5	115
5	4	65	66.5	68	4	147	147.0	147

To the left is a fairly typical TABULATE table. The callout numbers on the table correspond to the callout descriptions above.

The following code was used to generate this example table. Notice that the RTS= option applies only to the LISTING destination. The ODS statements are not shown here, but are included in the sample code for this book. See http://support.sas.com/authors.

```
proc tabulate data=advrpt.demog;
   class race;
   var ht wt;
   table race,
         (ht wt)*(n*f=2. min*f=4. median*f=7.1 max*f=4.)
         /rts=6;
   run;
```

The STYLE= option can be used to control virtually all of the same attributes that can be set by the ODS style. Some of these attributes can be dependent on the ODS destination, OS, or printer; however, the most commonly used attributes are generally available. Some of these common attributes include:

Controls	Attribute	Possible Values
Font	font_face=	times, courier, other fonts supported by the OS
Text size	font_size=	6, 8, 10 (sizes appropriate to the font)
Text style	font_style=	italic, roman
Text density	font_weight=	bold, medium
Text width	font_width=	narrow, wide
Foreground color	foreground=	color (color printers or displays)
Background color	background=	color (color printers or displays)

The STYLE= option uses either curly braces or square brackets to contain the list of attributes and their values. This step demonstrates the use of the STYLE override in a variety of statements. The callout numbers refer back to the previous table, as well as to the code that follows.

```
proc tabulate data=advrpt.demog;
   class race / style={font_style=roman};  ❷
   classlev race / style={just=center};  ❸
   var ht wt  / style={font_weight=bold  ❹
                       font_size=4};
   table race='(encoded)',
         (ht wt)*(n*f=2.*{style={font_weight=bold  ❻
                                 font_face='times new roman'}}
                 min*f=4. median*f=7.1 max*f=4.)
         /rts=6
          box={label='Race'  ❶
               style={background=grayee}};
   keyword n / style={font_weight=bold};  ❺
   run;
```

❶ The background color and a label of the RTS box are changed. Notice that the label has been removed from RACE in the TABLE statement and placed in the box using the LABEL= option.

❷ The heading for RACE is to be written without italic (the default). For the JOURNAL style, which is used in this example, italic is the default for the heading; consequently, this option has no effect. For other styles, such as PRINTER, italic is not the default and this style override would make a difference.

❸ The labels of the individual levels of RACE are centered. The STYLE= option on the CLASSLEV statement applies to the individual levels.

Race	height in inches❹ weight in pounds							
❶ ❷	N ❺Min	Median	Max	N	Min	Median	Max
1	42 62	69.0	74	42	109	177.0	215
2	18 62	67.0	72	18	98	158.0	240
❸3	8 64	64.0	68	8	105	105.0	133
	❻						
4	4 64	64.5	65	4	112	113.5	115
5	4 65	66.5	68	4	147	147.0	147

8.1.3b TABULATE Using the Journal Style Various STYLE= Attributes Have Been Changed

❹ On the VAR statement the STYLE= option changes the attributes associated with the variable headings.

❺ Adjust the label for the N statistic by bolding it. Notice that the headings for the other statistics remain unchanged.

❻ Cell attributes associated only with the N statistic are bolded.

8.1.4 Controlling Table Content with the CLASSDATA Option

The content of the table formed by the TABULATE procedure is influenced a great deal by the levels of classification variables in the data. Through the use of the CLASSDATA option we can identify a secondary data set to further influence the table appearance.

For the examples in this section the data set SYMPLEVELS contains only the variable SYMP, which takes on only the values '00', '01', and '02'. It should be noted, however, that in the data to be analyzed (ADVRPT.DEMOG) the variable SYMP never takes on the value '00', but otherwise ranges from '01' to '10'.

Using CLASSDATA with the EXCLUSIVE Option

The behavior and application of the CLASSDATA= option and the EXCLUSIVE option is very similar in the TABULATE step as it is in the MEANS and SUMMARY procedures (see Section 7.9). The CLASSDATA= option specifies a data set containing levels of the classification variables. These levels may or may not exist in the analysis data and can be used to either force levels into the table or to exclude levels from the table.

When the CLASSDATA= option is used with the EXCLUSIVE option, as in the following example, only those levels in the CLASSDATA= data set (including any levels not in the analysis data set) are displayed.

```
proc tabulate data=advrpt.demog
              classdata=symplevels exclusive;
    class symp;
    var ht wt;
    table symp,
          (ht wt)*(n*f=2. min*f=4. median*f=7.1 max*f=4.);
    run;
```

8.1.4a Using CLASSDATA= with EXCLUSIVE

	height in inches				weight in pounds			
	N	Min	Median	Max	N	Min	Median	Max
symptom code								
00	0	.	.	.	0	.	.	.
01	4	64	67.5	71	4	115	138.5	162
02	10	66	67.0	67	10	131	155.0	155

The symptom code '00' does not exist in the analysis data, but is included in the table. Symptom codes '03' through '10' are excluded from the table as they do not appear in the data set SYMPLEVELS.

When the CLASSDATA= option is used without the EXCLUSIVE option, all levels of the classification variable from either the CLASSDATA= data set or the analysis data are included in the table.

The EXCLUSIVE option can also appear on the CLASS statement; however, it will work with the CLASSDATA= option only when it is used on the PROC statement.

Using CLASSDATA without the EXCLUSIVE Option

When the EXCLUSIVE option is not used, the levels of the CLASSDATA data set can still be used to add rows to the resulting table. Here the EXCLUSIVE option has been removed from the previous example.

```
proc tabulate data=advrpt.demog
            classdata=symplevels;
   class symp;
   var ht wt;
   table symp,
         (ht wt)*(n*f=2. min*f=4.
               median*f=7.1 max*f=4.);
   run;
```

In this example the SYMP= '00' level has been added to the table; however, no rows have been excluded.

MORE INFORMATION

Section 12.1.2 discusses the use of pre-loaded formats with PROC TABULATE to accomplish similar results.

8.1.4b Using CLASSDATA= without EXCLUSIVE

	height in inches				weight in pounds			
	N	Min	Median	Max	N	Min	Median	Max
symptom code								
00	0	.	.	.	0	.	.	.
01	4	64	67.5	71	4	115	138.5	162
02	10	66	67.0	67	10	131	155.0	155
03	4	65	66.5	68	4	147	154.5	162
04	13	62	68.0	74	13	98	187.0	195
05	8	63	69.0	69	8	163	177.0	201
06	11	63	64.0	65	11	105	105.0	177
09	2	68	68.0	68	2	133	133.0	133
10	13	62	68.0	72	13	158	160.0	240

8.1.5 Ordering Classification Level Headings

Like many procedures that use classification variables, the default order for the level headings is ORDER=INTERNAL. Unlike the REPORT procedure the default order does not change for formatted variables.

```
proc format;
   value $SYMPTOM
      '01'='Sleepiness'
      '02'='Coughing'
      '03'='Limping'
      '04'='Bleeding'
      '05'='Weak'
      '06'='Nausea'
      '07'='Headache'
      '08'='Cramps'
      '09'='Spasms'
      '10'='Shortness of Breath';
   run;
```

The format $SYMPTOM., which is shown here, is used with the variable SYMP. Whether or not

```
proc tabulate data=advrpt.demog
                order=formatted;
   class symp sex;
   var wt;
   table sex*wt=' '*n=' '
         ,symp
         /box='Patient Counts'
          row=float
          misstext='0';
   format symp $symptom.;
   run;
```

8.1.5 Controlling Order
ORDER=FORMATTED

Patient Counts	symptom code							
	Bleeding	Coughing	Limping	Nausea	Shortness of Breath	Sleepiness	Spasms	Weak
patient sex								
F	5	6	2	7	5	2	0	2
M	8	4	2	4	8	2	2	6

the format is applied, the heading values reflect the INTERNAL order of the values of SYMP. Only if the format is assigned and the ORDER=FORMATTED is specified will the headings be placed in formatted order.

When dealing with date values the internal order or the order of the date values is often preferred over the formatted order. In the following example the visit dates are counted within months; however, we want to view the monthly totals in chronological (INTERNAL) order. In this example if we had used either the MONNAME. or MONTH. formats, the months for the two years would have been confounded.

```
proc tabulate data=advrpt.lab_chemistry;
   class labdt /order=internal;
   table labdt,n*f=2.;
   format labdt monyy.;
   run;
```

8.1.5 Controlling Order *Ordering Month Name*	
	N
LAB *TEST* *DATE*	
JUL06	19
AUG06	7
SEP06	11
OCT06	14
NOV06	7
DEC06	15
JAN07	7
FEB07	8
MAR07	21
APR07	12
MAY07	13
JUN07	15
JUL07	9
AUG07	6
SEP07	3
OCT07	1
NOV07	1

MORE INFORMATION
The ORDER= option is discussed in detail in Section 2.6.2. The VALUE statement in PROC FORMAT has the option NOTSORTED, which allows you to both format a variable and control the value order, is described in Section 12.4.

SEE ALSO
Formatting a TABULATE prior to copying it to Excel is discussed in a sasCommunity.org article at http://www.sascommunity.org/wiki/Proc_Tabulate:_Making_the_result_table_easier_to_copy_to_Excel. Indenting row headers is discussed in a SAS Forum thread, which contains links to other papers as well, at http://communities.sas.com/message/45339.

8.2 Expanding PROC UNIVARIATE

The capabilities of this procedure have been expanded in each of the last several releases of SAS and it is not unusual for even seasoned programmers to be only partially aware of all that it can now do. This section is a survey of some of those newer or less commonly known capabilities.

8.2.1 Generating Presentation-Quality Plots

A number of presentation-quality graphics, such as those produced by SAS/GRAPH, can also be produced by PROC UNIVARIATE. Some of the plotting capabilities require the presence of SAS/GRAPH even though a SAS/GRAPH procedure is not being called. Graphics are implemented through a series of statements which include:

- HISTOGRAM   builds histograms
- INSET   adds legends and text to the graph
- PROBPLOT   creates probability plots
- QQPLOT   creates quantile-quantile plots

The following example shows some of the flexibility of these statements by building three histograms that are overlaid by the normal distribution. In this example the plot generated by UNIVARIATE will be written to a file.

```
filename out821a "&path\results\g821a.emf"; ❶
goptions device=emf ❷
         gsfname=out821a ❸
         noprompt;
title1 '8.2.1a Plots by PROC UNIVARIATE';
proc univariate data=advrpt.demog;
   class race sex;
   var ht;
   histogram /nrows=5 ncols=2 ❹
             intertile=1 cfill=cyan ❺ vscale=count ❻
             vaxislabel='Count'; ❼
   inset ❽ mean='Mean Height: ' (5.2) / noframe position=ne
                                        height=2 font=swissxb;

   run;
   quit;
```

❶ The plot is to be saved as an EMF file. EMF and the older CGM files are generally considered best if the plot is to be imported into a word processing document as it has been here. The EMF file type has the further advantage of the capability of modifying and editing the graph in the Microsoft Image Editor.

❷ The DEVICE graphics option specifies the type of file to be created.

❸ The GSFNAME graphics option identifies the *fileref* that points to the file that is to be generated.

❹ The classification variables RACE and SEX form 10 combinations based on 5 values of RACE and 2 for SEX. These form the rows and columns for the plot.

❺ In a color representation of the histogram, the vertical bars are cyan.

❻ The scale of the vertical axis will be based on the patient counts. Other choices for the scale could include: PERCENT and PROPORTION.

❼ A label is specified for the vertical axis of each of the five histograms.

❽ The INSET statement inserts text, including various statistics, into the graph. Here the MEAN of HT is written in the upper right (NorthEast) corner of each graph using the 5.2 format and SWISSB, a SAS/GRAPH font. Notice that the default font for the title and the selected font for the INSET (SWISSB) are not particularly good choices. Under Windows most Windows fonts are available for use in graphics such as this. Alternatively the font could have been specified as `font='Arial Narrow /b'`. ARIAL is used in the next example.

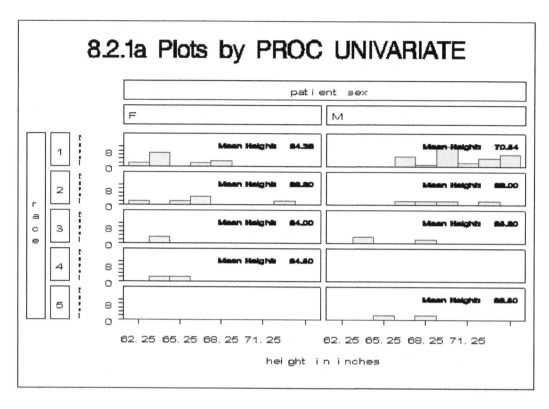

The histograms that are generated by the HISTOGRAM statement can be overlaid with one of several different statistical distributions. These distributions include:

- normal
- lognormal
- gamma
- Weibull

In this example a normal distribution is overlaid on a histogram of the data.

```
title1 f=arial
        '8.2.1b Normal Plots by PROC UNIVARIATE';
proc univariate data=advrpt.demog;
   var wt;
   histogram /midpoints=100 to 250 by 15 ❾
           cfill=cyan vscale=count
           vaxislabel='Count'
           normal (l=2 color=red); ❿

   inset mean='Mean: ' (6.2)/position=nw
                          height=4 font=arial;
   run;
   quit;
```

❾ The MIDPOINTS option is used to specify both the range of the values to be plotted and the widths of the individual bins represented by the histogram bars. This MIDPOINTS option is the same as is used in PROC GCHART and the syntax is similar to an iterative DO loop.

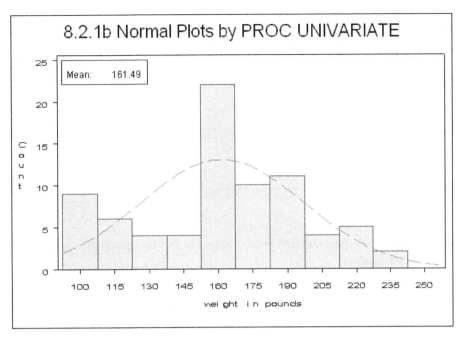

8.2.1b Normal Plots by PROC UNIVARIATE

❿ The normal distribution (based on the mean and variance of the data) is to be overlaid on the histogram. The line type is to be dashed (L=2) and the line color is set to RED.

The INSET statement is used to write the mean in the upper left (NorthWest) corner.

Notice the quality of the title font. The TITLE statement has specified the ARIAL font, which renders better than SWISSB in an EMF file (see Section 9.1 for more on adding options to TITLE statements).

Although not shown in this example, you can collect the actual and predicted percentage of observations for each midpoint by using the OUTHISTOGRAM= option. This option names a data set that will contain the predicted percentage for each distribution.

Although UNIVARIATE is not a SAS/GRAPH procedure, the graphics that it produces can take advantage of some of the other SAS/GRAPH capabilities. It will recognize several SAS/GRAPH statements, including AXIS and SYMBOL statements. Additionally it respects the ANNO= option so that it can utilize the ANNOTATE facility of SAS/GRAPH.

MORE INFORMATION
Section 9.2 discusses other SAS/GRAPH options and statements that can be used outside of SAS/GRAPH. Under some conditions the default font selection for portions of the graph results in virtually unreadable text. Portions of the text in the plot in Section 8.2.1a are very hard to read. This can be mitigated by using the FTEXT option, which is also discussed in Section 9.2.

8.2.2 Using the CLASS Statement

As is the case with a number of other summary and analysis procedures, multiple CLASS statements and CLASS statement options are supported (see Section 7.1). However, unlike other summary procedures, you can only specify up to two classification variables.

One of the CLASS statement options used specifically with UNIVARIATE is the KEYLEVEL= option. This option can be used to control plot order by specifying a *primary* or *key* value for the classification variable. The selected level will be displayed first.

```
title1 f=arial
       '8.2.2 KEYLEVEL Plots by PROC UNIVARIATE';
proc univariate data=advrpt.demog;
class race sex/keylevel=('3' 'M');
var ht;
histogram /nrows=5 ncols=2
          intertile=1 cfill=cyan vscale=count
          vaxislabel='Count';
run;
quit;
```

The single CLASS statement used here could have been rewritten as two statements, one for

```
class race / keylevel='3';
class sex  / keylevel='M';
```

each classification variable.

When using a CLASS statement, the printed output is also broken up into each combination of classification variables.

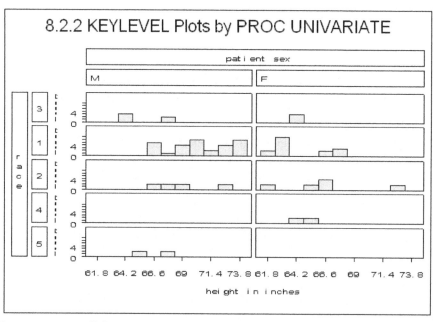

In the plot, notice that RACE level '3' and SEX level 'M' are positioned first—they have been designated as the KEYLEVELs.

Some of the other text in this graphic is very hard to read, and not only because of the size of the graph on this page. When fonts are not explicitly declared, default hardware fonts are sometimes selected that do not render well for all devices. The FTEXT= option, which is discussed in Section 9.2, can be used to explicitly specify default fonts.

8.2.3 Probability and Quantile Plots

In addition to histograms, UNIVARIATE has the capability of generating probability and quantile-quantile plots. The syntax and resulting graphics are similar for each of these types of plots. Typically these plots are used to compare the data to a known or hypothetical distribution. Probability plots are most suited as a graphical estimation of percentiles, while the quantile-quantile plots (also known as QQplots) are better suited for the graphical estimation of distribution parameters.

Probability Plots

Probability plots can be generated by use of the PROBPLOT statement.

```
title1 f=arial '8.2.3a Probability Plots';
symbol1 v=dot c=blue;  ❶

proc univariate data=advrpt.demog;
   var wt;
   probplot /normal(mu=est sigma=est  ❷
                    color=red l=2 w=2);  ❸

   inset mean='Mean: ' (6.2)
         std ='STD: ' (6.3) / position=nw
                             height=4 font=arial;
   run;
```

❶ The SYMBOL statement can be used to control the plot symbols for the percentiles. Here the requested plot symbol is a blue dot.

❷ The probability plot is to be compared to a normal distribution. The mean and standard deviation can be specified (MU and SIGMA), or they can be estimated from the data, as was done here.

❸ The estimated distribution is to be depicted with a dashed (L=2) red line with a thickness of 2.

As the distribution of the data approaches the theoretical distribution, the data percentile points should fall on the dashed line.

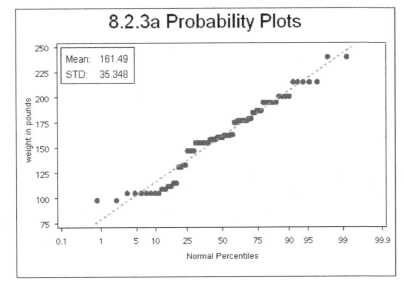

QQ Plots

Rather than using the percentiles as the horizontal axis, the quantile plots break the horizontal axis into quantile ranges.

```
proc univariate data=advrpt.demog;
   var wt;
   qqplot /normal(mu=est sigma=est
                  color=red l=2 w=2);

   inset mean='Mean: ' (6.2)
         std ='STD: ' (6.3) / position=nw
                                height=4 font=arial;
   run;
```

This QQPLOT statement uses the same options as were used in the percentile probability plots.

The resulting plots are generally very similar.

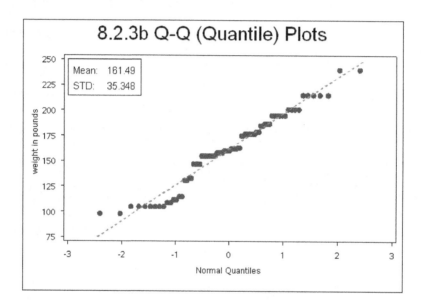

8.2.4 Using the OUTPUT Statement to Calculate Percentages

UNIVARIATE now supports the use of the OUTPUT statement. The syntax is essentially the same as is used in the MEANS and SUMMARY procedures; however, there are a number of statistics that are *only* available in the UNIVARIATE procedure. These statistics include several test statistics that are not included in the printed output. It is also possible to generate a wider range of percentiles. When used in conjunction with a CLASS statement, the output data set contains one observation per combination of classification levels. There are no rollup levels; the results are essentially what you would expect when using the NWAYS option with the MEANS or SUMMARY procedures.

The example shown here uses the OUTPUT statement to create a data set containing a series of percentiles that cannot be easily obtained in the MEANS or SUMMARY procedures.

```
proc univariate data=advrpt.demog
               noprint;
   class sex; ❶
   var wt;
   output out=unistats ❷
          mean = wt_mean
          pctlpre=wt_ ❸
          pctlpts=0 to 10 by 2.5, ❹
                 50,
                 90 to 100 by 2.5;
   run;
```

❶ The output data set will contain one observation for each level of the classification variable.

❷ The data set UNISTATS will be created by the OUTPUT statement.

❸ The PCTLPRE= option provides a prefix for the names of the variables containing the percentiles which are requested by the PCTLPTS option.

❹ The percentile request uses syntax similar to an iterative DO, which can include a compound list as is shown here.

```
8.2.4 Using the OUTPUT Statement in UNIVARIATE

Obs sex     wt_mean  wt_0  wt_2_5  wt_5   wt_7_5  wt_10

  1   F     144.548    98      98     98     105     105
  2   M     172.913   105     105    105     105     133

Obs   wt_50  wt_90  wt_92_5  wt_95  wt_97_5  wt_100

  1    155    187     187     187     215      215
  2    177    215     215     215     240      240
```

8.3 Doing More with PROC FREQ

Although PROC FREQ has been a part of SAS from its inception, it has changed very little. There are, however, a few newer and underutilized options that are now available.

8.3.1 OUTPUT Statement in PROC FREQ

In addition to the ODS OUTPUT destination, the OUTPUT statement can be used in a PROC FREQ step to write statistics generated by the procedure to a data set. The list of available statistics, as is shown in the documentation, is fairly extensive and corresponds to the statistics that can be generated by the TABLES statement.

The desired statistics must be requested on the TABLES statement if they are to be available in the OUTPUT statement.

```
proc freq data=advrpt.demog
        (where=(race in('1','2')));
    table race*sex/chisq; ❶
    output out=FreqStats ❷
         all; ❸
    run;
```

Statistics for Table of race by sex

Statistic	DF	Value	Prob
Chi-Square	1	2.5926	0.1074
Likelihood Ratio Chi-Square	1	2.5636	0.1093
Continuity Adj. Chi-Square	1	1.7493	0.1860
Mantel-Haenszel Chi-Square	1	2.5494	0.1103
Phi Coefficient		-0.2079	
Contingency Coefficient		0.2035	
Cramer's V		-0.2079	

Fisher's Exact Test	
Cell (1,1) Frequency (F)	14
Left-sided Pr <= F	0.0935
Right-sided Pr >= F	0.9706
Table Probability (P)	0.0642
Two-sided Pr <= P	0.1517

Sample Size = 60

❶ The CHISQ option requests a series of contingency table statistics for the table defined by RACE and SEX.

❷ The data set WORK.FREQSTATS will be created.

❸ Rather than select specific statistics by name, all the statistics generated by the CHISQ option on the TABLE statement are to be included in the data set. This option generates a warning as the ALL keyword requests statistics that are not included in the CHISQ option. The warning would not have been issued had either the ALL been replaced with CHISQ or if the additional statistics requested by the ALL had been included in the TABLES statement (MEASURES, CMH, N).

The portion of the table generated by the CHISQ option is shown here.

Many of these same values can be found in the data set, which has been printed below.

```
8.3.1 Using the OUTPUT Statement in FREQ

Obs  N  _PCHI_  DF_PCHI  P_PCHI  _LRCHI_  DF_LRCHI  P_LRCHI  _AJCHI_  DF_AJCHI  P_AJCHI  _MHCHI_

1    60  2.59259   1    0.10736  2.56359    1      0.10935  1.74934    1      0.18596  2.54938

Obs  DF_MHCHI   P_MHCHI    XPL_FISH   XPR_FISH   XP2_FISH    _PHI_     _CONTGY_    _CRAMV_

1       1       0.11034    0.093531   0.97063    0.15174   -0.20787   0.20352    -0.20787
```

8.3.2 Using the NLEVELS Option

The NLEVELS option on the PROC FREQ statement generates a summary table that shows the number of levels, including missing, for each of the classification variables included in the

```
proc freq data=advrpt.demog
        nlevels;
   table _all_ /noprint;
   run;
```

TABLES statement. In this example the TABLE statement requests a count of levels for all variables in the data set. The NOPRINT option prevents the printing of all tables, except the one generated by the NLEVELS option.

8.3.2 Using the NLEVELS Option on the FREQ Statement

The FREQ Procedure

Number of Variable Levels

Variable	Label	Levels	Missing Levels	Nonmissing Levels
subject		77	0	77
clinnum	clinic number	27	0	27
lname	last name	75	0	75
fname	first name	61	0	61
ssn	social security number	76	0	76
sex	patient sex	2	0	2
dob	date of birth	74	1	73
death	date of death	9	1	8
race	race	6	1	5
edu	years of education	9	1	8
wt	weight in pounds	22	0	22
ht	height in inches	13	0	13
symp	symptom code	9	1	8
death2		9	1	8

The resulting table shows all the variables in ADVRPT.DEMOG and the number of distinct values of each.

Knowing the number of distinct levels of a variable can be helpful when writing automated programs. The data contained in the NLEVELS table can be written to a data set using the ODS OUTPUT destination, and once there it can be harvested as metadata for use by the macro language.

```
ods output nlevels=varcnts;
proc freq data=advrpt.demog
        nlevels;
   table _all_ /noprint;
   run;
```

MORE INFORMATION
The ODS OUTPUT destination is discussed in more detail in Section 11.1.

SEE ALSO
SAS Tip number 30867 discusses the NLEVELS option at http://support.sas.com/kb/30/867.html.

8.4 Using PROC REPORT to Better Advantage

Like PROC TABULATE, the REPORT procedure is underutilized by many analysts. The details of its usage can be found in *Carpenter's Complete Guide to the SAS REPORT Procedure* (Carpenter, 2007a). This section will not provide an introduction to the procedure, but will instead cover a few topics that have been known to be problematic.

Much of the confusion is centered on the use of the compute block, which is unique to PROC REPORT. Consequently, most of the examples in this section involve the use of the compute block. Carpenter (2007a) has a number of more detailed examples on the syntax, use, and operating sequencing of the compute block. This book also includes a Microsoft PowerPoint copy of Russ Lavery's "Animated Guide to the REPORT Procedure's Compute Block."

MORE INFORMATION
The use of pre-loaded formats with PROC REPORT to create report subsets can be found in Section 12.1.1.

8.4.1 PROC REPORT vs. PROC TABULATE

Both the REPORT and TABULATE procedures can create summary reports and each has basically the same access to the standard suite of summary statistics.

Unlike TABULATE, the REPORT procedure can provide detail reporting as well as summary reporting capabilities. REPORT has the added flexibility to calculate and display columns of information based on other columns in the report.

Because of the unique way that the TABULATE procedure structures the report table it has a great deal more flexibility to present the groups, sub-groups, and statistics as either rows or columns. This is especially true for vertically concatenated reports, which are very straightforward in TABULATE and difficult in REPORT (see Carpenter, 2007a, Section 10.1 for details on creating a vertically concatenated report using PROC REPORT).

8.4.2 Naming Report Items (Variables) in the Compute Block

Compute blocks are unique to the REPORT procedure. While they have a number of similarities to the DATA step, there are sufficient differences, not only in syntax, but more importantly in how they are executed, which is why they warrant a discussion in this book.

In the DATA step you name the items (variables) on the Program Data Vector , PDV, explicitly by name. Although the term *variable* is often used to address values in the compute block, they are more accurately known as *report items*. In the compute block the rules for naming report items are not nearly as straightforward as in the DATA step. In the compute block there is no PDV, and the compute block can be used to address report items that are not actually variables.

There are four different situations that determine how a report item is to be named in the compute block. These result in three distinct report item naming conventions.

SEE ALSO
The topic of naming report items in compute blocks is specifically addressed in Carpenter (2006a and 2007a).

Explicitly by Name

In the DATA step variable names are used explicitly. While this same naming convention will *sometimes* work in the compute block, you need to understand when it will and will not work. The name is used explicitly when the variable or report item:

- has a define usage of GROUP, ORDER, COMPUTED, or DISPLAY
- is a temporary variable, which is created and used only in a compute block
- is the automatic temporary variable named _BREAK_
- is a report item alias (see below)

Using a Compound Name

Compound variable names are needed when a report item has a define usage of ANALYSIS, which means that is has been used to calculate a statistic (SUM is the default). The compound name is a combination of the variable name and the statistic that it has been used to calculate. The general form is *variablename.statistic*, and in the compute block you might address the mean of the variable WT as shown to the left.

```
compute wt;
   wt.mean = wt.mean/2.2;
endcomp;
```

Directly When Using an Alias

Creating an alias of a report item allows you to use that report item in multiple ways. The following COLUMN statement generates a series of aliases for the HT analysis variable. Each of these aliases will be used to calculate a different statistic.

```
column region ht
       ht=htmin   ht=htmax
       ht=htmean ht=htmedian;
```

When an alias is used in a compute block, it is named explicitly. Here the HTMIN alias of HT is used in a LINE statement.

```
compute after;
   line @3 'Minimum height is ' htmin 6.1;
endcomp;
```

Indirectly When Using the Absolute Column Number

The define type of ACROSS creates a series of columns. These columns, and indeed any column in the report, can be referenced by using the column number as an indirect column reference.

This pseudo variable name is always of the form, _Cxx_, where the *xx* is the column number as read from left to right on the report. The column count even includes any columns that will ultimately not be printed, e.g., those columns defined with NOPRINT or NOZERO.

When one or more report items have a define usage of ACROSS, it is not possible to address the transposed columns by name. To address these columns it is necessary to use absolute column numbers.

MORE INFORMATION

Absolute column references are used in the example in Section 8.4.3.

8.4.3 Understanding Compute Block Execution

In the following example a compute block is used to convert the mean weight from pounds to kilograms. Since WT is nested under SEX, which has a define usage of ACROSS, there will be two columns associated with WT. As a consequence, absolute column numbers must be used in the compute block that performs the conversion ❷.

Although this step executes without any syntax errors, it contains a *huge* logic error. In order to understand the error and what causes it we need to first understand the execution of the compute blocks.

```
proc report data=advrpt.demog nowd;
    column edu sex,wt ❶ wt=allwtmean;
    define edu / group 'Years/Ed.';
    define sex / across order=formatted;
    define wt  / mean 'Mean' format=5.1;
    define allwtmean / mean 'Mean' format=5.1;

    compute wt; ❷
        _c2_ = _c2_/2.2; ❸
        _c3_ = _c3_/2.2;
    endcomp;

    compute allwtmean; ❹
        allwtmean = allwtmean/2.2;
    endcomp;
run;
```

❶ Patient weight is nested within SEX, which has a define usage of ACROSS. An alias for WT, ALLWTMEAN, is also declared.

❷ In the compute block for WT, the values for each of the two genders are converted from pounds to kilograms.

❸ _C2_ holds the mean weight of female patients and is the second column in the report (counting from left to right).

❹ ALLWTMEAN (mean weight ignoring SEX) is a computed report item and is named directly.

```
8.4.3a Showing ACROSS With
Compute Blocks
Convert LB to KG

          patient sex
Years      F      M
  Ed.    Mean   Mean   Mean
   10      .    88.2   88.2
   12    31.4   81.7   76.4
   13    44.4   86.8   89.5
   14    22.7   47.7   49.1
   15    34.0   60.5   70.7
   16    32.6   75.9   75.1
   17    29.5   70.5   66.0
   18      .    79.1   79.1
```

In this example there are two compute blocks, one associated with WT and one for the alias ALLWTMEAN. Since there are two columns associated with WT within SEX (one for each of the two genders), the compute block for WT will execute twice for each row in the report. As a matter of fact, counting the one for ALLWTMEAN, *three* compute block executions take place for each report row.

Since the compute block for WT will execute *twice* for each report row, this causes a *very* nasty error. Notice that in the mean weights for females, the values have been divided by 2.2 *twice*. The problem goes away if the calculations for _C2_ and _C3_ are placed in the compute block for ALLWTMEAN, which is executed only once for each row.

❺ The compute block for ALLWTMEAN will execute only once. Any given compute block can reference any report item to its left on the COLUMN statement, so there is no issue with placing references to all three columns in this single compute block.

```
proc report data=advrpt.demog nowd;
   column edu sex,wt wt=allwtmean;
   define edu / group 'Years/Ed.';
   define sex / across order=formatted;
   define wt  / mean 'Mean' format=5.1;
   define allwtmean / mean 'Mean' format=5.1;

   compute allwtmean;  ❺
      _c2_ = _c2_/2.2;
      _c3_ = _c3_/2.2;
      allwtmean = allwtmean/2.2;
   endcomp;
run;
```

The resulting report now shows that the mean weight for the females has been successfully converted to kilograms.

```
8.4.3b Showing ACROSS With Compute Blocks
             Convert LB to KG

             patient sex
     Years    F      M
      Ed.   Mean   Mean   Mean
       10     .    88.2   88.2
       12   69.1   81.7   76.4
       13   97.7   86.8   89.5
       14   49.9   47.7   49.1
       15   74.8   60.5   70.7
       16   71.8   75.9   75.1
       17   64.9   70.5   66.0
       18     .    79.1   79.1
```

8.4.4 Using a Dummy Column to Consolidate Compute Blocks

In the previous section we were able to solve a nasty problem by taking advantage of a compute block associated with a report item that lay to the right of the columns with the problem. When there is no compute block 'to the right', a compute block that is based on a dummy column can be used to calculate all three mean values. In this example, ALLWTMEAN is to the left of the other columns. Consequently, its compute block could not be used in calculations of report items to its right in the COLUMN statement.

```
proc report data=advrpt.demog nowd;
   column edu wt=allwtmean sex,wt dummy ❶;
   define edu / group 'Years/Ed.';
   define allwtmean / mean 'Overall Mean' format=7.1;
   define sex / across order=formatted;
   define wt   / mean 'Mean' format=5.1;
   define dummy / computed noprint ❷;

   compute dummy;
      _c4_ = _c4_/2.2; ❸
      _c3_ = _c3_/2.2;
      allwtmean = allwtmean/2.2;
   endcomp;
run;
```

❶ The DUMMY column must be the furthest to the right on the COLUMN statement. Or at least it must be to the right of any columns used in the compute block.

8.4.4 Consolidating Compute Blocks Using a DUMMY Column

| Years Ed. | Overall Mean | patient sex | |
		F Mean	M Mean
10	88.2	.	88.2
12	76.4	69.1	81.7
13	89.5	97.7	86.8
14	49.1	49.9	47.7
15	70.7	74.8	60.5
16	75.1	71.8	75.9
17	66.0	64.9	70.5
18	79.1	.	79.1

❷ The NOPRINT option appears on the DEFINE statement for DUMMY as we are not interested in having this column displayed.

❸ The column numbers for the male and female values have now changed ($\_C3\_$ is now the mean weight of the females). Since these three conversions are independent of each other, they can be performed in any order.

8.4.5 Consolidating Columns

Sometimes we want to show the information contained in multiple report items within a single column. Doing so provides us with additional control over the appearance of the report items. In the following example we want to display the mean along with its standard error, and we want the values to be displayed as *mean (se)*.

```
proc report data=advrpt.demog nowd;
   column edu sex,(wt wt=wtse meanse ❶);
   define edu    / group 'Years/Ed.';
   define sex    / across order=formatted; ❷
   define wt     / mean noprint; ❸
   define wtse   / stderr noprint;
   define meanse / computed 'Mean (SE)' format=$15.; ❹

   compute meanse/char length=15; ❺
      _c4_ = cat(put(_c2_,5.2),' (',put(_c3_,5.2),')'); ❻
      _c7_ = cat(put(_c5_,5.2),' (',put(_c6_,5.2),')');
   endcomp;
   run;
```

❶ WT, its alias WTSE, and the computed report item MEANSE, are all nested under SEX ❷, which has a define usage of ACROSS.

❸ The values for WT and WTSE are not to be printed. They are used only to form the concatenated value (MEANSE).

❹ The computed report item MEANSE is defined.

❺ The computed variable MEANSE is defined as character with length of 15.

❻ The mean ($\_C2\_$) and the SE ($\_C3\_$) for females are concatenated into a single value ($\_C4\_$).

```
    8.4.5 Consolidating Columns within an ACROSS Variable
                  Weight Within Gender

                          patient sex
         Years  F                    M
          Ed.   Mean (SE)            Mean (SE)
           10     .   (  .  )        194.1 ( 5.75)
           12   152.0 ( 9.71)        179.8 ( 6.73)
           13   215.0 (  .  )        191.0 (12.00)
           14   109.9 ( 1.78)        105.0 ( 0.00)
           15   164.6 (13.72)        133.0 ( 0.00)
           16   158.0 ( 0.00)        167.0 ( 7.70)
           17   142.8 ( 9.84)        155.0 ( 0.00)
           18     .   (  .  )        174.0 (15.59)
```

The computed report item MEANSE is constructed in the compute block by concatenating the MEAN and SE, neither of which is printed individually. This also allows us to add the parentheses.

Because SEX has the define usage of ACROSS, absolute column references must be used in the compute block.

8.4.6 Using the STYLE= Option with LINES

When writing to destinations such as PDF, RTF, and HTML, the STYLE= option can be used to override values in the ODS style without using PROC TEMPLATE to redefine the style itself. This option is available for use with the REPORT, TABULATE, and PRINT procedures. In REPORT, it can be used with the LINE statement in the compute block.

```
proc report data=advrpt.demog nowd;
   column edu sex,(wt wt=wtse) wt=n wt=allwt;
   define edu / group 'Years/Ed.';
   define sex / across order=formatted;
   define wt  / mean 'Mean' F=5.1;
   define wtse / stderr 'StdErr' f=5.2;
   define n   / n noprint;
   define allwt / mean 'Overall/Mean' f=5.1;

   compute after/style(lines)={just=center
                              font_face=Arial
                              font_style=italic
                              font_size=10pt};
      line ' ';
      line @10  'Overall Statistics:';
      line @15 n 3. ' Subjects had a mean weight of
'
         allwt 5.1 ' pounds';
   endcomp;
run;
```

Here the style override option is used to change the justification, font, font style, and font size of the text written by the LINE statements.

In the LINE statement, the @10 and @15 control the left most starting position for the text. These values are ignored for destinations other than LISTING, and the STYLE option is ignored in the LISTING destination.

8.4.6 Using STYLE on the COMPUTE statement
Patient Weight

| Years Ed. | patient sex | | | | Overall Mean |
| | F | | M | | |
	Mean	StdErr	Mean	StdErr	
10	.	.	194.1	5.75	194.1
12	152.0	9.71	179.8	6.73	168.1
13	215.0	.	191.0	12.00	197.0
14	109.9	1.78	105.0	0.00	108.1
15	164.6	13.72	133.0	0.00	155.6
16	158.0	0.00	167.0	7.70	165.2
17	142.8	9.84	155.0	0.00	145.2
18	.	.	174.0	15.59	174.0

Overall Statistics:
76 Subjects had a mean weight of 160.5 pounds

Inline formatting can also be used in the compute block with the LINE statement; however, there are a couple of things that you should be aware of as the formatting becomes more complicated. Since LINE statements are consolidated before execution, you may not be able to change style attributes at the LINE statement level within a compute block. In this example the STYLE option will be applied to each of the LINE statements.

If you do need to change attributes on individual lines, the inline formatting will probably have to be done in a separate compute block. This can pose a problem if you are working with the COMPUTE AFTER (end of report) compute block.

In the following example an artificial variable PAGEFLAG is introduced. Since it is a constant, the COMPUTE AFTER PAGEFLAG block and the COMPUTE AFTER block will both take place at the end of the report. This will allow LINE statements with two different styles to be used.

```
* Show the use of the inline formatting;
ods rtf file="&path\results\E8_4_6b.rtf";
ods escapechar='~';  ❶
title1 '8.4.6b Using Inline Formatting';
title2 '~S={just=r} Patient Weight';  ❷
data demog;
set advrpt.demog;
pageflag=1;  ❸
run;

proc report data=demog(where=(sex='F')) nowd;
   column pageflag edu sex,(wt wt=wtse) wt=n wt=allwt;
   define pageflag / group noprint;  ❹
      . . . . define statements not shown . . . .
   compute after pageflag;  ❺
   line "~S={just=l background=pink } Females Only";
   endcomp;
   compute after/style(lines)={just=center  ❻
                       font_face=Arial
                       font_style=italic
                       font_size=10pt};
      line ' ';
      line @10  'Overall Statistics:';
      line @15 n 3. ' Subjects had a mean weight of '
                 allwt 5.1 ' pounds';
   endcomp;
run;
ods rtf close;
```

❶ An escape character is specified for use with the inline formatting sequences.

❷ Inline formatting is used to right justify the title.

❸ A constant variable is created that will allow us to have a second compute block at the end of the report.

❹ This report item is not printed, but since it has a define usage of GROUP, a compute block can be associated with it.

❺ A COMPUTE AFTER block is defined for the constant report item.

❻ Effectively there are now two compute blocks that will be executed at the end of the report.

SEE ALSO
More detail on the use of the style override option (see Sections 11.4.1 and 11.5 for more examples) and inline formatting (see Section 8.6) can be found in *Carpenter's Complete Guide to the SAS REPORT Procedure* (Carpenter 2007a).

8.4.7 Setting Style Attributes with the CALL DEFINE Routine

Unique to PROC REPORT, the CALL DEFINE routine can be used in the compute block to set various attributes. Unlike the STYLE= option shown in Section 8.4.6, as a routine CALL DEFINE can be conditionally executed. This highly flexible routine can be used to set or reset a number of attributes including formats, links, and styles.

In the following example the DEFINE routine is used to form a visual boundary by changing the background color for a dummy column. The PDF destination is used to create the report, and a gray vertical band is generated through the use of a computed variable, DUMMY, and the CALL DEFINE routine.

```
proc report data=advrpt.demog nowd;
   column edu sex,(wt wt=wtse) dummy ❶
          wt=allwt wt=allwtse;
   define edu / group 'Years/Ed.';
   define sex / across order=formatted;
   define wt   / mean 'Mean' F=5.1;
   define wtse / stderr 'StdErr' f=5.2;
   define dummy / computed ' ' ; ❷
   define allwt / mean 'Overall/Mean' f=5.1;
   define allwtse / stderr 'Overall/StdErr' f=5.2;

   compute dummy/char length=1;
      call define(_col_,'style', ❸
                     'style={background=cxd3d3d3 ❹
                            cellwidth=1mm}'); ❺
      dummy = ' '; ❻
   endcomp;
run;
```

❶ A computed column is created to hold the visual separator.

❷ The label for the computed column is set to blank.

❸ _COL_ indicates that the result of the routine is to be applied to the entire column. The second argument, STYLE, indicates that this is to be a style override. The third argument is the style attribute that will be overridden.

❹ The background color is set to a light shade of gray.

❺ Although the cell width is set to 1mm, you will probably need to experiment to obtain the desired width as this is only a nominal value.

❻ The computed variable is assigned a missing value.

8.4.7 Creating a Vertical Space Using CALL DEFINE

| | patient sex | | | | | |
| | F | | M | | Overall | Overall |
Years Ed.	Mean	StdErr	Mean	StdErr	Mean	StdErr
10	.	.	194.1	5.75	194.1	5.75
12	152.0	9.71	179.8	6.73	168.1	6.36
13	215.0	.	191.0	12.00	197.0	10.39
14	109.9	1.78	105.0	0.00	108.1	1.32
15	164.6	13.72	133.0	0.00	155.6	11.12
16	158.0	0.00	167.0	7.70	165.2	6.19
17	142.8	9.84	155.0	0.00	145.2	7.93
18	.	.	174.0	15.59	174.0	15.59

SEE ALSO

Section 7.5 of *Carpenter's Complete Guide to the SAS REPORT Procedure* (Carpenter 2007a) discusses the CALL DEFINE routine in detail.

8.4.8 Dates within Dates

When a report item is nested within itself, the resulting table is generally less than satisfactory unless you take some precautions.

Processing dates can be especially problematic as they can fall into several ranges at the same time. A given date is specific to a year, to a quarter, and to a month. When you want to create

```
data visits;
   set advrpt.lab_chemistry
      (keep=visit labdt sodium);
   year=year(labdt);
   qtr = qtr(labdt);
   run;
```

summarizations for more than one date level at the same time, you *could* create dummy variables for each level and then summarize using these levels as classification variables. This requires an extra step, such as the one shown to the left, which we can avoid when using PROC REPORT.

In the previous DATA step, the variable LABDT is used to create two different summary levels. The date is being used two different ways at the same time. You can conduct the same type of summarizations in a REPORT step by creating an alias; however, whenever you nest a variable under itself, you should be aware of some of the pitfalls of the technique.

The following report counts the number of patients that were seen for each visit type within quarter and year. The same variable, LABDT, is used for both the quarter and year summary.

```
proc report data=advrpt.lab_chemistry nowd;
   column visit ('Patient Counts Within Quarter' ❶
               labdt=year, labdt,sodium,n); ❷
   define visit  / group'Visit';
   define year   / across format=year. order=formatted ' ' ❸;
   define labdt  / across format=yyq6. order=internal ' '; ❹
   define sodium / display ' ';
   define n      / ' ' format=2. nozero ❺
               style={just=center}; ❻
   run;
```

❶ Spanning text is defined for the report.

❷ The lab date is nested under an alias of lab date (YEAR). The N statistic is nested under SODIUM, which is in turn nested under date.

❸ The dates will be consolidated into each represented year. The order will be determined by the formatted value.

❹ The quarters are nested within year, and every quarter—regardless of year—will appear under each year. This means that '2007Q1' will appear without any values under year 2006. We can eliminate these empty columns through the use of the NOZERO option.

❺ Any column that is always empty is completely eliminated by the NOZERO option.

8.4.8 Dates within Dates
Using the Original Date Variable

Patient Counts Within Quarter						
2006		2007				
2006Q3	2006Q4	2007Q1	2007Q2	2007Q3	2007Q4	

Visit	2006Q3	2006Q4	2007Q1	2007Q2	2007Q3	2007Q4
1	6	4	3	3	1	.
2	6	4	3	2	1	.
4	6	4	3	2	1	.
5	3	6	3	2	1	.
6	4	6	3	2	1	.
7	4	4	5	2	1	.
8	3	3	6	.	.	.
9	5	1	5	2	1	.
10	.	4	5	3	2	.
11	.	.	.	5	1	.
12	.	.	.	5	1	.
13	.	.	.	3	3	.
14	.	.	.	3	2	1
15	.	.	.	3	2	1
16	.	.	.	3		

❻ The STYLE= option is used to center the counts within each quarter.

The use of the NOZERO option is a key technique when nesting variables such as a date within itself. Without the use of the NOZERO option there would necessarily be a number of empty columns.

This example counts the number of patients with non-missing values of SODIUM. While the N statistic can be used for the lab date, the NOZERO option will not work on a grouping variable. Consequently, an intermediate analysis variable, SODIUM, is needed. SODIUM is a variable that in this case we are not particularly interested in, but it allows us to use the N statistic.

In this example the quarters are ordered appropriately and the ORDER=INTERNAL option ❹ is not necessary. If, instead of using quarters the dates had been grouped using month name, this option could have been used to place the columns in date order rather than alphabetical order.

8.4.9 Aligning Decimal Points

Unlike PROC PRINT the REPORT procedure does not by default align decimal points within a column of the output. This can be seen in the following example, which prints the values of SODIUM in the LAB_CHEMISTRY data set.

```
ods pdf file="&path\results\e8_4_9a.pdf"
        style=journal;
title2 'Unaligned Decimals';
proc report data=advrpt.Lab_chemistry
            nowd;
   column subject visit labdt sodium;
   run;
ods pdf close;
```

8.4.9 Aligning Decimal Points
Unaligned Decimals

PATIENT ID	VISIT NUMBER	LAB TEST DATE	sodium
200	1	07/06/2006	14
200	2	07/13/2006	14.4
200	1	07/06/2006	14
200	4	07/13/2006	14
200	4	07/13/2006	14
200	5	07/21/2006	14.2

There are a couple of easy ways to align the decimal points in the SODIUM column. When you are writing to the RTF or PDF destination, as we are here, the JUST= style attribute can be used on the DEFINE statement.

```
proc report data=advrpt.Lab_chemistry nowd;
   column subject visit labdt sodium;
   define sodium / style(column)={just=d};
   run;
```

8.4.9 Aligning Decimal Points
Aligned Decimals

PATIENT ID	VISIT NUMBER	LAB TEST DATE	sodium
200	1	07/06/2006	14
200	2	07/13/2006	14.4
200	1	07/06/2006	14
200	4	07/13/2006	14
200	4	07/13/2006	14
200	5	07/21/2006	14.2

While the columns are aligned, the decimal point is not always shown. When the use of a format is an option, and it generally is, the format will not only cause the decimal points to be aligned, but the decimal point will be displayed.

Here a format is used to align the decimal points instead of the style override.

```
proc report data=advrpt.Lab_chemistry
nowd;
   column subject visit labdt sodium;
   define sodium / f=4.1;
   run;
```

8.4.9 Aligning Decimal Points Aligned Decimals Using a Format			
PATIENT ID	VISIT NUMBER	LAB TEST DATE	sodium
200	1	07/06/2006	14.0
200	2	07/13/2006	14.4
200	1	07/06/2006	14.0
200	4	07/13/2006	14.0
200	4	07/13/2006	14.0
200	5	07/21/2006	14.2

8.4.10 Conditionally Executing the LINE Statement

Unlike in the DATA step where we can conditionally execute the PUT statement, the analogous LINE statement in a PROC REPORT compute block cannot be conditionally executed. However, we can conditionally assign values to write with the LINE statement. The first attempt, which is shown below, demonstrates this problem.

In this example we would like to write a message following each level of SEX. If the count is 35 or more, we want to display the mean weight; however, for counts under 35, we just want a note stating the low count.

```
proc report data=advrpt.demog nowd;
column sex race wt wt=meanwt;
define sex  / group;
define race / group;
define wt     / analysis n 'Patient Count';
define meanwt/ analysis mean 'Mean Weight' f=5.1;
compute after sex;  ❶
   if wt.n ge 35 then do;  ❷
      line 'Overall mean weight is: ' meanwt 5.1;  ❸
   end;
   else line 'Patient Count Below 35';  ❹
endcomp;
run;
```

❶ The text will be written after each grouping of the report item SEX.

❷ If the total N is greater than 34, we want to write the mean ❸ using a LINE statement.

❹ For small numbers, we just want this constant text to be written.

Clearly the LINE statements have not been executed conditionally. In fact both statements have been executed for each level of the report item SEX! This is obviously not what we intended, but what actually happened?

8.4.10 LINE Statements Attempted Conditional Execution			
patient sex	race	Patient Count	Mean Weight
F	1	14	159.0
	2	10	148.6
	3	3	105.0
	4	4	113.5
Overall mean weight is: 144.5 Patient Count Below 35			
M	1	28	184.7
	2	8	188.5
	3	5	116.2
	5	4	147.0
Overall mean weight is: 174.4 Patient Count Below 35			

During the process that evaluates the statements in a compute block, the LINE statements are effectively moved to the end of the step. The compute block from above essentially becomes the one shown here. This behavior is very different from anything that we see in the DATA step.

```
compute after sex; ❶
   if wt.n ge 35 then do; ❷
   end;
   else;
   line 'Overall mean weight is: ' meanwt 5.1; ❸
   line 'Patient Count Below 35'; ❹
endcomp;
```

Consequently we cannot conditionally execute the LINE statement; we can, however, conditionally build what will be displayed by the LINE statement.

```
compute after sex;
   if wt.n ge 35 then do;
       text= 'Overall mean weight is: '||put(meanwt,5.1); ❺
   end;
   else text = 'Patient Count Below 35'; ❺
   line text $31.; ❻
endcomp;
```

❺ Here we create a temporary variable (TEXT) that will take on the desired value to be displayed.

❻ The LINE statement is then executed.

SEE ALSO
The conditional execution of the LINE statement in PROC REPORT is discussed in SAS Sample #37763 at http://support.sas.com/kb/37/763.html.

8.4.10 LINE Statements Conditional Preparation

patient sex	race	Patient Count	Mean Weight
F	1	14	159.0
	2	10	148.6
	3	3	105.0
	4	4	113.5
Patient Count Below 35			
M	1	28	184.7
	2	8	188.5
	3	5	116.2
	5	4	147.0
Overall mean weight is: 174.4			

8.5 Using PROC PRINT

PROC PRINT is one of those procedures that everyone uses on a regular basis. It is designed to dump the data and is generally not used to generate *pretty* output. However, there are some things that you can do with PRINT that can make even this standard procedure more useful.

8.5.1 Using the ID and BY Statements Together

Although the PRINT procedure does not have a CLASS statement you can offset groups with a combination of the BY and ID statements. Variables that are common to these two statements will cause two changes to the standard report generated by PROC PRINT.

```
title1 '8.5.1 PRINT with BY and ID Statements';
proc print data=advrpt.clinicnames;
   by region;
   id region;
   var clinnum clinname;
   run;
```

```
8.5.1 PRINT with BY and ID Statements

region    clinnum    clinname

   1      011234     Boston National Medical
          014321     Vermont Treatment Center

  10      107211     Portland General
          108531     Seattle Medical Complex

   2      023910     New York Metro Medical Ctr
          024477     New York General Hospital
          026789     Geneva Memorial Hospital
          . . . . portions of this table are not shown . . . .
```

In the PRINT step to the left, both the BY and ID statements use the variable REGION. When used together this way, the value for REGION is written only once for each region (this is the default behavior for PROC REPORT for GROUP and ORDER variables). Also a blank line has been inserted after each REGION.

This specialized layout for PROC PRINT is generated when all of the variables in the BY statement also appear in the same order at the start of the ID statement.

8.5.2 Using the STYLE= Option with PROC PRINT

The STYLE= option, which is discussed in Sections 8.1.3 and 8.4.6, can also be used with PROC PRINT. This is a style override option and it is used to change the attributes generated by the selected ODS style.

In the general syntax shown here, notice that the attributes are surrounded by curly brackets.

```
style<(location)>={attribute=attribute_value}
```

In the current releases of SAS you are able to use the square bracket instead of the curly braces.

Specification of the *location* is optional, since there is a default assignment when it is left off. However, you generally will want to specify the location as it is used to control where the attribute assignment is to be applied. Supported locations include:

- DATA — cells (also COLUMNS or COL)
- TOTAL — sub-total (used on the SUM statement)
- GRANDTOTAL — overall total (used on the SUM statement)
- HEADER — column header (also HEAD and HDR)
- N — used when the N option is specified
- OBS — cells in the OBS column
- OBSHEADER — header for the OBS column
- TABLE — controls table structure such as cell width

The STYLE= option can be applied on the PROC PRINT statement as well as on other procedure step statements. A combination of the specified location and the statement containing the option will determine what portion of the table is modified by the option.

Not all style option locations are appropriate for all PRINT statements. The following table shows the available locations and the statements to which they can be applied.

PROC Statements	Supported Style Locations
PROC PRINT	data header, n, obs, obsheader, table
BY	none
ID	header, data
VAR	header, data
SUM	header, data, total, grandtotal

Some of the style attributes that can be modified include:

- BACKGROUND
- BORDERCOLOR
- BORDERCOLORDARK
- BORDERCOLORLIGHT
- FONT_FACE
- FONT_WEIGHT
- FOREGROUND

In the PROC Statement

When the STYLE= option is used on the PROC statement, the attributes tend to control the overall appearance of this particularly attractive table.

```
title1 'Using STYLE= with PRINT';
title2 '8.5.2a on the PROC Statement';
proc print data=advrpt.demog(obs=5)
      style(col)= [background=cyan] ❶
      style(header)= [background=yellow ❷
                      font_weight=bold]
      style(obs)= [background=pink] ❸
      style(obsheader)= [background=cyan] ❹
      ;
   var clinnum subject sex dob;
   run;
```

❶ The background color is reset for all the data values in each column.

❷ Two attributes for the column headers are reset.

❸ The background color for the OBS column is set to pink.

❹ The background color for the OBS column header is changed to CYAN.

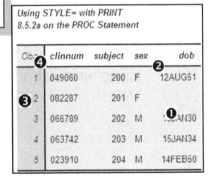

Using STYLE= with PRINT
8.5.2a on the PROC Statement

Obs	clinnum	subject	sex	dob
1	049060	200	F	12AUG51
2	082287	201	F	
3	066789	202	M	13JAN30
4	063742	203	M	15JAN34
5	023910	204	M	14FEB50

Supporting Statements

Although style attributes used with the PROC statement generally apply to the table as a whole, STYLE= options that are applied on supporting statements give you more specific control. Additionally these attributes tend to override those set on the PROC statement.

Like the CLASS statement (see Section 7.1), which can be split into multiple statements, several of the PRINT statements that allow lists of variables can also be split into multiple statements.

```
proc print data=advrpt.demog(obs=5)
     style(col)= [background=cyan] ❶
     style(header)= [background=yellow
                     font_weight=bold]
  ;
  id clinnum / style(hdr data ❷)={background=blue
                                  foreground=white};
  var subject / style(header)={background=red ❸
                               foreground=white}
              style(column)={background=red
                             foreground=white};
  var sex dob edu; ❹
  sum edu / style(grandtotal)={font_weight=bold ❺
                               background=blue
                               foreground=white};
  run;
```

This allows you to specify different options for different variables. We can take advantage of this ability when applying the STYLE= option ❸❹.

❶ The STYLE= options on the PROC statement override the defaults associated with the ODS style, JOURNAL.

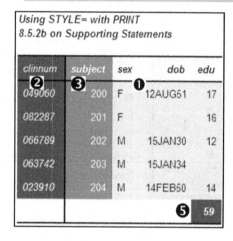

Using STYLE= with PRINT
8.5.2b on Supporting Statements

clinnum	subject	sex	dob	edu
049060 ❷	200 ❸	F ❶	12AUG51	17
082287	201	F	.	16
066789	202	M	15JAN30	12
063742	203	M	15JAN34	.
023910	204	M	14FEB50	14
			❺	59

❷ Two locations are specified for the ID variable. Notice that the HEADER location has been abbreviated as HDR.

❸ Two STYLE= options, each with its own location, are specified. Since both have the same attributes, they could have been combined as in ❷.

❹ The VAR statement has been split into two statements. These variables do not have a style override and will utilize the attributes specified in the PROC statement ❶.

❺ The column total receives three attribute overrides.

MORE INFORMATION

The style override option is used to produce traffic lighting effects is discussed in Section 11.5.4.

8.5.3 Using PROC PRINT to Generate a Table of Contents

The PRINT procedure is designed to work with lists of information. When the list of values contains links to other files, PRINT can create a table of contents to a report.

In this example HTML anchor tags are formed as data values. When they are displayed by ODS in the HTML destination these values become links to other files.

```
data clinlinks(keep=region clinnum clinic);
   set clinicnames;
   length clinic $70;
   clinic = catt("<a href='cn",     ❶
                 clinnum,            ❷
                 ".html'>",
                 clinname,           ❸
                 "</a>");
   run;

proc print data=clinlinks;
   var region clinnum clinic;        ❹
   run;
```

The character variable CLINIC is used to hold the HTML anchor tag. The CATT function ❶ is used to concatenate the pieces of the anchor tag statement. The clinic number ❷ is used in the name of the file to which the tag will link. The name of the clinic ❸ will be displayed. In the PROC PRINT step ❹ all that is required is to display the data. The LISTING destination does not know what to do with an HTML anchor tag and will therefore show the data as it is stored ❺.

```
8.5.3 Clinics in the Study

Obs    region    clinnum                      clinic

 1       4        049060      <a href='cn049060.html'>Atlanta General Hospital</a>  ❺
 2       6        066789      <a href='cn066789.html'>Austin Medical Hospital</a>
 3       5        051345      <a href='cn051345.html'>Battle Creek Hospital</a>
 4       3        031234      <a href='cn031234.html'>Bethesda Pioneer Hospital</a>
                          .... portions of the table are not shown ....
```

8.5.3 Clinics in the Study			
Obs	region	clinnum	clinic
1	4	049060	Atlanta General Hospital
2	6	066789	Austin Medical Hospital
3	5	051345	Battle Creek Hospital
4	3	031234	Bethesda Pioneer Hospital

When the table is displayed using the HTML destination, the value is interpreted as an anchor tag and is displayed as a linkable item. Here the first four items in the PROC PRINT are shown.

MORE INFORMATION

The creation of links is discussed in more detail in Section 11.4.

SEE ALSO

The REPORT procedure is even more flexible for creating this type of display. See Carpenter (2007b) for more on creating links in your table.

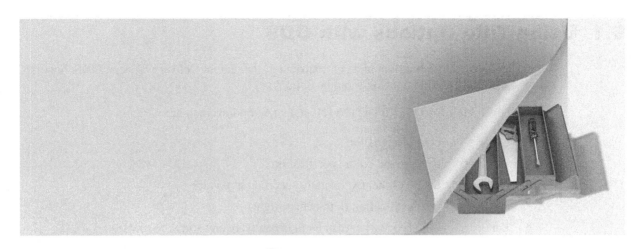

Chapter 9

SAS/GRAPH Elements You Should Know—Even if You Don't Use SAS/GRAPH

The Output Delivery System, ODS, gives us a great deal of the kind of control that we must have in order to produce the kinds of reports and tables that are expected of us. Although we will often include graphical elements in our tables, it turns out that a number of options, statements, and techniques that are associated with SAS/GRAPH can be utilized to our benefit even when we are NOT creating graphs. In this chapter you will learn how to take advantage of these graphical elements even when you are not using SAS/GRAPH.

Some of the options and statements described in this chapter are not available if your site has not licensed SAS/GRAPH. If SAS/GRAPH is not available to you, try to get your site to license it; otherwise, experiment. A lot of the things shown in this chapter will work anyway, but not everything!

SEE ALSO
Carpenter (2010b) contains a number of other related examples.

9.1 Using Title Options with ODS

For destinations that support font and color attributes, the Output Delivery System, ODS, honors many of the SAS/GRAPH title and footnote options.

A few of the traditional TITLE/FOOTNOTE statement options include:

- Color=   color designation
- BColor=   background color specification
- Height=   height of the text (usually specified in points)
- Justify=   text justification (left, center, right)
- Font=   font designation (can include hardware and software fonts)

Most of these options can be abbreviated. For the options shown above, you can use the uppercase letters in the option name as an abbreviation.

There are also a few font modification options. These include:

- BOLD   boldface the text
- ITALIC   italicize the text
- UNDERLINE   underline the text

Colors can include most standard color names as well as any of the RGB or gray-scale colors that are appropriate for the output destination.

These options are listed in the Base SAS TITLE statement documentation, as well as in the SAS/GRAPH documentation; however, a number of the SAS/GRAPH TITLE statement options are not supported outside of the graphics environment. The following example demonstrates some of these TITLE statement options using titles associated with an RTF report.

```
title1   f='times new roman' ❶
         h=15pt c=blue ❷
         bc=yellow ❸
         '9.1a Using TITLE Options';

ods rtf file="&path\results\E9_1a.rtf"
         style=rtf;

title2 f='Arial' h=13pt c=red
       j=l ❹
       bold ❺
       'English Units';

proc report data=advrpt.demog nowd split='*';
    ... portions of the REPORT step are not shown ....
```

❶ You may use any font available to your system. Fonts consisting of more than one word must be enclosed in quotation marks.

❷ The font size is set to 15 points. This can be a fairly nominal size, as actual size can depend on the destination and how it is displayed.

❸ The background color is set to yellow.

❹ JUSTIFY=LEFT has been abbreviated.

❺ The font is boldfaced.

RTF   PDF   HTML

9.1a Using TITLE Options	9.1a Using TITLE Options	9.1a Using TITLE Options
English Units	English Units	English Units
Mean Weight	Mean Weight	Mean Weight

When using the RTF style in the RTF destination, changing the background color (BC=) adds a box around the title.

In RTF, by default, the titles and footnotes are added to the HEADER and FOOTERS of the document when the table is imported. Footers are at the bottom of the physical page, and not necessarily at the bottom of the table. The titles/footnotes can be made a part of the table itself through the use of the BODYTITLE option. For shorter tables this can move the footnote to the base of the table.

```
ods rtf file="&path\results\E9_1b.rtf"
        style=rtf
        bodytitle;
```

Through the use of the background color option (BCOLOR) you can change the color behind the title's text. This option can also be used to create colorful horizontal lines. The BCOLOR option specifies the background color. Some text, if only a blank space, must be specified.

```
title1 '9.1c Horizontal Lines';
title2  h=5pt bcolor=blue ' ';  ❻
footnote h=5pt bcolor=blue ' ';  ❼
ods html file="&path\results\E9_1c.html";
proc print data=sashelp.class(obs=4);
run;
ods html close;
```

There are quite a few other SAS/GRAPH TITLE statement options. Most of these options are ignored outside of SAS/GRAPH. Depending on the destination and style, some SAS/GRAPH TITLE statement options are occasionally not ignored (when you think that they should be). In these cases they tend to yield unanticipated results.

9.1c Horizontal Lines

Obs	Name	Sex	Age	Height	Weight
1	Alfred	M	14	69.0	112.5
2	Alice	F	13	56.5	84.0
3	Barbara	F	13	65.3	98.0
4	Carol	F	14	62.8	102.5

SEE ALSO

The horizontal lines example originated from a tip supplied by Don Henderson on sasCommunity.org at http://www.sascommunity.org/wiki/Tip_of_the_Day:April_26.

TITLE / FOOTNOTE options are also documented at http://support.sas.com/documentation/cdl/en/lrdict/64316/HTML/default/viewer.htm#a000220968 .htm (see Example 3 for specific usages of these options).

9.2 Setting and Clearing Graphics Options and Settings

Most procedures that have graphics capabilities can also take advantage of many graphics options and settings. Not all of the graphics options will be utilized outside of the SAS/GRAPH environment, so you may need to do some experimenting to determine which graphics options are used for your OS, version of SAS, ODS destination, and the procedure of interest.

Graphics options are set through the use of the GOPTIONS statement. Like the OPTIONS statement, this global statement is used to set one or more graphics options. Because there are a great many aspects to the preparation and presentation of a high-resolution graphic, there are necessarily a large number of graphics options.

A few of the more commonly used options are shown here.

Option	Example Value	What It Does
htext=	2.5	sets the size for text
ftext=	Arial	sets a default font for text characters
border	noborder border	determines if a border is to be placed around the graphic
device=	emf	identifies the instruction set for the rendering of the graphic
gsfname=	*fileref*	the graphic is written to the file at this location

Because these options and settings have a scope for the entire session, if you are in an interactive session and execute two or more programs that use or change some of these options, it is not uncommon to have the options from one program interfere with the options of the next program. You can mitigate this interference by setting or resetting the options to their default values at the start of each program by using the RESET= option.

```
goptions reset=symbol;
```

The RESET= graphics option can be used to reset a number of different groups of graphic settings. The following table shows some of these groups.

RESET=	What It Does
all	resets all graphics options and settings. Resets values from some other statements as well (see below).
goptions	resets only graphics options to their default values
symbol	clears all symbol statement definitions
legend	clears all legend statement definitions
title	clears all title definitions; same as `title1;`
footnote	clears all footnote definitions; same as `footnote1;`

The following is a rather typical set of GOPTION statements.

```
FILENAME fileref "&path\results\FinalReport.emf";

goptions reset=all border  ❶
         ftext=simplex;  ❷
GOPTIONS GSFNAME=fileref   GSFMODE=replace  ❸
         DEVICE=emf;  ❹
*goptions device=win  ❺
         targetdevice=emf;  ❻
```

❶ The RESET=all option clears all graphics options and sets them to their default values. Borders around the graphs are then turned on.

❷ The FTEXT= option is used to set the default font for graphics text. SIMPLEX is similar to ARIAL; however, this SAS/GRAPH font may not be available if your site does not license SAS/GRAPH.

❸ Graphics Stream File options, GSF, are used to route the graph to a file.

- GSFNAME= points to the destination of the graphic (in this case a *fileref* named FILEREF).

- GSFMODE= if the graphic file already exists, REPLACE indicates that the graphic is to be replaced.

❹ The DEVICE= option is used to structure the graph for the appropriate physical or virtual destination. EMF is a good device when the graphic is to be included in a word processing document.

❺ During program development you will want to see the graph displayed on the monitor (DEVICE=WIN); however, you may want to view it as it will ultimately be displayed on the final destination. The TARGETDEVICE= option ❻ attempts to show you the graph on the display device (DEVICE=) using the constraints of the eventual final device (TARGETDEVICE=). In this production example this development statement has been commented out.

FTEXT= can be especially important when generating graphics using procedures such as PROC UNIVARIATE. In the example plot appearing in Section 8.2.1a, and shown here as well, some of the text is virtually unreadable. This can be a result of the automatic default selection of hardware fonts that do not scale appropriately. The FTEXT= option ❼ can be used to specify

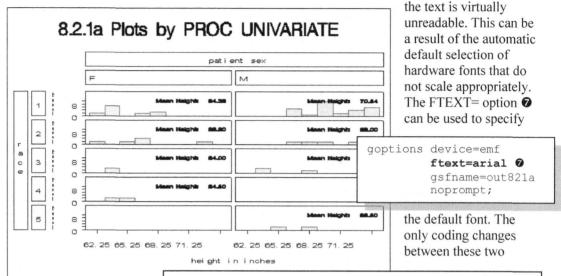

```
goptions device=emf
         ftext=arial  ❼
         gsfname=out821a
         noprompt;
```

the default font. The only coding changes between these two versions of this graphic is the use of the FTEXT= option. Notice that there are several changes including text orientation as well as readability. Note that the font used on the interior graphics was not changed.

When these options are being used with ODS you may want to control whether or not they should override the selected style. Starting in SAS 9.2 the application of some of

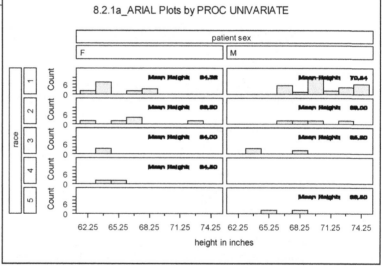

these options can be restricted with the use of the ODS USEGOPT statement. When USEGOPT is in effect, the settings for the following graphics options will take precedence over the ODS style and will affect all of your ODS output, including tables as well as graphics GOPTIONS. Affected graphics options include:

FTEXT=, CTEXT=, HTEXT=, FTITLE=, CTITLE=, HTITLE=.

When ODS NOUSEGOPT is in effect, the settings for these graphics options will not override the value in the style definition in your ODS output.

MORE INFORMATION

Although not generally shown in the code included in the text of this book, graphics options are used in examples throughout Chapters 9 and 10. Examine the sample code for more examples of the use of graphics options.

9.3 Using SAS/GRAPH Statements with Procedures That Are Not SAS/GRAPH Procedures

There are a number of other procedures that although not part of SAS/GRAPH are none-the-less able to take advantage of SAS/GRAPH statements when generating high-resolution graphs.

A few of the more common procedures that I have found to be useful that also have high-resolution graphics capabilities include:

Base

PROC UNIVARIATE (see Section 8.2)

SAS/QC

PROC CAPABILITY

PROC SHEWART (see Section 10.1.3)

SAS/STAT

PROC BOXPLOT (see Section 10.1.1)

PROC PROBIT

PROC REG

The remainder of this section is a very brief introduction to some of the statements that can be used outside of SAS/GRAPH. Better and more complete introductions to SAS/GRAPH can be found in numerous papers, as well as in several books.

CAVEAT

If you do not have access to SAS/GRAPH and depending on your release of SAS, some of the techniques and capabilities described in this section may not be available to you. This is true even if you are not using a SAS/GRAPH procedure.

MORE INFORMATION

The probability and QQ plots generated by PROC UNIVARIATE in Section 8.2.3 can take advantage of the SYMBOL statement (see Section 9.3.1).

Section 9.1 demonstrates the use of SAS/GRAPH options in TITLE and FOOTNOTE statements as they can be applied to output generated by the Output Delivery System.

SEE ALSO

Books that specifically provide introductions to SAS/GRAPH include Carpenter and Shipp (1995), Carpenter (1999), and Miron (1995).

9.3.1 Changing Plot Symbols with the SYMBOL Statement

The SYMBOL statement is used to control the appearance of items within the graphics area. As you would suspect this includes plot symbols, but it also controls the appearance of lines, and how points are joined with these lines. All plot symbols and lines have attributes, e.g., color, size, shape, thickness. These attributes are all controlled with the SYMBOL statement.

There can be up to 99 numbered SYMBOL statements. Attributes to be controlled are specified through the use of options. Options are

```
symbol1 color = blue h = .8 v=dot;
```

specified through the use of their names and, in most cases, the names can be abbreviated.

This SYMBOL statement requests that the plot symbols (a dot) be blue, with a size (height) of .8 units.

Fortunately, since the SYMBOL statement is heavily used, it is usually fairly straightforward to apply. A quick study of the documentation will usually serve as a first pass instruction. Like so many things in SAS however, there are a few traps that you should be aware of when applying the SYMBOL statement in more complex situations.

A few of the numerous SYMBOL statement options are shown in the following table.

Option	Option Abbreviation	Example Value	What It Does
color=	c=	blue	sets the color of the symbol or line
height=	h=	1.5	specifies the size of the symbol
value=	v=	star	identifies the symbol to be used in the plot
interpol=	i=	join	indicates how plot symbols are to be connected
line=	l=	1	assigns line numbers; 1, 2 , and 33 are the most useful
width=	w=	2.1	identifies line width; the default is usually 1

SYMBOL Definitions Are Cumulative

Although SYMBOL statements, like TITLE and FOOTNOTE statements, are numbered, that is about the only similarity with regard to the way that the definitions are established. When a TITLE3 statement is specified, the definition for TITLE3 is completely replaced. Not only is a given TITLE statement the complete definition for that title, but that same TITLE3 statement automatically clears titles 4 through 10. SYMBOL statement definitions, on the other hand, are cumulative, and each numbered statement is independent of statements with other numbers.

The two SYMBOL statements on the left could be rewritten as a series of statements.

```
symbol1 color = blue v=none
        i=box10 bwidth=3;
symbol2 color = red   v=dot
        i=join line=2 h=1.2;
```

```
symbol2 v=dot  i=join;
symbol1 color = blue;
symbol2 color = red;
symbol1 v=none i=box10 bwidth=3;
symbol2 line=2 h=1.2;
```

The graphics option RESET can be used in the GOPTIONS statement to clear SYMBOL statement definitions.

```
goptions reset=symbol;
```

SYMBOL Definition Selection Is NOT User Directed

When symbols or lines are to be used in a graph, the procedure first checks to see if there are any user defined symbol definitions (of course, there are defaults for everything when SYMBOL statements have not been used). The procedure then selects the next available symbol definition. This means that if SYMBOL2 was just used, the procedure will look for a SYMBOL3 definition.

Unfortunately it is not generally possible to directly tie a given symbol statement to a given line or symbol. This means that you will need to have at least a basic understanding of symbol definition selection for the procedure that you are planning on using.

The following example uses PROC REG to perform a regression analysis on HT and WT in the DEMOG data set. The PLOT statement can be used to create a plot of the results of the analysis.

```
goptions reset=all; ❶

title1 f=arial bold 'Regression of HT and WT'; ❷
title2 '9.3.1a No SYMBOL Statement';

proc reg data=advrpt.demog;
model ht = wt; ❸
plot ht*wt/conf; ❹
run;
quit;
```

❶ All graphics options are set to their defaults.

❷ TITLE statement options are used to select boldface ARIAL as the font for the first title.

❸ HT is used as the dependent variable.

❹ The CONF option is used to request the plotting of the confidence intervals and predicted values.

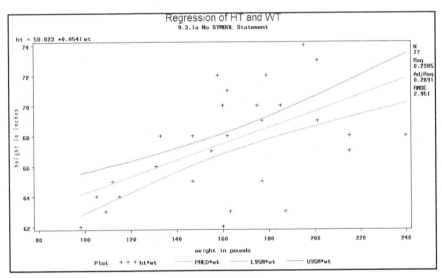

Although the procedure selects colors and line types for the predicted value line and for the confidence intervals, the data is plotted using the plus '+' symbol.

We can use the SYMBOL statement to gain control of the plot symbol.

❺ For the data points, the SYMBOL1 statement is used to select the plot symbol attributes. In this case the color and the symbol (V=).

```
title2 '9.3.1b With SYMBOL Statements';
symbol1 c=blue  v=dot; ❺
symbol2 c=red; ❻
symbol3 c=green r=2; ❼
```

❻ The color for the estimated line is specified. You will need to experiment to determine which SYMBOL statement will be used by which aspect of the graph.

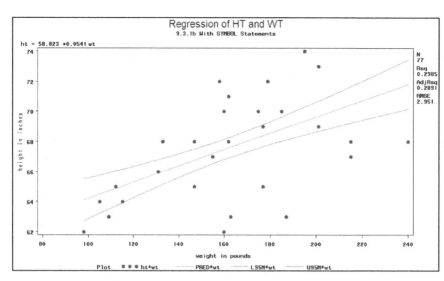

❼ The confidence lines are colored green. The R= option causes this symbol definition to be reused a second time. Otherwise, we could have specified the SYMBOL4 definition to be the same as the SYMBOL3 definition.

The colors and plot symbols are shown in the legend at the bottom of the graph. You can take control of the legend though the use of the LEGEND statement (see Section 9.3.2). Although not supported by PROC REG, for some procedures you can eliminate the legend altogether using the NOLEGEND option.

9.3.2 Controlling Axes and Legends

Control of any and all aspects of the horizontal and vertical axes can be obtained through the use of the AXIS statement. This global statement can be one of the most complex statements in SAS/GRAPH, if not within SAS itself, and it is clearly outside of the scope of this book to do much more than just partially describe this statement. Closely related to the AXIS statement in syntax is the LEGEND statement, which is used to control the appearance of the graph's legend. The following is a brief introduction to these two statements.

Like the SYMBOL statement, you can have up to 99 numbered AXIS and LEGEND statements.

```
goptions reset=axis;
goptions reset=legend;
```

Also like the SYMBOL statement, the axis and legend definitions are cumulative. Both axis and legend definitions can be cleared with the RESET= option.

AXIS Statement

The AXIS statement can be used to control the axis lines, tick marks, tick mark text, and axis labels. You can specify fonts and color for all text. For any of the lines you can control the styles (type of line), thickness, color, and length.

The axis definition is built through a series of options. Some of these options will themselves have options, and the layers of options with options can often be three deep. To make things even more interesting, some options will appear in multiple ways and their effect will depend on position and usage. Clearly just knowing how to apply the options and how to nest them can be complicated.

Most of the options that can appear in several different aspects of the statement are text appearance options. Most options are similar to those used as TITLE and FOOTNOTE statement options (see Section 9.1), and there is also some overlap with those options used in the SYMBOL statement. Some of the more common text appearance options include:

Option	Option Abbreviation	Example Value	What It Does
height=	h=	10pct	sets size to 10 percent
color=	c=	cxdedede	sets color to a shade of gray
font=	f=	Arial	sets the font to ARIAL
'text string'		'Units are mg'	assigns a text string to the option

The first layer of options control major aspects of the axis. These include such things as:

- ORDER=   range of values to be included
- LABEL=   axis label
- VALUE=   tick mark control
- MAJOR=   major tick marks (the ones with text)
- MINOR=   minor tick marks

When building an AXIS statement, parentheses are used to form groups of sub-options, and indenting to each level of option can be helpful in keeping track of which options go with what.

```
axis2 order =(3 to 6 by 1) ❶
      label =(h=2 ❷
              font='Times New Roman' ❸
              "Potassium Levels")❹
      minor =(n=1) ❺
             angle=90 ❻
             rotate=0 ❼ ;
axis1 minor=(n=4)    ❺
      color=black
      label=( "BMI")
      order=(15 to 40 by 5);
```

This is a fairly typical AXIS statement. Notice that the values of options are in parentheses. This allows you to specify the sub-options.

❶ ORDER= Restricts the axis range; data can be excluded. Here the range of the axis is limited to values between 3 and 6 with major tick marks at the integers. The VALUE= option (shown in the LEGEND example) specifies the attributes of the major tick marks.

9.3.2a Initial Visit BMI and Potassium
Using an AXIS Statement

❷ H= sets the height of the label's text to 2 units (the default units are cells).

❸ FONT= specifies the font. Fonts with multiple words should be quoted.

❹ *'text'* specifies the text for the label, which overrides the variable's label.

❺ MINOR= specifies the number of minor tick marks. The keyword NONE can be used to turn off minor tick marks.

❻ ANGLE= rotates the entire label 90 degrees (from horizontal to vertical). Angle=0 is horizontal.

```
proc gplot data=bmi;
    plot potassium*bmi/haxis=axis1 ❽
                       vaxis=axis2;
    run;
```

❼ ROTATE= rotates the letters within the line of text individually.

❽ The individual AXIS definitions are assigned to an axis on the plot or graphic through options on the PLOT statement.

LEGEND Statement

The general syntax, statement structure, and even many of the options of the AXIS statement are shared with the LEGEND statement. In the regression plots in Section 9.3.1, the legend appears at the bottom of the graph. We can change its location as well as its appearance.

```
legend1 position=(top left inside) ❶
        value=(f='arial' t=1 'Height' ❷
                         t=2 'Predicted'
                         t=3 'Upper 95'
                         t=4 'Lower 95')
        label=none ❸
        frame ❹
        across=2; ❺

proc reg data=advrpt.demog;
model ht = wt;
plot ht*wt/conf
        legend=legend1; ❻
run;
```

❶ The legend can be INSIDE or OUTSIDE of the graphics area. It can also be moved vertically and horizontally.

❷ The VALUE= option controls the text associated with the four individual items in the legend.

❸ The LABEL=NONE option turns off the legend's label.

❹ The FRAME option adds a box around the legend. Other options allow you to change the width, color, and shadowing of the frame.

❺ The ACROSS=2 option allows at most 2 items for each row in the legend.

❻ The LEGEND= option identifies the appropriate legend statement.

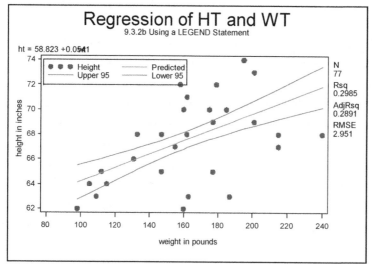

For this graph, especially when displayed in black and white, the legend is fairly superfluous. Many procedures have an option (NOLEGEND) that can be used to prevent the display of the option. PROC REG does not support the NOLEGEND option. Consequently, there is no way to prevent the legend from appearing when any of the PLOT statement options that cause multiple items to be displayed (such as CONF) are used. The following example uses a LEGEND statement to minimize the impact of the legend.

```
legend2 value= none  ❼
        label=none  ❽
        shape=symbol(.001,.001);  ❾

proc reg data=advrpt.demog;
model ht = wt;
plot ht*wt/conf
          legend=legend2;  ❿
run;
```

❼ The text for the individual values is turned off.

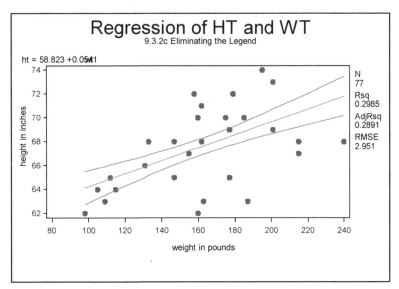

MORE INFORMATION
The AXIS statement is used in an example in Section 10.1.1.

❽ The label is turned off.

❾ The individual symbol elements cannot be turned off (SHAPE does not support NONE); therefore, the values are made very small.

❿ The LEGEND2 definition is selected for use.

9.4 Using ANNOTATE to Augment Graphs

The annotate facility gives us the ability to customize the output generated by the procedure. The huge advantage is that the customization can be data dependent, without recoding. The key to the process is the annotate data set. This data set contains the instructions that are to be passed to the annotate facility. Each observation in the data set is one instruction, and very often the instruction is fairly primitive, e.g., pick up a pen.

The instructions in the data set are passed to the procedure that is generating the graphic through the use of the ANNOTATE= option. You can tell if a procedure can take advantage of the annotate facility when it supports this option or its abbreviation, which is ANNO=.

Since the annotate facility interprets each observation of the annotate data set as an instruction, it uses the values of specific variables to form the intent of the instruction. You do not get to choose the names of the variables, but you have a great deal to do with the values that the variables take on. In order for the instruction to provide a valid instruction to annotate, the variables and their values have to provide answers to three primary questions.

Questions to Be Asked	Possible Variables Used to Answer the Questions
WHAT is to be done?	FUNCTION (this variable will always be present)
WHERE is it to be done?	X, Y, XSYS, YSYS
HOW is it to be done?	COLOR, SIZE, STYLE, POSITION

The value of the variable FUNCTION is *always* specified, and *this* value determines what other variables will be used by annotate when the instruction is executed. FUNCTION should be a character variable with a length of 8. There are over two dozen possible values for FUNCTION; three of the commonly used values are shown here.

Value of FUNCTION	What It Does
label	adds a text label to the graphic
move	moves the pointer to another position on the graphic without drawing anything
draw	draws a line from the current position to a new position on the graphic

Other commonly used annotate functions include tools for:

- generating polygons, bars, and pie slices
- drawing symbols, arrows, and lines
- including text and images

For annotate operations that are associated with a location on a graphic, variables are used to specify the location. In order to identify a location, you will need to specify the coordinate system (e.g., XSYS, YSYS) and a location within that coordinate system (e.g., X, Y).

```
data bmilabel(keep=function ❶
                   xsys ysys x y
                   text color style
                   position size);
   set advrpt.demog;

   * Define annotate variable attributes;
   length color function $8; ❷
   retain function 'label' ❸
          xsys ysys '2'
          color 'red'
          style 'arial'
          position '2'
          size .8;

   * Calculate the BMI. Note those outside of
   * the range of 18 - 26;
   bmi = wt / (ht*ht) * 703;
   if bmi lt 18 or bmi gt 26 then do; ❹
      * Create a label;
      text = put(bmi,4.1); ❺
      x=wt;
      y=ht;
      output bmilabel; ❻
   end;
   run;
```

In this example the Body Mass Index, BMI, is calculated and then added using the annotate facility to the regression plot generated by PROC REG (see Section 9.3).

❶ The annotate data set is named (WORK.BMILABEL), and the variables that it will contain are specified.

❷ The annotate variables FUNCTION and COLOR are assigned a length in order to avoid truncation. This is always a good idea.

❸ The annotate variables that are constant for all the instructions (observations) are assigned values with the RETAIN.

❹ Create an annotate instruction (a label) only for those observations with a BMI outside of the stated range.

❺ The variables X and Y contain the coordinates of this data point on the graph. TEXT, the annotate label, contains the value of the variable BMI.

❻ Write this annotate instruction to the annotate data set.

❼ The ANNO= option is used to name the data set ❶ that contains the annotate instructions.

```
proc reg data=advrpt.demog;
   model ht = wt;
   plot ht*wt/conf legend=legend1
            anno=bmilabel;  ❼
   run;
```

The BMI values that are outside of the selected range are added to the plot as annotate labels. The location of the label is based on the data values that are also used to generate the plot and the regression.

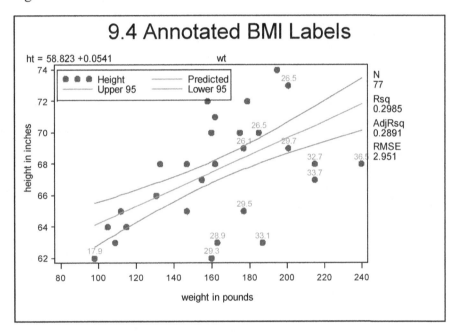

SEE ALSO

An introduction to the annotate facility can be found in Carpenter (1999).

Values can also be added to points using the SYMBOL statement option POINTLABEL. See http://communities.sas.com/message/100627#100627.

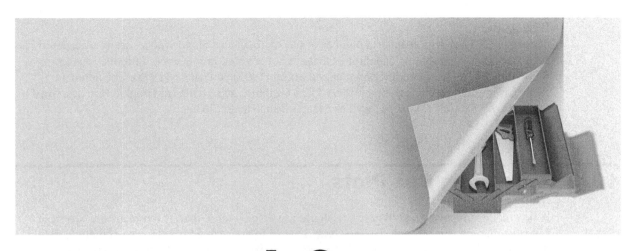

C h a p t e r 10

Presentation Graphics—More than Just SAS/GRAPH

SAS/GRAPH software has had the ability to create presentation-quality graphs since its introduction. Currently within SAS there are several graphing systems, with some of the newest innovations associated with ODS Statistical Graphics. Outside of these two graphics systems there are a number of procedures that have plotting and graphing capabilities that are comparable to SAS/GRAPH. A well-rounded programmer will be aware of each of these systems and will be able to take advantage of the strengths of each. This chapter briefly discusses the plotting capabilities of some procedures that are not part of SAS/GRAPH.

MORE INFORMATION
The plotting capability of PROC UNIVARIATE (see Sections 8.2.1 through 8.2.3) and PROC REG (see Sections 9.3 and 9.4) are demonstrated in other sections of the book.

A review of the SYMBOL, AXIS, and LEGEND statements, which are used throughout this chapter, can be found in Chapter 9.

SEE ALSO

Carpenter (2010b) demonstrates a number of these procedures along with a variety of capabilities. SAS has extensive graphing capabilities that are not covered in this book. Introductions to SAS/GRAPH can be found in Carpenter and Shipp (1995) and Miron (1995). The annotate facility is introduced in Carpenter (1999). ODS Graphics and Statistical Graphics are described in the books by Kuhfeld (2010), as well as Matange and Heath (2011).

10.1 Generating Box Plots

A box plot is a type of graph that has been used to display more than two dimensions worth of information on a single graph. Unlike some other graphics techniques that also attempt to display more than two dimensions, the box plot can do so without creating visual distortions that otherwise can mislead the reader (Carpenter 1994 and Carpenter 1995, Section 5.7). Although used heavily in some disciplines, they are unfortunately ignored in others.

Traditionally, the number of ways to generate a box plot within SAS was fairly limited. User-written programs were common with some of the more sophisticated published examples presented by Michael Friendly (Friendly, 1991). The SYMBOL statement within SAS/GRAPH can also be used to generate box plots; however, even with the addition of recent options this is still a limited technique. More recent additions to SAS provide procedures that can be used to generate box plots. Of these, the only procedure dedicated to the generation of box plots is PROC BOXPLOT (see Section 10.1.1), which is part of SAS/STAT software. Other procedures that generate variations of this type of data display include:

- PROC GPLOT   SAS/GRAPH   (using the SYMBOL statement, see Section 10.1.2)
- PROC MIXED   SAS/STAT
- PROC SHEWART   SAS/QC   (see Section 10.1.3)

SEE ALSO

The programs and macros found in Michael Friendly's book (Friendly, 1991) are well written and well explained. The techniques described provide a flexibility that is hard to beat even in the newer procedures.

10.1.1 Using PROC BOXPLOT

PROC BOXPLOT, a fairly recent addition to SAS/STAT software, is used to create a variety of types of box plots. The PLOT statement is used to provide primary control, and PLOT statement options in addition to those shown in the example below include:

- BOXSTYLE   indicates the type of box to be displayed
- BOXWIDTH   used to control the box width
- BOXWIDTHSCALE   allows box width to vary according to a function of the group size
- NOTCHES   draws the boxes with notches
- SYMBOLLEGEND   attaches a legend statement (see Section 9.3.2)

Although this is a SAS/STAT procedure, it has the capability of utilizing statements that are normally associated with SAS/GRAPH. These include the SYMBOL, AXIS, and LEGEND statements (see Section 9.3).

```
symbol1 color = blue h = .8 v=dot;
axis1 minor=none color=black
      label=(angle=90 rotate=0);

proc boxplot data=demog;
   plot wt*symp/ cframe   = cxdedede  ❶
                 boxstyle = schematicid  ❷
                 cboxes   = red
                 cboxfill = cyan
                 vaxis    = axis1  ❸
                 ;
   id race;  ❹
   run;
```

❶ The background color inside of the frame is set to gray using the color CXdedede.

❷ There are several styles of boxes. SCHEMATICID causes points outside of the whiskers to be labeled using the ID variable, which is RACE ❹, in this graph.

❸ The AXIS statement is used to control various aspects of the vertical axis. AXIS statements can be applied to either the horizontal axis, HAXIS=, or the vertical axis, VAXIS= (see Section 9.3).

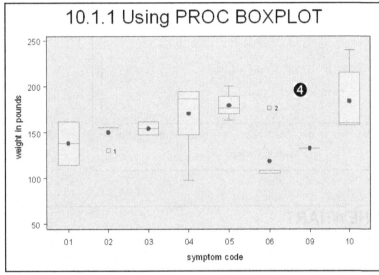

❹ The ID statement names the variable(s) used to identify the outlying points.

In this example the SYMBOL statement has been used to control the color and symbol designating the mean (a blue dot). The median is designated with a horizontal line and the upper and lower limits of the box are the 25th and 75th percentiles.

10.1.2 Using PROC GPLOT and the SYMBOL Statement

The SYMBOL statement (see Section 9.3.1) can be used to generate box plots directly in the GPLOT procedure. The control is through the use of the INTERPOL= option, which is usually abbreviated as I=. When I= takes on the value of BOX, data are condensed into a box plot for constant values of the horizontal variable (SYMP).

```
symbol1 color = blue v=none i=box10 bwidth=3;  ❶
symbol2 color = red  v=dot  i=none h=1.2;

axis1 minor=none color=black
      order=(50 to 250 by 50)
      label=(angle=90 rotate=0);
axis2 order = ('00' '01' '02' '03' '04' '05'  ❷
               '06' '07' '08' '09' '10' '11')
      value = (t=1  ' ' t=12 ' ');  ❸

proc gplot data=demog;
   plot wt*symp/ haxis   = axis2
                 vaxis   = axis1
                 ;
   run;
```

❶ The boxes that form the box plot are defined using this SYMBOL statement. The options form the characteristics of the boxes:

- Color=   identifies the outline color
- Value=   indicates that plot symbols are not needed
- Interpol=   requests box plots with the BOX option BOX10 whiskers are at the 10th and 90th percentiles
- bwidth=   specifies the width of the boxes

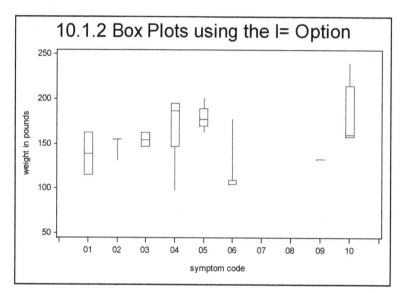

10.1.2 Box Plots using the I= Option

❷ The major tick marks for the horizontal axis are declared. '00' and '11' do not appear in the data and their labels are set to blank using the VALUE option ❸. They provide horizontal spacing control.

10.1.3 Using PROC SHEWHART

PROC SHEWHART, which is available in SAS/QC software, can be used to generate a number of process control charts, and box plots are one of the supported chart forms.

The following code can be used to create box plots using PROC SHEWHART. Note that the horizontal plot variable must be character.

```
symbol1 color = blue v=dot i=box10 bwidth=3;
symbol2 color = red  v=dot  i=none h=1.2;

axis1 minor=none color=black
      order=(50 to 250 by 50)
      label=(angle=90 rotate=0);
axis2 order = ('00' '01' '02' '03' '04' '05'
               '06' '07' '08' '09' '10' '11')
      value = (t=1  ' ' t=12 ' ');

proc shewhart data=demog;
   boxchart wt*symp/
      haxis   = axis2
      vaxis   = axis1
        stddeviations nolimits
      ;
   run;
```

The AXIS statements are used to augment the axes. Although SHEWHART and CAPABILITY are not SAS/GRAPH procedures, they support AXIS, SYMBOL, and PATTERN statements.

You can also use the ANNOTATE facility with these procedures.

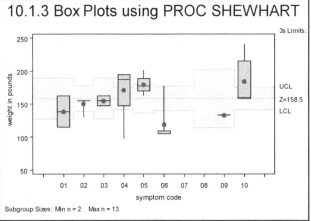

10.1.3 Box Plots using PROC SHEWHART

10.2 SAS/GRAPH Specialty Techniques and Procedures

SAS/GRAPH software has been a product of SAS software for quite a long time. During that time not only has its capabilities continued to expand, but so have the technologies to which it is delivering. As a result there are a great many graphics problems that have been solved over the years. Here are a few of the more interesting.

10.2.1 Building Your Own Graphics Font

Although the plot symbols available through SAS/GRAPH software are generally adequate for our graphing needs, occasionally you may want to tailor plot symbols for specific needs. The GFONT procedure can be used to create plot symbols. The procedure is used to draw the symbol shape in much the same way as the annotate facility is used to draw, that is by drawing from one coordinate to the next.

In this example we are not satisfied with the 'lumpy' appearance of the dot symbol, and would like to create a smoother symbol. This portion of a graph was generated using the standard DOT symbol.

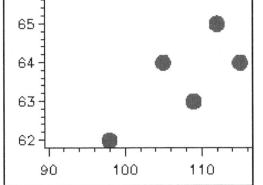

```
symbol1 color = blue v=dot h=2;
```

We can smooth out the circle by generating our own symbol using PROC GFONT.

A control data set that draws the font characters (in this case just a circle) is first generated in a DATA step. Like annotate data sets, specific variables are used to pass information to the GFONT procedure. Here we are drawing the outline of a circle using 721 short line segments. This control data set is then passed to GFONT.

```
data fontdata(keep=char seg x y lp ptype);
retain char 'c'  ❶
       seg   1   ❷
       lp    'p'; ❸
ptype='w'; x=-10; y=110; output fontdata; ❹
ptype='v'; x=100; y=50;  output fontdata; ❺
do deg = 0 to 360 by .5;
    rad = deg*arcos(-1)/180;
    x=50*cos(rad)+50;
    y=50*sin(rad)+50;
    output fontdata;  ❻
end;
run;
libname gfont0 "&path\data";  ❼
proc gfont data=fontdata
           name=mydot  ❽
           filled  ❾
           resolution=3;  ❿
run;
symbol1 f=mydot ❽ c=blue v='c' ❶ h=1;
```

❶ The variable CHAR is used to hold the keyboard character 'c' used to designate this symbol within the MYDOT font ❽.

❷ Symbol segments allow you to create symbols with disconnected parts (segments). This symbol has only one segment so this value is a constant.

❸ This is to be a polygon figure.

❹ Width of the character (PTYPE='w'). The symbol will have a width of 100 units. Allow about 10% extra for character spacing.

❺ This is the first coordinate of the plot symbol (PTYPE='v').

❻ The segment is written to the control data set.

❼ Generated fonts are stored in a catalog named FONTs. SAS/GRAPH software searches for this catalog in numbered *librefs* whose names start with GFONT.

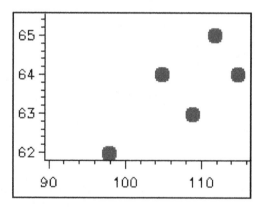

❽ Name the font.

❾ We want the shape to be filled.

❿ Increase the font resolution. You may notice little or no difference between a value of 1 and 3.

The resulting dots are a bit smoother. In this screen capture some pixelation has taken place when the image was copied into this document.

SEE ALSO
Carpenter (1995) uses PROC GFONT to create a logo and sunflower symbols that self adjust according to the plotted value. Plot symbol resolution was further discussed with alternate solutions in the SAS Forum thread at
http://support.sas.com/forums/thread.jspa?threadID=12547&start=0&tstart=0.

10.2.2 Splitting a Text Line Using JUSTIFY=

Within SAS/GRAPH software you can use the text justification option to split lines of text. Very often this technique can be used where you otherwise would be unable to split the text line. All that you need to do is to repeat the justification option. The second occurrence will cause a line split.

```
title1 f=arial h=1.2 justify=c '10.2.2'
              j=center 'Splitting Text'; ❶
symbol1 c=blue v=dot h=1.5;
symbol2 c=red v=dot h=1.5;
axis1 reflabel = (h=1.5
              t=1 c=red  j=left 'Overweight' ❷
                         j=l ' '
                         j=l c=blue  'Normal');
axis2 order=(1920 to 1970 by 10)
      label=(j=c 'Birthyear'j=c 'All Subjects'); ❸
```

❶ The justification option can be written as either JUSTIFY or abbreviated as a J. The position (Left, Center, or Right) can also be abbreviated. Here the repeated justification option causes a split in the text of the title.

❷ The reference label on the vertical reference line has been split into three text lines.

❸ The horizontal axis label has been split into two lines.

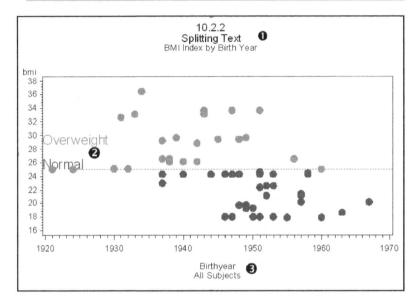

10.2.3 Using Windows Fonts

SAS/GRAPH software has a number of built-in fonts. In addition if you are executing in the Windows environment, you may also use Windows TrueType fonts.

The GOPTIONS statement (see Section 9.2) is used to specify graphics options, and one of these options, the FTEXT= option, can be used to specify fonts. On titles the FONT= option (see Section 9.1) can also be used to specify fonts.

```
goptions reset=all noborder
         device=emf
         gsfname=image gsfmode=replace
         ftext='Arial'; ❶

title1 f=arial 'This is a Title'; ❷
```

❶ Arial has been selected as the default font for all graphics text. As a general rule, fonts that are not SAS/GRAPH fonts should be quoted when they are named as a graphics option. The name *must* be quoted when it has more than one word.

❷ The font for a title can also be specified (see Section 9.1 for more on TITLE statement options). It is not necessary to quote the font name in the TITLE statement, unless the font name has more than one word, e.g., 'Times New Roman'.

You can see a list of available fonts, and their alternate designations, by selecting the FONTS entry in the Windows control panel. Select Start → Control Panel → Fonts.

Some fonts have named variations that include bold and italics ('Arial Bold'). You may also want to specify font modifiers. Of the three available font modifiers, bold (/bold or /bo ❸) and italic (/italics or /it), are the most commonly used.

```
axis1 label=(f='arial/bo ❸' angle=90 rotate=0 'Patient Weight');
```

There are a number of symbol sets that are included with SAS/GRAPH software. For symbol sets other than the default symbol set, the name of the symbol set is specified with the FONT= (or F=) option on the SYMBOL statement. For special characters that are not SAS/GRAPH special characters (characters that do not have a value mapped to the keyboard), the character can often be inserted from the character map (shown to the right). Select Start → Programs → Accessories → System Tools → Character Map. ❹ Select the character of interest; as shown on the right, a starburst design from the Wingdings 2 font has been selected. ❺ Press the copy button to place the code in the paste buffer. ❻ Paste the value into the appropriate location (the V= option on the SYMBOL statement is shown here). Notice that the symbol is likely to appear differently in the SAS program than it will in the graph.

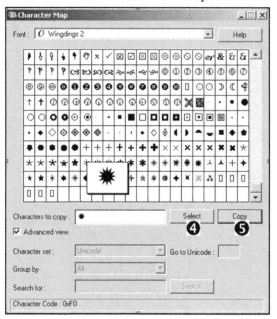

```
symbol1 c=blue
        f='wingdings 2' v='ð' ❻
        i=box10 bwidth=3;
```

CAVEAT: If you choose a non-standard font that is only available on your local machine, your SAS program becomes less transportable.

10.2.4 Using PROC GKPI

The GKPI (Key Performance Indicator) procedure is fairly new to SAS/GRAPH. This procedure allows you to quickly build indicator bars or dials that show the relative status of a value within a range that you have specified. There are several types of performance indicators available through this procedure.  The horizontal slider (HBULLET) is shown here. You can specify the range of values, colors to separate ranges, and the current value.

```
goptions reset=all noborder
        device=javaimg ❶
        xpixels=130 ypixels=50
        ftext='Arial';
```

The JAVAIMG device ❶ is required for the construction of the indicator. Here the GOPTION statement also sets the horizontal (XPIXELS) and vertical (YPIXELS) size in pixels.

Each execution of the procedure generates a single indicator graphic; therefore, for practical applications the GKPI procedure step will need to be called within a macro loop of some kind.

```
%macro slider(gname,bmi);
proc gkpi mode=raised;
hbullet actual=&bmi ❷ bounds=(0 18.5 25 30 50)❸ /
        noavalue nobvalue
        target=. colors=(blue,green,yellow,red) ❹
        name="c:\temp\&gname" ❺;
run;
quit;
%mend slider;
```

❷ The indicator value, which determines the length of the horizontal line, is passed into the macro and is treated as a constant in the GKPI procedure.

❸ The value range endpoints are specified.

These will set the zones for the colors ❹ as well as the placement of the ACTUAL= and TARGET= values.

❹ The colors of the individual segments are specified. In this example these are constants, but they too could be declared using macro parameters.

❺ The NAME= option names the file that will contain the individual indicators.

We would like to report on each subject's body mass index, BMI, value. In that report we need to show the indicator calculated for each specific subject. This means that we need to run the GKPI procedure once for each subject in the data set after first calculating the BMI value. The CALL EXECUTE routine allows us to create a series of macro calls; one for each observation in the DATA step.

```
title;
ods html file="c:\temp\slider.gif"; ❻
data bmi(keep=subject ht wt gname bmi);
    set advrpt.demog(obs=8);
    length gname $4;
    bmi = wt / (ht*ht) * 703; ❼
    gname=cats('G',subject); ❽
    call execute('%slider('||gname||','||put(bmi,4.1)||')'); ❾
    run;
ods html close;
```

❻ It is necessary to have an ODS destination open. We will not use this GIF file, but it will also contain the individual indicators.

❼ The BMI value is calculated.

❽ The subject number will become a part of the name of the file that contains the indicator that will be imported by PROC REPORT.

❾ The CALL EXECUTE routine is used to build a series of macro calls—one for each incoming observation. The macro %SLIDER is passed the parameters needed for each specific subject (the subject number and the associated BMI value). This is sufficient information for PROC GKPI to generate the subject specific indicator.

```
%slider(G200,24.3)
%slider(G201,21.4)
%slider(G202,25.1)
%slider(G203,36.5)
%slider(G204,18.0)
%slider(G205,25.1)
%slider(G206,23.0)
%slider(G207,24.3)
```

After all the calls to the macro %SLIDER have been executed, a series of PNG files will have been generated ❺. These files have been named ❾ so that they can be imported by PROC REPORT ❿ in the correct order.

```
ods pdf file="&path\results\E10_2_4.pdf" style=default;
title font=arial '10.2.4 Using GKPI';
proc report data=bmi nowd;
column subject gname ht wt bmi slider;
                          .... code not shown ....
define slider / computed ' ';

compute slider/char length=62;
slider=' ';
imgfile = "style={postimage='c:\temp\"||trim(left(gname))||".png'}";
call define ('slider','style',imgfile); ❿
endcomp;
run;
ods pdf close;
```

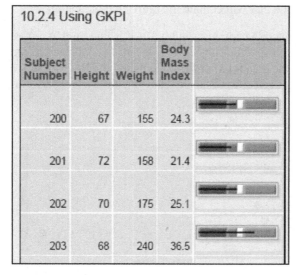

❿ The REPORT step creates a computed report item (SLIDER) that will hold the indicator image. The temporary variable IMGFILE contains the pointer to the PNG file ❺ that contains the image. For subject 200, IMGFILE contains:
```
style={postimage='c:\temp
\G200.png'}.
```

A portion of the resultant PDF file is shown here.

10.3 PROC FREQ Graphics

The FREQ procedure has now been included in the list of base procedures that can produce graphics through the ODS Statistical Graphics routines. The TABLE statement supports the PLOT= option, which can be used to generate a number of graphs.

A number of different types of plots are available, especially if you are calculating test statistics, such as those generated with the CHISQ option. The PLOTS= option is used to make the plot

```
ods graphics on;
ods pdf file="&path\results\E10_3.pdf";
proc freq data=advrpt.demog;
    table wt / plots=cumfreqplot(scale=freq); ❶
    table sex*race/plots=freqplot(scale=percent); ❷
    run;
ods pdf close;
```

requests. You may specify specific types of plots, as is done below, or you may request all plots (plots=all).

❶ The CUMFREQPLOT can be used to generate frequency and cumulative frequency histograms. There are a number of modifier options that further refine the plot requests. Here we specifically request that the vertical axis be frequencies.

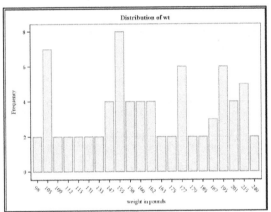

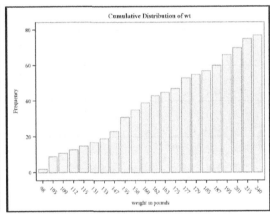

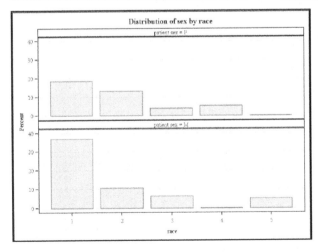

❷ For two-way tables we often need to see the relationship between the frequencies of the combinations of values. In this plot request we ask for a frequency plot of the relationship between the RACE and SEX. The plot request has been modified to have the vertical axes show percentages.

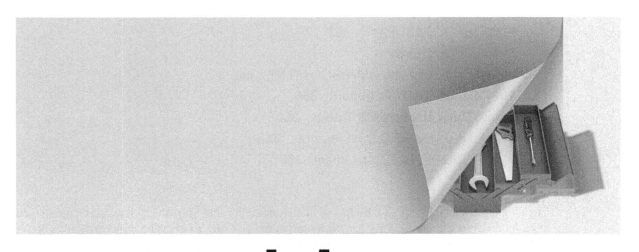

Chapter **11**

Output Delivery System

The Output Delivery System, ODS, has so many intricacies that only a very few can be mentioned here. Indeed, more than one book has been written where ODS is either the primary topic or an important secondary topic. This chapter is aimed at highlighting a few useful topics.

SEE ALSO

The classic go-to document for ODS is *Output Delivery System: The Basics and Beyond* by Haworth, Zender and Burlew (2009). Find tip sheets for ODS at http://support.sas.com/rnd/base/ods/scratch/ods-tips.pdf

Lund (2006) covers a great many of the topics found in this chapter and has a very nice summary of ODS attributes and the destinations to which they apply.

11.1 Using the OUTPUT Destination

While most procedures have one or more options that can be used to route procedural results to data sets, not all values can be captured this way. The OUTPUT destination allows us to capture procedure results as data. This destination is especially useful when there is no option available to write a specific statistic to a data set, or when a procedure does not have the capability of generating output data sets.

The output from each procedure is organized into one or more objects. These objects have a series of properties including a name and a label. This name (or the label) can be used on the ODS OUTPUT statement as an option to create an output data set.

The examples in this section use PROC UNIVARIATE; however, most of the discussion applies to most other procedures as well.

11.1.1 Determining Object Names

In its simplest, the UNIVARIATE procedure creates five output objects, and we will need at least the object name to make use of the OUTPUT destination.

```
ods trace on; ❶
proc univariate data=advrpt.demog;
   var ht wt;
   run;
ods trace off; ❷
```

The labels of the five basic objects produced by PROC UNIVARIATE for each of the analysis variables can be seen in the RESULTS window. If you right click on the label you can examine the objects attributes, including the name.

Since the ODS TRACE statement with the ON option ❶ was used, these attributes will also be displayed in the LOG. The portion of the LOG shown to the right shows the attributes of two of the five objects. The TRACE statement is turned off with the OFF option ❷.

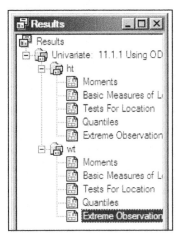

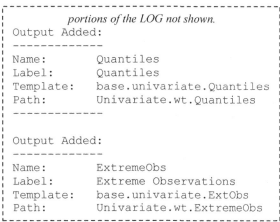

```
                portions of the LOG not shown.
Output Added:
-------------

Name:        Quantiles
Label:       Quantiles
Template:    base.univariate.Quantiles
Path:        Univariate.wt.Quantiles
-------------

Output Added:
-------------

Name:        ExtremeObs
Label:       Extreme Observations
Template:    base.univariate.ExtObs
Path:        Univariate.wt.ExtremeObs
```

11.1.2 Creating a Data Set

The name of the output object (or its label) is used as an option on the ODS OUTPUT statement to name the data set that is to be created. In this example we want to collect information on the observations with the extreme values. By default PROC UNIVARIATE identifies the five observations containing the maximums and minimums of the analysis variables.

```
ods listing close; ❶
title1 '11.1.2a Naming the OUTPUT Data Set';
ods output extremeobs=maxmin; ❷
proc univariate data=advrpt.demog;
   id lname fname; ❸
   var ht wt; ❹
   run;

ods listing; ❺
proc print data=maxmin;
   run;
```

❶ In this case we only want PROC UNIVARIATE to create a data set (no printed output), so all destinations other than OUTPUT are closed. We could not just use the NOPRINT option, because it also blocks the ability of the OUTPUT destination to create a data set.

❷ The name of one or more output objects (EXTREMEOBS) is used as an option on the ODS OUTPUT statement to name the data set that is to be created (WORK.MAXMIN).

❸ The ID statement names one or more variables useful in identifying the selected observations.

❹ Two analysis variables are specified.

❺ The LISTING destination has been turned back on and PROC PRINT is used to show the data set built by the OUTPUT destination.

The LISTING output of the resulting data set (WORK.MAXMIN) shows the observation number and identification variables for the five maximum and minimum values for each analysis variable.

```
11.1.2a Naming the OUTPUT Data Set

       Var            lname_      fname_                     lname_      fname_      High
Obs    Name    Low    Low         Low      LowObs    High    High        High        Obs

  1     ht      62    Moon        Rachel       51      74    Lawless     Henry        38
  2     ht      62    Karson      Shawn        36      74    Mercy       Ronald       50
  3     ht      62    Cranston    Rhonda       18      74    Nabers      David        53
  4     ht      62    Carlile     Patsy        11      74    Panda       Merv         56
  5     ht      63    Temple      Linda        72      74    Taber       Lee          70
  6     wt      98    Karson      Shawn        36     215    Mann        Steven       43
  7     wt      98    Carlile     Patsy        11     215    Marks       Gerald       44
  8     wt     105    Stubs       Mark         69     215    Rose        Mary         63
  9     wt     105    Maxwell     Linda        49     240    Antler      Peter         4
 10     wt     105    Leader      Zac          39     240    King        Doug         37
```

Using CLASS and BY Variables

When CLASS or BY variables are added to the PROC step, the resulting data set is expanded to include them.

```
title1 '11.1.2b CLASS Variable Present';
ods output extremeobs=maxclass(keep=sex varname high ❻
                                    lname_high fname_high);
ods listing close;
proc univariate data=advrpt.demog;
   class sex; ❼
   id lname fname;
   var ht wt;
   run;
```

❻ Data set options (see Section 2.1) can be included when naming the new data set.

❼ When one or more classification variables are used, they are added to the new data set for each combination of levels. The BY statement yields a similar result; however, the order of both the observations and the variables will be different.

```
11.1.2b CLASS Variable Present

       Var                        lname_      fname_
Obs    Name    sex    High        High        High
                ❼
  1     ht      F      68    East         Jody
  2     ht      F      68    Rose         Mary
  3     ht      F      68    Wills        Norma
  4     ht      F      72    Adamson      Joan
  5     ht      F      72    Olsen        June
  6     ht      M      74    Lawless      Henry
  7     ht      M      74    Mercy        Ronald
        .... portions of this listing are not shown ....
```

Using the Object's Label

The object's label can be used instead of the object name on the ODS OUTPUT statement. Here the example from 11.1.2a is repeated using the quoted label (extreme observations) ❽ instead of the object name.

```
ods output ❽'extreme observations'=extobs;
```

Driving an Automated Process

Any data set or even any information arranged in rows and columns can be used as the driving information for automating a process. The SAS macro language is especially powerful when it comes to creating applications and programs that rely on external information to make decisions.

In this example that external information will be a data set created through the use of ODS and the OUTPUT destination. The example process shown here, one that we would like to execute many times, is a simple PROC PRINT, but in reality it could be any number of DATA and PROC steps.

```
%macro process(dsn=,whr=);
    proc print data=&dsn;
        where &whr;  ❶
        run;
%mend process;
```

In this case we want to execute the %PROCESS macro once for every level of a classification variable (in this example the classification variable must be character). The control will be accomplished using a WHERE clause ❶ that will be constructed in the controlling macro (%DOPROCESS) ❹.

The controlling macro, %DOPROCESS, uses PROC FREQ with the OUTPUT destination ❷ to form a data set containing one level for each distinct value of the classification variable (&CVAR).

```
%macro doprocess(dsn=, cvar=);
ods output onewayfreqs=levels;  ❷
proc freq data=&dsn;
    table &cvar;
    run;
data _null_;
    set levels;  ❸
    whr = cats("&cvar='",&cvar,"'");  ❹
    call execute('%nrstr(%process(dsn='
                ||"&dsn"||',whr='
                ||whr||'))');  ❺
    run;
%mend doprocess;

%doprocess(dsn=advrpt.demog, cvar=sex)  ❻
```

❷ ODS OUTPUT and a PROC FREQ are used to create a data set (WORK.LEVELS) that will contain one row for each unique value of the classification variable (&CVAR).

❸ The data set created by the OUTPUT destination is read as input for the DATA _NULL_ step.

❹ The WHERE criteria is constructed and placed in the variable WHR, which will be added as text to the macro call ❺.

❺ CALL EXECUTE is used to build a series of calls to the macro %PROCESS; one for each level of the classification variable.

❻ The %DOPROCESS macro is called with the data set and classification variable of interest.

MORE INFORMATION

In this example an SQL step could have also been used to create the data set WORK.LEVELS, but this is not always the case. PROC SQL is used to create a distinct list of values in the second example in Section 11.2.2.

11.1.3 Using the MATCH_ALL Option

Obviously once the data set has been created, a variety of subsetting techniques can be used to break it up into distinct slices. If you know that the data set is to be broken up using BY/CLASS values, you can save the subsetting step(s) by using the MATCH_ALL option ❶.

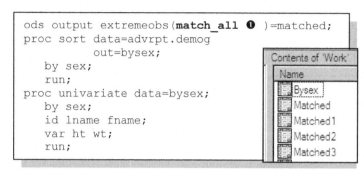

```
ods output extremeobs(match_all ❶ )=matched;
proc sort data=advrpt.demog
          out=bysex;
   by sex;
   run;
proc univariate data=bysex;
   by sex;
   id lname fname;
   var ht wt;
   run;
```

Notice the placement of the option, within parentheses following the object name.

In this step there are four combinations of SEX (M/F) and the analysis variables (HT/WT), and the MATCH_ALL option generates four data sets - one for each combination. The first is named WORK.MATCHED, the second WORK.MATCHED1, and so on. The listing for the first combination (WORK.MATCHED) is shown below. Notice that in this example a BY statement was used, rather than a CLASS statement as in the example in Section 11.1.2b. This allows you to observe the differences in the order of the variables and observations. Here the first of the data sets (WORK.MATCHED) is printed.

```
11.1.3 Using MATCH_ALL
matched

           Var        lname_    fname_                      lname_    fname_   High
Obs   sex  Name  Low   Low       Low      LowObs   High   High      High     Obs

 1     F   ht    62   Moon      Rachel      19      68    East      Jody       6
 2     F   ht    62   Karson    Shawn       12      68    Rose      Mary      22
 3     F   ht    62   Cranston  Rhonda       5      68    Wills     Norma     30
 4     F   ht    62   Carlile   Patsy        4      72    Adamson   Joan       2
 5     F   ht    63   Temple    Linda       28      72    Olsen     June      21
```

11.1.4 Using the PERSIST= Option

The PERSIST option is typically used to modify selection lists, and what we have essentially done in the ODS OUTPUT statement (Section 11.1.3) is specify a list of selected objects. The PERSIST= option determines how long the specified object should remain on the selection list. For the OUTPUT destination the list of selected objects is by default cleared (EXCLUDE ALL) at the step boundary.

Although the OUTPUT destination will by default generate data sets from a single PROC step, you may want to create a data set based on the results of two or more steps (generally of the same procedure). Rather than first creating the data sets individually and then later combining them in a secondary DATA step, they can be combined directly using the PERSIST= option ❶.

```
title1 '11.1.4 Using the Persist Option';
ods output extremeobs(persist=proc ❶)=pmatched;
ods listing close;


proc univariate data=advrpt.demog;
   class sex;
   id lname fname;
   var ht wt;
   run;

proc univariate data=advrpt.demog;
   class edu;
   id lname fname;
   var ht wt;
   run;
ods output close; ❷
```

❶ Using PERSIST=PROC maintains the selection list past the step boundary. The data set remains open until the destination is closed ❷ or the list is otherwise cleared. This allows ODS to write the results of more than one procedure to the same data set (WORK.PMATCHED).

❷ This ODS destination is closed. When the PERSIST= option is used it is important to close this destination. In this example you could have also used CLEAR instead of CLOSE; however, CLEAR merely resets the SELECT/EXCLUDE list to EXCLUDE=ALL.

The list of variables generated by the two PROC UNIVARIATE steps is almost the same. Since the classification variable differs, each step will contribute its classification variable as a column in the new data set (WORK.PMATCH). Effectively the data sets created by the two procedure steps individually have been concatenated. Adding the MATCH_ALL option would have resulted in separate data sets (see Section 11.1.5).

SEE ALSO
The PERSIST= option is discussed by Bryant, Muller, and Pass (2003).

11.1.5 Using MATCH_ALL= with the PERSIST= Option

When the MATCH_ALL= option and the PERSIST= option are used together, a series of related data sets with similar Program Data Vectors can be generated.

```
ods output extremeobs(match_all=series ❶
                      persist=proc ❷)=HT_WT ❸;
ods listing close;

proc univariate data=advrpt.demog;
   class sex;
   id lname fname;
   var ht wt;
   run;

proc univariate data=advrpt.demog;
   class edu;
   id lname fname;
   var ht wt;
   run;
ods output close;
```

In this example the MATCH_ALL= results in a series of data sets in the form of ❸ WORK.HT_WT, WORK.HT_WT1, WORK.HT_WT2, etc. Since we are using multiple procedures ❷ along with classification variables, the number of data sets may not be easily known. ❶ The list of the names of the data sets generated by the use of the MATCH_ALL=option can be stored in a macro variable (&SERIES) ❶.

```
64    %put &series;
HT_WT HT_WT1 HT_WT2 HT_WT3 HT_WT4 HT_WT5 HT_WT6 HT_WT7 HT_WT8
HT_WT9 HT_WT10 HT_WT11 HT_WT12 HT_WT13 HT_WT14 HT_WT15
HT_WT16 HT_WT17 HT_WT18 HT_WT19
```

For this example we could write this macro variable's value to the LOG using the %PUT statement.

In earlier versions of SAS this list was often used with a SET statement to concatenate the data sets. This gives us some additional control. In the DATA step that concatenates these data sets an IF statement ❹ has been used to subset the data. A WHERE statement could not be used because the variable EDU is present only in those data sets generated by the second PROC UNIVARIATE step. The SET statement could also have been written without the macro variable by using a data set list abbreviation.

```
data HT_WT_all;
   set &series;
   if edu < '13';  ❹
   run;
```

When using the current versions of SAS, the MATCH_ALL= option is no longer necessary to produce a concatenated data set. In this example including the PERSIST= ❷ without the MATCH_ALL= ❶ (see Section 11.1.4), would result in a single data set (WORK.HT_WT) that included the output from both PROC UNIVARIATE steps.

```
set ht_wt:;
```

11.2 Writing Reports to Excel

There are several ways to write reports and procedure output directly to EXCEL tables. The results vary and method selection should depend on the desired result.

Destination	File Type	File Characteristics
HTML	HTML	Uses the HTML4 tagset to generate an HTML 4.0 file. Not all style attributes are transferred to EXCEL.
HTML3	HTML	HTML 3.2 standard file. Was the only HTML destination under SAS 8. Attribute handling is different than the HTML destination under SAS®9.
MSOFFICE2k	HTML	(tagset) Supports importation of SAS/GRAPH images. Optimized for MSOffice 2k environment.
EXCELXP	XML	(tagset) Emphasis is on the data not the text. Supports writing to EXCEL Workbooks and multiple worksheets.

By far the most flexible approach and the only one that supports the XML standard is through the use of the EXCELXP tagset. This tagset is under constant development with new features being added on a regular basis. The latest version of this tagset along with a number of supporting papers and examples can be downloaded at http://support.sas.com/rnd/base/ods/odsmarkup/.

```
ods tagsets.excelxp ❶
           style=default
           path="&path\results"
           body="E11_2.xls";
title1 '11.2 Using the EXCELXP Tagset';
title2 "Using the ExcelXP Tagset";

proc report data=advrpt.demog
           nowd split='*';
       portions of the PROC REPORT step not shown
```

❶ EXCELXP is a tagset within the MARKUP destination. You may also address the tagset as an option on the ODS MARKUP statement ❷ .

```
❷ ods markup tagset=excelxp . . .
```

The emphasis for this tagset is on the data and not necessarily on the text. Notice that, using the defaults, the two titles do not even appear in the Excel spreadsheet. Of course the titles can be included through the use of the embedded `_titles='yes'` option. This is just one of the many available options. To see the current list of available options use the DOC='help' option ❸, which writes a list of options to the LOG.

	A	B	C	D
1		Mean Weight in Pounds		
2		Gender		
3		F	M	
4	Symptom			Ratio F/M
5	1	115	162	1.409
6	2	147	155	1.054
7	3	162	147	0.907
8	4	151	183	1.209
9	5	163	185	1.135
10	6	127	105	0.829
11	9		133	
12	10	170	193	1.135
13		147	168	1.144

```
ODS tagsets.excelxp file="&path\results\test.xml"
                ❸ options(doc="help");
```

SEE ALSO
The EXCELXP tagset is further discussed in (Andrews, 2008). Eric Gebhart (Gebhart, 2010) has written a number of papers on this tagset. Vince DelGobbo has written over a dozen papers on the EXCELXP tagset. An overall index to Vince's papers can be found at http://www.sas.com/events/cm/867226/ExcelXPPaperIndex.pdf with a full list of his and other SAS author's papers found at http://support.sas.com/rnd/papers/index.html.

11.2.1 EXCELXP Tagset Documentation and Options

The operation of the EXCELXP tagset is controlled through the use of options. These are implemented using the OPTIONS option with the options themselves enclosed in parentheses

```
ODS tagsets.excelxp
    path="&path\results"
    body="E11_2_1.xls"
    options(doc="help");
```

which follow the OPTIONS keyword. Here the DOC= option is used to write the tagset's full documentation to the LOG. If you want to learn about recent changes to the tagset, the CHANGELOG option `options(doc="changelog")` will show you the timing of changes and summary of new features.

SEE ALSO

The EXCELXP tagset is constantly being updated and refined. If you are using the version of the EXCELXP tagset that was shipped with SAS it is unlikely to be current. You can learn how to download and install the latest version of the EXCELXP tagset by reading SAS Note #32394 at http://support.sas.com/kb/32/394.html.

11.2.2 Generating Multisheet Workbooks

When writing a report or data set to Excel, it is not uncommon to need to break it up into portions that are written to individual sheets in the workbook. This can be accomplished in a couple of different ways. The primary difference between the two techniques shown here is whether you are sending a report or just the data to the spreadsheet.

When writing a report to Excel the EXCELXP tagset is the most flexible choice. This is demonstrated in this example with a PROC PRINT. This technique utilizes BY-group processing to break up the report.

```
%macro multisheet(dsn=,bylist=);
ods tagsets.excelxp ❶
          style=default
          path="&path\results"
          body="E11_2_2a.xls"
          options(sheet_name='none' ❷
                  sheet_interval='bygroup' ❸
                  embedded_titles='no'); ❹

proc sort data=&dsn out=sorted;
   by &bylist;
proc print data=sorted;
   by &bylist; ❺
   run;
ods tagsets.excelxp close;
%mend multisheet;
%multisheet(dsn=advrpt.demog,bylist=race)
```

❶ The EXCELXP tagset is selected for use.

❷ Let the tagset determine the sheet name.

❸ The SHEET_INTERVAL option determines how to break up the sheets.

❹ Titles will not be included on the report. This is the default.

❺ The BY line must be specified when using the SHEET_INTERVAL of 'BYGROUP' ❸.

For this macro call one sheet will be created for each level of RACE, including any missing values of RACE.

When breaking up a data set to multiple sheets we do not need to invoke the power of the EXCELXP tagset. Instead we can use PROC EXPORT. Here the EXPORT step is inside of a %DO loop that will execute once for each level of the selected classification variable.

```
%macro multisheet(dsn=,cvar=);
  %local varcnt type string i;
  proc sql noprint;  ❻
    select distinct &cvar
        into :idvar1 - :idvar&sysmaxlong
          from &dsn;
      %let varcnt = &sqlobs;
      quit;

data _null_;  ❼
    if 0 then set &dsn;
    call symputx('type',vtype(&cvar),'l');
    stop;
    run;

%do i = 1 %to &varcnt;
      %if &type=N %then %let string=&&idvar&i;  ❽
      %else %let string="&&idvar&i";
      proc export
            data=&dsn(where=(&cvar=&string))  ❾
            outfile="&path\results\E11_2_2b.xls"
            dbms= excel
            replace;
        sheet = "&cvar._&&idvar&i";  ❿
        run;
  %end;
%mend multisheet;
%multisheet(dsn=advrpt.demog,cvar=race)
```

❻ PROC SQL is used to create a series of macro variables to hold the distinct levels of the classification variable (&CVAR).

❼ A DATA _NULL_ step is used to determine if the classification variable is numeric or character.

❽ The WHERE criteria is established differentially for numeric and character variables.

❾ The data set is subsetted using the WHERE criteria established at ❽.

❿ The sheet is named using a combination of the variable's name and its value.

SEE ALSO

The must-read paper on this topic is by Vince DelGobbo (2007). The PROC EXPORT example shown here was adapted to a sasCommunity.org article which shows a less generalized program to break up a data set into separate EXCEL sheets http://www.sascommunity.org/wiki/Automatically_Separating_Data_into_Excel_Sheets.

11.2.3 Checking Out the Styles

In addition to customized styles that you or your company may have created, SAS ships with over 40 predefined styles. With so many styles to choose from and since not all style attributes are carried over to the Excel spreadsheet when reports are written to Excel using the EXCELXP tagset, it becomes important to be able to visualize your report for each of the currently defined styles.

The %SHOWSTYLES macro uses the macro language to write a report to Excel once for each available style.

```
title1 '11.2.3 Showing Style in Excel';

%macro showstyles;
%local i stylecnt;
proc sql noprint;
    select scan(style,2,'.')  ❶
        into: style1-:style&sysmaxlong
            from sashelp.vstyle;
    %let stylecnt = &sqlobs;  ❷
    quit;

%do i = 1 %to &stylecnt;
    ods markup tagset=excelxp
            path="&path\results"
            file="&&style&i...xml"  ❸
            style=&&style&I  ❹
            options(sheet_name="&&style&i"  ❺
            embedded_titles='yes');

        title2 "Using the &&style&i Style";
        proc report data=sashelp.class nowd;  ❻
            column name sex age height weight;
            define age    / analysis mean f=4.1;
            define height / analysis mean f=4.1;
            define weight / analysis mean f=5.1;
            rbreak after /summarize;
            run;
    ods markup close;
%end;
%mend showstyles;
ods listing close;

%showstyles  ❼
```

❶ An SQL step is used to create a macro variable for each of the unique style names.

❷ The number of styles found is saved in &STYLECNT.

❸ Inside the %DO loop the style name will be contained in the macro variable reference &&STYLE&I. The filename, therefore, contains the style name.

❹ The style of interest is specified on the STYLE= option.

❺ The sheet name will be the name of the style.

❻ The report code will be exactly the same for each of the styles.

❼ The %SHOWSTYLES macro is called.

11.3 Inline Formatting Using Escape Character Sequences

There are some types of formatting that is difficult or impossible to implement directly using ODS styles and options. However, through the use of an escape character it is possible to pass destination-specific commands directly to the destination that builds the output. The escape character alerts ODS that the associated formatting sequence of characters are not to be used directly by ODS, but rather are to be passed to the receiving destination.

The escape character should be one that you do not otherwise use in your SAS programs. Since I

```
ods escapechar='~';
```

tend to not use the tilde as a negation mnemonic, it makes a good escape character. The escape character is designated using the ODS ESCAPECHAR option.

The escape character is used to note an escape sequence that may contain one or more destination commands or functions. The syntax varies by the kind of command.

Type	General Form	Used to	Section
Formatting Functions	~{*function* text}	Control pagination, superscripts, subscripts.	11.3.1 11.3.2
Style Modification	~S={*attribute characteristics*} ~{style *elements and attributes*}	Assign style attributes.	11.3.3
Sequence Codes	~*code*	Manipulate line breaks, wrapping, and indentations.	11.3.4
Raw Text Insertion	~R/*destination* "*rawtext*" ~R "*rawtext*"	Insert destination-specific codes.	11.3.5

SEE ALSO

Carpenter (2007a, Section 8.6) discusses inline formatting in the context of PROC REPORT steps.

Haworth, Zender, and Burlew (2009) discuss the use of the escape character and inline formatting sequences in a variety of usages.

Zender (2007) covers all the basics in this easy-to-read SAS Global Forum paper on inline formatting. If you want to know more, this should be the first paper that you read on this topic.

11.3.1 Page X of Y

A common requirement for multipage reports is to indicate the current page as well as the total number of pages, something like `page 2 of 6`. This can be accomplished in several ways, and the appropriate methodology depends on the destination, the placement of the text, and how the text is to be written out.

The RTF Destination - Using PAGEOF

When writing to the RTF destination, the *pageof* sequence can potentially be used to add page numbering. Designed to be used in a title or footnote, this formatting sequence can have unintended consequences when used elsewhere or if the BODYTITLE option is also used.

In this example a PROC REPORT step is executed with a BY statement causing a new page to be generated for each value of the BY variable.

```
ods escapechar='~';  ❶
options nobyline;

ods rtf file="&path\results\E11_3_1a.rtf"
        style=rtf;
proc sort data=advrpt.demog
          out=demog;
   by symp;
   run;
proc report data=demog nowd split='*';
   title2 '#byvar1 #byval1';  ❷
   title3 '~{pageof}';  ❸
   by symp;
   column sex wt ht;
   define sex      / group 'Gender' order=data;
   define wt       / analysis mean format=6.1 ' ';
   define ht       / analysis mean format=6.1 ' ';
   compute after;
      line @3 'Page ~{pageof}';  ❹
   endcomp;
   run;
ods _all_ close;
```

❶ The tilde is designated as the escape character.

❷ Because the value of the BY variable has been placed in the title using the #BYVAL and #BYVAR options, the BYLINE is turned off. These options may not fully work if the TITLE statement containing the options is outside the PROC step.

❸ The *pageof* formatting sequence is designated in the title. Notice that the sequence is enclosed in braces that follow the escape character. The page numbering appears correctly in the title line ❺.

❹ For demonstration purposes the page numbering has also been requested through the LINE statement, which will write at the bottom of the report. Here the numbering is calculated incorrectly ❻. Remember that the *pageof* formatting sequence is designed to be used in the TITLE or FOOTNOTE statements and not in a LINE statement.

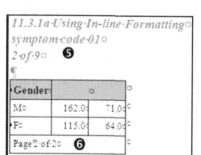

When using RTF with MSWORD, the page numbers are sometimes not shown until the document is either printed or viewed in the print preview window.

MORE INFORMATION
The #BYVAL and #BYVAR title and footnote options are also used in Sections 7.4.3 and are discussed in Section 15.1.2.

SEE ALSO
Usage note 15727 discusses the use of the BODYTITLE option with the PAGEOF sequence.

The PDF and RTF Destinations - Using THISPAGE and LASTPAGE
The PAGEOF sequence is only available for the RTF destination; however, you can create even more flexible paging for both the RTF and PDF destinations by using the THISPAGE and LASTPAGE ❼ formatting sequences. Like the PAGEOF sequence, these are designed to be used in titles and footnotes, and unanticipated results can be expected when using the BODYTITLE option with the RTF destination.

```
ods pdf file="&path\results\E11_3_1b.pdf"
        style=printer;
ods rtf file="&path\results\E11_3_1b.rtf"
        style=rtf;

title1  '11.3.1b Using In-line Formatting';
title2 '#byvar1 #byval1';
title3 'THISPAGE and LASTPAGE';
title4 h=10pt
        'This is Page ~{thispage} of a Total of ~{lastpage} Pages';  ❼
proc means data=demog n mean;
   by symp;
   class sex;
   var wt ht;
   run;
ods _all_ close;
```

11.3.1b Using In-line Formatting
symptom code 01
THISPAGE and LASTPAGE
This is Page 2 of a Total of 9 Pages ❼

The MEANS Procedure

patient sex	N Obs	Variable	Label	N	Mean
F	2	wt ht	weight in pounds height in inches	2 2	115.0000000 64.0000000
M	2	wt ht	weight in pounds height in inches	2 2	162.0000000 71.0000000

The behavior of the LASTPAGE sequence is similar to PAGEOF in the RTF destination. You may not be able to observe the total number of pages until you either do a print preview or scroll down a few pages in the table after it has been imported.

11.3.2 Superscripts, Subscripts, and a Dagger

In addition to paging information you can also draw attention to specific text by adding superscripts, subscripts, and a dagger symbol using inline formatting. Inline formatting functions include:

~{super 1}	makes the 1 a superscript
~{sub 14}	the number 14 becomes a subscript
~{dagger}	the dagger symbol may be used instead of numbers.

```
title1  '11.3.2 In-line Formatting';

ods escapechar='~';

ods pdf file="&path\results\E11_3_2.pdf"
        style=printer;
title2 'Superscripts and a Dagger ~{dagger}'; ❶
proc report data=advrpt.demog nowd split='*';
   column symp wt ht;
   define symp / group 'symptom' order=data;
   define wt   / analysis mean format=6.1
                 'Weight~{super 1}' ❷;
   define ht   / analysis mean format=6.1
                 'Height~{super 2}' ❷;
   compute after;
      line @1❸ '~{super 1} ❷Pounds';
      line @1❸ '~{super 2} ❷Inches';
      line @1❸ '~{dagger} ❶ Using inline formatting';
   endcomp;
   run;
ods _all_ close;
```

❶ The dagger symbol is placed in the title. The symbol itself does not have the same appearance in all destinations.

❷ Superscripts are used to annotate the units of measure. In this PROC REPORT step they are applied in both the DEFINE statement and the LINE statement.

❸ The column placement notation (@1) is not reliable in most destinations, especially when using proportional fonts. The @1 will reliably left justify the text; however, using something like @5 will not necessarily align text across rows.

11.3.2 In-line Formatting
Superscripts and a Dagger †

symptom	Weight[1]	Height[2]
02	150.2	66.8
10	184.4	68.5
06	118.8	64.0
04	170.8	68.7
03	154.5	66.5
09	133.0	68.0
05	179.5	67.5
01	138.5	67.5

[1]Pounds

[2]Inches

† Using inline formatting

MORE INFORMATION

Aligning text across rows can be accomplished using inline formatting sequence codes; see Section 11.3.4. This same example demonstrates the use of the dagger as well.

11.3.3 Changing Attributes

Attributes associated with text can be changed using the inline style modifier. Most of the standard attributes that are set by the ODS style can be modified. The style modifier can be used virtually anywhere that you specify text. This includes not only titles and footnotes, but the labels of formatted values, and even the data itself. There are two general forms of the style modifier.

Prior to SAS 9.2 the only available form was:

```
~S={attribute=value}
```

Notice that an uppercase S= follows the escape character and precedes the curly braces. A typical use could be to change attributes of the text in a title.

```
title2 '~S={font_face="times new roman"} ❶ Initial'
       '~S={font_style=roman} ❷ Coded'
       '~S={} ❸ Symptoms';
```

11.3.3a In-line Formatting
Initial **Coded Symptoms**

❶ The default title2 font is changed to Times New Roman.

❷ Change the default font style from italics.

❸ Changes are turned off and the defaults are restored.

Starting with SAS 9.2 a more flexible form of style modifier was introduced. The general form is:

```
~{style [element=attribute] text}
```

The style element and its attribute are enclosed within square brackets and together they precede the text to which the element/attribute pairs are to be applied. Notice that the braces enclose the text as well as the style elements.

The following TITLE statement generates the same title line as the TITLE statement of the previous example. Since modified attributes apply only to the text

```
title2 '~{style [font_face="times new roman"]Initial}'
       '~{style [font_style=roman] Coded}'
       ' Symptoms';
```

11.3.3b In-line Formatting
Initial **Coded Symptoms**

within the braces, the default attributes are applied to 'Symptoms'.

Either type of style modification sequence can be used outside of the TITLE and FOOTNOTE statements.

```
compute after;
  line @3 '~{style [font_weight=bold font_size=3] ~{super 1} Pounds}';
  line @3 '~{style [font_weight=light] ~{super 2}Inches}';
endcomp;
```

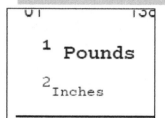

In fact they can be used most places where text is displayed. This includes data, labels, and formatted values. Here style modification sequences are applied in a LINE statement within a PROC REPORT compute block.

SEE ALSO
Haworth, Zender and Burlew (2009, pg 268) creates several style modification sequences based on data values.

11.3.4 Using Sequence Codes to Control Indentations, Spacing, and Line Breaks

For some destinations, a series of sequence codes can be used with the escape character to control line breaks, indentations, and spacing. These codes may not work equally well for each of the primary destinations, and performance may be dependent on the version of SAS being used.

Code Sequence	What it does	Note
~m	Indentation location marker.	
~-2n	Forces a line break (takes ~m into consideration).	
~xn	Forces x line breaks (does not take ~m into consideration).	x= # of line feeds
~w	Suggested location for an optional line break.	
~_	Creates a non-breaking space.	underscore
~xz	Inserts one of four error codes.	x=1, 2, 3, or 4

```
proc format;
   value $genttl   ❶        ❷
      'f','F'='Fe~mmale~-2nSubjects'
      'm','M'='Ma~mle~-2nSubjects';
   run;              ❶      ❷

title1 "Controlling Line Breaks";
proc report data=advrpt.demog nowd;
   columns sex ht wt;
   define sex   / group format=$genttl.
                  'Subject~w Gender'; ❸
   define ht    / analysis mean
                  format=5.2
                  'Height~{dagger}❹';
   define wt    / analysis mean
                  format=6.2
                  'Weight~{dagger}';
   rbreak after / summarize;

   compute after;❹       ❶
    line @1 '~{dagger} Eng~mlish Measures'
           '~-2nHeight(in.)~-2nWeight(lbs.)';
    line @1 'All su~mbjects were screened during '
           '~-2nthe intake session at visit one.';
   endcomp;      ❷
   run;
```

❶ The ~m is used to mark the alignment location (indentation) for the line breaks specified with the ~-2n sequence.

❷ Line breaks are forced using the ~-2n notation. Without forcing these breaks the width of the table would be driven by the LINE statement.

❸ The ~w indicates an optional line break. This means that this will be the preferred location for a line break, if one is needed. In this table the text will break at the space with or without using the ~w.

❹ The dagger symbol is added to associate the column label with the units footnote.

Controlling Line Breaks

Subject Gender	Height†	Weight†
Female Subjects	65.06	144.55
Male Subjects	69.24	172.91
	67.56	*161.49*

† English Measures
    Height(in.)
    Weight(lbs.)
All subjects were screened during
    the intake session at visit one.

11.3.5 Issuing Raw RTF Specific Commands

An RTF table or report is generated using a series of commands, control words, and field codes that are specific to the RTF destination (Section 15.3 goes into more detail on RTF code). Normally we do not need to know anything about those commands because they are written for us by ODS. However, when formatting sequences are not available for a specific task **and** you know the appropriate underlying command for the destination, you can pass the raw RTF destination-specific control code from SAS for execution at the destination.

You can issue these raw destination-specific commands using one of the following inline formatting functions. While the syntax used in SAS 9.1.3 will work in SAS 9.2, the newer preferred syntax is available starting in SAS 9.2. The escape character (here a tilde is used) must be declared using the ODS ESCAPECHAR option:

SAS 9.1.3 (and earlier)

```
~R/destination 'command'
```
especially useful when multiple destinations are open ~R 'command'

SAS 9.2

```
~{raw 'command'}
~{dest[destination] 'command'}
```

In the following example, raw RTF commands are passed both through a format and in TITLE statements. The commands themselves are preceded by a back slash and followed by a space. Multiple raw commands can be chained together, and they are turned off by following the control code with a 0 (zero).

```
ods escapechar = '~';
proc format;
   value $gender
      'f','F'='~{raw \b F\b0\i emale}' ❶
      'm','M'='~{raw \b M\b0\i ale}' ❶;
   run;
title1  ~{raw '11.3.5 \i0 Using \b\ul RTF\b0\ul0  Codes'}; ❷
title2 ~{raw '\i0 Italics off'};
proc report data=advrpt.demog nowd;
   columns sex ht wt;
   define sex   / group format=$gender.;
                   ... code not shown ...
```

11.3.5 Using <u>RTF</u> Codes Italics off		
patient sex	**Height**	**Weight**
Female	*65.06*	*144.55*
Male	*69.24*	*172.91*
	67.56	*161.49*

❶ Bolding is turned on and off with the \b and \b0, while italics are turned on with \i. Notice that the text in this format has control down to the letter (first letter bolded – remaining letters in italics).

❷ The default for the RTF style turns italics on for titles. These have been turned off (except for the example number; 11.3.5).

SEE ALSO
A few common RTF commands can be found in Haworth, Zender, and Burlew (2009, pg 128). The same section discusses an alternative approach using a style rather than inline formatting.

11.4 Creating Hyperlinks

Text, reports, graphs, and tables within electronic documents can automatically be connected by creating hyperlinks. SAS can create these links almost anywhere that text is displayed, as well as within graphic objects. Links can be established between portions of tables, graphs, other locations within a table, and between tables of different types. Within a table links can be established within data values, formats, titles, header text, and graphic symbols.

MORE INFORMATION
Links are created in a PROC PRINT example in Section 8.5.3.

SEE ALSO
A number of the following techniques are presented in PROC REPORT examples by Carpenter (2007b).

11.4.1 Using Style Overrides to Create Links

In the TABULATE, REPORT, and PRINT procedures the style overrides can be especially useful for creating links. The application of the style override option (STYLE=) is very similar in all three procedures; however, in PROC REPORT links can also be generated through the use of the CALL DEFINE routine.

In this example a report summarizes the data for each of the symptoms (SYMP). It is constructed so that clicking on the symptom number links to a report for that symptom.

```
%macro sympRPT;
title1  '11.4.1 Hyperlinks Using Style Overrides';

ods html file="&path\results\E11_4_1.htm" ❶
        style=journal;
title2 'Symptoms';
proc report data=advrpt.demog nowd split='*';
   column symp wt ht;
   define symp / group 'Symptom' order=internal
                 missing;
   define wt   / analysis mean format=6.1 'Weight';
   define ht   / analysis mean format=6.1 'Height';
   compute symp;
      stag = 'E11_4_1_'||trim(left(symp))||'.htm'; ❷
      call define(_col_,'url',stag);
   endcomp;
   run;
ods _all_ close;

proc sql noprint;
   select distinct symp
      into: sym1-:sym999 ❸
         from advrpt.demog;

%do s=1 %to &sqlobs; ❹
ods html file="&path\results\E11_4_1_&&sym&s...htm"
        style=journal;
title2 "Symptom &&sym&s";
proc report data=advrpt.demog(where=(symp="&&sym&s")) ❺
            nowd split='*';
   column sex wt ht;
   define sex / group 'Sex' order=internal
                style(header)={url='e11_4_1.htm'}; ❻
   define wt   / analysis mean format=6.1 'Weight';
   define ht   / analysis mean format=6.1 'Height';
   run;
ods _all_ close;
%end;
%mend symprpt;

%symprpt
```

❶ The primary HTML file is named and then created using PROC REPORT.

❷ Each individual value of SYMP is associated with a file that will contain the summary for that symptom. The CALL DEFINE is used to associate the URL with the symptom.

❸ An SQL step is used to assign the distinct values of SYMP to macro variables.

❹ A macro %DO loop is used to cycle through the individual values of symptom (&&SYM&S).

❺ The macro variable holding a specific symptom (&&SYM&S) is used to subset the data.

❻ The style override option is used to assign the URL attribute to the header, which will now link back to the primary table.

11.4.1 Hyperlinks Using Style Overrides Symptoms		
Symptom	Weight	Height
_	177.8	69.5
01	138.5	67.5
02 ❷	150.2	66.8
03	154.5	66.5
04	170.8	68.7
05	179.5	67.5
06	118.8	64.0
09	133.0	68.0
10	184.4	68.5

11.4.1 Hyperlinks Using Style Overrides Symptom 02		
Sex ❻	Weight	Height
F	147.0	66.7
M	155.0	67.0

MORE INFORMATION

Style overrides are introduced and discussed further in Sections 8.1.3 (TABULATE), 8.4.6 (REPORT), and 8.5.2 (PRINT). The CALL DEFINE routine is introduced in Section 8.4.7.

11.4.2 Using the LINK= TITLE Statement Option

In the TITLE and FOOTNOTE statements the LINK= option can be used to specify the file to which you want to link. The option can point to an internal anchor location, a local file, or may even contain a fully qualified path.

```
title1  '11.4.2 LINK= Option';

ods pdf file="&path\results\E11_4_2.pdf"
        style=journal;
title2 'Patient List Report';
title3 link='E11_4_1.htm' ❶
        'Symptom Report';
proc print data=advrpt.demog;
   var lname fname sex dob symp;
   run;
ods _all_ close;
```

In this example the LINK= option ❶ is used to point back to the primary file created in Section 11.4.1.

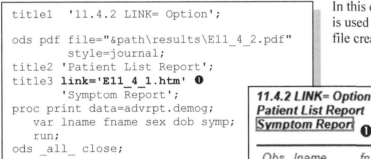

11.4.2 LINK= Option
Patient List Report
Symptom Report ❶

Obs	lname	fname	sex	dob	symp
1	Adams	Mary	F	12AUG51	02
2	Adamson	Joan	F	.	10
3	Alexander	Mark	M	15JAN30	

MORE INFORMATION

Other TITLE and FOOTNOTE statement options are described in Section 9.1.

11.4.3 Linking Graphics Elements

Most graphics elements can be created so that the viewer can click on them and then 'drill down' to another graph or report. Linkable graphic elements include, but are not limited to, histogram bars, scatter plot lines and symbols, maps, pie slices, and legend elements. Linkable elements can be enabled through the annotate facility as well as the techniques shown here.

While reports can be linked among the three primary destinations (MARKUP, PDF, RTF), linking from a graphics element requires a file designed for web viewing, such as GIF or PNG, with a markup overlay, such as HTML. This necessarily means that these techniques will not work for the LISTING destination.

The key is the generation of a character variable that contains the name of the file to which the element is to link. The following program creates a vertical histogram. Clicking on any one of the bars will display a report for the study participants with the selected number of years of education.

```
filename e1143 "&path\results\e11_4_3.png";  ❶

* Initialize graphics options;
goptions reset=all border
         ftext=swiss
         htext=1;
goptions device=png
         gsfname=E1143;  ❶

data demog;
   set advrpt.demog(keep=edu wt);
   drilledu = catt('href=E11_4_3.pdf#_',
                 left(put(edu,2.)));  ❷
   run;

              . . . . code not shown . . . .

* Create a chart that links to the summary report;
ods html path="&path\results" (url=none)  ❸
         body='E11_4_3.html';

PROC GCHART DATA=demog;
   VBAR edu / type=mean sumvar=wt
              discrete
              patternid=midpoint
              html=drilledu  ❹
              raxis=axis1
              ;
   run;
   quit;
ods html close;
```

❶ A PNG histogram will be created by PROC GCHART. By itself this file will not contain linkable elements.

❷ The variable DRILLEDU contains the name of the file to which we will be linking. In this example all the reports are in a single file with internal anchor point labels ❸, therefore, the paths are all relative to each other. This variable could contain a fully qualified path.

❸ Create the overlay file that contains the linkable elements that are associated with the graph ❶.

❹ The variable that contains the 'link to' filename ❷ is identified using the HTML option.

The HTML_LEGEND option can be used when elements of the legend are to be made linkable.

The bars of the histogram will link to a series of reports, one for each vertical bar (value of EDU).

```
%macro BldRpt;
ods pdf file="&path\results\e11_4_3.pdf" ❺
        style=journal;

proc sql noprint; ❻
    select distinct edu
        into :edu1 - :edu99
            from advrpt.demog(keep=edu);
    %let educnt=&sqlobs;
    quit;

%do i = 1 %to &educnt; ❼
    ods pdf anchor="_&&edu&i"; ❽
    ods proclabel 'Symptom Summary'; ❾
    title3 "&&edu&i Years of Education";
    proc report data=advrpt.demog
                    (where=(edu=&&edu&i ❼))
                contents="_&&edu&i Years" ❿
                nowd;
        columns symp sex,wt;
        define symp / group;
        define sex  / across 'Gender';
        define wt   / analysis mean;
        run;
%end;
ods pdf close;
%mend bldrpt;
%bldrpt
```

❺ The individual reports, to which we are linking from the histogram, are to be stored in this one PDF file.

❻ The individual values of EDU are determined and stored in a list of macro variables using an SQL step.

❼ A macro loop creates a series of reports – one for each value of EDU.

❽ The drill down variable ❷ contains an anchor label reference within the PDF file. This label is created using the ANCHOR= option. The ANCHOR points are coordinated by using the macro variable reference &&EDU&I.

❾ The ODS PROCLABEL statement is used to replace the procedure name in the PDF bookmarks with user-specified text.

❿ The PDF Bookmark value is changed using the CONTENTS= option.

Clicking on the vertical bar associated with subjects with 12 years of education (second from left), opens the indicated PDF file at the appropriate anchor point.

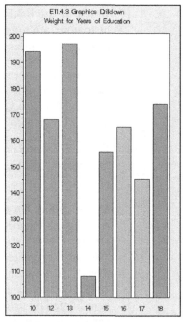

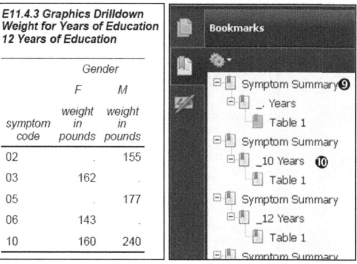

**E11.4.3 Graphics Drilldown
Weight for Years of Education
12 Years of Education**

| | Gender | |
| | F | M |
symptom code	weight in pounds	weight in pounds
02	.	155
03	162	.
05	.	177
06	143	.
10	160	240

MORE INFORMATION
Internal links are discussed further in Section 11.4.4.

11.4.4 Creating Internal Links

While hyperlinks will usually point to the beginning of a file, it is not unusual for the link to point to a location within a file. This can be a location within a file other than the one being viewed or to another location within the same file. This is known as an internal link. In the example in Section 11.4.3 (see ❷) one of the links formed is a internal link.

Internal links are designated using a pound or hash sign (#). A pointer to an internal location within a file in the local directory might be named: `E11_4_3.pdf#_15`. The internal location is

```
ods pdf anchor=_15;
```

marked using the ANCHOR= option.

In the following example a PDF document is created that contains the output from a PROC TABULATE and two PRINT procedure steps. The three reports are linked using internal locations.

```
            . . . . code not shown . . . .
 proc format;
    value $genlnk ❶
        'M' = '#Males'
        'F' = '#Females';
    run;

 ods pdf anchor='Master';
 ods proclabel='Overall';
 proc tabulate data=tabdat.clinics;
    class sex ;
    classlev sex/ style={url=$genlnk. ❷
                      foreground=blue};
    var wt;
    table sex=' ',
          wt*(n median min max)
          / box='Gender';
    run;

 ods pdf anchor='Males'; ❸
 ods proclabel='Males'; ❹
 title2 link='#Master' ❺□'Return to Master';
 title3 c=blue 'Males';
 proc print data=tabdat.clinics;
    where sex='M'; ❻
    var lname fname ht wt;
    run;
            . . . . code not shown . . . .
```

❶ A format is being used to assign the link location. The pound sign (#) identifies the link location as internal to the current file. Internal locations are specified with the ANCHOR= option ❸.

❷ The name of the file or, in this case, the internal location to which we will link is assigned using the URL attribute. Since this STYLE override option is on the CLASSLEV statement, the levels of this classification variable will form the links.

❸ The ANCHOR= option marks an internal location in the current document to which we can link.

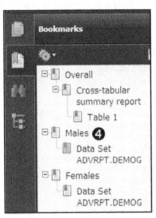

Linking Documents Master Table				
Gender	weight in pounds			
	N	Median	Min	Max
F ❶-❷	31	155.00	98.00	215.00
M	46	177.00	105.00	240.00

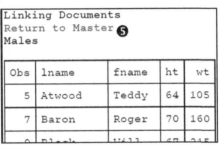

Linking Documents Return to Master ❺ Males				
Obs	lname	fname	ht	wt
5	Atwood	Teddy	64	105
7	Baron	Roger	70	160

❹ The ODS PROCLABEL statement changes how the results of this procedure are labeled in the PDF bookmarks.

❺ The LINK= option can be used to create links in titles and footnotes. Notice the use of the # sign to designate the internal link

❻ A WHERE statement is used to create the table for the males.

CAVEAT
When using SAS 9.2 under Windows, internal link locations and pointers for PDF documents should all be designated in lowercase. In SAS 9.1.3 mixed case is acceptable. This behavior may change in future releases of SAS. The problem stems from the conversion of all the anchor locations to lowercase when the document is rendered.

11.5 Traffic Lighting

Traffic lighting is a technique that allows the programmer to automatically change attributes of a report based on the values that are presented. Traditionally the background color is changed to red/yellow/green – hence the name. However any of the attributes can be changed; and most often a style override is used to change the attribute value. The key to the process is a user-defined format that is used to map the display value to the attribute that is to be changed.

MORE INFORMATION
Style overrides are introduced and discussed further in Sections 8.1.3 (TABULATE), 8.4.6 (REPORT), and 8.5.2 (PRINT). The CALL DEFINE routine is introduced in Section 8.4.7.

11.5.1 User-Defined Format

A user-defined format is used to associate an ODS attribute value with the values that are to be displayed in the table. Once established the format can be used by the TABULATE, REPORT, or PRINT procedures.

```
proc format;
   value $serious_f
      'YES','yes' = 'white';
   value $serious_b
      'YES','yes' = 'red';
   value $severity_f
      '3' = 'black'
      '4','5'= 'white';
   value $severity_b
      '3' = 'yellow'
      '4','5'= 'red';
   run;
```

Although traditionally *traffic lighting* techniques are used to change colors, the extension of the concept allows us to change multiple attributes, including cell attributes such as links, font, font size, and bolding.

The VALUE or INVALUE statement is specified as always, however, the format's label (text to the right of the equal sign) is the attribute value. The format is then used along with the attribute to which it is to be applied. Generally the association will be created using a STYLE override option and takes the form of *attribute=format*.

11.5.2 PROC TABULATE

Although when using PROC TABULATE the STYLE override options can be applied to all aspects of the table, for traffic lighting we will generally want to highlight values of statistics or values derived from the data. To do this the STYLE override option will be nested under the statistic to which it is to be applied.

```
proc format;
   value MaxWT_f  ❶
      235-high  = 'white';  ❷
   value MaxWT_b  ❸
      235-high  = 'red';  ❹
   value MinWT_f
      low-<100  = 'white';
   value MinWT_b
      low-<100  = 'red';
   run;

title1 '11.5.2 Traffic Lighting: TABULATE';
title2 'Weight Compliance';
ods listing close;
ods pdf file="&path\results\e11_5_2.pdf"
        style=journal;
proc tabulate data=advrpt.demog
              (where=(clinnum in:('05','06')));
   class clinnum;
   var wt;
   table clinnum,
         wt*(min*{style={background=minwt_b.
                         foreground=minwt_f.}}
             max*{style={background=maxwt_b.  ❷
                         foreground=maxwt_f.  ❶}});
   run;
ods pdf close;
```

In the protocol for this study the subject's initial weight should be between 100 and 235 pounds. This table highlights those clinics that have enrolled subjects that are out of compliance.

❶ The MAXWT_F. format will be used to alter the color of the text (foreground) ❷ for the maximum weights. The color of the background (red ❹)is altered by the MAXWT_B. ❸ format.

11.5.2 Traffic Lighting: TABULATE
Weight Compliance

clinic number	weight in pounds	
	Min	Max
051345	215.00	215.00
054367	160.00	160.00
057312	158.00	158.00
059372	98.00	98.00
063742	201.00	240.00
063901	187.00	187.00
065742	147.00	155.00
066789	175.00	175.00

The resulting table shows that at least two patients have been enrolled that do not meet the weight criteria. You will generally not be pleased with dark text on a dark background, which is why the foreground color was changed to white, when the background color was to be changed to red.

SEE ALSO
Carpenter (2010a) has a PROC TABULATE example of traffic lighting using the STYLE override option.

11.5.3 PROC REPORT

When using PROC REPORT the style override option can be applied in two ways. First, statement options such as the ones used in the example in Section 11.5.2 are applied directly. Second, in PROC REPORT the CALL DEFINE routine allows us to conditionally apply style overrides.

The examples in this section make use of the formats established in Section 11.5.1. In this report we would like to highlight adverse events which are either serious (SER='YES') or have a severity level which is greater than 3.

```
proc report data=advrpt.ae(where=(sev>'1')) nofs;
    column subject aestdt ser sev aedesc;
    define subject / order;
    define aedesc / order;
    define aestdt / display;
    define ser / display
                style(column) ={background=$serious_b. ❶
                                foreground=$serious_f.};
    define sev / display
                style(column) ={background=$severity_b. ❷
                                foreground=$severity_f.};
    run;
```

11.5.3a Traffic Lighting: REPORT Adverse Event List				
PATIENT ID	AE START DATE	AE SERIOUS	AE SEVERITY	
201	09/20/2006	NO	2	BILATERAL SHOULD
	09/06/2006	NO	3	LEFT GROIN PAIN
	09/10/2006	NO	4 ❷	LEFT HYDRONEPHR
	07/25/2006	NO	2	LEG CRAMPS (INCR
203	09/26/2006	NO	2	BILAT. ANKLE EDEM
	09/26/2006	NO	2	GROSS HEMATURIA
	11/21/2006	NO	2	INCREASED WEAKN
	09/13/2006	NO	2	UTI (URINE-BACTER
204	11/21/2006	N(❶	2	CIRCUMFERENTIAL
	11/06/2006	YES	3	DVT-RIGHT LEG (SA
	10/24/2006	NO	2	INCREASED NOCTU
	11/08/2006	NO	2	MENTAL CONFUSIC
205	11/17/2006	NO	2	PAIN LEFT MID-BAC
206	01/30/2007	YES	5	RIGHT SUBDURAL F
208	01/23/2007	YES	3	BACTEREMIA

Formats for the foreground and background colors are applied to the display of the values of the seriousness (SER ❶) and severity (SEV ❷) of the event.

In the figure for example 11.5.3a (to the left), the severity codes 3 and greater are highlighted regardless of the seriousness of the event. These two formats are independent of each other and both are applied. Notice that subject 201 has two events highlighted (SEV>2), however, neither of the events is rated as serious (SER=NO).

Because the CALL DEFINE is executable we can use IF-THEN/ELSE processing to conditionally assign a format. In this next example the traffic lighting style override for the severity is moved to a CALL DEFINE and is only applied for serious events (SER=YES). ❸ The CALL DEFINE results are applied to the current column ❹ which is SEV since we are executing the SEV compute block. The style override is selected ❺ and the style attributes that are to be applied are the same ones as were applied in the previous example ❷.

```
        define ser / display
                    style(column) ={background=$serious_b. ❶
                                    foreground=$serious_f.};
        define sev / display;
        compute sev;
            if ser='YES' then ❸
              call define(_col_, ❹
                          'style', ❺
                          'style ={background=$severity_b. ❷
                                   foreground=$severity_f.}');
        endcomp;
```

A portion of the PROC REPORT step is shown here.

❸ The format for the severity has only been applied when the event is serious (SER='YES').

11.5.3b Traffic Lighting: REPORT Adverse Event List

PATIENT ID	AE START DATE	AE SERIOUS	A SEVE
201	09/20/2006	NO	2
	09/06/2006	NO	3
	09/10/2006	NO	4
	07/25/2006	NO	2
203	09/26/2006	NO	2
	09/26/2006	NO	2
	11/21/2006	NO	2
	09/13/2006	NO	2
204	11/21/2006	NO	2
	11/06/2006	YES	3

SEE ALSO
The use of traffic lighting with PROC REPORT is covered in detail in Carpenter (2006b).

11.5.4 Traffic Lighting with PROC PRINT

Traffic lighting with PROC PRINT, like with PROC TABULATE and PROC REPORT (see Sections 11.5.2 and 11.5.3), is applied using the style override option. As was discussed in Section 8.5.2, which introduced the style override option for the PRINT procedure, the option can be applied on the VAR statement. The following example utilizes the formats created in Section 11.5.1 and the output mimics the first output generated by the REPORT procedure in Section 11.5.3.

```
ods pdf style=journal file="&path\results\E11_5_4a.pdf";
proc print data=advrpt.ae(where=(sev>'1'));
   by subject;
   id subject;
   var aestdt ;
   var ser / style(column)={background=$serious_b. ❶
                            foreground=$serious_f.};
   var sev / style(column)={background=$severity_b. ❷
                            foreground=$severity_f.};
   run;
ods pdf close;
```

Notice that the style override option is applied as an option on the VAR statement ❶ ❷. By splitting the VAR statement into two statements you can apply the traffic lighting formats differentially.

11.5.4 Traffic Lighting Using PROC PRINT
Severe Adverse Events

SUBJECT	AESTDT	SER	SEV
201	07/25/2006	NO	2
	09/06/2006	NO	3
	09/10/2006	NO	4
	09/20/2006	NO	2
203	09/13/2006	NO	2
	09/26/2006	NO	2
	09/26/2006	NO	2
	11/21/2006	NO	2
204	11/06/2006	YES	3
	11/08/2006	NO	2

Like the figure in Section 11.5.3a this table applies the formats on the two variables independently. In PROC REPORT we were able to use the CALL DEFINE routine to conditionally apply the format for severity. In PROC PRINT the CALL DEFINE routine is not available and formats cannot be conditionally assigned.

SEE ALSO
Carpenter (2006b) discusses traffic lighting in detail for the PROC REPORT step.

11.6 The ODS LAYOUT Statement

The ODS LAYOUT statement, which is available but was not yet production for SAS 9.2, can be used to place the results of multiple procedures, including graphics, on one page. The page is defined in terms of regions with a given procedure's output being placed into a specific region. The user gets to define the number, size, and placement of the regions.

There are two basic types of layouts ABSOLUTE and GRIDDED. Absolute layouts, shown below, have fixed region sizes, while gridded layouts can be more dynamically allocated.

The ability to define and use the output regions is made available by turning on LAYOUT with the START option on the ODS LAYOUT statement. The general form of a program that utilizes LAYOUT will be something like the following:

```
ods layout start;
        ods region . . .
        proc . . . .
        ods region . . .
        proc . . . .
ods layout end;
```

Layout is turned on with the START option and off with the END option. Between the start and end will be one or more region definition followed by the procedure that will write to that region. Regions are rectangular and within the layout page regions are defined with the ODS REGION statement.

The primary options are:

- X= horizontal position of the upper-left corner (measured from the left).
- Y= vertical position of the upper-left corner (measured from the top).
- width= width of the region.
- height= height of the region.

You get to select the size of the region; however, you do need to be careful. If the region is too small this warning may appear in the LOG when using layout.

```
WARNING: THE ABSOLUTE REGION WAS TOO SMALL TO ACCOMMODATE THE TEXT
SUPPLIED. OUTPUT WAS LOST.
```

```
%let text1 = ~S={font_face=arial   ❶
font_weight=bold}11.6 Using ODS
LAYOUT~S={font_face=arial}~nMean Weight and Height;
%let text2 = ~nfor Symptom and Years of Education;
title1;

ods pdf file="&path\results\E11_6.pdf" ❷
        style=journal
        startpage=never; ❸
ods escapechar='~';
ods layout start width=7in height=10in; ❹

ods region x=1in y=1in width=7in height=.5in; ❺
ods pdf text="&text1&text2"; ❻

ods region x=0.5in y=1.5in width=7in height=4in; ❼

proc report data=advrpt.demog nowd;
        .... REPORT code not shown ....
ods region x=1.5in y=1.5in width=3in height=4in; ❽

proc report data=advrpt.demog nowd;
        .... REPORT code not shown ....
ods _all_ close; ❾
ods layout end; ❿
```

The layout to the left establishes three regions. One for a title written by the TEXT= option, and two for side-by-side PROC REPORT tables.

❶ The title for the page is saved in macro variables for use with the ODS PDF TEXT= option ❻. Inline formatting sequences (~n, ~S) are described in Section 11.3.

❷ The PDF destination is opened.

❸ Restrain PDF from starting a new page when going to a new procedure.

❹ Open the layout space with an overall size definition.

❺ Define the first region, which will hold the title text stored in the macro variables &TEXT1 and &TEXT2.

❻ The text is written to the first region.

❼ and ❽ The regions used by the two PROC REPORT steps are defined.

❾ The PDF destination is closed. Actually here all destinations are closed.

❿ The ODS LAYOUT section is closed.

11.6 Using ODS LAYOUT
Mean Weight and Height
for Symptom and Years of Education

Symptom	Weight	Height
01	138.5	67.5
02	150.2	66.8
03	154.5	66.5
04	170.8	68.7
05	179.5	67.5
06	118.8	64.0
09	133.0	68.0
10	184.4	68.5

	Gender			
	F		M	
Years of Education	Weight	Height	Weight	Height
10	.	.	194.1	71.4
12	152.0	64.5	179.8	68.9
13	215.0	68.0	191.0	70.7
14	109.9	64.3	105.0	64.0
15	164.6	64.2	133.0	68.0
16	158.0	72.0	167.0	70.0
17	142.8	64.8	155.0	67.0
18	.	.	174.0	69.0

SEE ALSO
Rob Nelson (2010) creates a similar, but more complex layout. Along with a number of other ODS options, Dan O'Conner and Scott Huntley (2009) discuss both absolute and gridded layout options in detail. Chen (2005) introduces gridded layout.

11.7 A Few Other Useful ODS Tidbits

There are just way too many options and statements to fully describe them here. However there are a couple of which you should be at least aware.

11.7.1 Using the ASIS Style Attribute

The HTML destination removes spacing to give a 'better' display. When you want to preserve spacing surrounding a value, the ASIS option can prevent this behavior.

```
proc print data=advrpt.demog(obs=5);
   id lname;
   var fname ht wt;
   format wt 8.1;
run;
```

lname	fname	ht	wt
Adams	Mary	67	155.0
Adamson	Joan	72	158.0
Alexander	Mark	70	175.0
Antler	Peter	68	240.0
Atwood	Teddy	64	105.0

Although the format used with WT has a width of 8, the space surrounding the numbers only reflects the cell margins. We can preserve that space by using the ASIS style attribute; however, this attribute may also affect the bottom margin attribute.

```
   var fname ht;
   var wt /style(data)={asis=yes}
          style(header)={just=c};
```

lname	fname	ht	wt
Adams	Mary	67	155.0
Adamson	Joan	72	158.0
Alexander	Mark	70	175.0
Antler	Peter	68	240.0
Atwood	Teddy	64	105.0

The VAR statement has been broken into two statements so that we can associate STYLE override options with the variable WT. The header as also been centered.

Notice that the ASIS attribute has also increased the bottom margin of the cell.

11.7.2 ODS RESULTS Statement

When we create RTF files under some combinations of operating systems and versions of Microsoft Office, a prompt is issued when the SAS viewer attempts to open the new RTF file. This can be especially troublesome when the files are generated as a part of an automated system or application.

One solution is to turn off the automatic viewing of the results. In the interactive environment this can be done manually by changing the preference settings. While running the SAS Display Manager go to:

Tools → Options → Preferences→ Results

Uncheck the "View results as they are generated" box.

The automatic viewing of results can also be controlled through the use of the ODS RESULTS statement. To override the preference that is selected through the Preferences Dialogue Box, you may use either:

ods results on;   Turn on automatic review of the results (typically the default).

ods results off;   Turn off automatic review of the results.

SEE ALSO

Alternative approaches and a deeper statement of the problem were discussed in the SAS Forum thread http://communities.sas.com/message/42066#42066.

Part **3**

Techniques, Tools, and Interfaces

Chapter **12**

Taking Advantage of Formats

The use of formats is essential to the process of analysis and reporting of data. The user must be well grounded not only in the basic application of formats, but the deeper, less commonly known aspects as well. The power of formats can be taken advantage of in a wide variety of situations. It is incumbent on the user that these capabilities are fully understood.

MORE INFORMATION
Section 3.5.1 discusses the difference between formats and informats and their uses in the conversion of character to numeric values.

SEE ALSO
Carpenter (2003a) introduces a number of topics on the use of user-written formats. Ron Cody (2008a) provides a number of examples of user-written formats which demonstrate a number of the options and techniques described in this chapter. Bilenas (2005) is a book that is devoted to the topic of SAS formats.

12.1 Using Preloaded Formats to Modify Report Contents

Generally when a level of a classification variable is not included in the data, that level will not appear in any reports or analyses based on that data. Preloaded formats provide a mechanism to not only force the inclusion of those absent classification levels, they can also be used to filter or remove unwanted levels.

Preloaded formats can be used with the REPORT, TABULATE, MEANS and SUMMARY procedures. Preloaded formats are invoked through the use of options, and the available options and their usage varies for each procedure. For each procedure it is the interaction of these options that determines the resulting table. For the examples in this section study the results of the various combinations of the options, then experiment using other combinations.

For each of these procedures the two primary options used are PRELOADFMT and EXCLUSIVE:

-   PRELOADFMT   Loads the format levels prior to execution. This option will always be present when using preloaded formats.

-   EXCLUSIVE   Only data levels that are included in the format definition are to appear in table.

```
proc format;
   value $regx
      '1'=' 1'
      '2'=' 2'
      'X'=' X' ;   ❶
   value $genderu
      'M'='Male'
      'F'='Female'
      'U'='Unknown';   ❶
   value $symp
      '00'= 'Unspecified'   ❶
      '01'= 'Sleepiness'
      '02'= 'Coughing'
      '03'= 'Limping';
   run;
```

As the name implies preloaded formats key off of what is generally a user-defined format. The formats $REGX., $GENDERU., and $SYMP., which are defined here, are used in the examples that follow. Each contains one level that is not in the data ❶, and both $REGX. and $SYMP. exclude levels that *are* found in the data. The format $REGX. is used with the variable REGION, which ranges from '1' through '10'. The format $SYMP. is used with the variable SYMP, which ranges from '01' through '10'. The format $GENDERU. is used with the variable SEX which takes on the values of 'F' and 'M'.

12.1.1 Using Preloaded Formats with PROC REPORT

When preloading formats with the REPORT procedure, the PRELOADFMT and, if used, the EXCLUSIVE options are applied on the DEFINE statement.

In addition to the PRELOADFMT and EXCLUSIVE options, the REPORT procedure can also take advantage of the options COMPLETEROWS and COMPLETECOLS:

- COMPLETEROWS   All rows representing format levels are to appear in the report.
- COMPLETECOLS   All columns representing format levels are to appear in the report.

The following examples show various combinations of these options. Although they do not discuss the use of COMPLETECOLS, its behavior is similar to COMPLETEROWS, which is discussed here. This can be easily demonstrated by applying these options to the report item SEX in these same examples.

Remember that you must apply the PRELOADFMT option on the DEFINE statement for each report item for which you want to preload a format.

Using PRELOADFMT with EXCLUSIVE

Preloading with the EXCLUSIVE option allows only those levels that are on the format *and* in the data. In PROC REPORT these options are applied on the DEFINE statement.

```
title2 'Using PRELOADFMT with EXCLUSIVE';
proc report data=demog nowd;
   column region sex,(wt=n wt);
   define region / group
                   format=$regx6.
                   preloadfmt exclusive;
   define sex     / across          format=$Genderu. 'Gender';
   define n       / analysis n    format=2.0 'N';
   define wt      / analysis mean format=6.2 'Weight';
   run;
```

```
12.1.1 Preloading Formats in PROC REPORT
Using PRELOADFMT with EXCLUSIVE

                  Gender
           Female         Male
  region   N  Weight    N  Weight
    1       .     .      4  195.00
    2       6  109.67    3  105.00
```

Using the format $REGX. along with these two options causes only regions 1 and 2 to appear in the report, as these are the only two regions that are both in the format and in the data.

Using COMPLETEROWS without EXCLUSIVE

The COMPLETEROWS option, which is used on the PROC statement, forces **all** levels that appear either in the data or in the format to appear on the report.

```
title2 'Using COMPLETEROWS with PRELOADFMT and without EXCLUSIVE';
proc report data=demog nowd completerows;
   column region sex,(wt=n wt);
   define region / group format=$regx6.
                   preloadfmt
                   order=data;
   define sex    / across           format=$Genderu. 'Gender';
   define n      / analysis n    format=2.0 'N';
   define wt     / analysis mean format=6.2 'Weight';
   run;
```

```
12.1.1 Preloading Formats in PROC REPORT
Using COMPLETEROWS with PRELOADFMT and without EXCLUSIVE

                  Gender
           Female          Male
  region   N  Weight     N  Weight
    1      0     .        4  195.00
    2      6  109.67      3  105.00
    X      0     .        0    .
    3      5  127.80      5  163.80
    4      4  143.00     10  165.60
    5      5  146.20      3  177.00
    6      3  187.00      6  205.33
    7      0     .        4  151.00
    8      4  160.00      0    .
    9      2  177.00      7  187.00
   10      2  163.00      4  177.00
```

In this example the ORDER=DATA option also appears on the DEFINE statement. Prior to SAS 9.2 the ORDER= option was expected, although not *always* required. Notice that all three format levels appear. Except for REGION 'X', which does not appear in the data, the output is now in data order (the data has been sorted by CLINNUM which nominally sorts the regions as well).

The default order for a formatted report item is ORDER=FORMATTED.

Using COMPLETEROWS with EXCLUSIVE

As was shown earlier, using EXCLUSIVE without the COMPLETEROWS option yields only those levels that are both in the format and in the data. Using the EXCLUSIVE option *with* the COMPLETEROWS option creates a report that contains each of the values that are in the format, regardless of whether or not they are in the data.

```
title2 'Using COMPLETEROWS with PRELOADFMT and EXCLUSIVE';
proc report data=demog nowd completerows;
   column region sex,(wt=n wt);
   define region / group format=$regx6.
                   preloadfmt exclusive;
   define sex    / across      format=$Genderu. 'Gender';
   define n      / analysis n   format=2.0 'N';
   define wt     / analysis mean format=6.2 'Weight';
   run;
```

```
12.1.1 Preloading Formats in PROC REPORT
Using COMPLETEROWS with PRELOADFMT and EXCLUSIVE

                  Gender
            Female        Male
   region   N  Weight    N  Weight
      1     0    .       4  195.00
      2     6  109.67    3  105.00
      X     0    .       0    .
```

Only those levels in the format, regardless of whether or not they are in the data, are included in the report.

12.1.2 Using Preloaded Formats with PROC TABULATE

When using preloaded formats with the TABULATE procedure the PRELOADFMT and EXCLUSIVE options are applied on the CLASS statement. As with the REPORT procedure these two options interact.

PRELOADFMT with EXCLUSIVE

In each of the examples in this section the user-defined format, $SYMP. is used. This format, which is defined in Section 12.1, contains only three of the 10 possible values that can occur in the data, and one additional value that does not occur in the data.

```
title2 'Using PRINTMISS With the EXCLUSIVE option';
proc tabulate data=advrpt.demog;
   class symp /preloadfmt exclusive; ❶
   var ht wt;
   table symp,
         (ht wt)*(n*f=2. min*f=4.
                  median*f=7.1 max*f=4.)
         / printmiss; ❸
   format symp $symp.; ❷
   run;
```

❶ The PRELOADFMT and EXCLUSIVE options appear on the CLASS statement associated with the classification variable that is to receive the preloaded format.

❷ The appropriate format is assigned to the classification variable.

❸ The PRINTMISS option allows the display of missing values in a PROC TABULATE table. Without including this option, levels added by the preloaded format, which would necessarily always be missing, would not be displayed.

*12.1.2 Using Preloaded Formats With TABULATE
Using PRINTMISS With the EXCLUSIVE option*

	height in inches				weight in pounds			
	N	Min	Median	Max	N	Min	Median	Max
symptom code								
Unspecified	0	.	.	.	0	.	.	.
Sleepiness	4	64	67.5	71	4	115	138.5	162
Coughing	10	66	67.0	67	10	131	155.0	155
Limping	4	65	66.5	68	4	147	154.5	162

The PRELOADFMT and EXCLUSIVE options used together eliminate all values of SYMP that are not on the format, while including values on the format that are not in the data.

Because the PRINTMISS option ❸ has been used, the "Unspecified" row appears in the table with the appropriate values for N.

Using PRELOADFMT without EXCLUSIVE

Using the PRELOADFMT option without the EXCLUSIVE option allows you to have all possible rows, including those without data and those that are not included in the format.

```
title2 'Without the EXCLUSIVE option';
proc tabulate data=advrpt.demog;
   class symp /preloadfmt;
   var ht wt;
   table symp,
       (ht wt)*(n*f=2. min*f=4.
               median*f=7.1 max*f=4.)
       /printmiss;
   format symp $symp.;
   run;
```

The resulting table shows all values of SYMP that are either in the data or in the format. This includes the level of SYMP that is not found in the data.

*12.1.2 Using Preloaded Formats With TABULATE
Without the EXCLUSIVE option*

	height in inches				weight in pounds			
	N	Min	Median	Max	N	Min	Median	Max
symptom code								
Unspecified	0	.	.	.	0	.	.	.
Sleepiness	4	64	67.5	71	4	115	138.5	162
Coughing	10	66	67.0	67	10	131	155.0	155
Limping	4	65	66.5	68	4	147	154.5	162
04	13	62	68.0	74	13	98	187.0	195
05	8	63	69.0	69	8	163	177.0	201
06	11	63	64.0	65	11	105	105.0	177
09	2	68	68.0	68	2	133	133.0	133
10	13	62	68.0	72	13	158	160.0	240

MORE INFORMATION

Preloading formats is also discussed in conjunction with the REPORT procedure (see Section 12.1.1) and the MEANS and SUMMARY procedures (see Section 12.1.3).

The EXCLUSIVE option can also be used with the CLASSDATA option; see Section 8.1.4.

SEE ALSO

Carpenter (2010a) discusses the use of preloaded formats with the TABULATE procedure.

12.1.3 Using Preloaded Formats with the MEANS and SUMMARY Procedures

As was the case with PROC TABULATE (see Section 12.1.2) the PRELOADFMT and EXCLUSIVE options appear on the CLASS statement, when they are used with the MEANS and SUMMARY procedures.

Using PRELOADFMT with EXCLUSIVE

Preloading with the CLASS statement options PRELOADFMT ❶ and EXCLUSIVE ❷ limits the levels of the classification variable to those that are both on the format and in the data. Essentially the format acts as a filter without resorting to either a subsetting IF statement or a WHERE clause.

```
title2 'Using the EXCLUSIVE Option';
proc summary data=advrpt.demog;
   class symp / preloadfmt ❶
                exclusive ❷;
   var ht;
   output out=withexclusive
          mean= meanHT;
   format symp $symp.;
   run;
proc print data=withexclusive;
   run;
```

Notice that unlike PROC TABULATE, this combination of CLASS statement options does NOT insert a row for the formatted value of SYMP that is not in the

12.1.3 Using Preloaded Formats With MEANS/SUMMARY Using the EXCLUSIVE Option

Obs	symp	_TYPE_	_FREQ_	meanHT
1		0	18	66.8889
2	Sleepiness	1	4	67.5000
3	Coughing	1	10	66.8000
4	Limping	1	4	66.5000

data (SYMP='00'). To add this level using the MEANS and SUMMARY procedures, the COMPLETETYPES option must also be included.

Using COMPLETETYPES with PRELOADFMT and EXCLUSIVE

The PROC statement option COMPLETETYPES (this option was introduced in Section 7.10) will interact with the PRELOADFMT and EXCLUSIVE options. As a result of this interaction, levels of the classification variable that are not in the data, but are on the format are now included in the summary. However, levels not on the format are not included in the table.

```
title2 'With EXCLUSIVE and COMPLETETYPES';
proc summary data=advrpt.demog
            completetypes;
   class symp / preloadfmt
                exclusive;
   var ht;
   . . . code not shown . . . .
```

12.1.3 Using Preloaded Formats With MEANS/SUMMARY With EXCLUSIVE and COMPLETETYPES

Obs	symp	_TYPE_	_FREQ_	meanHT
1		0	18	66.8889
2	Unspecified	1	0	.
3	Sleepiness	1	4	67.5000
4	Coughing	1	10	66.8000
5	Limping	1	4	66.5000

The 'Unspecified' level for SYMP now appears in the report even though it is not in the data (_FREQ_=0).

Using COMPLETETYPES without EXCLUSIVE

When the COMPLETETYPES option is used without the EXCLUSIVE option, all levels of the classification variable appear whether it is in the data or if it is only in the preloaded format.

```
title2 'Two Classification Variables';
title3 'COMPLETETYPES Without EXCLUSIVE';
proc summary data=advrpt.demog completetypes;
   class symp sex / preloadfmt ;
   var ht;
   output out=twoclass mean= meanHT;
   format symp $symp.
          sex  $genderu.;
   run;
proc print data=twoclass;
   run;
```

A portion of the table generated by the two classification variables SYMP and SEX, which both have preloaded formats, is shown here. Notice that each format

level not associated with data appears as do the unformatted levels (only SYMP='04' is shown in this partial table).

12.1.3 Using Preloaded Formats With MEANS/SUMMARY
Two Classification Variables
COMPLETETYPES Without EXCLUSIVE

Obs	symp	sex	_TYPE_	_FREQ_	meanHT
1			0	65	67.2000
2		Female	1	29	65.0690
3		Male	1	36	68.9167
4		Unknown	1	0	.
5	Unspecified		2	0	.
6	Sleepiness		2	4	67.5000
7	Coughing		2	10	66.8000
8	Limping		2	4	66.5000
9	04		2	13	68.6923

12.2 Doing More with Picture Formats

Although generally speaking picture formats are only nominally an advanced topic, they are far underutilized and there are some concepts that are unique to picture formats that are commonly misunderstood.

The PICTURE statement is used to build what is essentially a template of zeros, nines, and other characters that are also to be displayed. The zeros and nines are used as placeholders. The nines indicate that a number must be placed at the location, even if it is a zero. A zero placeholder indicates that a number is placed at the location if it is non-zero (embedded zeros are displayed).

SEE ALSO

The book *The Power of PROC FORMAT* (Jonas Bilenas, 2005) is a good source for some of the more introductory picture format topics.

The documentation for the PICTURE statement gives a clear description of the application of the format to the value that is to be formatted. The application process should be well understood before working with fractional values.

12.2.1 Date Directives and the DATATYPE Option

The DATATYPE= option on the PICTURE statement can be used to build date-specific formats. This option allows the use of 'directives', which tell the PICTURE statement how to further structure or format the value relative to the type of data that is to be interpreted.

The directives are individual case-sensitive letters that indicate a specific portion of the DATE, TIME, or DATETIME value. There are over 15 directives and the case of the letters used as directives is important:

Y	Year	M	Minute
m	Month	S	Second
d	Day	b	Month abbreviation
H	Hour	B	Month name

In the PICTURE statement the directive is preceded by a percent sign (%) which acts like an escape character. Single quotes are used to prevent the interpretation of the % as a macro language trigger.

The following format is used to display a SAS datetime value in a format that can be used in DB2.

```
proc format;
   picture dbdate
      other = '%Y-%0m-%0d:%0H:%0M:%0S ❶' (datatype=datetime) ❷;
   run;

data _null_;
   now = '11sep2010:15:05:27'dt;
   put now=;
   put now= dbdate.;
   call symputx('selldate',now); ❸
   run;
```

❶ As in other PICTURE formats a zero may be used as a placeholder.

❷ The DATATYPE option determines how the incoming value is to be interpreted. Option values indicate the type of data that the format will be used with. These data types include:

- DATE   SAS date value
- TIME   SAS time value
- DATETIME   SAS datetime value
- DATETIME_UTIL   SAS datetime value specific for the utilities industry (SAS 9.3)

The LOG shows:
```
now=1599836727
now=2010-09-11:15:05:27
```

In the DATA step above the macro variable &SELLDATE was created ❸. Once formatted this macro variable could be used to write this DATETIME value into SQL code that is to be passed through to a DB2 server. The %SYSFUNC calls the PUTN function which will render the formatted value.

```
%put %sysfunc(putn(&selldate,dbdate.));
```

Because of the variety of directives and the availability of the DATATYPE option, there is a great deal of flexibility as to the resulting formats. This means that a format can be generated for any of the three datetime data value types. This flexibility can be demonstrated by creating alternate formats for the MONNAME. format (which can be used only with DATE values). The MONTHNAME. and MONTHABB. formats created here are used with DATETIME values.

```
proc format;
   picture monthname
      other = '%B ❹      ' (datatype=datetime);
   picture monthabb
      other = '%b ❺' (datatype=datetime);
   run;

data _null_;
   now = '11sep2010:15:05:27'dt;;
   put now=;
   put now= monthname.;
   put now= monthname3.; ❻
   put now= monthabb.;
   run;
```

❹ The %B directive returns the month name. When defining the format be sure to leave sufficient space for the longest month (September), otherwise truncation of the month's name could take place.

❺ The lowercase %b directive returns the first three letters of the month in uppercase. Again allow three spaces (including the directive and the escape character).

❻ If an abbreviated month name is desired in mixed case, a width value can be included with the format for the full name.

```
now=1599836727
now=September ❹
now=Sep ❻
now=SEP ❺
```

SAS 9.3 Date Directive Enhancements

Prior to SAS 9.3 fractional seconds were truncated (even when the ROUND option was applied to the PICTURE format). Starting in SAS 9.3 fractional seconds are rounded. Although generally of minor concern this can change the date for time values within a half second of midnight.

```
proc format;
 picture myDayT (round)
   low - high = '%0d%b%0Y:%0H:%0M:%0S'(datatype=datetime)
     ;
run;
```

	Display Value	
DateTime Constant	**MYDAYT. (prior to SAS 9.3)**	**MYDAYT. (SAS 9.3 and after)**
'01apr2011:12:34:56.7'dt	01APR2011:12:34:56	01APR2011:12:34:57
'01apr2011:23:59:59.7'dt	01APR2011:23:59:59	02APR2011:00:00:00

The utility industry often wants to reference a midnight date to be 24:00:00 instead of 00:00:00. The new DATATYPE= value DATETIME_UTIL allows this.

```
proc format;
  picture ymdtime (default=19)
    other='%Y-%0m-%0d %0H:%0M:%0S' (datatype=datetime_util);
   run;
```

DateTime Constant	YMDTIME. Display Value
'01nov2008:00:00:00'dt	2008-10-31 24:00:00
'01nov2008:00:00:01'dt	2008-11-01 00:00:01

```
proc format;
  picture durtest(default=27)
    other='%n days %H hours %M minutes'
          (datatype=time);
  run;

 data _null_;
    start = '01jan2010:12:34'dt;
    end = '01feb2010:18:36'dt;
    diff = end - start;
    put diff=durtest.;
    run;
```

The %n directive allows you to count the number of complete days within an interval. This allows you to return the duration in days/hours/minutes/seconds between two datetime values. The %D directive cannot be used because it returns the day of the month.

The LOG shows the result of the PUT statement.

```
diff=31 days 6 hours 2 minutes
```

12.2.2 Working with Fractional Values

Picture formats do not automatically handle fractional values well. This is especially true for values between zero and 1, and the value of zero itself must also be taken into consideration or it will not display correctly.

The problem with fractions is demonstrated in the following example. The variable VAL ranges from 0 to 3 by .25, and three picture formats have been created to display these values.

```
proc format;
   picture showval
      other = '0000'; ❶
   picture withdec
      other = '00.0'; ❷
   picture twodec
      other = '09.00'; ❸
   run;

data vallist;
   do val = 0 to 3 by .25;
      val2 = val;
      val3 = val;
      val4 = val;
      output;
   end;
   format val2 showval.
          val3 withdec.
          val4 twodec.;
   run;
```

```
12.2.2 Picture Formats
Showing Decimals
                      ❶      ❷      ❸
Obs      val     val2    val3    val4

  1     0.00                    0.00
  2     0.25              2      0.25
  3     0.50              5      0.50
  4     0.75              7      0.75
  5     1.00      1      1.0     1.00
  6     1.25      1      1.2     1.25
  7     1.50      1      1.5     1.50
  8     1.75      1      1.7     1.75
  9     2.00      2      2.0     2.00
 10     2.25      2      2.2     2.25
 11     2.50      2      2.5     2.50
 12     2.75      2      2.7     2.75
 13     3.00      3      3.0     3.00
```

❶ The format SHOWVAL. does not contain an explicit decimal point. The displayed values are only the integer portions. Fractional values are not displayed and the 0 is not displayed.

❷ The WITHDEC. format contains a decimal point which allows a single decimal value (tenths). Values below 1 are not displayed correctly, and the 0 is not displayed at all.

❸ Adding a 9 in the digits place forces the format to write a value in that location. The values less than 1 are now displayed correctly. The 0 is also now being displayed.

When working with values that are less than one be sure to force at least one significant digit by using the 9 as the placeholder in the format label.

Truncation is another area that warrants extra consideration when dealing with picture formats. Values that do not fit into the picture template (the format label) are truncated. The variable X in this example ranges from a small fraction to a value over one thousand.

```
proc format;
   picture showdec
      other = '09.00';  ❸
      run;
data x;   ❹                        ❺        ❻
do x = .007,.017,.123,1.234, 12.345, 1234;
   y=x;
   output;
end;
format y showdec. x 8.3;
run;
```

❸ The format is defined with two decimal places and 9 in the unit's digit. Notice that the implied format used for X does not have sufficient range to display both the largest and smallest value. Something like the

```
12.2.2 Picture Formats
Showing Decimals

Obs            x          y

1          0.007       0.00  ❹
2          0.017       0.01
3          0.123       0.12
4          1.234       1.23
5         12.345      12.34  ❺
6       1234.000      34.00  ❻
```

format 8.3 would have had the range to show all the values.

❹ The value less than .01 (the smallest permitted by the format) is not surprisingly truncated.

❺ Only two of the three least significant digits can be displayed. The display value for X is rounded but for Y the thousandths are truncated.

❻ We have only allowed for values up to 99.99. Values over 100 have the most significant digits truncated.

The rounding and truncation issues can be addressed in the format definition. The SHOWDECR.

```
proc format;
   picture showdecr (round)  ❼
      other = '00009.00';  ❽
      run;
```

format shown here allows both rounding and numbers up to 99999.99.

❼ The ROUND option causes the format to round fractional values.

```
12.2.2 Working with Fractional Values
Showing Decimals

Obs        x         y

1       0.007      0.01
2       0.017      0.02
3       0.123      0.12
4       1.234      1.23
5      12.345     12.35
6    1234.000   1234.00
```

❽ Be sure to include a sufficient number of placeholders to accommodate the largest number.

12.2.3 Using the MULT and PREFIX Options

While text that is to be included, either within or following the formatted value can be included as part of the picture definition, text that is to precede the display value is ignored. When you want preceding text to be a part of the displayed value, the PREFIX= option can be used.

The MULT= option allows the application of a multiplier. This value is multiplied against the incoming value and the result is displayed.

A common alternative solution to handling fractional values involves the use of the MULT= and PREFIX= options.

```
proc format;
   picture showdec
      0          = '9' ❶
      0<  - <.01 = '9'(prefix='<.01') ❷
      .01 - <1   = '99'(prefix='0.' mult=100) ❸
      other      = '00000.00'; ❹
   run;
data x;                                    ❺
do x = 0,.001,.012,.123,1.234, 12.345, 1234;
   y=x;
   output;
end;
format y showdec.;
run;
```

❶ The zero is handled separately.

❷ Values that are smaller than .01 add a prefix value that includes the <.01 text.

❸ For values from .01 up to 1, multiply the value by 100 and add the '0.' prefix.

❹ Make sure that there are sufficient significant digits.

❺ Since we did not use the (ROUND) option this value has been truncated.

```
12.2.3a Picture Formats
Using MULT and PREFIX

Obs          x          y

1         0.00          0   ❶
2         0.00       <.010  ❷
3         0.01        0.01
4         0.12        0.12
5         1.23        1.23
6        12.35       12.34   ❺
7      1234.00     1234.00   ❹
```

Since the numbers in this example have two decimal places, multiplying them by 100 (10^n where n is the number of significant digits to be displayed) turns the number into an integer. The decimal is then inserted via the picture format.

In the following example dollars are being converted to British pounds (the conversion constant used for this example is 0.635, which is almost certainly not the current currency conversion constant).

```
title2 'Using The MULT and PREFIX Options';
proc format;
   picture pounds
      1   - 10  = '9,00'❻(mult=63.5 ❼prefix='£'❽)
      10< - 100 = '09,0'(mult=6.35 ❾prefix='£')
      100<- high= '000.000.000' (mult=.635 prefix='£');
   run;

data money;
   do dollars =  1.23, 12.3, 123, 1230, 12300;
      pounds = dollars;
      output;
   end;
   format dollars dollar10.2 pounds pounds.;
   run;

proc print data=money;
   run;
```

❻ The display value template specifies two decimal places (using the European style with a comma separating the whole numbers from the fractions.

❼ The multiplication factor is specified using the MULT= opton.

❽ The PREFIX= option is used to designate one or more leading symbols. Here the option is applied to each value/label pair; however, it does not need to be constant, as it is in this example.

```
12.2.3a Picture Formats
Using The MULT and PREFIX Options

Obs       dollars      pounds

1          $1.23        £0,78
2         $12.30        £7,8
3        $123.00         £78
4      $1,230.00        £781
5     $12,300.00      £7.810
```

❾ The multiplication factor changes to accommodate the number of decimal places.

The symbol for pounds, £, which was used in this example, does not appear on most US keyboards. Since special characters such as this one, exist in most fonts, utilizing them in SAS is fairly straightforward. From your word processor insert the character or symbol of choice into a document. From the document copy it and then paste it into the SAS editor, where it can now be used in your code.

In the following expansion of the POUNDS. format, we allow for the inclusion of negative values as well as values between 0 and 1.

```
picture pounds
   -1   - <0  = '99'   (mult=63.5 prefix='£-0,')  ❿
    0        = '9'    (prefix='£')
    0 < - <1 = '99'   (mult=63.5 prefix='£0,')
    1   - 10 = '9,00' (mult=63.5 prefix='£')
   10< - 100 = '99,0' (mult=6.35 prefix='£')
  100<- high= '000.000.000' (mult=.635 prefix='£');
```

❿ Because the absolute value is taken on mapped values, the minus sign must also be added as a prefix character.

```
12.2.3b Using the PREFIX and MULT Options
Including Negative Values

Obs       dollars      pounds

1         $-0.12       £-0,07
2          $0.00          £0
3          $0.12        £0,07
4          $1.00        £0,63
5          $1.23        £0,78
6         $12.30        £07,8
7        $123.00         £78
8      $1,230.00        £781
9     $12,300.00      £7.810
```

SEE ALSO

Chapman (2003) has a number of examples of PICTURE formats that use the MULT= option.

12.2.4 Display Granularity Based on Value Ranges – Limiting Significant Digits

Sometimes we want the precision of the displayed value to self-adjust as the size of the numbers change. This was shown, but not really commented on, in the examples in the previous section. In the following example we want to limit the displayed value to no more than 6 digits; however, the values themselves range from 0 to billions.

```
title2 'Limiting Significant Digits';
proc format;
  picture Tons
            0         =   '9'
          0< -     <1 =   '99' (prefix='0.' mult=100)
           1 -    <10 =   '9.99'
          10 - <1000 =   '000.9'
        1000 - <1e06 =   '000,000'  ❶
        1e06 - <1e09 =   '000.999M'  ❷ (mult=1e-03)  ❸
        1e09 - <1e12 =   '000.999B' (mult=1e-06);  ❹
  run;

data imports;
    do tons = 0, .15, 1.5,1.5e2, 1.5e4, 1.5e7, 1.5e10;
        fmttons = tons;
        output;
    end;
    format fmttons tons.;
    run;
proc print data=imports;
    run;
```

❶ Values between 1000 and 1,000,000 (1e6) are shown without any decimal places.

❷ For values in the millions, an 'M' is displayed following the 6 most significant digits.

❸ We have shifted the decimal six places; however, only three have been lost.

❹ The decimal shift is now 9 digits and of these only three are shown before the decimal point. The multiplication factor is therefore 1e-06.

This type of format can be especially useful on graphs where we need to control the width of the tick mark values on the axis.

```
12.2.4 Picture Formats
Limiting Significant Digits

Obs                    tons      fmttons

1                      0.00            0
2                      0.15         0.15
3                      1.50         1.50
4                    150.00        150.0
5                  15000.00       15,000
6               15000000.00      15.000M
7            15000000000.00      15.000B
```

12.3 Multilabel (MLF) Formats

When creating formats, overlapping format ranges are generally not allowed. Multilabel formats overcome this limitation; however, only selected procedures are able to utilize this special type of format. Multilabel formats are created using the MULTILABEL option on the VALUE statement.

12.3.1 A Simple MLF

In the following example we summarize the years of education into high school and college. At the same time we want to see the graduate school subset of those attending college. This can easily be accomplished using a multilabel format.

```
proc format;
    value edlevel (multilabel) ❶
        9-12 = 'High School'
        13-high='College'
        17-high='Graduate Studies'; ❷
    run;
```

```
title1 '12.3.1 Multi-label Formats';
proc tabulate data=advrpt.demog;
    class edu / mlf ❸;
    class sex;
    var wt;
    table edu=' ' all,
          sex*wt*(n*f=2.
                  mean*f=5.1
                  stderr*f=6.2)
          /box=edu;
    format edu edlevel.;
    run;
```

❶ The MULTILABEL option sets up the format to be used with the MLF option in PROC TABULATE.

❷ The ranges for 'College' and 'Graduate Studies' overlap (are not distinct). Without the MULTILABEL option, PROC FORMAT would generate an error and the format would not be created.

❸ The MLF option appears on the CLASS statement associated with the formatted variable. Without this option 'Graduate Studies' will not be displayed as a level of EDU.

❹ PROC TABULATE correctly counts

and totals the number of persons in each education category. Notice that the count for 'Graduate Studies' does not contribute to the overall total.

The MLF option can also be used with CLASS variables in PROC MEANS. In future releases of SAS it may be implemented in other procedures as well.

12.3.1 Multi-label Formats

years of education	patient sex					
	F			M		
	weight in pounds			weight in pounds		
	N	Mean	StdErr	N	Mean	StdErr
College	23	142.0	7.02	23	156.6	6.96
Graduate Studies	8	142.8	9.84	6	167.7	10.64
High School	8	152.0	9.71	22	187.0	4.59
All ❹	31	144.5	5.76	45	171.4	4.75

Procedures that do not utilize overlapping format values (do not support the MLF option) will only use the primary range of the format.

MORE INFORMATION
Very often we would like to have better control of the order of the formatted values (in this example High School comes after College). Section 12.4 discusses the use of the NOTSORTED option. Multilabel formats are used to calculate a moving average in Section 12.3.2.

12.3.2 Calculating Rolling Averages

There are a number of ways to calculate a rolling or moving average within SAS. The use of multilabel formats provides a quick and easy programming solution to this often taxing problem. In this example we would like to calculate a three-visit rolling average of the patient's potassium levels. The variable VISIT will take on the values of $1 - 16$.

```
data control(keep=fmtname start end label hlo);
     retain fmtname 'avg' ❶
            hlo 'M'; ❷
     do start=1 to 14;
        end=start+2; ❸
        label=cats('VisitGrp', put(start,z2.)); ❹
        output Control;
     end;
     hlo='O'; ❺
     label='Unknown';
     output;
run;

proc format cntlin=control; ❻
   run;

proc summary data=advrpt.lab_chemistry;
   by subject;
   class visit / mlf; ❼
   format visit avg.; ❽
   var potassium;
   output out=rollingAVG
          mean= Avg3Potassium;
   run;
```

❶ A format control data set containing the value pairs, labels, and format name (AVG.) is created.

❷ The HLO variable is used to designate this as a multilabel format.

❸ START is the lower bound of the moving average and END is the upper bound. In this example the width will contain up to three visits.

❹ The label is assigned a value. For the group starting with visit 4, LABEL will be VisitGrp04, which will be the average of visits 4, 5, and 6.

❺ Although not needed here, it is always a good idea to specify the 'OTHER' group.

❻ PROC FORMAT creates the format using the CONTROL data set.

❼ The overlapping ranges of a multilabel format are ignored unless the MLF option is specified.

❽ The format is associated with the classification variable.

The format AVG., which is generated above, is effectively defined as shown to the right.

```
proc format;
   value avg (multilabel)
       1 - 3 = 'VisitGrp01'
       2 - 4 = 'VisitGrp02'
       3 - 5 = 'VisitGrp03'
       4 - 6 = 'VisitGrp04'
       5 - 7 = 'VisitGrp05'
       6 - 8 = 'VisitGrp06'
       7 - 9 = 'VisitGrp07'
       8 - 10= 'VisitGrp08'
       9 - 11= 'VisitGrp09'
      10- 12= 'VisitGrp10'
      11- 13= 'VisitGrp11'
      12- 14= 'VisitGrp12'
      13- 15= 'VisitGrp13'
      14- 16= 'VisitGrp14'
      other = 'Unknown';
   run;
```

While this technique is a fast and easy way to generate rolling averages, you should be careful to fully understand how it will work with your data. In the portion of the data shown here for SUBJECT 201, there is no visit 3 and there is a duplicate observation for visit 2.

```
12.3.2 Using MLF for Calculating Rolling Averages

 Obs    SUBJECT    VISIT        LABDT    potassium

  15      201        1      07/07/2006      3.8
  16      201        2      07/14/2006      4.2
  17      201        2      07/14/2006      4.2
  18      201        4      07/26/2006      4.2
  19      201        5      07/21/2006      4.7
  20      201        6      07/29/2006      4.5
  21      201        7      08/04/2006      4.0
  22      201        8      08/11/2006      4.0
  23      201        9      09/12/2006      4.2
  24      201       10      10/13/2006      3.9
```

After applying the format in the PROC SUMMARY step, we notice in the portion of the data set listing for this subject (201) that there is an average for visit 3 (it contains visits 4 & 5 only). Also notice that because the duplicate observation (visit 2) precedes the missing visit 3 the means for the first two visits have been distorted.

This subject also did not have any follow-up visits after visit 10. This is reflected in the N

```
12.3.2 Using MLF for Calculating Rolling Averages

 Obs   SUBJECT     VISIT       _FREQ_   Avg3Potassium

  12     201                     10        4.17000
  13     201      VisitGrp01      3        4.06667
  14     201      VisitGrp02      3        4.20000
  15     201      VisitGrp03      2        4.45000
  16     201      VisitGrp04      3        4.46667
  17     201      VisitGrp05      3        4.40000
  18     201      VisitGrp06      3        4.16667
  19     201      VisitGrp07      3        4.06667
  20     201      VisitGrp08      3        4.03333
  21     201      VisitGrp09      2        4.05000
  22     201      VisitGrp10      1        3.90000
```

associated with the last two rolling averages.

MORE INFORMATION

Section 3.1.7 uses an ARRAY to calculate a running average. Section 12.7 discusses the process of creating a format from the data in more detail.

SEE ALSO

The first time that I learned about this technique was from Liang Xie who suggests using a multilabel format to create a rolling window in a sasCommunity.org tip http://www.sascommunity.org/wiki/Tips:Summarize_data_in_a_rolling_window.

Moving averages are calculated using PROC EXPAND by Vora (2008).

12.4 Controlling Order Using the NOTSORTED Option

Normally when a user-defined format is created, the format is internally placed into sorted order. Consequently it does not particularly matter what order the value/label pairings are specified in the value statement. However, this reordering can be prevented by using the NOTSORTED option on the VALUE statement ❶. When applied, the internal order of the format remains as it is defined. When an ORDER=FORMATTED option is applied to a format that is created using the NOTSORTED option, the order of the pairings in the format definition is used.

In the PROC TABULATE example in Section 12.3 'College' appears first because it is first alphabetically. Here the format is created using the NOTSORTED option. Since the level for 'High School' is listed first in the following PROC FORMAT, 'High School' will appear before 'College' in the report.

```
proc format;
   value edlevel (notsorted)  ❶
      9-12 = 'High School'
      13-high='College';
   run;

proc tabulate data=advrpt.demog;
   class edu sex;
   var wt;
   table edu all,
         sex*wt*(n*f=2. mean*f=5.1 stderr*f=6.2);
   format edu edlevel.;  ❷
   run;
```

❶ When the NOTSORTED option is used on the VALUE statement, the order that the item pairs are defined in the VALUE statement is preserved.

❷ The format is used as usual.

CAVEAT

When a format is created it is optimized internally to make the assignment process as efficient as possible. Using the NOTSORTED option negates some of that optimization; however, for formats with fewer than a dozen or so value pairs it generally makes little practical difference. For very large formats (hundreds or thousands of items) there may be some performance access issues when using the NOTSORTED option. Be sure to experiment with your data and OS when using this option for larger formats.

12.4 Formats Defined with NOTSORTED

	patient sex					
	F			M		
	weight in pounds			weight in pounds		
	N	Mean	StdErr	N	Mean	StdErr
years of education						
High School	8	152.0	9.71	22	187.0	4.59
College	23	142.0	7.02	23	156.6	6.96
All	31	144.5	5.76	45	171.4	4.75

12.5 Extending the Use of Format Translations

A user-defined format can be used to point to another secondary format or, in SAS 9.3, even a function. This allows us to create customized formats that retain characteristics of either another format or a function. The format or function call appears unquoted on the right side of the equal sign (in the label area) in square brackets.

12.5.1 Filtering Missing Values

```
proc format;
   value missdate
      . = 'Unknown'  ❶
      other=[date9.];  ❷
   run;

proc print data=advrpt.demog;
   var lname fname dob;
   format dob missdate.;  ❸
   run;
```

We need to create a format that will handle missing date values differently than non-missing values. In the MISSDATE. format to the left, missing values are mapped to the word 'Unknown' ❶, while all other values are formatted using the DATE9. format.

❷ The label for the 'other' category is the secondary or nested format. It is enclosed in square brackets instead of quotes.

❸ The nested format is used as is any other format.

```
12.5 Nested Formats

Obs      lname          fname          dob

  1      Adams          Mary        12AUG1951
  2      Adamson        Joan        Unknown
  3      Alexander      Mark        15JAN1930
  4      Antler         Peter       15JAN1934
       . . . . portions of the listing not shown . . . .
```

The first few lines of the listing generated by PROC PRINT show that the date of birth has been formatted. Since Joan Adamson's date of birth is missing, it has been displayed as 'Unknown'.

```
proc format;
   value pctzero
      .='0.00'
      other=[6.2];
   run;
```

Similar nested formats can be used to check for valid data or data that can be formatted. This is the case with the PCTZERO. format which is used to map missing values to special characters for reporting purposes. Here missing values are mapped to '0.00', while the non-missing values are displayed using the 6.2 format.

12.5.2 Mapping Overlapping Ranges

Another use of nested formats is to define ranges or groups with multiple ranges. Notice here that the 'Secondary' and 'Out of Range' groups span the 'Primary' group. A change to one group definition requires a change to another's as well.

```
proc format;
   value agegrps
      low - 40  = 'Out of Range'
      40 - <48  = 'Secondary'
      48 -  52  = 'Primary'
      52<-  65  = 'Secondary'
      65<- high = 'Out of Range'
      other = 'Unknown';
   run;
```

Instead we can create a series of nested formats that call each other. Although potentially more work to set up initially, these formats offer more flexibility because the range for each group is totally self-contained. The format for the inner most range (PRIMARY.) is called first, and if the value is outside of the primary range, the secondary range format (SECOND.) is called. Notice that the secondary range spans the primary range. In automated systems this can be a huge advantage.

In an assignment statement the format is used as any other format might be used:

```
proc format;
   value primary
      48 -  52  = 'Primary'
      other = [second.];
   value second
      40 - 65   = 'Secondary'
      other = [OOR.];
   value oor
      low - high = 'Out of Range'
      other = 'Unknown';
   run;
```

```
agegroup = put(startage,primary.);
```

12.5.3 Handling Text within Numeric Values

SAS supports up to 28 types of numeric missing values (see Section 2.10.1). These sometimes need to be associated with codes in the data. When importing data from raw text files, the codes themselves may be inserted as non-numeric values into the numeric fields. We need the ability to read the column as a numeric value while differentiating among the various codes.

```
proc format;
   invalue inage
      y, yz = .y
      s, ss = .s
      other = [2.];
   run;
data surveyAge;
   input patcode $
         age inage.;
   datalines;
1 45
2 yz
3 36
4 ss
5 y
   run;
```

The informat INAGE. converts selected codes into the specific missing values .S and .Y. All other non-numeric codes will map to the standard numeric missing (.). The data set SURVEYAGE contains the following values. Remember that special numeric missing values print without the period.

Obs	patcode	age
1	1	45
2	2	Y
3	3	36
4	4	S
5	5	Y

MORE INFORMATION
The use of special numeric missing values is discussed in Section 2.10.1. A numeric informat similar to this one is created from a data set in Section 6.7.3.

SEE ALSO
The SAS Forum thread found at http://communities.sas.com/message/48729 discusses various methods for avoiding errors when reading mixed fields.

12.5.4 Using Perl Regular Expressions within Format Definitions

Starting in SAS 9.3 the REGEXPE option can appear on the value side of the equal sign for a value/label pair. This option follows a quoted Perl regular expression and causes the format to effectively act like the PRXCHANGE function.

The $ABC2DEF. informat shown here has been created to convert the letter combination 'abc' to 'def'.

```
proc format;
    invalue $abc2def (default=20)  ❶
        's/abc/def/' (REGEXPE)❷ = _same_❸;
    run;

data _null_;
x=input('abc',$abc2def.);   put x=; ❹
x=input('xabcx',$abc2def.); put x=; ❹
x=input('xyz',$abc2def.);   put x=; ❺
x=input('def',$abc2def.);   put x=; ❺
run;
```

❶ The informat $ABC2DEF. is defined with a default length of 20.

❷ The PRX string is defined and followed by the REGEXPE option, which causes the PRX string to be interpreted as a Perl Regular eXpression.

❸ The special _SAME_ operator is used to pass the result of the PRX string to the format's label.

❹ The LOG shows that both a stand-alone occurrence and an embedded occurrence of 'abc' are converted to 'def'.

❺ Letter combinations other than 'abc' are not changed and are therefore passed through the format 'as-is'.

```
x=def     ❹
x=xdefx   ❹
x=xyz     ❺
x=def     ❺
```

12.5.5 Passing Values to a Function as a Format Label

Starting with SAS 9.3 it is possible to pass values into functions via formats. The huge advantage is that the functions, which are embedded into the format, become available outside the DATA step, and can be used wherever formats are used.

```
proc format;
  value fmtname (default=10)
        other=[myfunc()];
run;
```

The function call occurs in the label portion of the value/label pair and is enclosed in square brackets. The specifications include:

- The function may take no more than one argument.
- Numeric functions return numeric values.
- Character functions return character values.
- The DEFAULT= option should be used to ensure proper widths.
- The function can be supplied by SAS or it can be user supplied through the use of PROC FCMP (see Section 15.2).

Using SAS Supplied Functions

The FIPSTATE function can be used to convert a FIPS state code into a two-letter state postal code abbreviation; however, there is no matching format to perform the same conversion. Here we use a format to pass the FIPS code to the FIPSTATE function. As is shown in the LOG, the FIPSTATE. format returns the 2-character abbreviation.

```
proc format;
  value fipstate
      other=[fipstate()];
  run;

data _null_;
    x=37; put x=fipstate.;
    run;
```

```
x=NC
```

The use of functions in formats opens a wide range of possibilities including the use of DATE, DATETIME, and TIME functions. In this example both numeric and character versions of a series of formats are created that execute various functions. Notice that conversion between numeric and character values will occur as needed.

```
proc format;
  value daten    (default=10) other=[date()];     ❶
  value $datec   (default=10) other=[date()];
  value dpartn   (default=10) other=[datepart()];  ❷
  value $dpartc  (default=10) other=[datepart()];
  value lenn     (default=10) other=[length()];    ❸
  value $lenc    (default=10) other=[length()];
  run;
data _null_;
  x=datetime();   ❹
  y=put(datetime(),best12.);   ❺
  z=put(date(),best12.);  ❻
  a=datepart(x);  ❼
  put x= y= z= a=;
        ❽          ❾          ❿
  put x=daten. x=dpartn. x=lenn.;
  put y=$datec. y=$dpartc. y=$lenc.;
  put z=$datec. z=$dpartc. z=$lenc.;
  run;
```

❶ The DATE function does not take an argument, but it can still be used within a format label.

❷ The DATEPART function will extract the date portion from a datetime value.

❸ The LENGTH function will provide the length of the argument. Although, the LENGTH function expects a character argument, a numeric value will be converted to a character value which will subsequently be passed to the LENGTH function.

A DATA _NULL_ step is used to create some numeric and character date and datetime values which are then used with the formats that were just created.

❹ X is a numeric value containing the current datetime value in seconds (there are 10 digits in the number of seconds). The execution shown here took place on day 18,946 (15nov2011).

❺ Y is a character variable of length 12 containing the number of seconds for the current datetime value.

❻ Z is a character variable of length 12 with the current date.

❼ A is the numeric date portion of the datetime value stored in X.

❸ The DATEN. and $DATEC. formats correctly return the current date value.

❾ The DPARTN. and DPARTC. formats return the date portion of the datetime value in days. The variable Z already contains a date value. The $DPARTC. format interprets this value as seconds and returns the date value of 0 (01jan1960). This shows that although the value is character it can be handled by the function as numeric.

```
x=1636995477.6  y=1636995477.6  z=18946  a=18946
      ❽              ❾           ❿
x=18946  x=18946  x=10
y=18946  y=18946  y=12
z=18946  z=0      z=12
```

❿ The LENGTH function is applied to the incoming value. Notice that for the variable X the numeric format (LENN.) is applied and only the number of whole numbers is counted (10). The character format ($LENC.) returns 12 when applied to the same number.

Using User-Defined Functions

When combined with user-defined functions, the ability to insert a value into a function through a format can be especially powerful.

In this example the user-defined functions C2FF() and F2CC() are used to convert between degrees Centigrade and degrees Fahrenheit. These two character functions add the scale symbols to the resultant value. Similar numeric functions that do not add the scale symbols are created in Section 15.2.2.

```
proc fcmp
      outlib=Advrpt.functions.Conversions;  ❶
   function c2ff(c) $;  ❷
      return(cats(((9*c)/5)+32,'°F'));  ❸
   endsub;  ❹

   function f2cc(f) $;
      return(cats((f-32)*5/9,'°C'));
   endsub;
   run;
options cmplib=(advrpt.functions);  ❺
data _null_;
   f=c2ff(100); put f=;  ❻
   c=f2cc(212); put c=;
   run;
```

❶ The OUTLIB= option specifies the data set and packet (CONVERSIONS) that will contain this function definition.

❷ The FUNCTION statement names the new function and its arguments. The $ is used to specify that this function returns a character value.

❸ The RETURN statement contains the value to be returned by the function. In this case the result of the conversion equation is concatenated to the temperature scale symbol.

❹ FUNCTION definitions are terminated with the ENDSUB statement.

❺ The CMPLIB option is used to point to the data set (ADVRPT.FUNCTIONS) that contains the function definition.

❻ A constant value (100°C) is converted using the C2FF() function, and the converted value is displayed using a PUT statement.

```
f=212°F  ❻
c=100°C
```

These two functions can be used wherever functions can be used; however, since most procedures will not accept the use of functions, they are not as generally usable as formats. Below these two functions are called by formats so that they can be used wherever formats can be used. This includes in the PUT statement where formats are anticipated and functions are not callable.

Since the C2FF() and F2CC() functions do not have more than one argument, they can be used directly in the label portion of a user-defined format. Although in this example the format name and the function name are the same, this is not in any way a requirement.

```
proc format;
   value c2ff (default=10)  ❼
      other=[c2ff()];  ❽
   value f2cc (default=10)
      other=[f2cc()];
      run;

data _null_;
      c=100; put c=c2ff.;  ❾
      f=212; put f=f2cc.;
      run;
```

❼ When using a function in the label it is best to set the default display width.

❽ The function is called from within the label by enclosing it in square brackets.

❾ The format is requested on the PUT statement. This request results in the execution of the function, and generates this text in the LOG.

```
c=212°F  ❾
f=100°C
```

The initial versions of these functions were written by Rick Langston, senior manager in software development at SAS.

Return the Quarter without the Year (Qq instead of yyQq)

In Section 15.2.1 the QNUM() function is created to remove the year portion of the value that is returned by the YYQ. format. In that example a second DATA step was required before we could use the results of the function in a PROC FREQ step. By using that function in a format, we can use the format directly in the PROC step and thereby avoid an additional pass of the data.

```
proc format;
   value qfmt other=[qnum()];
run;

options cmplib=(advrpt.functions);
proc freq data=advrpt.lab_chemistry
         order=formatted;
   table visit*labdt;
   format labdt qfmt.;
   run;
```

Here the QNUM() function is used in the format label. The format is then used in the PROC FREQ step. The elimination of steps will almost always improve processing efficiency.

12.5.5 Functions as Format Labels

The FREQ Procedure

Frequency Percent Row Pct Col Pct	Table of VISIT by LABDT				
		LABDT(LAB TEST DATE)			
VISIT(VISIT NUMBER)	Q1	Q2	Q3	Q4	Total
1	3	3	7	4	17
	1.78	1.78	4.14	2.37	10.06
	17.65	17.65	41.18	23.53	
	8.33	7.50	12.73	10.53	
2	3	2	7	4	16
	1.78	1.18	4.14	2.37	9.47
	18.75	12.50	43.75	25.00	
	8.33	5.00	12.73	10.53	

Informats can also be used with functions-as-labels. This is a case where the user wanted the feature of the TRAILSGN informat, but that informat does not handle implied decimal specifications. In this example the numbers in the data have been entered without the decimal point (the value 12 should be .12). The function will use the TRAILSGN informat and then divide by 100 and return the result.

```
proc fcmp outlib=work.functions.smd;
    function tsgn(text $);
        put 'in tsgn: ' text=;
        x = input(text,trailsgn10.);
        x = x/100;
        return(x);
    endsub;
run;
options cmplib=(work.functions);
proc format;
    invalue tsgn(default=10)
        other=[tsgn()];
data _null_;
    input x: tsgn.;
    put x=;
cards;
1
1-
12-
123-
123+
1+
0
run;
```

```
1     x=.01
1-    x=-0.01
12-   x=-0.12
123-  x=-1.23
123+  x=1.23
1+    x=0.01
0     x=0
```

MORE INFORMATION
Details on the use of PROC FCMP to create user-defined functions can be found in Section 15.2.

12.6 ANYDATE Informats

The ANYDATE informats (available starting in SAS 9) are designed to allow you to read in a variety of mixed date forms including:

- DATE, DATETIME, and TIME
- DDMMYY, MMDDYY, and YYMMDD
- JULIAN, MONYY, and YYQ.

There are various forms of these informats:

- ANYDTDTE.   extracts the date portion
- ANYDTDTM.   extracts the datetime portion
- ANYDTTME.   extracts the time portion

12.6.1 Reading in Mixed Dates

This example demonstrates the flexibility of these informats. Here the ANYDTDTE10. informat is applied to a number of different date forms. The DATESTYLE system option ❶ is used to resolve some of the possible ambiguities by declaring a default ordering for the month/day/year portions of the dates.

```
options datestyle=mdy; ❶
data new;
input date anydtdte10.;
format date date9.;
datalines;
01/13/2003
13/01/2003
13jan2003 ❷
13jan03
13/01/03 ❸
01/02/03
03/02/01
run;
```

```
12.6 ANYDATE Informats

Obs          date

 1      13JAN2003
 2      13JAN2003
 3      13JAN2003 ❷
 4      13JAN2003
 5      03JAN2013 ❸
 6      02JAN2003
 7      02MAR2001
```

A PROC PRINT of the data set shows that these date values have been read into the data set as SAS dates. The supported date forms are quite varied and include DATE9. ❷. You should note that the DATESTYLE= option cannot fully resolve all ambiguities and, consequently, some dates may be misinterpreted. Since 13 is not a valid number for a month ❸, the informat assumes that the order of the values has been changed and it guesses that the correct order is YMD. Since the informats can detect invalid values and make informed guesses as to the correct order of the date portions, it becomes *very* important for the user to either know the data well, or to at the very least, understand exactly how the incoming values can be converted. In this example it is likely that ❸ should have been read as dd/mm/yy and not yy/mm/dd.

12.6.2 Converting Mixed DATETIME Values

When the incoming string contains not only dates, but time values as well, the conversion process becomes even more complicated. For the time portion (and the following applies to SAS time values as well), the hours can be specified using the 24 hour clock or the 12 hour clock with AM/PM also included.

Datetime strings might include values such as those shown to the right. With the exception of the ambiguous date with a two-digit year ❶ the ANYDTDTM informat will correctly interpret the date time portion of these strings. It is unable, however, to utilize the AM/PM codes, which are ignored.

```
10/13/2011:15:45:12
2011-03-01T15:20:45
9/13/2011 11:52:54 AM
9/13/2011 11:52:54 PM
13/09/2011 11:52:54 PM
11/09/12 11:52:54 AM ❶
2011/09/12 11:52:54 PM
```

Fortunately the MDYAMPM. informat is available. This informat correctly interprets the AM/PM portion of the datetime value; however, it requires that the date portion be in MDY order.

Since there is no informat that will effectively combine the flexibility of the ANYDTDTM. informat with the ability to interpret the AM/PM, the following adjustment can be used. ❷ The

```
dt_plus= input(daytime,anydtdtm.) ❷
        +(43200 ❸
        *(^^index(upcase(daytime),'PM')❹));
```

```
12.6 ANYDATE Informats
Converting Mixed DATETIME Values

Obs          daytime                        dt_plus

 1    10/13/2011:15:45:12        13OCT2011:15:45:12
 2    2011-03-01T15:20:45        01MAR2011:15:20:45
 3    9/13/2011 11:52:54 AM      13SEP2011:11:52:54
 4    9/13/2011 11:52:54 PM      13SEP2011:23:52:54
 5    13/09/2011 11:52:54 PM     13SEP2011:23:52:54
 6    11/09/12 11:52:54 AM       09NOV2012:11:52:54
 7    2011/09/12 11:52:54 PM     12SEP2011:23:52:54
```

ANYDTDTM. informat is used to convert the datetime string. The INPUT function returns the datetime value in seconds. ❸ If a 'PM' is present we need to add 12 hours worth of seconds (43,200=12*60*60) to the datetime value. ❹ The INDEX function searches for an occurrence of 'PM' and the location is converted to a binary 0/1 value which is multiplied by the number of seconds to add.

12.7 Building Formats from Data Sets

The VALUE, INVALUE, and PICTURE statements are usually used to create a user-defined format or informat. As the number of value pairs becomes large, coding these statements becomes inconvenient (for me large can be less than a dozen). Fortunately you can also define formats and informats using a data set.

PROC FORMAT accepts a data set to control the definition of the format or informat. The procedure expects specific variables, and the observations of the data set are used to form the value pairs. As a minimum the data set used to control the formation of the format must contain the variables FMTNAME, START, and LABEL ❶. It may also contain over twenty other variables that can be used to define the format.

In this example the data set CNTRLFMT will be used create a character format ($CL_REG.) that will map the clinic number (CLINNUM) into a region (REGION).

```
data cntrlfmt(keep=fmtname start label)❶;
    set advrpt.clinicnames(rename=( clinnum=start ❷
                                    region=label)); ❷
    retain fmtname '$cl_reg'; ❸
    run;
proc format cntlin=cntrlfmt; ❹
    run;
```

❶ You may keep variables that will not be used by PROC FORMAT. Extraneous variables will be ignored.

❷ The variable START contains the data value that is to be mapped (left side of the value pair), while LABEL is the value that will appear as a result of the mapping.

❸ The character variable FMTNAME contains the name of the format that is to be created.

❹ The CNTLIN= option is used to specify the data set that contains the format definition.

The control data set may define more than one format definition; however, if it does the data set must be sorted by format name (or at least grouped by format name). In the following example

```
data cntrlfmt(keep=fmtname start label);
   set advrpt.clinicnames(rename=( clinnum=start));
   length fmtname $8 label $40;
   fmtname = '$cl_reg';
   label = region;
   output cntrlfmt;
   fmtname = '$cl_name';
   label = clinname;
   output cntrlfmt;
   run;
proc sort data=cntrlfmt;
   by fmtname start;  ❺
   run;
proc format cntlin=cntrlfmt;
   run;
```

each incoming observation is used to build two formats ($CL_REG. and $CL_NAME.). ❺ Notice that the control data set has been sorted by the format name prior to passing it to the FORMAT procedure.

The list of potential variables in the control data set that can be utilized by PROC FORMAT is

```
proc format cntlout=control(where=(fmtname='CL_NAME'));
   run;
proc print data=control;
   run;
```

quite extensive, and while they are well documented, you can use

PROC FORMAT and the CNTLOUT= option to surface the format definition by writing it to a data set. This will reveal the variable names that can be used in a format definition and can give you a good idea about their usage.

```
12.7 Building Formats from Data

                                                                    D L
       F                                        D              D A A
       M                                        E L  P   N     D I T N
       T   S        L                           F E  R    O  S E E G A G
       N   T        A                           A N F E M F E T E E C 3 T U
 O  A  A   E  B                             M M U G U F U I D Y X X H S S Y A
 b  M  R   N  E                             I A L T Z I L L I P C C L E E P G
 s  E  T   D  L                             N X T H Z X T L T E L L O P P E E

 1 CL_NAME 011234 011234 Boston National Medical     1 40 27 27 0  0  0 C N N
 2 CL_NAME 014321 014321 Vermont Treatment Center     1 40 27 27 0  0  0 C N N
 3 CL_NAME 023910 023910 New York Metro Medical Ctr   1 40 27 27 0  0  0 C N N
 4 CL_NAME 024477 024477 New York General Hospital    1 40 27 27 0  0  0 C N N
 5 CL_NAME 026789 026789 Geneva Memorial Hospital     1 40 27 27 0  0  0 C N N
```

The HLO variable can be especially useful because it allows you to specify not only the HIGH and LOW open ended ranges, but the keyword OTHER ❻ as well. Another less known usage of HLO is to specify a nested format (HLO='F') ❼. The format STUDYDT., created here, sets the acceptable range of dates for the study. All other dates will be displayed as 'Out of Compliance'.

```
data intervals(keep=fmtname start end label hlo);
   retain fmtname 'studydt';
   length label $20;
   start = '12jan2006:01:01:01'dt;
   end = '24nov2007:11:12:13'dt;
   label = 'datetime18.';
   hlo = 'F';  ❼
   output intervals;
   start=.;  ❽
   end=.;  ❽
   hlo='O';❻
   label= 'Out of Compliance' ;
   output intervals;
   run;

proc format cntlin=intervals;
   run;
```

❽ Although set to missing in this example, the START and END variables do not need to be cleared for the observation containing `HLO='O'`.

MORE INFORMATION
An example in Section 6.5 creates a format from data in order to perform a table lookup. The example in Section 12.3.2 creates a multilabel format based on data generated in a DATA step.

SEE ALSO
The following SAS Forum thread contains an example of a format built from a data set http://communities.sas.com/message/39814.

12.8 Using the PVALUE Format

When displaying values between zero and one, especially values close to zero, it is often difficult to determine the number of decimal values needed. The PVALUE. format was designed to display small probability values. The number of decimal points (4 in this table) ❶ designates the smallest number that can be displayed by the format. Smaller numbers will be displayed with a < sign. ❷

This format was designed to work with probability values that are necessarily constrained to be between 0 and 1, consequently this format does not handle negative values or even 0 well. All values less than the minimum specified (for PVALUE6.4 this is .0001) will be displayed the same (as <.0001). ❸ Even the minimum (.0001) is displayed as <.0001.

X	PVALUE6.4 ❶
0.000000	<.0001
0.000006	<.0001 ❷
0.000050	<.0001
0.000100	<.0001 ❸
0.000400	0.0004
0.003000	0.0030
0.020000	0.0200
0.100000	0.1000

If you want to take advantage of the capabilities of the PVALUE format, but suspect that some numbers will be equal to or less than zero or even greater than one, then you may want to create a format that incorporates, but does not solely depend on PVALUE.

Here the RANGE. format is created using a combination of formats.

```
proc format;
  value range
  low - <0 = [best7.]
  0        = [1.]
  0<  -  1 = [pvalue6.4]
  1<  -high= [best6.];
run;
```

The table to the right shows how various values (I) are displayed using the PVALUE7.4 format (J) and the RANGE. format (K).

12.8 Using the PVALUE Format			
Obs	i	j	k
1	-4.1	<.0001	-4.1
2	-0.0001	<.0001	-0.0001
3	-0.00001	<.0001	-1E-5
4	0	<.0001	0
5	0.00001	<.0001	<.0001
6	0.00001	<.0001	<.0001
7	0.0003	0.0003	0.0003
8	0.02	0.0200	0.0200
9	0.1	0.1000	0.1000
10	1	1.0000	1.0000
11	12.34	12.3400	12.34
12	3456.789	3456.79	3456.8

12.9 Format Libraries

User-defined formats and informats are saved in a catalog, which by default will have the name FORMATS. When a format library is not named on the PROC FORMAT statement, the format definition is written to the catalog WORK.FORMATS. The entry name is the same as the format. Here the definition of the TONS. format will be stored in WORK.FORMATS.TONS.FORMAT.

```
proc format;
  picture Tons
    0 = '9'
.... code not shown ....
```

The catalog entry type will depend on the type of format that is created:

- FORMAT   Numeric format
- FORMATC   Character format
- INFMT   Numeric informat
- INFMTC   Character informat

Because each of the four types has a different catalog entry type, the same format name can be used up to four times.

12.9.1 Saving Formats Permanently

Formats are stored permanently by using the LIBRARY= option on the PROC FORMAT statement. The LIBRARY= option is used to specify the *libref* that is to contain the FORMATS catalog. Any formats created by this PROC FORMAT will be stored in a catalog with the name of FORMATS in the *libref* ADVRPT.

```
proc format library=advrpt;
.... code not shown ....
```

Because of the way that SAS searches for format catalogs (see Section 12.9.2), formats that are stored in a catalog named FORMATS in a *libref*

```
libname library 'c:\myfmts';

proc format library=library;
     .... code not shown ....
```

named LIBRARY will by default be included in the search path. The code at first seems a bit odd, but this FORMAT step will create or add to the catalog LIBRARY.FORMATS.

Although I find it to be generally a good idea, you are not required to store the format definitions in a catalog named FORMATS. You can specify the catalog name as a second level on the

```
proc format library=advrpt.projfmt;
     .... code not shown ....
```

LIBRARY= option. Here the catalog will be named PROJFMT.

12.9.2 Searching for Formats

When requesting a format SAS first checks in WORK.FORMATS and then, if the *libref* LIBRARY is defined, SAS will look in LIBRARY.FORMATS. Since format libraries are not usually conveniently located in these two locations, we need to be able to search for formats in a variety of places and in catalogs named something other than FORMATS.

The FMTSEARCH= system option is used to identify not only the *librefs*, but also the order for the search. In this example SAS will look for the requested format in the catalog ADVRPT.PROJFMT, and then in catalogs named FORMATS in the *librefs* WORK and LIBRARY in that order. Since WORK appears in the FMTSEARCH list, the default catalog is no longer

```
options fmtsearch=(advrpt.projfmt work library);
```

WORK.FORMATS, and it is not searched first.

12.9.3 Concatenating Format Catalogs and Libraries

When your formats are spread among multiple catalogs, the search can be simplified by concatenating the catalogs.

Catalogs with the same name will be implicitly concatenated when they reside within concatenated libraries. In the following example formats are being written to two different libraries (the *librefs* are OLDFMT and NEWFMT).

```
libname oldfmt 'c:\temp1';
libname newfmt 'c:\temp2';
libname allfmt (newfmt oldfmt); ❶

proc format library=oldfmt;
  value yesno 1 = 'Yes' ❷
              0 = 'No';
  value generation
   low - <1950 = 'Greatest'
   1950 - high = 'Boomer';
  run;

proc format library=newfmt;
  value gender 1 = 'Female'
               0 = 'Male';
  value yesno 1 = 'No' ❷
              0 = 'Yes';
  run;

title1 12.9.3 Display Format names;
proc catalog cat=allfmt.formats; ❸
  contents;
  quit;

options fmtsearch=(allfmt work); ❹
```

❶ Any references to the *libref* ALLFMT will point to both of the other two locations.

❷ The YESNO. format appears in both format libraries. Notice that the two definitions are *not* the same.

❸ PROC CATALOG is used here to show the locations of the formats. The column LEVEL refers to the library containing the catalog that contains the format.

❹ Before any of the formats in the ALLFMT library can be used, the library must be included on the search path. Formats in ALLFMT will be found before any formats with the same name in the WORK library. Within ALLFMT the catalogs are searched from left to right. The version of the YESNO. format in the NEWFMT library will be used.

The output generated by the CATALOG procedure shows that the various formats can be found in the concatenated library ALLFMT. It also shows that the YESNO. format, which is defined in both the OLDFMT and NEWFMT catalogs is being picked up from the first catalog in the list (NEWFMT – LEVEL=1).

12.9.3 Display Format names

		Contents of Catalog ALLFMT.FORMATS				
#	Name	Type	Level	Create Date	Modified Date	Description
1	GENDER	FORMAT	1	11Dec11:14:40:49	11Dec11:14:40:49	
2	GENERATION	FORMAT	2	11Dec11:14:40:49	11Dec11:14:40:49	
3	YESNO	FORMAT	1	11Dec11:14:40:49	11Dec11:14:40:49	

Chapter 13

Interfacing with the Macro Language

A great deal has been written on the macro language. The documentation provided by SAS Institute is good and two SAS Press books have been written on the subject, *Carpenter's Complete Guide to the SAS Macro Language, 2nd Edition* (Carpenter, 2004), and *SAS Macro Programming Made Easy, Second Edition* (Burlew, 2006). The treatment of the macro language in this book, therefore, must necessarily be limited to a few topics.

SEE ALSO

Russ Tyndall, a Principal Technical Support Analyst for the DATA Step and Macro Language at SAS Institute, has written TS739 (Tyndall, 2005), which contains a number of advanced tips as well as newer features of the macro language.

13.1 Avoiding Macro Variable Collisions—Make Your Macro Variables %Local

The rule of thumb that a macro variable created within a macro will be local to that macro is a very dangerous rule simply because it is usually correct. In the world of programming the problem that only shows up occasionally is one of the worst to detect and fix. Simply knowing the rules (and there is nothing simple about knowing the rules) for symbol table assignment is not enough. Most symbol table assignments depend on circumstances that are usually unknown, and even more often unknowable, to the macro programmer. The rules for table assignment are described in Carpenter, 2004, Section 13.6, page 395.

Macro variable collisions occur when a macro variable assignment is written to the unintended symbol table and inadvertently overwrites an existing macro variable's value. Very often this happens when the same macro variable name is inadvertently used in more than one of a series of nested macros.

```
%macro primary;
 .... code not shown....

   %do i = 1 %to &dsncnt;  ❷
    %chksurvey(&&dsn&i)
   %end;

  .... code not shown....
%mend primary;
```

A subtle example of a macro variable collision, and one that can cause horrid errors while leaving the programmer blissfully unaware, is contained in the program fragments shown here. In this case a secondary macro (%CHKSURVEY) is called from within a %DO loop in %PRIMARY.

The %DO loop in %PRIMARY seems to work without error. But a closer inspection of the inner macro %CHKSURVEY reveals a hidden problem.

```
%macro chksurvey(dset);
 .... code not shown....

   %do i = 1 %to 5;  ❶

  .... code not shown....
%mend chksurvey;
```

❶ The %DO loop in %CHKSURVEY uses &I as the index variable. This variable would *usually* be local to %CHKSURVEY; however, since &I already exists in the higher table of the calling macro, it will not be local. When %CHKSURVEY executes, it will modify the value of &I in the higher table of %PRIMARY. ❷

In this example, if &DSNCNT ❷ is less than or equal to 6, the loop in %PRIMARY will execute only once. If the programmer is lucky, &DSNCNT will be greater than 6 and an infinite loop will be created. This is lucky because then the programmer will at least know to look for the problem!

Unfortunately, we cannot protect the macro variables in a symbol table from being overwritten by a macro that is called by the outer macro. However, we can protect higher symbol tables by forcing all of our macro variables onto the LOCAL symbol table. This is done through the use of

```
%macro chksurvey(dset);
   %local i;  ❸
 .... code not shown....

   %do i = 1 %to 5;

  .... code not shown....
%mend chksurvey;
```

the %LOCAL statement. When using nested macro calls, macros that call macros, *ALWAYS* use %LOCAL to prevent collisions!! In the previous example the collision would have been avoided by simply adding the %LOCAL statement to %CHKSURVEY. ❸ As an aside we do not know for sure that the macro variable &I in %PRIMARY will not also cause problems to an even higher table, and a %LOCAL statement should also have been

included in that macro.

In the previous paragraph, I suggest that you *ALWAYS* use the %LOCAL statement. Others have suggested that this is too strident, that the %GLOBAL statement and the global symbol table exist for a reason—so that we can use them. After all, the argument goes, there are situations when you want to pass a value out of a local symbol table and into the global environment. While I concede the desire, and *might* even admit to having used the global symbol table in this way, it is my admittedly biased opinion that this should not be a first choice, and that it is generally a solution employed when one of the alternative techniques discussed in this chapter are either not possible or more likely are not fully understood. That said, there is nothing wrong with using the global symbol table to pass values as long as the programmer fully understands the risks and has correctly and successfully mitigated them.

The problem with using the global symbol table is that it does not exist in the parent-child hierarchy implied by nested macros. The calling macro is the parent and the called macro is the child macro. We can take advantage of this relationship.

The best way to avoid collisions is to take direct control of symbol table placement. This is what we did by placing the %LOCAL statement in our macro definitions. Another method of doing this, without using the global symbol table, is to define your macro variables, and the symbol tables that they use, in such a way as to allow their values to flow to a higher symbol table. This technique also allows a macro function to return more than one value without resorting to global macro variables.

While this method is not particularly flexible, it can be useful in some circumstances. The key to its success is to remember that if a macro variable already exists in a higher symbol table, and NOT the most local table, a macro variable assignment will be written to the higher table. This technique is demonstrated in the example that follows.

The name of the analysis data set (&DSN) is determined in the macro %GETDATANAME. In this case, for some reason, %GETDATANAME cannot be written as a macro function (see Section 13.7 for more on writing macro functions), and consequently it passes the data set name out of the macro using the macro variable &DSN. If %GETDATANAME does not explicitly have a %LOCAL statement for &DSN, the macro variable will be written to the next higher table in which it already exists; in this case this is the local table for %PRIMARY ❹.

```
%macro primary;
   %local dsn;  ❹

   %getdataname  ❺

proc print data=&dsn  ❻
   .... code not shown....
%mend primary;

%macro getdataname;
   .... code not shown....
   %let dsn = biomass;  ❺
   .... code not shown....
%mend getdataname;
```

❹ The macro variable &DSN is added to the local symbol table for %PRIMARY (with a null value).

❺ %GETDATANAME assigns a value to the macro variable &DSN. Assuming that &DSN does not already exist on the local symbol table for %GETDATANAME (there is no %LOCAL statement), and since it already exists in a higher table ❹, its value is written to the higher table (the local table for %PRIMARY). ❻ The data set name generated in %GETDATANAME is available during the execution of the remainder of the macro %PRIMARY, because it resides on the local table for %PRIMARY. The value of &DSN flows from the child macro (%GETDATANAME) to the higher parent macro (%PRIMARY).

MORE INFORMATION
Section 13.6 has another example which purposefully passes a macro variable through a higher table. The macro in Section 13.2.2 passes macro variables out of a macro by using the global symbol table.

SEE ALSO
Carpenter (2005) goes into detail on the subject of macro variable collisions and how they can be avoided. You can also read more about collisions and macro variable referencing scopes in Carpenter (2004, Section 5.4.2).

13.2 Using the SYMPUTX Routine

Starting in SAS®9 the SYMPUTX routine is offered as alternative to the SYMPUT routine for building macro variables from within the DATA step. My preference is to always use the SYMPUTX routine. When I am modifying existing programs that contain a call to SYMPUT, I will whenever possible, convert it to a SYMPUTX.

13.2.1 Compared to CALL SYMPUT

The SYMPUTX routine has two major advantages over SYMPUT, and one minor disadvantage. The only disadvantage of SYMPUTX relative to SYMPUT is that its name has one more letter to type, and that letter is an x! And actually this is not so much of a disadvantage.

Advantages of SYMPUTX over SYMPUT:

- Automatic conversion of numeric input (without a note in the LOG). ❶
- Uses a field width of up to 32 characters when it converts a numeric second argument to a character value. CALL SYMPUT uses a field width of up to 12 characters. ❷
- The value's leading and trailing blanks are removed prior to assignment. ❸
- Ability to force the macro variable onto the local or global symbol table. ❹

The advantages of SYMPUTX over SYMPUT are great enough that generally all new coding is being done using SYMPUTX in preference to SYMPUT. The advantages are shown in this example that writes the value of the numeric variable EDU for subject 205 into three macro variables. ❶ The use of the numeric variable

```
data _null_;
  set advrpt.demog(where=(subject=205));
  call symput('EDU205a',edu);  ❶
  call symput('EDU205b',left(put(edu,3.)));  ❷
  call symputx('EDU205c',edu);  ❸
  run;
%put |&edu205a|  |&edu205b|  |&edu205c|;
```

EDU generates a conversion note in the LOG as well as the storage of a right justified character string.

❷ When using SYMPUT numeric values must be converted using the PUT function to avoid the note in the LOG. Here a TRIM function could have been used to avoid the storage of the trailing blank.

❸ The SYMPUTX routine solves both of the issues shown in ❶ and ❷.

The LOG shows:

```
NOTE: Numeric values have been converted to character values at the
places given by: ❶
    (Line):(Column).
    47:26
NOTE: There were 1 observations read from the data set ADVRPT.DEMOG.
      WHERE subject=205;
NOTE: DATA statement used (Total process time):
      real time          0.06 seconds
      cpu time           0.00 seconds

51  %put |&edu205a|  |&edu205b|  |&edu205c|;
|       12|  |12 |❷  |12|  ❸
```

The optional third argument to the SYMPUTX routine can be used to place the macro variable onto the local ('l' or 'L') or global ('g' or 'G') symbol table ❹. The ability to control the symbol table assignment is especially

```
call symputx('EDU205c',edu,'l' ❹);
```

important when attempting to avoid macro variable collisions (see Section 13.1). When the name of the macro variable is known as it is in this example, it is just as easy to use the %LOCAL (or %GLOBAL) statement. However, when the macro variable's name is derived during the program's execution, a %LOCAL statement is not always possible.

MORE INFORMATION

The SYMPUTX examples in Sections 13.2.2 and 13.2.3 use the third argument to make the symbol table assignment.

13.2.2 Using SYMPUTX to Save Values of Options

The macro %SCALEPOS is used to rescale the SAS/GRAPH VPOS and HPOS graphics options. The values of these options are stored in macro variables generated by a call to the SYMPUTX routine.

```
%macro ScalePos(hvscale=2.5);
data _null_;
      set sashelp.vgopt(keep=optname setting);
      where optname in('HPOS','VPOS');
      call symputx(optname,setting,'G'); ❺
      run;

goptions hpos=%sysevalf(&hpos * &hvscale)
      vpos=%sysevalf(&vpos * &hvscale);
%mend scalepos;
goptions reset=all dev=win;
%scalepos(hvscale=1.5)

    . . . . code not shown . . . .
* Reset the HPOS and VPOS graphics options;
goptions hpos=&hpos vpos=&vpos; ❻
```

❺ The name of the macro variable to be created is stored in the variable OPTNAME, and consequently the first argument to the SYMPUTX routine is not a constant. The third argument, which can be either uppercase or lowercase, allows us to place these macro variables onto the global symbol table.

❻ These macro variables can be used later to reset the graphics options back to their original values.

This technique of passing variables to the global symbol table assumes that we know that the macro variables &HPOS and &VPOS either do not already exist or that it is OK for the macro %SCALEPOS to change their values. If this is not the case then we are at risk for having a macro variable collision (see Section 13.1).

SEE ALSO

Other methods of retrieving, storing, and reestablishing options and their values can be found in Carpenter (2004, Section 10.3.1).

13.2.3 Using SYMPUTX to Build a List of Macro Variables

It is very common to work with lists of items within the macro language, and there are several ways to create and process the items in these lists (Fehd and Carpenter, 2007). One common way of creating a list of macro variables is through the use of SYMPUTX.

The following DATA step is part of a macro that creates a list of all variables within a SAS data set that are of one type (numeric or character). The name of the macro variables to be created will take on the values of &VARNAME1, &VARNAME2, etc.

```
data _null_;
  set sashelp.vcolumn(where=(libname="%upcase(&lib)" &
              memname="%upcase(&mem)" &
              type="&type")) end=eof;
  call symputx('varname'||left(put(_n_,9.)),name,'L');
  if eof then call symputx('varcnt',_n_);
  run;
```

The CALL SYMPUTX could have been somewhat simplified by using the CATS function to perform the concatenation.

```
call symputx(cats('varname',_n_),name,'L');
```

Using a DATA step is not the only way, nor even necessarily the easiest, to create a list of numbered macro variables. See the sample code associated with this section for a PROC SQL example.

MORE INFORMATION
A list of macro variables is created using a PROC SQL step in Section 13.5.

SEE ALSO
In the macro language there are four primary ways of handling lists. These are discussed in Fehd and Carpenter (2007). Rozhetskin (2010) gives a number of clear and straightforward examples of the use of list processing for a variety of tasks. Crawford (2006) introduces a macro to simplify list processing.

13.3 Generalized Programs—Variations on a Theme

The macro language is first and foremost a code generator. As such, one of its strengths is to create and store reusable code. The next two sections discuss the process of generalization. As you become stronger in the macro language and more comfortable with the process itself, you may find that some of the steps shown in this section will become compressed or even eliminated.

13.3.1 Steps to the Generalization of a Program

Because macro programs can be difficult to debug, it is often easier to start with a working (non-macro) step or program. Then examine your code and modify it using these steps:

- Identify those things that change from use to use.
- Convert these items to macro language elements.
- Use named parameters with reasonable defaults.

```
proc means data=advrpt.demog noprint;
  class sex;
  var ht wt;
  output out=stats
    n=
    mean=
    stderr= / autoname;
  run;
```

Consider the following simple PROC MEANS step. We want to generalize the step to allow processing against any data set and any list of classification and analysis variables. We may also want to allow the user to choose whether or not the procedure will generate printed output. The bolded sections of code are those things that we will need to control using macro language elements. These are the items that will be dependent on run-time conditions.

```
proc means data=&dsn &print;
  class &classlst;
  var &varlst;
  output out=&outdsn
    n=
    mean=
    stderr= / autoname;
  run;
```

These dependencies are then converted to macro language elements, in this case macro variables. The values of these macro variables can be supplied in a number of different ways. For the simple, most straightforward case, the values can be supplied as parameters in a macro call.

The macro %MYMEANS uses keyword parameters to specify the macro parameters. In addition the macro also performs some logic checks.

```
%macro mymeans(dsn=advrpt.demog,
               classlst=sex,
               varlst=ht wt,
               outdsn=stats,
               print=noprint ❶);
proc means data=&dsn
 %if &outdsn = %then print; ❷
 %else &print;; ❸
%if &classlst ne %then %do;class &classlst;%end; ❹
%if &varlst ne %then %do; var &varlst; %end; ❹
%if &outdsn ne %then %do; ❺
 output out=&outdsn n= mean= stderr= / autoname;
%end;
run;
%mend mymeans;
```

❶ By default no printed output will be written; however, if there is no summary data set specified ❷ printed output is automatically generated.

❸ The user can request both printed output and a summary data set by setting &PRINT to PRINT. The second semicolon closes the PROC statement.

❹ The CLASS and VAR statements are only written if one or more variables have been specified. The %DO blocks are not really needed here, but they eliminate the need to have a double semicolon, such as was used on the %ELSE ❸.

❺ When the name of a summary data set is provided the OUTPUT statement is written.

The following calls of the %MYMEANS macro demonstrate its flexibility.

This call to %MYMEANS will change the data set and analysis variables, use the default classification variable, and will produce no printed output.

```
%mymeans(dsn=sashelp.class,varlst=height weight)
```

```
%mymeans(outdsn=)
```

Here only printed output is generated using all the standard defaults for the macro.

SEE ALSO
Carpenter (2009) describes these steps to generalization in more detail.

13.3.2 Levels of Generalization and Levels of Macro Language Understanding

Another way of looking at the generalization steps described in Section 13.3.1 is to think of the process that one must go through as they learn the macro language. One could divide the learning process into three primary steps:

- Code substitution
- Use of macro statements and macro logic
- Creation of dynamic applications using the macro language

As your macro language skills increase, and your understanding of the macro language process solidifies, you will find that you will be able to write more complex programs.

SEE ALSO
Stroupe (2003) uses the term *Text Substitution* in a very nice introduction to the macro language.

Code Substitution
Typically programs and macros written at this level expect that the user will supply all the information needed by the macro. These programs are characterized by a lack of macro logic, and the use of macro variables that contain single items of information.

```
%let dsn = advrpt.demog;
%let vars= subject ht wt;
proc print data=&dsn;
  var &vars;
  run;
```

There is only a very short learning curve for these techniques, which can usually be quickly applied even by programmers fairly new to SAS.

Macro Statements and Macro Logic
In this stage the user gives more control to the macro, and the macro can determine some information generalization from its incoming parameters. At this level the programmer starts to take advantage of macro logic and utilizes the macro functions.

```
%macro printit;
.... code not shown....
%let dsn = advrpt.demog;
%let vars= subject ht wt;
proc print data=&dsn;
  %if &vars ne %then var &vars;;
  run;
.... code not shown....
%mend printit;
```

This level of learning takes longer to master and requires a more thorough understanding of the basic programming aspects of SAS. Many very good macro programmers never venture beyond this level of learning.

Dynamic Programming
Characteristics of applications and programs written using dynamic macro programming techniques include:

- A minimum of information is passed to the macro.
- A macro is adept at determining what it needs.
- Macro logic utilizes information outside of that passed into the macro.

- Macros that call macros are typical.
- Utility macros and macro functions are common.

It is common for dynamic macros to build and process lists of values (see Section 13.2.3 for one

```
%let dsn = advrpt.demog;
proc print data=&dsn;
  var %varlst(&dsn);
```

method for building a list of macro variables). Remember the steps to generalization (see Section 13.3.1) as you begin the process of converting your program from one that is controlled manually to one that builds its code dynamically.

13.4 Utilizing Macro Libraries

If you write more than the occasional macro, or if you share macros with colleagues, or if you *ever* define the same macro in different programs/places, you should be using macro libraries. Macro libraries provide the ability to remove the macro definition (%MACRO to %MEND statements) from your programs. By placing the macro definitions in a library, other programmers in your group can have access to the same macro definitions. Libraries allow you to effectively share your macro definitions without copying and storing them in multiple locations.

There are three basic types of macro libraries:

- %INCLUDE   these are not true macro libraries.
- Autocall   macro definitions (%MACRO to %MEND) are stored as code.
- Stored Compiled   compiled macros are stored in permanent catalogs.

Each of the three forms has value and each is worth knowing; however, if you only learn one type, learn to use the autocall macro library. This library is most often used, and it has a number of advantages over the other two forms of libraries.

A macro is defined through the use of the %MACRO and %MEND statements. When these statements are executed, the macro facility performs a macro compilation. This is not really a true compilation, and is little more than a check on macro syntax. The compiled macro definition is then written to a SAS catalog with an entry type of MACRO. The default catalog is WORK.SASMACR and the entry name will be the name of the macro itself.

SEE ALSO
A full treatment of the use of macro libraries can be found in Carpenter (2001a and 2004). Extensive use is made of Stored Compiled Macro Libraries in Section 13.9 and in Sun and Carpenter (2011).

13.4.1 Establishing an Autocall Library

By default an autocall macro library is automatically made available. This library contains a fairly extensive collection of macros that are provided with SAS. These include macros such as %LEFT, %VERIFY, and %QTRIM.

Two system options, MAUTOSOURCE and SASAUTOS=, are used to control the use of the

```
* Default option settings;
options mautosource
    ❶sasautos=sasautos❷;
```

autocall library. The ability to access an autocall
library is turned on with the MAUTOSOURCE
system option (by default this option is on). The
physical location of the autocall library is specified
using the SASAUTOS= ❶ system option.

By default the SASAUTOS= option's value is an automatic composite *fileref* also named
SASAUTOS ❷. This *fileref* points to various locations (which locations and how many depends
to some extent on your release of SAS , and the products that you lease). These locations are used
to house the autocall macro definitions that are supplied by SAS.

You may add your own macro definitions to the autocall library by storing them in one or more

```
filename mymacs "<phys path to my macro definitions>";
filename prjmacs "<phys path to the project macro definitions>";
filename COmacs "<phys path to the company wide macro definitions>";
options mautosource
    ❶sasautos=(mymacs prjmacs comacs sasautos ❷);
```

locations and then by adding those locations to the SASAUTOS= option. Notice the use of the
FILENAME statement, *not* a LIBNAME ❸ statement. Under directory-based systems the *fileref*
will point to the directory level and under zOS to a partitioned data set.

The only further constraint is that the macro name must match the name of the file that contains
the definition. If you were to create the definition for the macro %ABC, the %MACRO ABC
statement through the %MEND ABC statement, would be stored in a file named ABC.SAS (or
under zOS, an ABC member name). On the UNIX OS, the name of the file that stores the macro
definition must be in all lowercase characters (abc.sas).

When the %ABC macro is called, SAS will search for the program ABC.SAS in the locations
(left to right) specified in the SASAUTOS= option. Once the file is found, the macro definition is
included, the %MACRO to %MEND macro definition is compiled, and then the %ABC macro is
executed.

While it is possible for the file containing the macro definition to contain code other than just the
%MACRO through %MEND statements, it is not a good idea to do so. By segregating the code so
that a given file contains only the definition for the macro for which it is named, macro definitions
become much easier to find, and control.

There are a couple of caveats to be aware of when using autocall libraries. First, be very careful to
include the automatic *fileref* SASAUTOS ❷. Failure to do so results in the loss of the ability to
use autocall macros supplied by SAS. Secondly, be sure to specify the library locations using
filerefs and not *librefs*. Use the FILENAME statement even though you are pointing to a location

```
* WRONG WAY TO SPECIFY THE LIBRARY!!!;
libname ❸ COmacs "c:\temp";
options mautosource
    sasautos=(comacs ❸ sasautos ❷);
%silly
```

and not to a specific file. Although no
error is issued when the SASAUTOS=
option is specified using a *libref*, the
use of a *libref* ❸ will cause problems
when the library is accessed.

SEE ALSO
Heaton and Woodruff (2009) discuss options for establishing a company-wide autocall library. SAS Sample Code 24-841, by Peter Crawford, discusses a macro that inserts *filerefs* into the SASAUTOS= list of locations http://support.sas.com/kb/24/841.html.

13.4.2 Tracing Autocall Macro Locations

As was discussed in Section 13.4.1, it is not uncommon to have an autocall library point to several locations. When a macro is called, SAS searches each location in turn and executes the first copy of the macro that is encountered. You may need to know which location contains the code for the called macro. The MAUTOLOCDISPLAY system option ❶, the default is NOMAUTOLOCDISPLAY, will write the physical location ❷ of the macro's definition, whenever a macro is retrieved from an autocall library and is subsequently used.

In this example the definition for the %OBSCNT macro resides in the directory shown for the *fileref* MYMACS. Each time the macro is called the LOG shows the path to the program (OBSCNT.SAS) containing the macro definition.

```
filename mymacs "&path\sascode\sasmacros";
options mautosource
    sasautos=(mymacs sasautos)
    mautolocdisplay;  ❶

%put There are %obscnt(sashelp.shoes) obs in sashelp.shoes;
```

```
451 %put There are %obscnt(sashelp.shoes) obs in sashelp.shoes;
MAUTOLOCDISPLAY(OBSCNT): This macro was compiled from the autocall file
            C:\AdvTechniques\sascode\sasmacros\obscnt.sas  ❷
There are 395 obs in sashelp.shoes
```

13.4.3   Using Stored Compiled Macro Libraries

Stored Compiled Macro Libraries are only available when turned on with the MSTORED system option. The SASMSTORE= option is then used to allocate the stored compiled macro library.

```
libname complib "&path\sascode\storedmacros";
options mstored
    sasmstore=complib;
```

Although the SASMSTORE option accepts only one *libref*, the library associated with that *libref* can be a concatenated or composite library.

If a stored compiled macro library is available, the /STORE option on the %MACRO statement can be used to direct the compiled macro to the permanent COMPLIB.SASMACR catalog.

```
%macro def / store;
  %put Stored compiled Version of DEF;
%mend def;
```

13.4.4 Macro Library Search Order

Understanding the macro library search order is crucial to understanding which version of a macro will be executed. When a macro, such as the %ABC macro, is called, SAS must search for the macro's definition. SAS first looks for the ABC.MACRO entry in the WORK.SASMACR catalog. Then, assuming that it is not found in the WORK catalog, and if stored compiled macro libraries are turned on, a search is made for the ABC.MACRO entry in each SASMACR catalog in the *libref* designated by the SASMSTORE= system option. Finally if a compiled entry has not yet been found, SAS starts a search in the autocall library locations for a program with the name of ABC.SAS.

In summary the search order is:

1. WORK.SASMACR
2. stored compiled macro libraries (COMPLIB.SASMACR in the above example)
3. autocall macro libraries

13.5 Metadata-Driven Programs

Metadata is data about the data. For the macro language the metadata very often contains the instructions that will be used to drive the macros. Instead of passing macro parameters the macros read data to determine the parameters.

13.5.1 Processing across Data Sets

In this example the researcher wants to print the key variables along with the critical variables for each of several data sets. The %PRINTALL macro has been written to make the listings; however, the macro obtains the information that it needs (data set name, BY variables, critical variables) from a SAS data set. This control file, which has one observation for each data set of interest, contains all the information needed by the %PRINTALL macro.

```
13.5.1 Using Metadata Across Data Sets
Meta-data Control File

Obs DSN            keyvars            critvars

 1  demog          subject            dob ht wt
 2  Lab_Chemistry  subject visit      labdt
 3  Conmed         subject mednumber  drug
```

```
%macro printall;
  %local i dsncount;
  * Build lists of macro vars;
  proc sql noprint;
  select dsn,keyvars,critvars ❶
    into :dsn1 - :dsn999, ❷
      :keyvar1 - :keyvar999,
      :critvar1 - :critvar999
    from advrpt.dsncontrol; ❸
  %let dsncount = &sqlobs; ❹

  %do i = 1 %to &dsncount; ❺
    title2 "Critical Variables for &&dsn&i";
    proc print data=advrpt.&&dsn&i;
      id &&keyvar&i; ❻
      var &&critvar&i;
      run;
  %end;
%mend printall;

%printall
```

We can use this table when we want to process across all data sets in the study or when we need data set specific information - such as the BY variables. The %PRINTALL macro needs the name of the data set, it's BY variables, and the list of its critical variables.

❶ The three variables that contain the metadata values of interest are selected from the control file.

❷ A list of macro variables is created. The SQL step does not support the SYMPUTX routine, instead it uses the INTO : clause to write macro variables to the symbol table.

❸ The control data set is read into the SQL step.

❹ The number of data sets, observations in the control data set, is saved in &DSNCOUNT.

❺ The %DO loop cycles through the &DSNCOUNT data sets with &I as the data set counter.

❻ The macro variable of the form &&VAR&I refers to the Ith element in the list. For &I=2, &&CRITVAR&I resolves to LABDT.

Notice the use of an asterisk style comment in the %PRINTALL macro. SAS recommends the use of the /* */ style of comments. Using asterisk style comments inside a macro to comment out macro language elements can cause problems (Carpenter, 2004, Section 13.3.5). A minimum rule should be to use macro comments to comment out or to annotate macro code. See the examples in Sections 13.5.2 and 13.7.

MORE INFORMATION
In this example and in the example in the following section, the metadata has been manually generated. Metadata can come from a number of sources, and some of these sources are available automatically. Section 13.8 discusses some of those sources of information.

13.5.2 Controlling Data Validations

In Section 2.3.3 there is a discussion of the use of a simple data set, which is used to populate the macro parameters of a data set specific error checking macro. Using similar techniques it is possible to build the checks themselves based on the data that is to be validated.

The metadata shown here contains the data set name and the check information associated with

```
13.5.2 Metadata Driven Field Checks

Obs  dsn             var          chkrating  chktype    chktext

 1   demog           subject      1          notmiss
 2   demog           RACE         2          list       ('1','2','3','4','5','6')
 3   conmed          medstdt_     4          datefmt    mmddyy10.
 4   lab_chemistry   potassium    2          maximum    6.7
```

that check. If the metadata is designed to contain sufficient information, it can easily be expanded to accommodate any number of checks on any number of data sets, and multiple checks can be performed on any given variable. The number of checks and the kinds of checks are only limited by the programmer's imagination.

In the examples in Section 2.3.3 a data validation and error reporting macro was developed that utilized metadata to perform simple data checks. We can expand on that macro by making use of the type of metadata shown here. In this example the checks are performed across all data sets, and the checks themselves are constructed from the information in the metadata.

Any number of different types of checks is possible; shown here are just a few to give you an idea of the possibilities (variable CHKTYPE in the metadata):

- notmiss   the variable may not contain missing values.
- list   the value must be in the list of values in CHKTEXT.
- datefmt   the formatted value of the variable (using the format in CHKTEXT) must not be missing.
- maximum   the value must be less than or equal to the value in CHKTEXT.

Using this approach any number of checks can be performed against a given variable or data set. Adding and changing checks does not require coding changes, unless a brand new check is introduced. Here a slight coding modification would be required if we wanted to introduce a check for minimum values.

```
%macro errrpt(dsn=, keyvars=subject);  ❶
%local i;
data _null_;  ❷
  set advrpt.fldchk(where=(upcase(dsn)=upcase("&dsn")));
  fldcnt+1;
  cnt = left(put(fldcnt,6.));
  call symputx('errdsn'||cnt,dsn,'l');
  call symputx('errvar'||cnt,var,'l');
  call symputx('errrating'||cnt,chkrating,'l');
  call symputx('errtype'||cnt,chktype,'l');
  call symputx('errtext'||cnt,chktext,'l');
  call symputx('chkcnt',cnt,'l');
  run;
data errrpt&dsn  ❸
    (keep=dsn
          &keyvars
          errvar errval errtxt errrating);
  length dsn    $25
     errvar   $15
     errval   $25
     errtxt   $25
     errrating 8;

set advrpt.&dsn;

%do i = 1 %to &chkcnt;  ❹
  %* Write as many error checks as are needed;
  if  ❺
  %* Determine the error expression;
  %if %upcase(&&errtype&i)   = NOTMISS %then  ❻
      missing(&&errvar&i);  ❼
  %else %if %upcase(&&errtype&i) = LIST %then
      &&errvar&i not in(&&errtext&i);  ❼
  %else %if %upcase(&&errtype&i) = DATEFMT %then
      input(&&errvar&i,&&errtext&i) eq .;  ❼
  %else %if %upcase(&&errtype&i) = MAXIMUM %then
      &&errvar&i gt &&errtext&i;  ❼
  then do;  ❽
   dsn = "&dsn";
   errvar = "&&errvar&i";
   errval = &&errvar&i;
   errtxt = "&&errtext&i";
   errrating= &&errrating&i;
   output errrpt&dsn;
   end;
  %end;
  run;

title2 "Data Errors for the &dsn data set";
proc print data=errrpt&dsn;
  run;
%mend errrpt;
```

❶ The %ERRRPT macro is passed the name of the data set to be checked and the key variables for that data set.

❷ A DATA _NULL_ step is used to read the error metadata appropriate for the data set to be checked.

❸ A data set containing the data errors is defined. Here it is written to the WORK directory and its name includes the name of the data set being checked.

❹ A macro %DO loop is used to cycle across the checks that have been requested in the metadata for this data set.

❺ The expression used to detect the data error will be written for the DATA step IF statement by a macro %IF and will be based on the metadata.

The IF statement terminates with a THEN DO/END at ❽.

❻ A %IF statement is used to determine the type of error comparison that is to be written.

❼ The error condition specified in the metadata is written.

❽ The error information is written to the error reporting data set. This THEN DO/END terminates the IF statement started at ❺.

The macro %ERRRPT is called once for each data set that is to have its data validated. The list of data sets used in the study, along with their key variables, can also be placed in metadata. For the study data being checked in this example, the metadata used to describe the study data sets can be found in ADVRPT.DSNCONTROL (see Section 13.5.1).

For the checks on the DEMOG data set the following DATA step is written by the %ERRRPT macro. Notice that only the two checks associated with this data set in the metadata have been included and that the appropriate variables have been used in the checks.

```
data errrptdemog (keep=dsn subject errvar errval errtxt errrating);
length dsn $25 errvar $15 errval $25 errtxt $25 errrating 8;
set advrpt.demog;
if missing(subject) ❼ then do;
dsn = "demog";
errvar = "subject";
errval = subject;
errtxt = "";
errrating= 1;
output errrptdemog;
end;
if RACE not in(('1','2','3','4','5','6')) ❼ then do;
dsn = "demog";
errvar = "RACE";
errval = RACE;
errtxt = "('1','2','3','4','5','6')";
errrating= 2;
output errrptdemog;
end;
run;
```

The ERRRPTDEMOG data set will contain any detected errors (in this case a single error was found—a missing value for the variable RACE).

```
13.5.2 Metadata Driven Field Checks
Data Errors for the demog data set

Obs   dsn   errvar   errval           errtxt              errrating   subject

 1    demog  RACE             ('1','2','3','4','5','6')      2          204
```

The macro %DATAVAL reads the DSNCONTROL metadata and then builds macro variable lists of data set names and BY variables. Those macro variable lists are then used to call the %ERRRPT macro (shown above) for each data set.

```
%macro dataval;
%local i;
* Determine list of data sets to check;
data _null_;  ❾
  set advrpt.dsncontrol;
  cnt = left(put(_n_,5.));
  call symputx('dsn'||cnt,dsn,'l');
  call symputx('keyvars'||cnt,keyvars,'l');
  call symputx('dsncnt',cnt,'l');
  run;

%* Perform data validation checks on
%* each data set;
%do i = 1 %to &dsncnt;
  %errrpt(dsn=&&dsn&i, keyvars=&&keyvars&i)  ❿
%end;
%mend dataval;
%dataval
```

❾ A DATA step is used to create the lists of macro variables. This type of list is generally easier to create using an SQL step.

❿ An iterative %DO loop is used to process across the list of data sets.

```
%macro dataval2;
* Determine list of data sets to check;
proc sql noprint;
  select dsn,keyvars
    into :dsn1-:dsn999,
       :keyvars1-:keyvars999
    from advrpt.dsncontrol;
  %let dsncnt=&sqlobs;
  quit;

%* Perform data validation checks;
%* on each data set;
%do i = 1 %to &dsncnt;
  %errrpt(dsn=&&dsn&i, bylst=&&keyvars&i)
%end;
%mend dataval2;
```

In some instances you may want to store the validation formula itself in the metadata. Although technically this can be more challenging, the methodology is an expansion of the techniques shown above.

The DATA _NULL_ step in the %DATAVAL macro could have been replaced with a PROC SQL step. The end result is the same with the exception that we cannot as easily control the symbol table for the derived macro variables.

The generation of the list of macro variables can be avoided altogether by using the CALL EXECUTE routine.

Since CALL EXECUTE is a DATA step routine, the %ERRRPT macro call can be generated directly for each observation using the DATA step variables. CALL EXECUTE places the macro call in a stack which executes after the execution of the DATA step.

```
%macro dataval3;
* Determine list of data sets to check;
data _null_;
  set advrpt.dsncontrol;
  call execute('%nrstr(%errrpt(dsn='||dsn||', keyvars='||keyvars||'))');
run;
%mend dataval3;
%dataval3
```

MORE INFORMATION
Section 2.3.3 introduces an example that uses metadata to drive a data validation macro. Section 13.5.1 introduces the use of the ADVRPT.DSNCONTROL data set.

SEE ALSO
Fehd and Carpenter (2007) and Rozhetskin (2010) discuss several different ways to process a list of metadata values.

Although not directly applicable to the use of metadata, discussions on how to store a formula as a data value and have it executed dynamically in a later DATA step can be found in the SAS Forum threads http://communities.sas.com/message/48498 and http://communities.sas.com/message/46975#46975.

13.6 ~~Hard Coding~~—Just Don't Do It

Hard coding takes place when study or data-specific information is inserted directly as code in our programs. Unfortunately this is an all too common practice that can cause a number of problems for the researcher:

- Code has embedded data dependencies.
- Changes to the dependencies requires coding changes in *all* programs, which have the dependency.
- Each modified program must be revalidated.

A simple example of a hard coded data dependency is the exclusion of a subject from an analysis. In this case we need to exclude subject 202 when data is read from ADVRPT.CONMED.

```
data conmed;
  set advrpt.conmed(where=(subject ne '202')); ❶
.. ..code not shown....
  run;
```

Creating a WHERE clause through the use of a WHERE= data set option to do the exclusion is quite easy ❶; however, for consistency the exclusion must take place in each program that utilizes data that contains that subject, and it is likely that the exclusion list will not remain constant. Keeping track of which programs utilize what data dependent exclusions can become tedious and error prone. By consciously developing tools for avoiding the use of hard coding, we can avoid the hard coding nightmare.

The macro language can be used to replace hard coded exception lists. The simplest solution is to just move the exception coding to a macro ❷ that can be called from any program that needs to account for the data exceptions ❸. While not very flexible, placing the macro in a macro library makes the exceptions available to all programs from a single, changeable source.

```
%macro exceptions;
where=(subject ne '202')  ❷
%mend exceptions;

data conmed;
  set advrpt.conmed(%exceptions  ❸);
  run;
```

Of course real life is rarely this simple Exceptions may only be appropriate for some data sets and there may also need to be data adjustments that need to be applied in only certain situations. Both of these cases lend themselves well to the creation and use of metadata to control the process. The use of metadata to drive a process is described in Section 13.5.

A simple extension of the previous example might include a metadata file such as the one shown here. This data set includes only the data set name and one or more exceptions. Other expressions are easily implemented using this approach.

```
data advrpt.DataExceptions;
  length dsn $12 exception $35;
  dsn='AE';     exception="(subject le '204')"; output;
  dsn='conmed'; exception="(subject ne '202')"; output;
  dsn='conmed'; exception="(subject ne '208')"; output;
  run;
```

In this version of the %EXCEPTIONS macro, the metadata are used to build a WHERE= data set option that can be used to subset the incoming data. ❹ The data exceptions metadata is read using an SQL step.

```
%macro exceptions(dsn=ae);
  * Build exception list;
  proc sql noprint;
    select exception into :explist separated by '&'  ❻
      from advrpt.dataexceptions  ❹
        where upcase(dsn)=upcase("&dsn");  ❺
    quit;
  %if &explist ne %then %let explist=where=(&explist);  ❼
%mend exceptions;

%let explist = ;  ❽
%exceptions(dsn=conmed)  ❾
%put &explist;
proc print data=advrpt.conmed(&explist)  ❿;
  run;
```

❺ Observations associated with the data set of interest are selected, and the values of the variable EXCEPTION are added to the macro variable &EXPLIST ❻.

❻ You need to be careful when using an ampersand within a macro variable, as was done here. In this macro the individual clauses are surrounded by parentheses, consequently the & will not be seen as a macro language trigger.

❼ The list of data exceptions are stored in the form of a WHERE= data set option. The macro variable &EXPLIST will NOT be written to the local symbol table for the %EXCEPTIONS macro, but will be written to the next higher table (where &EXPLIST already has been established with a null value) ❽.

❽ The macro variable is initialized to a null value. This not only ensures that the macro variable does not contain a value from a previous execution of %EXCEPTIONS, it also adds the macro variable to the most local symbol table (given the code that we see here, this may be the global symbol table). The value of &EXPLIST generated within %EXCEPTIONS will, therefore, be written to this higher symbol table and not to the local table for %EXCEPTIONS. This helps to control the possibility of a macro variable collision with the value of the macro variable &EXPLIST in a higher symbol table.

❾ The %EXCEPTIONS macro is executed and a value is assigned to the macro variable &EXPLIST. The LOG shows the resulting WHERE clause.

```
15   %put &explist;
where=((subject ne '202')&(subject ne '208'))
```

❿ The WHERE= data set option with the exceptions is added when the data set is used. When there are no exceptions for a given data set, &EXPLIST will have a null value and no observations will be excluded.

MORE INFORMATION
Section 13.1 specifically discusses macro variable collisions.

13.7 Writing Macro Functions

A macro function (Chung and Whitlock (2006) use the terminology *Function-Style macros*) is a macro that is written so that it mimics the behavior of a function. Several of the Autocall macros supplied by SAS (including %LEFT, %QTRIM, and %VERIFY) are actually macro functions. It is not all that difficult to write a macro function, but there are three rules that you need to follow to successfully cause the macro to work like a function.

A function returns a specific value of interest, and only that value of interest. Since the macro language is first and foremost a code generator, we want to make sure that the only code generated by our function is the value that is to be returned. Of course we would also like our macro to be robust and to not interfere with any other code that we use in conjunction with our macro function. The following three rules ensure that your macro will operate like a function and will not interfere with other code.

Your macro function should:

- Use only macro language elements: no DATA steps or PROC steps.
- Create no macro variables that are not local to the macro.
- Resolve to the value that is to be passed out of the function.

When written following these rules, your macro function can be used in both DATA steps and with macro language elements. Here is a fairly classic macro function, which is a slightly modified version of a macro of the same name that appears in Carpenter, 2004 (Section 11.5.1). This macro function returns the number of observations in a SAS data set by opening and examining the data set's metadata.

Notice that the %OBSCNT macro contains only macro statements, that all the macro variables

```
%macro obscnt(dsn);
%local nobs dsnid rc;  ❶
%let nobs=.;

%* Open the data set of interest;
%let dsnid = %sysfunc(open(&dsn));

%* If the open was successful get the;
%* number of observations and CLOSE &dsn;
%if &dsnid %then %do;
   %let nobs=%sysfunc(attrn(&dsnid,nlobs));
   %let rc =%sysfunc(close(&dsnid));
%end;
%else %do;
   %put Unable to open &dsn - %sysfunc(sysmsg());
%end;

%* Return the number of observations;
&nobs  ❷
%mend obscnt;
```

created in the macro are forced onto the local symbol table ❶, and that the value of the number of observations, &NOBS, is passed out of the macro as resolved text ❷. As occurs here, it is quite common that the value to be passed back stands alone as a macro language element–not as a complete statement. Here the macro variable &NOBS ❷ will resolve to the number of observations during the execution of the macro. This becomes the only non-macro language element in the macro and, therefore, becomes the resolved value of the macro. If we assume that the data set WORK.CLINICS has 88 observations, the %IF statement in the code box on the left resolves to the one in the code box on the right.

```
%if %obscnt(clinics) > 5 %then %do;
```

```
%if 88 > 5 %then %do;
```

The first rule requires all statements to be macro language statements. This includes comments. In this example macro comments have been used; however, they could have been replaced with the SAS recommended /* */ style comments. Although the /* */ style comment is not a macro language element, as is the %* style comment, the /* */ style comment is stripped out even earlier in the parsing process and, consequently, will not interfere with the macro function as would an asterisk style comment.

Very often the macro function can be written to contain only a single macro language phrase. This code segment is executed and the result is passed out of the macro to the calling program. It is important to remember that code segments are handled differently in the macro language than are code segments in the DATA step. Macro variables are commonly resolved without being a part of a complete macro statement ❷, and a macro function, especially %SYSFUNC, can be a complete element in and of itself ❸.

```
%macro wordcount(list);
  %sysfunc(countw(&list,%str( )))  ❸
%mend wordcount;

%let list = a  Bb c d;
%put %wordcount(&list);
```

The %WORDCOUNT macro function, shown here only contains a %SYSFUNC macro function. Notice that the %SYSFUNC ❸ is not a part of a complete statement, and is not followed by a semicolon. The %SYSFUNC function call will resolve to the number of words in &LIST (4 in the example shown here).

The following macro, %AGE, which was written by Ian Whitlock and appears in Chung and Whitlock (2006), is a macro function that returns a person's age in years. This macro function is designed to be used either in a DATA step or in a PROC step WHERE clause. The macro

```
%macro age(begdate,enddate);
  (floor((intck('month',&begdate,&enddate)-(day(&enddate)<day(&begdate)))/12))
%mend age;
```

assumes that &BEGDATE and &ENDDATE are either SAS date values or variables that hold the date values. Here the %AGE macro is used to list those subjects over 45 as of the specified date.

```
proc print data=advrpt.demog;
  * select subjects over 45 as of Feb 18, 1998;
  where %age(dob,'18feb1998'd) gt 45;
  var fname lname dob;
  run;
```

The value returned by this function is the expression itself (*not its resolved value*—which is determined at the time of execution). As in the previous example, %WORDCOUNT, the macro contains only an expression and not a complete statement.

MORE INFORMATION
Alternative methods of calculating age can be found in Section 3.2.

Because a macro function can also be used to return a value, the macro call itself can be used as a part of a macro statement. The macro %NEXTDOG determines the next available macro variable name that starts with the letters DOG.

```
%let dog=scott;
%let dog1=bill;
%let dog2=george;
%let dog3=notsue;  ❹

%macro nextdog;
%local cnt;
%let cnt=;
%do %while(%symexist(dog&cnt));  ❺
  %let cnt=%eval(&cnt+1);  ❻
%end;
&cnt  ❼
%mend nextdog;

%put nextdog is %nextdog;
%let dog%nextdog=Johnny;  ❽
%put nextdog is %nextdog;
```

❹ The macro variable &DOG3 has been defined. The next available name will be &DOG4.

❺ The %SYMEXIST function is used to determine if a given macro variable currently exists on the global symbol table.

❻ Increment the counter and check for the next macro variable.

❼ When a given macro variable is not found ❺ the %DO %WHILE loop terminates and the next available value is passed out of the macro.

❽ The next available macro variable is automatically assigned using the next available number.

SEE ALSO
Carpenter (2002) and Carpenter (2004: Section 7.5.2) both cover the rules associated with the creation of macro functions.

The original age formula used by Chung and Whitlock (2006) was devised by Kreuter (2004).

The %NEXTDOG macro was used to demonstrate a concept in a SAS Forum thread http://communities.sas.com/thread/14805.

13.8 Macro Information Sources

Macros that are sophisticated enough to seek out and utilize information that they need without resorting to user input have a great advantage in both flexibility and power. In order for us to write these macros, we must be aware of these information sources and how and when to use them.

Fortunately, there is a great deal of information that is easily accessible to the macro language.

13.8.1 Using SASHELP and Dictionary tables

A series of views and on-demand tables have been constructed to provide a great deal of information about SAS and the environment in which it is running. These come in two basic flavors:

- DICTIONARY tables   available only within an SQL step.
- SASHELP views   can be used anywhere a data set or a view can be used.

The full list of SASHELP views and DICTIONARY tables can be found in the SAS documentation. The following list is selection of some of these that I have found to be most helpful.

DICTIONARY Tables and Associated SASHELP Views		
DICTIONARY Table	**SASHELP View**	**Description**
CATALOGS	VCATALG	Contains information about known SAS catalogs.
COLUMNS	VCOLUMN	Contains information about columns in all known tables.
DICTIONARIES	VDCTNRY	Contains information about all DICTIONARY tables.
ENGINES	VENGINE	Contains information about SAS engines.
EXTFILES	VEXTFL	Contains information about known external files.
FORMATS	VFORMAT VCFORMAT	Contains information about currently accessible formats and informats.
GOPTIONS	VGOPT VALLOPT	Contains information about currently defined graphics options (SAS/GRAPH software). SASHELP.VALLOPT includes SAS system options as well as graphics options.
INDEXES	VINDEX	Contains information about known indexes.
LIBNAMES	VLIBNAM	Contains information about currently defined SAS libraries.
MACROS	VMACRO	Contains information about currently defined macro variables.

(continued)

DICTIONARY Table	SASHELP View	Description
DICTIONARY Tables and Associated SASHELP Views (continued)		
MEMBERS	VMEMBER VSACCES VSCATLG VSLIB VSTABLE VSTABVW VSVIEW	Contains information about all objects that are in currently defined SAS libraries. SASHELP.VMEMBER contains information for all member types; the other SASHELP views are specific to particular member types (such as tables or views).
OPTIONS	VOPTION VALLOPT	Contains information about SAS system options. SASHELP.VALLOPT includes graphics options as well as SAS system options.
STYLES	VSTYLE	Contains information about known ODS styles.
TABLES	VTABLE	Contains information about known tables.
TITLES	VTITLE	Contains information about currently defined titles and footnotes.
VIEWS	VVIEW	Contains information about known data views.

To learn more about a given SASHELP view simply explore it like you would any data set. A quick look at the view with PROC CONTENTS or VIEWTABLE is generally sufficient for you to understand what the view contains.

The DICTIONARY tables must be explored using an SQL step. The DESCRIBE statement can be used to write the column names and attributes to the LOG, while the SELECT statement will write the contents of the table to any open ODS destinations.

```
proc sql;
describe table dictionary.tables;
```

```
select * from dictionary.members;
```

The information in these tables can then be transferred to macro variables for processing by the macro language. This can be done in either the DATA or SQL steps. In this PROC SQL step a comma-separated list of the names of all of the data sets in the ADVRPT library is written to the macro variable &TABLELIST.

```
title2 'Build a list of data sets';
proc sql noprint;
select memname
  into :tablelist separated by ','
  from dictionary.members
    where libname='ADVRPT';
%put &tablelist;
quit;
```

MORE INFORMATION

Examples of the use of SASHELP.VTABLE and DICTIONARY.MEMBERS can be found in Section 1.1.5.

13.8.2 Retrieving System Options and Settings

It is not unusual for a macro to need to adjust the value of a system option or setting during macro execution. If a macro changes system settings, such as options, during execution, these settings should be returned to their original values at the completion of the macro's execution. This means that the original settings must be captured, saved, and restored.

The SASHELP views and DICTIONARY tables mentioned in Section 13.8.1 are one source of this type of information. Sources for SAS system option settings include:

- SASHELP.VOPTIONS
- SASHELP.VALLOPT
- SASHELP.VGOPT
- DICTIONARY.GOPTIONS
- DICTIONARY.OPTIONS
- GETOPTION function

The portion of the macro %SECURECODE shown here grabs the current settings for the system

```
%macro securecode;
data _null_;
  set sashelp.voption
    (where=(optname in('MPRINT','MLOGIC','SYMBOLGEN'))); ❶
  call symputx('hold'||left(optname),optname, 'l'); ❷
  run;
options nomprint nomlogic nosymbolgen; ❸

/* secure code goes here*/❹

options &holdmprint &holdmlogic &holdsymbolgen; ❺
%mend securecode;
```

options MPRINT, SYMBOLGEN, and MLOGIC ❶, saves them in macro variables whose names start with HOLD ❷, and then turns these options off ❸ so that the code used in the macro ❹ will not be revealed in the LOG. At the conclusion of the macro these three options are reset to their original values ❺.

It is also possible to collect setting values through the use of functions. There are a great many functions that can be used to obtain this kind of information and it is essential for an advanced macro programmer to be well versed in which ones can be useful.

The documentation groups functions by category and some of these categories contain functions that are especially useful for obtaining information about system settings. These include:

- External files
- SAS File I/O
- Special

The following example detects the location of a file using the PATHNAME function ❻ (the assumption being that the macro programmer only knows the *fileref*—and not the actual physical location when calling the macro). The macro uses this information to create a new *fileref* using the FILENAME ❼ function, which points to a different file (&NEWNAME) in the same location. Notice that the first argument of the FILENAME function is expected to be a macro variable when used with %SYSFUNC and that the & is not used.

```
filename super 'c:\temp\super.pdf';

%macro newref(locref=, newlocref=, newname=);
 %local origref origname nameloc newloc rc;
 %let origref = %sysfunc(pathname(&locref));  ❻
 %let origname= %scan(&origref,-1,\);
 %let nameloc = %sysfunc(indexw(&origref,&origname,\));
 %let newloc = %substr(&origref,1,&nameloc-1)&newname;
 %let rc    = %sysfunc(filename(newlocref,&newloc));  ❼
 %put %sysfunc(fileexist(&newlocref));
%mend newref;
%newref(locref=super,newlocref=silly,newname=freqplot.pdf)
```

Once the physical path has been retrieved ❻, it can be dissected. The filename is extracted from the end of the string (&ORIGNAME) and its starting location noted (&NAMELOC). Using this location the new name (&NEWNAME) can then be appended onto the location portion of the path. Once the new path has been constructed, the new *fileref* (&NEWLOCREF) can be established using the FILENAME function ❼.

The GETOPTION function (in the Special Category of the list of functions), can be used to retrieve current system option settings. The %FINDAUTOS macro shown here is used to retrieve the physical locations of the autocall macro libraries and to write them to the LOG. The *filerefs* associated with the autocall library are stored in the SASAUTOS system option. Because the %QSCAN is used to parse the list of locations, this macro assumes that *filerefs*, and not physical names of files (which could contain characters that would be interpreted as word delimiters), are used in the definition of the SASAUTOS system option.

```
%macro findautos;
%local autoref i ref refpath;
%let autoref = %sysfunc(getoption(sasautos));❽
%let i=0;
%do %until(&ref eq);
  %let ref = %qscan(&autoref,&i+1);
  %if &ref eq %then %return;
  %let refpath=%qsysfunc(pathname(&ref));  ❾
  %let i = %eval(&i + 1);
  %put &i &ref &refpath;  ❿
%end;
%mend findautos;
```

❽ The value of this option, or any other system option, can be retrieved with the GETOPTION function.

❾ The list of *filerefs* can then be passed, one at a time, through the PATHNAME function.

❿ In this macro the resulting path, including in this case a composite location, is written to the LOG.

```
%put %sysfunc(getoption(sasautos));
%findautos
```

```
44  %put %sysfunc(getoption(sasautos));
(advmac, sasautos)
45  %findautos
1 advmac C:\AdvTechniques\sascode\sasmacros
2 sasautos (
'C:\Program Files\SAS\SASFoundation\9.2\core\sasmacro'
'C:\Program Files\SAS\SASFoundation\9.2\accelmva\sasmacro'
          .... portions of the SASAUTOS definition are not shown ....
```

MORE INFORMATION
The DATA step SCAN function is used in a related example in Section 3.6.6. Examples related to determining the location of executing programs can be found in Section 14.6.

SEE ALSO
Carpenter (2008b) demonstrates various ways to retrieve the physical location of a file, even when it is on a server with a mapped drive.

13.8.3 Accessing the Metadata of a SAS Data Set

Several of the SASHELP views and DICTIONARY tables described in Section 13.8.1 provide information about the attributes of data sets (e.g., variable names, formats, variable type). Much of this same information can be obtained through the use of either PROC CONTENTS or DATA step functions. Depending on what you intend to do there can be decided performance differences between these approaches. Experiment—results will likely vary from one situation to another.

Using PROC CONTENTS
The OUT= option on the CONTENTS procedure can be used to create a data set that contains the information of interest. Like the examples in Section 13.8.1, once this information is in data set form it can be harvested and used by a number of techniques.

```
%macro varlist(dsn=sashelp.class, type=1);
%* TYPE 1=numeric
%*       2=character;
%local varlist;
proc contents data=&dsn
              out=cont(keep=name type ❶
                         where=(type=&type))
              noprint;
  run;
proc sql noprint;
  select name
    into :varlist separated by ' ' ❷
      from cont;
  quit;
  %put The list of type &type variables is:
    &varlist; ❸
%mend varlist;
%varlist(dsn=advrpt.demog,type=1)
```

In the macro %VARLIST, the user may request that a list of either numeric or character variables be written to the LOG. The CONTENTS procedure returns the column TYPE as 1 for numeric variables and 2 for character variables.

❶ The OUT= option is used to write a data set containing the metadata. The data set is in the form of one observation for each variable in the data set.

❷ The list of space separated variable names that meet the numeric/character attribute is written to the macro variable &VARLIST.

❸ The list of variable names is then written to the LOG.

If you want to create a summary of all of the data sets in a library the keyword _ALL_ can be

```
proc contents data=advrpt._all_
       out=cont
       noprint;
   run;
```

used. This PROC CONTENTS step will create a table (CONT) that contains one observation for each variable by data set combination in the library ADVRPT.

PROC CONTENTS is a fast method for generating a list of metadata attributes. However, there are limitations. Because of the use of the two PROC steps it is not possible to use this technique in a macro that will mimic a macro function (see Section 13.7). The use of DATA step functions along with the %SYSFUNC macro function can eliminate this limitation, improve performance, and generally simplify the macro coding.

Using DATA Step Functions

The metadata of a SAS data set can be accessed directly using DATA step functions. Although virtually never used within a DATA step, these functions are extraordinarily helpful when accessing metadata from within the macro language.

These functions allow us to open and close the data set as well as to query all sorts of things about the metadata itself. We can even manipulate the data itself; however, that is rarely necessary.

In the %MAKELIST macro shown here, we again need to return a list of either numeric or character variables. Rather than use either a PROC or DATA step to access the metadata, this macro goes directly to the source.

```
%macro makelist(dsn=sashelp.class, type=N);
%* TYPE = N for numeric
%*        C for character;
%local dsid i varlist rc;
%let dsid = %sysfunc(open(&dsn));  ❹
%do i = 1 %to %sysfunc(attrn(&dsid,nvar));  ❺
  %if %sysfunc(vartype(&dsid,&i))=%upcase(&type)  %then  ❻
    %let varlist=&varlist %sysfunc(varname(&dsid,&i));  ❼
%end;
%let rc = %sysfunc(close(&dsid));  ❽
&varlist  ❾
%mend makelist;
%put Char vars are: %makelist(dsn=advrpt.demog,type=c);  ❿
```

❹ The data set of interest is opened for inspection. The opened data set is assigned a non-zero identification number (saved here in &DSID), which is used by a number of other functions. Once opened a series of functions can be applied to the metadata or even to the data itself.

❺ The ATTRN function is especially useful. It retrieves numeric attributes from the metadata. Its first argument is the identification number of the opened data set, and the second argument is used to select the attribute of interest. Here the NVAR argument is used to select the number of variables stored in the data set. This number becomes the upper bound for the %DO loop which processes across the variables in the data set.

❻ The VARTYPE function returns the type (N=numeric, C=character) of the &i[th] variable. This value is compared to the requested variable type.

❼ The VARNAME function returns the name of the &i[th] variable.

❽ After the information has been retrieved from the metadata, the data set is closed.

❾ This is a macro function and the list of variable names is returned.

❿ The %MAKELIST macro is called with a request for the names of character variables (type=c).

MORE INFORMATION
DATA step functions that return variable characteristics can be found in Section 3.6.5.

13.9 Macro Security and Protection

At times, such as when executing macros under a controlled environment or as part of a larger application, it may be necessary to limit the user's access to various aspects of the coding of our macros. Sometimes we need to prevent the dissemination of proprietary code. Other times we need to force the use of a particular version of a macro.

When control-related issues are discussed by experienced SAS programmers, common topics include:

- What version of a macro is being executed?
- How can we control for the correct version?
- How do we avoid macro variable collisions and protect our macro variables, compiled macros, and the source code?

MORE INFORMATION
Sherman and Carpenter (2007) discuss the protection of user IDs and passwords when accessing external databases from within SAS. This topic is also discussed in less detail in Section 5.4.2.

SEE ALSO
Sun and Carpenter (2011) discuss a number of aspects of the control and protection of macro code, macro operation, and macro variables.

13.9.1 Hiding Macro Code

When using stored compiled macros, the SOURCE option on the %MACRO statement can be used to store your macro's definition, the code itself, in the catalog along with the compiled macro. This code can then be reclaimed from the catalog using the %COPY

```
%macro abc/store source;
```

statement. Of course if your users can see the source code for your macro, they can then re-engineer your macro. Obviously this is definitely not the best way to hide your macro source code.

Regardless of whether or not the SOURCE option was used on the %MACRO statement it is still

```
filename maccat catalog 'advrpt.sasmacr.abc.macro';
data _null_;
  infile maccat;
  input;
  list;
  run;
```

to some extent possible to reclaim some of the original macro definition. A DATA _NULL_ step can be used to write the hex codes that are

associated with the macro's compiled definition to the LOG. Another technique which has been attributed to Ian Whitlock uses the %QUOTE function to surface the macro code.

```
%put %quote(%abc);
```

The SECURE option on the %MACRO statement prevents even this partial recovery of the code by encrypting the compiled macro definition.

```
%macro def/store secure;
```

Use of the SECURE option (with or without the STORE option) causes the system option values

```
options mprint symbolgen mlogic;
%macro dtest/secure;
proc print data=sashelp.class;
  run;
%mend dtest;
%dtest
```

of NOMPRINT, NOMLOGIC, and NOSYMBOLGEN to be temporarily set during the execution of the macro. The SECURE option in the example to the left resets the MPRINT system option to NOMPRINT during the execution of %DTEST. After %DTEST has completed execution, the value of MPRINT is restored as the system option value.

The SECURE option allows us to keep the source code of our validated macro out of the hands of those who may want to re-engineer our code. However, since one of the primary functions of a macro is to serve as a code generator any code generated by the macro, even with the SECURE option in effect, can still be seen by those executing the macro.

13.9.2 Executing a Specific Macro Version

In addition to hiding the macro source code, we may also want to control which version of a given macro is to be executed. If we have written and validated a given macro, we may need to make sure that our version is the one executed by our users. Nominally we do this by placing our version of the macro in the Autocall library and/or in the stored compiled macro library.

The user could still circumvent the use of our version of the macro by writing his/her own macro from scratch and then force its use in preference to the validated version using any of several techniques. Each of these techniques results in the compiled version of their macro being written to the WORK.SASMACR catalog. Since the WORK.SASMACR catalog is always searched first, their version will then be seen (and executed) in preference to ours.

The user's version may also be *inadvertently* written to either the autocall library or to the stored compiled macro catalog. You can provide some protection by making each of these locations READ ONLY to all but the developers of the macros. Still the user can compile and execute a macro from the WORK.SASMACR catalog.

The system options NOMCOMPILE and NOMREPLACE are partial solutions to these circumventions. However, they are not without side effects, and of course like other system options, protection of the option's value itself is outside of our control. NOMCOMPILE prevents the compilation of new macro definitions and NOMREPLACE restricts the storage of a new compiled version of a macro that has already been compiled.

NOMCOMPILE prevents the compilation of new macros, but does not prevent the use of macros that are already stored in a library. This option is a solution to the user's macro which was written to replace one of ours ❶ ❸. However, it also prevents the user from compiling any new macro definitions. Notice also that macro definitions stored in the autocall library ❷ are not affected by NOMCOMPILE.

```
options nomcompile;

%* attempt to compile an autocall macro (%OBSCNT);
%* The macro compiles!!!; ❷
%put OBS count for DEMOG is %obscnt(advrpt.demog);

* Because of NOMCOMPILE
* macro MYGHI does not compile; ❶
%macro myghi;
%put compile from within program;
%mend myghi;
%myghi

* Because of NOMCOMPILE
* Included macro definitions do not compile; ❶
%inc "&path\sascode\Chapter13\frominc13_9_2.sas";
%frominc13_9_2

* Because of NOMCOMPILE
* The macro is not stored or compiled; ❸
%macro storeghi / store;
%put macro was compiled and stored;
%mend storeghi;
%storeghi
```

A portion of the LOG shows that, because of the NOMCOMPILE option, the macro %MYGHI was not compiled, and was not available for execution.

```
1563 %macro myghi;
ERROR: Macro compilation has been disabled by the NOMCOMPILE option. Source code
     will be discarded until a corresponding %MEND statement is encountered.
1564 %put compile from within program;
1565 %mend myghi;
1566 %myghi
     -
    180
WARNING: Apparent invocation of macro MYGHI not resolved.
```

```
options mcompile nomreplace;

* Compile the autocall macro GHI;
* Definition is stored in WORK.SASMACR; ❷
* (entry GHI.MACRO does not already exist);
%ghi

* Unauthorized version of GHI
* does not replace version from the
* autocall library;
%macro ghi; ❶
%put Unauthorized version of GHI;
%mend ghi;
%ghi
```

The NOMREPLACE system option is used to prevent the replacement of a macro definition that has already been compiled and resides in WORK.SASMACR. It has no affect on stored compiled macro libraries and, therefore, offers us no protection against a scenario where the user overwrites the official version of the macro in the stored compiled macro library ❸.

When used together these two system options give us some protection for our authorized macro versions from the three cases shown above. Unfortunately the use of NOMCOMPILE can severely limit the use of user written macros.

```
ERROR: The macro GHI will not be compiled because the NOMREPLACE option is set. Source code
     will be discarded until a corresponding %MEND statement is encountered. ❶
```

Since one of the biggest issues is that the WORK.SASMACR catalog is searched first, this is the catalog on which we need to focus. If we were to copy all the macro definitions in our stored

```
%macro copysasmacr/store;
proc catalog catalog=advrpt.sasmacr;
  copy out=work.sasmacr;
  quit;
%mend copysasmacr;
```

compiled macro library to WORK.SASMACR at the very start of the program/application, the NOMREPLACE option would be protecting our authorized macro definitions! The macro %COPYSASMACR can be placed in our

library and then called in the autoexec.sas program, **before WORK.SASMACR exists**. With the NOMREPLACE option in effect, but not NOMCOMPILE, users will be free to create their own macros, but not macros with the same name as any of our validated macros.

The %COPYSASMACR macro will not work if the WORK.SASMACR catalog already exists. Since compiling any macro, even one that is to be saved as a stored compiled macro creates WORK.SASMACR, the call to %COPYSASMACR needs to be one of the first things executed in the AUTOEXEC program.

As an aside, SAS uses an internal pointer to track whether or not WORK.SASMACR exists. Using the %COPYSASMACR does not reset this pointer. This generally does not matter, because any macro executed by the user will exist in the stored compiled library as well, and it is the version (ADVRPT.SASMACR) that will actually be executed (not the version in the WORK.SASMACR catalog). As soon as the user compiles or attempts to compile any macro, the pointer is reset and the macro facility will start checking WORK.SASMACR first for macro definitions.

Another approach is to purge the WORK.SASMACR catalog of any macros with a name of one of your protected macros prior to running your application. Although the WORK.SASMACR

```
%macro purgework(macname=);
proc catalog cat=work.sasmacr
       entrytype=macro;
  delete &macname;
  quit;
%mend purgework;
%purgework(macname=abc def ghi myghi)
```

catalog cannot be deleted its members can be deleted. The macro %PURGEWORK deletes one or more entries from the WORK.SASMACR catalog. The list of entries to be deleted may contain macros that are not in the catalog, these generate an ERROR in the LOG; however, the remaining macro entries are still deleted. It should be noted that while this

technique seems to work, deleting macros from the WORK.SASMACR catalog is NOT a technique that is supported by SAS Institute. Starting in SAS 9.3 there is a new statement called %SYSMACDELETE that allows you to delete macros from the WORK.SASMACR catalog.

Incorporation of the FORCE option on the PROC CATALOG statement allows you to overwrite

```
%macro vercopy(verlist=)/store;
proc catalog c=complib.sasmacr
       force
       et=macro;
  copy out=work.sasmacr ;
  select &verlist;
  quit;
%mend vercopy;
```

an existing compiled macro in the WORK.SASMACR catalog. Here the %COPYSASMACR macro has been enhanced so that one or more selected macro entries can be copied from the stored compiled macro library to the WORK.SASMACR catalog. Unlike %COPYSASMACR, the %VERCOPY macro can be executed after the WORK.SASMACR catalog has been established. If it is called just

before using one of your controlled macros, you can be sure that you will be using your own version of the macro that is to be called.

Here the %BIGSTEP macro will be calling %CLEANUP and %GHI. In order to make sure that our versions of these two macros exist in the WORK.SASMACR catalog, we call %VERCOPY.

```
%macro bigstep;
  %vercopy(verlist=cleanup ghi)
  %cleanup
  %* do other things;
  %ghi
%mend bigstep;
```

SEE ALSO

Sun and Carpenter (2011) offer more detail on a number of additional techniques for the protection of macros and macro code.

13.10 Using the Macro Language IN Operator

The IN comparison operator can be used in the DATA step to form lists of values that are to be checked against. For the macro language, the IN operator was briefly available in the initial release of SAS®9, and has returned with some differences in SAS 9.2.

Although the syntax is a bit different, the macro language IN comparison operator is similar in function to the IN comparison operator that can be used elsewhere, as in the DATA or SQL steps. The operator symbol is the pound sign (#), and the mnemonic IN can also be used. By default this comparison operator is not available in SAS 9.2, but can be made available through the use of the MINOPERATOR option.

SEE ALSO

Usage Note 35591 http://support.sas.com/kb/35/591.html discusses the IN operator, as does Usage Note 31322 http://support.sas.com/kb/31322, which shows how to use the IN operator with a NOT. Examples in the documentation can be found at http://support.sas.com/documentation/cdl/en/mcrolref/61885/HTML/default/viewer.htm#a003092 012.htm.

Warren Repole has an example of the IN operator as a part of his series of articles titled "Don't Be a SAS Dinosaur: Modernize Your SAS Programs," http://www.repole.com/dinosaur/separatedby.html.

13.10.1 What Can Go Wrong

One of the reasons that the IN operator was unavailable in SAS 9.1 was because of a confusion between the mnemonic IN and the postal code abbreviation for the state of Indiana, which is also

```
%macro BrokenIN;
%let state=in;
%if &state=CA %then %put California;
%else %put Not California;
%mend brokenin;
%brokenin
```

IN. In the initial release of SAS®9 the macro %BROKENIN shown here would fail, because &STATE resolves to IN before the expression is evaluated. During the evaluation of the expression the IN is seen as mnemonic for the IN operator. In SAS 9.1 there is no confusion, because the IN operator is not available. In SAS 9.2 the operator returns, but with options that help to remove the confusion. In SAS 9.2 the IN operator is, by default, not available, and %BROKENIN will execute correctly.

Regardless of the release of SAS or the status of the MINOPERATOR option, the IN can be

```
%if %bquote(&state)=CA %then %put California;
```

masked from being interpreted as the mnemonic for the IN operator by using a quoting function. Here %BQUOTE is used to mask the resolved value of &STATE until after the evaluation of the expression when it will not cause a problem. For this particular example the %BQUOTE will also prevent parsing errors for the states of Oregon (OR) and Nebraska (NE).

13.10.2 Using the MINOPERATOR Option

Starting in SAS 9.2 you can choose whether or not to make the IN operator available. The selection is done using the MINOPERATOR option (the default is NOMINOPERATOR). This option can be applied as a system option or, and this is my recommendation, as an option on the macro statement.

In Section 13.10.1 the macro %BROKENIN was shown to work correctly in SAS 9.2; however, it

```
option minoperator;
%macro BrokenIN;
%let state=in;
%if &state=CA %then %put California;
%else %put Not California;
%mend brokenin;
%brokenin
```

will fail in SAS 9.2 when the MINOPERATOR option is turned on. Instead of applying the MINOPERATOR at the system level, if it is applied at the macro level you gain a finer degree of control. In the remaining examples in this section this option will always be set at the macro level and the assumption will be that the system option remains turned off (NOMINOPERATOR).

Although we commonly refer to the letters (IN) as the operator symbol, the IN is actually the mnemonic, and the pound sign (#) is the actual symbol for the operator. Some of the confusion discussed above would go away if we could turn off the mnemonic and just use the # sign, but that choice is not currently an option.

The macro %TESTIN can be used to demonstrate the use of the IN operator. Here the NOMINOPERATOR system option ❶ has been set to mimic the default value (SAS 9.2). Since

```
* system option default value;
option nominoperator;  ❶

%macro testIN(dsn=demog)/minoperator;  ❷
%if %upcase(&dsn) # AE CONMED DEMOG ❸ %then %do;
  %put &dsn count %obscnt(advrpt.&dsn);
%end;
%mend testin;
%testin(dsn=ae)
```

we want to use the operator in the macro it is turned on, just for the execution of %TESTIN, through the use of the MINOPERATOR option ❷ on the %MACRO statement. In this comparison, the expression checks to see if the resolved value of

&DSN is in the list of acceptable values ❸ to the right of the operator. The expression is often easier to read when the list of values is enclosed in parentheses. The # can be replaced with the mnemonic. The IN operator resolves to true or false (1 or 0) and, therefore, uses an implied %EVAL function.

```
%upcase(&dsn) in (AE CONMED DEMOG)
```

13.10.3 Using the MINDELIMITER= Option

By default the IN operator expects the list of values to be space separated; however, it is possible to use the MINDELIMITER= option to specify an alternate delimiter for the list. Like the MINOPERATOR option, the MINDELIMITER= option can be specified at the system option level or at the macro level, and again my recommendation is to always apply it on the macro statement and not at the system level. Here the MINDELIMITER= option has been added to the %MACRO statement ❹ to allow a comma separated list ❺ of values.

```
%macro testIN(dsn=demog)/minoperator
         mindelimiter=','; ❹
%if %upcase(&dsn) in(AE,CONMED,DEMOG) ❺ %then %do;
   %put &dsn count %obscnt(advrpt.&dsn);
%end;
%mend testin;
%testin(dsn=demog)
```

By enclosing the list in parentheses and by using the IN mnemonic, as was done here, the macro expression more closely mimics the syntax used with the DATA step's IN operator.

13.10.4 Compilation vs. Execution for these Options

When using the system option versions of MINOPERATOR and MINDELIMITER=, it is important to understand the difference between compilation and execution of the macro. If you intend to use these options on the %MACRO statement (as I have suggested), rather than as system options, then the issues discussed below will not affect you, as the values declared on the %MACRO statement will override the system options during both compilation and execution.

The MINDELIMITER= System Option

The value of the MINDELIMITER= option is set when the macro is compiled. Subsequent changes to this option will not affect the execution of the macro.

```
options minoperator mindelimiter=','; ❻
%macro testIN(dsn=demog);
%if %upcase(&dsn) in(AE,CONMED,DEMOG) %then %do;
   %put &dsn count %obscnt(advrpt.&dsn);
%end;
%else %PUT &DSN not on the list;
%mend testin;
. . . .
* change in mindelimiter does not break the macro;
options minoperator mindelimiter=' '; ❼
%testin(dsn=conmed)
```

❻ The MINDELIMITER= option has been set prior to the compilation of the macro %TESTIN.

❼ Prior to execution, the value of the MINDELIMITER= option has been changed to a blank. Although the value of the MINDELIMITER= option has been changed ❼, the %TESTIN macro will still operate correctly. This means that before you compile a macro that depends on the MINDELIMITER= system option, you must either know the current value of the option or explicitly set this system option prior to compiling the macro. You do not, however, need to know the current setting when executing the macro.

The MINOPERATOR System Option

For the MINOPERATOR system option, it is the value that is in effect at the time of macro execution that is applied. It does not matter whether MINOPERATOR or NOMINOPERATOR is

```
77  options nominoperator; ❽
78  %testin(dsn=conmed)
ERROR: Required operator not found in expression:
%upcase(&dsn) in(AE,CONMED,DEMOG)
ERROR: The macro TESTIN will stop executing.
```

set during the compilation of the macro; however, if NOMINOPERATOR is in effect when the macro is called, the macro will fail. The LOG shown here shows

that the macro %TESTIN, which executed successfully in the previous example ❼, fails with the MINOPERATOR option turned off ❽.

Remember you will not need to worry about the current settings of these system options if you always set their values as options on the %MACRO statement.

13.11 Making Use of the MFILE System Option

Generally when debugging a macro the options MPRINT, SYMBOLGEN, and MLOGIC are sufficient to find the problem with the code. You can also use the MLOGICNEST and MPRINTNEST options when the macros are nested. Sometimes, however, you just need a bigger hammer. The MFILE system option gives us another way of looking at the results of the macro.

The macro language is primarily a code generator, and the MFILE system option allows us to save the code generated by the macro. This saved code will be completely free from all macro references. Here for demonstration purposes the macro %PRINTIT performs a simple PROC PRINT.

```
%Macro PRINTIT(dsn=,varlist= ,obs=);
proc print data=&dsn
  %if &obs ne %then (obs=&obs);;
  %if &varlist ne %then var &varlist;;
  run;
%mend printit;

options mprint mfile; ❶
filename mprint ❷ "&path\sascode\Chapter13\E13_11_mfile.sas" ❸;
%printit(dsn=advrpt.demog,varlist=subject lname fname dob,obs=4)
```

❶ In order to use the MFILE system option, the MPRINT option must also be turned on.

❷ The resulting SAS code is written to the file associated with the *fileref* MPRINT (it must be MPRINT; you do not get to choose the name for this *fileref*).

The generated SAS code is written to the file (E13_11_mfile.sas ❸), and contains the PROC PRINT after all the macro references and logic have been resolved and executed.

```
proc print data=advrpt.demog (obs=4);
var subject lname fname dob;
run;
```

13.12 A Bit on Macro Quoting

Outside of the macro language quote marks (double and single) are used by the parser to distinguish character strings from things like options and variable names. In the macro language quotes are not seen as parsing characters because the & and % characters serve as macro language triggers. While quote marks are not parsing characters, they and other special characters can have special meaning in the macro language. As a consequence when we must work with these special characters; which may also include the comma, semicolon, colon, and Boolean operators, we need to take precautions to make sure that they are interpreted correctly.

Fortunately the macro language includes a number of functions that mask these special characters. However, these various functions do not all mask all of the same special characters in the same way. There are not only several quoting functions that are supplied with the macro language, but there are also a number of text functions that can be used to return quoted text.

It is not always obvious when macro quoting is going to be required. In this simple example all we want to do is write the macro variable &LIST to the LOG and we would like to make sure that the list is left justified.

Here the %LEFT function fails because it has too many positional arguments. The message comes from the fact that %LEFT is really a macro with a single positional parameter. But the real problem is that the commas in &LIST are being interpreted as parameter separators. The commas have special meaning to the macro parser. This meaning can be masked by using a quoting function. Here the %STR function is used

```
%let list = butter, cheese, milk;
%put %left(&list);
```

to mask the commas. The %LEFT now executes as we would expect.

```
%let list = %str(butter, cheese, milk);
%put %left(&list);
```

Internally the macro quoting functions insert an invisible character into the text string that allows the parser to ignore the special characters. Effectively the quoted text stored in &LIST becomes something like is shown here. Where the symbol ☐ stands for the invisible masking character. Under Windows it is sometimes possible to surface this invisible character in the LOG, and it shows up as the ☐ symbol.

```
%let list =butter☐,☐ cheese☐,☐ milk;
```

In the following TITLE statement we would like to insert the run date using the WORDDATE18. format. Since this numeric format will be right justified we want to left justify it in the title. This fails for the same reason as the %PUT failed in the previous example. On May 24th the %SYSFUNC returns a date string which is to be left justified. The

```
title1 "13.12 %left(%sysfunc(date(),worddate18.))";
```

```
title1 "13.12 %left(   May 24, 2011)";
```

comma inserted by the WORDDATE18. format is the culprit.

A quoting function is needed to mask the comma in the date. In this case the %SYSFUNC function has a quoting analogue %QSYSFUNC which

```
title1 "13.12 %left(%qsysfunc(date(),worddate18.))";
```

can be used to mask the comma.

There are several macro quoting functions. Each is slightly different and the differences are often quite subtle. Fortunately you do not need a full understanding of each of the various quoting functions in order to mask special characters. Almost all of your quoting needs can be met with the %BQUOTE and %NRSTR quoting functions. The following table gives a brief summary of some of these functions.

Quoting Function	This Function's Super Power
%STR	Easy to type—quotes most of the usual suspects.
%BQUOTE	Quotes even more characters than %STR.
%NRSTR	Masks & and %, and prevents their resolution or execution.
%UNQUOTE	Removes macro masking characters.

Each of the macro functions that returns text (such as; %LEFT, %UPCASE, %TRIM, %SYSFUNC) has a quoting analog (same name preceded by a Q) that returns text with special characters masked. This includes the macro language triggers & and %. For instance the function call %LEFT(A&P) would attempt to resolve the &P as a macro variable, whereas the quoting analog (%QLEFT) will not. Regardless of whether or not characters have been masked before the use of these functions, they will be unmasked in the returned text, unless a quoting analog is used.

```
%let p = proc;
%let store=
   %nrstr( My favorite store is the A&P. );  ❶
%put |&store|; ❷
%put %left(&store);  ❸
%put %qleft(&store);  ❹
```

❶ The %NRSTR allows the preservation of leading and trailing blanks and also masks the &P so that no attempt is made to resolve it as a macro variable.

```
7   %put |&store|;
| My favorite store is the A&P. |  ❷
8   %put %left(&store);
My favorite store is the Aproc  ❸
9   %put %qleft(&store);
My favorite store is the A&P.  ❹
```

❷ The value of &STORE contains leading and trailing blanks as well as a masked &P.

❸ The %LEFT left justifies the text and allows the resolution of &P. (&P. is replaced by the letters `proc`). Notice that the period is seen as a part of the macro variable name and is also replaced.

❹ The %QLEFT left justifies the text without removing the masking characters around the A&P.

For my work I usually tend to use the %BQUOTE and %NRSTR quoting functions almost to the exclusion of the others. For text functions my general rule is to use the Q analogue version unless there is a reason not to do so.

SEE ALSO
Whitlock (2003) introduces and discusses the ins and outs of macro quoting—start here if you are new to macro quoting. Rosenbloom and Carpenter (2011b) specifically address issues associated with macro variables that contain special characters. Macro quoting is discussed in detail in Carpenter (2004, Section 7.1).

Chapter 14

Operating System Interface and Environmental Control

Very often a significant portion of our use of SAS interfaces with the Operating System, OS. Fortunately there are a number of tools and techniques that can be used to help us make that interface smoother. Usually it is a matter of passing information between the OS and SAS. Other times it is helpful to be able to have SAS execute OS commands directly. Some of the techniques that I have found to be useful are included in this chapter.

An important interface that is not discussed in this chapter is SAS Enterprise Guide. This topic is thoroughly covered by Slaughter and Delwiche (2010). Other less frequently used interfaces, which are also not discussed in this chapter include: SAS/ASSIST, SAS Desktop, and SAS/INSIGHT.

MORE INFORMATION

A great deal of information on SAS's interface with the operating system is available in various forms of metadata (see Section 13.8).

SEE ALSO

Peter Crawford (2006b) discusses a number of Display Manager techniques.

14.1 System Options

Most system options either control the SAS environment or they control the way that SAS interfaces with the operating system. While most of these options are fairly straightforward to use, you need to at least be aware of them. In addition some have proven to be especially useful. A few of the options that fall into one or more of these categories are included here.

14.1.1 Initialization Options

Initialization options are specified when SAS is invoked and usually along with the same script

that calls the SAS execution file. Although initialization options are available for all operating systems, the syntax and application varies by OS. Most of the examples shown here are for the Windows OS; however, the implementation with other operating systems should be fairly straightforward.

Under the Windows OS, when executing interactively, these options are declared on the target line of the properties of the SAS shortcut. The options themselves are preceded by a dash (-). Here the -CONFIG initialization option is specified on the TARGET line. Notice that all paths are enclosed in *double* quotes (double quotes are required by Microsoft for all paths that have any embedded blanks).

Most of the options shown in this section can also be specified in the configuration file (see Section 14.3).

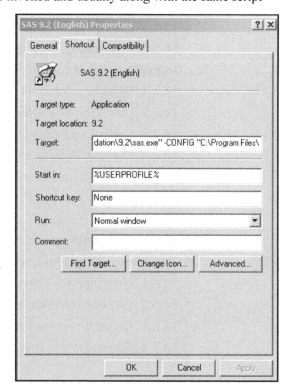

SEE ALSO
Hurley (2007) introduces the use of initialization options.

The -SASINITIALFOLDER Option
This initialization option controls the default location that SAS first opens when the Display Manager brings a SAS program into the editor. Simply follow the option with the path to the desired folder.

```
-sasinitialfolder "folder location"
```

For the sample code associated with this book the -SASINITIALFOLDER option might contain:

```
-sasinitialfolder "C:\InnovativeTechniques\SASCode"
```

The 'Start In:' location on the PROPERTIES dialog box has limited utility in this context. It can be used to designate the base path information, but not necessarily the location of the SAS code of interest.

The -SYSPARM Option
The -SYSPARM option is used to pass information into a SAS program where it can be accessed using the automatic macro variable &SYSPARM. Here a portion of a path is passed into the SAS session when SAS is initialized ❶. Within the SAS session the

```
-sysparm "c:\temp" ❶
```

automatic macro variable &SYSPARM ❷ can be used to access this information. The quotes used with the initialization option ❶ are not stored with

```
libname advrpt v9 ❷ "&sysparm\InnovativeTechniques\Data";
```

&SYSPARM ❷.

The -SPLASHLOC Option
This option is more fun than practical. It allows you to point to a bit mapped image, BMP file, that replaces the normal splash screen shown while SAS is being loaded. Instead of seeing the SAS logo, you can view a cute picture of your favorite dog, cat, or moose. Of course the image is not displayed for long. The only practical application that I have seen, other than on April 1, is for use with a specific tool or program that you want to identify visually.

The -AUTOEXEC Option
The -AUTOEXEC option identifies the path and name of a SAS program that is to be automatically executed when SAS is started (see Section 14.2). This program executes after the configuration file and can be extremely useful when setting up a SAS environment that is to be tailored for a specific application.

The -CONFIG Option
The -CONFIG option identifies the path and name of a SAS configuration file (see Section 14.3). This is not a SAS program, but is used during the initialization process to set up the base SAS environment. Many SAS installations restrict the modification of this file.

The -ALTLOG Option
Using this option you can write a separate copy of the LOG to a file of your choice. This becomes an alternative to using PROC PRINTTO.

The -SYSIN Option
When running SAS programs in batch mode, a specific SAS program can be named to be executed using the -SYSIN option. When you use this option, it becomes unnecessary to change

the location for the launching of SAS. You only need to specify the path to the program that is to be executed.

SEE ALSO

The sasCommunity.org article by Ron Fehd "Batch processing under Windows" has examples of the use of -SYSIN and other initialization options http://www.sascommunity.org/wiki/Batch_processing_under_Windows.

The -RTFCOLOR Option

I do not remember where I heard about this option and I have not found it documented, but it seems to work under Windows during an interactive session. Using this initialization option allows you to write your log to a RTF file with preserved colors (notes, warnings, and errors).

During your SAS session or from within your SAS program, issue a DM statement with the WRTFSAVE option. Alternatively you could issue the WRTFSAVE command from the command line when the LOG window is active.

```
DM 'log; WRTFSAVE "c:\temp\mylog3.rtf"';
```

Even with -RTFCOLOR specified, logs generated with PROC PRINTTO and the -ALTLOG initialization option do not retain their color.

CAVEAT Use this option with care, do not write applications that depend on it, because this option is undocumented and there is no guarantee or even likelihood that it will continue to even exist, let alone be supported in future releases of SAS.

The -INITSTMT and -TERMSTMT Options

The -INITSTMT and -TERMSTMT options are generally used together, and provide a mechanism to automatically execute a SAS statement at the beginning and end of a SAS session. The -INITSTMT option designates a statement that is to execute immediately following the execution of the AUTOEXEC, while the -TERMSTMT option specifies a statement that will execute just before the session closes. When the SAS statement is a macro call, the entire macro is executed at the appropriate time.

Since macro variables are stored in memory, their values are not retained from one SAS session to the next. This example uses the -INITSTMT and -TERMSTMT options to save and restore macro variable values between SAS sessions. These options designate macro calls to %GETGLOBAL and %SAVEGLOBAL, both of which are autocall macros.

```
-initstmt='%getglobal' -termstmt='%saveglobal'
```

The %GETGLOBAL macro recovers the saved values from the data set

```
%macro GetGlobal;
data _null_;
   set advrpt.globalvars(where=(name ne 'PATH'));
   call symputx(name,value,'g');
   run;
%mend getGlobal;
```

ADVRPT.GLOBALVARS and uses the SYMPUTX routine to reestablish them as global macro variables. Because %GETGLOBAL macro definition is in the autocall library, which was established by the AUTOEXEC, the macro can be immediately executed by the -INITSTMT option.

The %SAVEGLOBAL macro will be executed as the SAS session terminates. All macro variables currently defined with a scope of global are written to the data set ADVRPT.GLOBALVARS, where they will be available for the next SAS session. For interactive sessions the ENDSAS statement must be executed in order for the -TERMSTMT option to be triggered.

```
%macro SaveGlobal;
data advrpt.GlobalVars(keep=name value);
   set sashelp.vmacro(where=(scope='GLOBAL'));
   run;
%mend saveglobal;
```

The -NOWORKINIT and -NOWORKTERM Options

By default the WORK location is initialized at the start of the SAS session and is cleared at the end of the SAS session. Usually this is exactly what you want to have happen; however, occasionally you may want to maintain the WORK location across SAS sessions. The -NOWORKINIT option prevents SAS from initializing the WORK location at the start of the SAS session, and -NOWORKTERM prevents SAS from clearing the WORK location when SAS terminates.

These two options are independent of each other but will generally be used together, as they allow you to maintain the WORK space across session boundaries.

14.1.2 Data Processing Options

There are a few SYSTEM options that are either misused or underutilized. This is by no means an exhaustive list, just a few that I have encountered. Some of these should be used only by the advanced user who has full knowledge of the ramifications of their use.

The MERGENOBY Option

Unless done in a very deliberate manner, performing a MERGE without a BY statement (one-to-one merge) can be very risky. When done inadvertently, the result is almost always not what is desired. Worse, the errors are often not obvious, and these may not produce any warnings, notes, or error messages in the LOG. The MERGENOBY option can be used to change this behavior. The MERGENOBY determines what action should be taken when the MERGE statement is used without a corresponding BY statement.

- NOWARN merge takes place without warning **(this is the default)**
- WARN   merge takes place and a warning is issued
- ERROR   merge does not take place and an error is written to the LOG

The MERGENOBY=WARN designation ❶ causes a warning to be issued to the LOG. It is my opinion that the MERGENOBY option should be set to ERROR, and only reset to WARN or NOWARN when a specific compelling need arises. The MERGENOBY=WARN produces the following warning in the LOG. Notice that although a warning is issued, the MERGE still takes place.

```
options mergenoby=warn;  ❶
data aemed;
   merge advrpt.ae
         advrpt.conmed;
   run;
```

```
WARNING: No BY statement was specified for a MERGE statement.
NOTE: There were 127 observations read from the data set ADVRPT.AE.
NOTE: There were 199 observations read from the data set ADVRPT.CONMED.
NOTE: The data set WORK.AEMED has 199 observations and 13 variables
```

MORE INFORMATION

A MERGE statement without a BY appears in a 'look-ahead' example in Section 3.1.4. The code associated with that example changes and then resets the current setting of the MERGENOBY option.

The DATASTMTCHK Option

This option protects the user from overwriting data sets due to dropped semicolons by restricting the use of certain data set names. The default is COREKEYWORDS which prevents you from naming a data set UPDATE, MERGE, RETAIN, or SET.

- COREKEYWORDS   data sets may not have a name corresponding to a core keyword **(default)**

- ALLKEYWORDS   also excludes all names that can start SAS DATA step statements (i.e., ARRAY, DO, OUTPUT)

- NONE   no restrictions (*defacto* default prior to V7)

In this DATA step the semicolon has been left off of the DATA statement ❷. Effectively we are

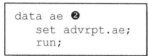

```
data ae ❷
   set advrpt.ae;
   run;
```

attempting to create three data sets: WORK.AE, WORK.SET, and ADVRPT.AE. Fortunately the COREKEYWORDS setting of the DATASTMTCHK option protects us. In V6 or with the option set to NONE, all three data sets would have been created with 0 observations and 0 variables – our permanent data set

ADVRPT.AE would have been wiped out!! In my opinion this option should not be reset to match V6 standards (DATASTMTCHK=NONE).

```
22   data ae
23      set advrpt.ae;
        ---
        56
ERROR 56-185: SET is not allowed in the DATA statement when option
DATASTMTCHK=COREKEYWORDS.
             Check for a missing semicolon in the DATA statement, or use
DATASTMTCHK=NONE
```

As an aside, neither the COREKEYWORDS nor the ALLKEYWORDS values will flag a data set, regardless of its name, which includes a *libref*. Consequently it would be possible, although of doubtful utility, to name a data set WORK.UPDATE, when naming it without the *libref*, UPDATE, would fail.

The VALIDVARNAME Option

The VALIDVARNAME option is used to determine valid variable naming conventions, and how the variable names are handled.

- V6   variable names must conform to V6 conventions (up to 8 characters). This option value is no longer documented, but still works for the current versions of SAS.

- V7   allows up to 32 character names, and the case at variable definition is remembered (the default).

- UPCASE   case is not remembered or used.

- ANY   allows non-standard naming conventions. Non-standard naming conventions have limited utility outside of Base SAS and SAS/STAT.

The VALIDVARNAME=V6 Option

```
option validvarname=v6;
data a;
    abcdefghig= 5;
    run;
```

Useful when you are creating data sets that must conform to V6 naming conventions. While acceptable for all other values of VALIDVARNAME, the following DATA step fails when VALIDVARNAME=v6, because the variable name has more than 8 characters. This value of the VALIDVARNAME= option may not be supported in future releases of SAS. In SAS 9.2 it has not been included in the documentation for the VALIDVARNAME option. Although the V6 value restricts the length of variable names it does not restrict the length of data set names, as was the case for SAS V6.

The VALIDVARNAME=ANY Option

This setting permits variable names that may come from other sources, such as ORACLE or EXCEL, that do not use the same naming conventions as SAS. We access these non-standard variables through the use of named literals that are needed whenever the name contains non-standard characters. Once the data have been imported into SAS, it is generally recommended that these variables are renamed to conform to standard SAS naming conventions. When a data set contains variables defined using named literals, in order to include non-standard characters, VALIDVARNAME= must always be set to ANY before the data set can be used. Also, whenever you access any variable containing non-standard characters you must always use named literals to name the variable.

A named literal is enclosed in quotes and the quoted string is followed by the letter n. Named literals are used just like any other variable name. Three variables are included in the VAR statement to the left.

```
options validvarnames=any;
proc print data=aexls;
   var subject 'ae-date'n 'ae#type'n;
   run;
```

The application of the use of VALIDVARNAMES=ANY can be shown using the EXCEL file E14_1_2AE.XLS. This spreadsheet has column headers that do not meet SAS naming standards.

	A	B	C
1	Subject	AE-date	AE#type
2	201	24-Apr-06	1
3	202	15-May-06	2

PROC IMPORT can be used to convert this EXCEL table to a SAS data set. Since the GETNAMES=YES option is used, SAS will attempt to use the column headers as variable names. Without first specifying VALIDVARNAME=ANY, non-standard characters would be converted to underscores.

In this example it would not have a large impact; however, naming conflicts can arise. By using VALIDVARNAME=ANY, we know what the columns will be named in the SAS data set (WORK.AEXLS). A look at the properties of WORK.AEXLS shows that the non-standard names are being used by SAS.

```
option validvarname=any;
PROC IMPORT OUT= WORK.AeXLS
          DATAFILE= "&path\data\E14_1_2AE.xls"
          DBMS=EXCEL REPLACE;
    SHEET="Sheet1$";
    GETNAMES=YES;
    MIXED=NO;
    SCANTEXT=YES;
    USEDATE=YES;
    SCANTIME=YES;
RUN;
```

Column Name	Type	Length
Subject	Number	8
AE-date	Number	8
AE#type	Number	8

Non-standard data set names are generally only honored in Base SAS and SAS/STAT, therefore it is generally a good idea to rename these variables as soon as is convenient. In this case the variable's label will reflect the original name.

When using a non-standard name, such as is done in this RENAME option, a named literal is used - the variable name is surrounded by quotes, which are immediately followed by the letter n.

```
data ae;
   set aexls(rename=('ae-date'n=AEDate
                     'ae#type'n=AEType));
   run;
```

14.1.3 Saving SAS System Options

Typically the settings for SAS System Options persist for the duration of the current job or the current SAS session. It is possible, however, to save the current system options settings in either the registry or in a data set. Once saved, they can be retrieved for a future session or job. Saving the options in a data set is more flexible as it allows other users to acquire the same option settings, which can include macro and format library controls.

Not all system options can be saved; most notably are initialization options and options that contain passwords.

The options can be saved and retrieved by the use of either procedures or commands.

MORE INFORMATION
The -INITSTMT and -TERMSTMT system options described in Section 14.1.1 could also be used to capture and save system options using similar techniques to those shown in that section to capture and save macro variable values.

SEE ALSO
The SAS Forum thread http://communities.sas.com/message/101546 discusses a number of ways to gather and reestablish system options.

Using PROCs OPTSAVE and OPTLOAD
The OPTSAVE procedure is used to write the current option settings to a SAS data set. The data set is like any other and consists of two columns (OPTNAME and OPTVALUE). A portion of the data set ADVRPT.CURRENT_SETTINGS is shown here. Notice the SASAUTOS option in line 155 ❶.

```
proc optsave out=advrpt.current_settings;
   run;
```

```
14.1.3 Saving System Options
Options saved using PROC OPTSAVE

Obs   OPTNAME            OPTVALUE

148   REPLACE            1
149   REUSE              NO
150   RIGHTMARGIN        0.000 IN
151   RSASIOTRANSERROR   1
152   S                  0
153   S2                 0
154   S2V                0
155   SASAUTOS           (advmac, sasautos) ❶
156   SASCMD
          ...portions of the table not shown...
```

In the code that follows, the SASAUTOS option is 'inadvertently' changed ❷. We can observe the change using PROC OPTIONS ❸ ❺, and we can recover the original setting of the option using PROC OPTLOAD ❹.

```
33   proc options option=sasautos; ❷
34      run;

     SAS (r) Proprietary Software Release 9.2   TS2M2

 SASAUTOS=adv        Search list for autocall macros ❸
NOTE: PROCEDURE OPTIONS used (Total process time):
      real time            0.00 seconds
      cpu time             0.00 seconds

35   proc optload data=advrpt.current_settings
                   (where=(optname='SASAUTOS')); ❹
36      run;

NOTE: PROCEDURE OPTLOAD used (Total process time):
      real time            0.12 seconds
      cpu time             0.01 seconds


37   proc options option=sasautos; ❺
38      run;

     SAS (r) Proprietary Software Release 9.2   TS2M2

 SASAUTOS=(advmac, sasautos) ❺
                   Search list for autocall macros
NOTE: PROCEDURE OPTIONS used (Total process time):
      real time            0.00 seconds
      cpu time             0.00 seconds
```

❺ PROC OPTIONS is used to view the current setting of the SASAUTOS option.

Here we have recovered the setting of a single option. If the WHERE clause had not been used ❹ all the option settings in the data set ADVRPT. CURRENT_ SETTINGS would have been used.

Using the DMOPTSAVE and DMOPTLOAD Commands

The DMOPTSAVE and DMOPTLOAD commands are similar to the PROC steps described above. As commands they are designed to be executed from the command line in the Display Manager, but this also means that they can either be assigned a key in the KEYS window (see Section 14.4.7), or that they can be executed from within the DM statement. Since the DM statement is generally the most flexible, the code is shown here.

```
dm 'dmoptsave advrpt.current_settings'; ❻
options sasautos=adv2; ❼
proc options option=sasautos; ❽
   run;
dm "dmoptload advrpt.current_settings ❾
             (where=(optname='SASAUTOS'))";
   run;
proc options option=sasautos;
   run;
```

❻ The current system option settings are saved using the DMOPTSAVE command.

❼ The SASAUTOS option is *accidentally* changed.

❽ The new SASAUTOS option value is ready to be used. All we do here is demonstrate that it has been changed.

❾ The *original* value for the SASAUTOS option is restored.

MORE INFORMATION

The DM statement is also discussed in Section 14.4.2.

14.2 Using an AUTOEXEC Program

When SAS initializes and starts execution, it automatically looks for and executes (if it exists) a program named AUTOEXEC.SAS. By default the AUTOEXEC.SAS program will only be found if it is in the !SASROOT directory. Through the use of the -AUTOEXEC initialization option (see Section 14.1.1), you can point to any SAS program in any location.

The AUTOEXEC.SAS can be any ordinary SAS program. It is commonly used to set up macro libraries, system options, *librefs*, and *filerefs*. Although the program can be named anything (Fred.jpg), AUTOEXEC.SAS is less obscure.

The AUTOEXEC.SAS shown here could be used to set up the *libref* used to access the data associated with this book. It expects that the automatic macro variable &SYSPARM ❶ was assigned the upper portion of the path structure when SAS was called (see Section 14.1.1 for more on the use of the -SYSPARM initialization option). ❷ The *libref* is then established and system options ❸ are declared, including the establishment of the autocall library ❹. This particular AUTOEXEC is short, but it could just as easily have contained any number of macro calls or even initiated an application.

```
* Autoexec.sas *;
%let path = ❶ &sysparm\InnovativeTechniques;
libname advrpt v9 "&path\Data"; ❷
filename advmac "&path\sascode\sasmacros";
options nodate nonumber nocenter; ❸
options sasautos=(advmac, sasautos); ❹
```

Typically my Windows desktop contains a different SAS icon for each active client/project. Each has its own unique autoexec program. This ensures that I use the correct data and the correct programs for each project. If you install and use the AUTOEXEC.SAS program that comes with the sample code for this book, the *libref* and autocall macro library used by the sample programs will automatically be available for your use.

14.3 Using the Configuration File

When SAS initializes, the configuration file is executed *before* the autoexec program. The configuration file is not a SAS program, and if you make changes, the editing must be done carefully. Some companies use the configuration file to provide common setup instructions to all the site's SAS installations and consequently do not allow the modification of this file.

If you do want to customize this file, I would suggest that you do not modify the original version, which can be found in the !SASROOT location. Instead you can copy and then edit the copy. You then point to the modified configuration file using the -CONFIG initialization option (see Section 14.1.1).

The default name for this file is SASV9.CFG. The location will vary according to OS and installation. Under a standard Windows setup, you may find it at one of the following locations:

- SAS 9.1   C:\Program Files\SAS\SAS 9.1\nls\en
- SAS 9.2   C:\Program Files\SAS\SASFoundation\9.2\nls\en
- SAS 9.3   C:\Program Files\SASHome\SASFoundation\9.3

Common customizations include:

- addition of a macro library location to SASAUTOS
- changing the location of the WORK directory
- modification of other default values like memory allocation and macro symbol table size

14.3.1 Changing the SASAUTOS Location

The following portion of a configuration file adds a directory ❶ to the list of SASAUTOS libraries. Adding this directory to the definition of SASAUTOS means that it will automatically be included in the autocall library, without specifying it in the SASAUTOS= system option.

```
/* Setup the SAS autocall library definition */
-SET SASAUTOS  ("\\groupserver\sascode\macros" ❶
               "!sasroot\core\sasmacro"
               "!sasext0\inttech\sasmacro"
               "!sasext0\access\sasmacro"
               "!sasext0\assist\sasmacro"
               "!sasext0\eis\sasmacro"
               "!sasext0\ets\sasmacro"
               "!sasext0\graph\sasmacro"
               "!sasext0\iml\sasmacro"
               "!sasext0\or\sasmacro"
               "!sasext0\qc\sasmacro"
               "!sasext0\share\sasmacro"
               "!sasext0\stat\sasmacro"
               )
```

In the configuration file the -SET keyword is used to establish an environmental variable, which will later be interpreted as a *fileref*.

Rather than modify the original CONFIG file (or a copy as described above), it is possible to invoke a second tailored configuration file that will augment or override selected portions of the original file. The configuration file shown below makes it unnecessary to modify the SASAUTOS option when defining your autocall library. It does this by inserting the locations directly into the SASAUTOS system option.

Inserting locations in the autocall library using the INSERT option in the CONFIG.CFG file has been problematic. Fortunately the following technique was worked out by Peter Crawford of Crawford Software Consultancy Limited.

```
-set advtech "C:\InnovativeTechniques\SASCode" ❷
-sasautos (sasautos)                            ❸
-insert sasautos !advtech\ProdMacros\           ❹
-insert sasautos !advtech\SASMacros\            ❺
```

❷ For convenience the ADVTECH environmental variable is created and subsequently used as a path abbreviation at ❹❺.

❸ SAS Institute's autocall library is reestablished using the SASAUTOS option. This step must be done first, because the original value of the SASAUTOS option is specified without using parentheses. This definition will replace the default value of SASAUTOS, with one that includes the parentheses. The parentheses are needed when more than one location is to be specified as is done at ❹❺.

❹ The physical path to the production autocall library is inserted into the SASAUTOS option. The -INSERT option writes the text at the beginning of the list. The APPEND option (which is not shown) can place the text at the end of the list. Notice the use of the environmental variable !ADVTECH which is designated as such using the exclamation point.

❺ A second autocall library is inserted into the SASAUTOS option. Because it is inserted second it will appear before the PRODMACROS library ❹.

❻ The tailored configuration file is pointed to using a second -CONFIG initialization option on

```
-config "C:\InnovativeTechniques\SASCode\Chapter14\E14_3_1.cfg" ❻
```

the execution line. This option will generally be in addition to the standard configuration file which will also use a -CONFIG option.

```
proc options option=sasautos;  ❼
   run;
%put %sysget(advtech);  ❽
```

❼ We can confirm the values in the SASAUTOS option by using PROC OPTIONS.

❽ The value of the ADVTECH environmental variable can be surfaced using the %SYSGET macro function.

The LOG shows that the SASAUTOS option has been modified and that the ADVTECH environmental variable has the anticipated value. Notice that the SASAUTOS *fileref* is not quoted ❼ in the LOG. If we had inserted the additional locations without first inserting the parentheses, this text would have been quoted and that portion of the autocall library would not have been available.

```
5     proc options option=sasautos;
6     run;

    SAS (r) Proprietary Software Release 9.2   TS2M2

  SASAUTOS=( '!advtech\SASMacros\' '!advtech\ProdMacros\' sasautos )  ❼
                  Search list for autocall macros
NOTE: PROCEDURE OPTIONS used (Total process time):
      real time             0.00 seconds
      cpu time              0.00 seconds
7
8     %put %sysget(advtech);
C:\InnovativeTechniques\SASCode   ❽
```

Other common initialization options that could be declared in this second configuration file could include:

- -SASINITIALFOLDER   (see Section 14.1.1)
- -AUTOEXEC   (see Sections 14.1.1 and 14.2)
- -VERBOSE   Show option setting in the log at initialization

SEE ALSO
SAS Problem Note 44791 discusses the problem associated with inserting an autocall location, but does not suggest the solution worked out by Peter Crawford http://support.sas.com/kb/44/791.html. Sample Code 42360 demonstrates the use of the APPEND option http://support.sas.com/kb/42/360.html.

14.3.2 Controlling DM Initialization

When the Display Manager is initialized you can, to a large extent, control the appearance and the available tools through the use of initialization options. This level of control is most generally of interest when you are executing an application, perhaps through SAS/AF or SAS/EIS, and you need to control your user's ability to access SAS.

❶ The entire top line can be turned off by using:

```
-AWSCONTROL NOTITLE
```

Alternatively you can change the title from the default (SAS) to one of your choosing:

```
-AWSTITLE "ABC Project"
```

❷ The list of drop-down menus can be turned off by using the -NOAWSMENU initialization option.

```
/* Prep for AF start up appliction */
-initcmd  "af c=control.Control.wrapper.frame af;
          toolclose;
          zoom;
          command close;
          wstatusln off;
          wwindowbar off;"
```

The -INITCMD option allows you to specify control commands when executing an application, such as SAS/AF. The initialization option shown here starts a SAS/AF application and closes most of the user's access to the rest of SAS.

14.4 In the Display Manager

The Display Manager is extremely customizable. Many of these possible customizations are a bit over the top, but there are a number that are very helpful.

SEE ALSO
Richard DeVenezia's Web site has a useful section on "actions" you might consider adding to the explorer window http://www.devenezia.com/downloads/sas/actions/.

14.4.1 Showing Column Names in ViewTable

By default the VIEWTABLE window displays variable labels as column headers. I find this to be

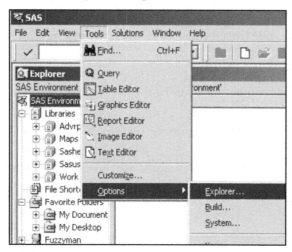

VIEWTABLE: Advrpt.Demog

	subject	clinic number	last name	first name	social security number	patient sex	date of birth
1		101 049060	Adams	Mary	079932455	F	12AUG51
2		102 082287	Adamson	Joan	011553218	F	.
3		103 066789	Alexander	Mark	743567875	M	15JAN30

very annoying as I almost always want to see the variable name. Fortunately the default can be changed to show the column names instead of the labels.

Although the following steps will be basically the same, the setup of some of the following dialog boxes will vary according to OS and version of SAS.

With the SAS Explorer Window active, select:
TOOLS → OPTIONS → EXPLORER.

This brings up the Explorer Options dialog box, select: MEMBERS → TABLE → EDIT

Edit the line with the ACTION of &OPEN and add colheading= name. The default for VIEWTABLE will now be to display variable names. Repeat the process for VIEWS.

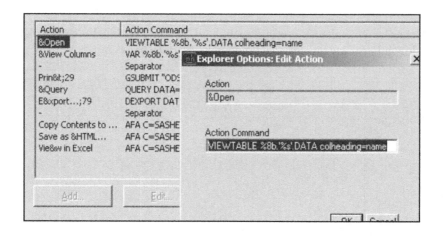

MORE INFORMATION

Rather than change the default behavior you can also change this setting when the ViewTable is invoked. Section 14.4.2 shows a DM command that invokes the ViewTable using column names rather than labels.

14.4.2 Using the DM Statement

The DM statement allows you to execute one or more Display Manager commands from within a SAS Program. These are the same commands that can be used in the command box, on the

```
DM <windowname> 'action' <windoname>;
```

command line, or from the KEYS window.

If you do not need to designate or change the active window all you need is the *action.* This is common when performing a DM task from within a batch program.

The following command can be used to clear the LOG window.

```
dm log 'clear';
```

Multiple DM commands can be included in a single DM statement by chaining them together with semicolons. This DM statement turns off the program editor and executes a SAS/AF

```
dm  af "pgm off; af cat = appls.allproj.passwd.program";
```

program making the SAS/AF window active.

You can also route the LOG much as you can by using PROC PRINTTO.

```
dm 'log; file "&path\logdump1.log"';
```

The POST command can be used to post a message box to the users screen. This can be a much stronger attention grabber than

```
dm 'post "this is a message"';
```

just a message to the LOG.

The enhanced editor is designated using the WEDIT command, and if followed by a filename, a specific file can be loaded and opened for editing.

```
dm 'wedit "C:\InnovativeTechniques\sascode\chapter14\e14_3_1.sas"' ;
```

The ViewTable can be invoked as well using either the VIEWTABLE or VT command. Here the data set ADVRPT.DEMOG is opened with the variable names shown as column headings

```
dm "viewtable advrpt.demog colheading=names";
```

(COLHEADING=LABELS is the default).

The DM statement can also be used to assign a specific command to a function key (see Section

```
dm 'keydef f12 "log;clear"';
```

14.4.7). The KEYDEF command is used to make the assignment. Here the F12 key is assigned to clear the log.

For combination keystrokes enclose the key definition in quotes as well. The SHIFT F9 key will

```
dm 'keydef "shf f9" "next VIEWTABLE:; end"';
```

now close the next open ViewTable window. Close a series of open ViewTable windows with successive selections.

MORE INFORMATION

The DM statement is used to execute the DMOPTSAVE and DMOPTLOAD commands in Section 14.1.3. Section 14.4.7 discusses the execution of DM commands through hot key assignments.

SEE ALSO

The DM statement and the WEDIT command are discussed in the SAS Forum thread http://communities.sas.com/thread/12520. Rosenbloom and Lafler (2011c) assign a macro call to a function key.

14.4.3 Enhanced Editor Options and Shortcuts

There are a number of options and shortcuts available for use with the Enhanced Editor. It is also possible to do a fair amount of customization.

Enhanced Editor Setup

There are only a few set up preferences that I would recommend that you change. Most of the defaults are fine for typical users. The options for the Enhanced Editor can be found when the editor is the active window. Use TOOLS → Options → Enhanced Editor. This brings up the Enhanced Editor Options dialog box. I like to select 'Show Line numbers' ❶, because it makes life easier for large programs. More importantly, be sure to check 'insert spaces for tabs' ❷ and 'replace tabs with spaces on file open' ❸. Both of these options help to make it easier to maintain the text formatting of a SAS program when it is transferred between programmers.

Enhanced Editor Keys

The editor has been set up with a number of shortcut key combinations. Depending on how you work and what things you tend to do, some of these key combinations can be very useful. You can see and learn more of these key combinations through the Enhanced Editor Keys pull-down menu. While the Enhanced Editor is the active window, go to TOOLS → OPTIONS → ENHANCED EDITOR KEYS.

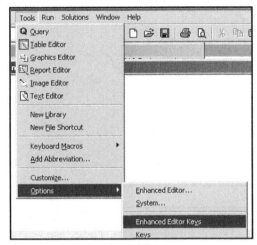

Explore the resulting dialog box to find the key combinations that are most useful to you. I especially like to use:

Ctrl+/	add comments
Ctrl+shift+/	remove comments
Ctrl+F2	mark a line in a program
F2	jump to the next marked line
Shift+F2	jump to the previous marked line

Not only are the defined key combinations very useful, but you can redefine the combinations and add new keyed operations. Notice that the BEEP command has no assigned key combinations and is therefore not available. While BEEP is probably not particularly useful, others can be; you can scroll down the list of available operations until you find one of interest. Here "Sort the selected lines" has been highlighted. Next press the "Assign keys..." button.

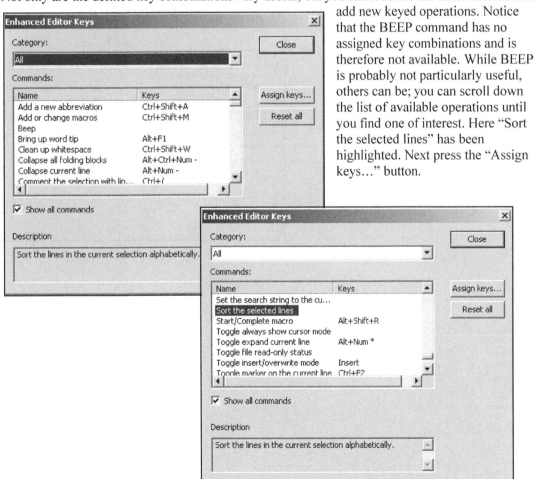

The 'Assign Keys' dialog box is used to assign a set of key strokes to this operation. Highlight the none in the 'Press new shortcut key:' box, and then press the desired keys. If you select a combination that is already in use, the keystroke combination is changed to be used with the new operation. In this example we are choosing to use Alt + Shift + R (this key combination will no longer be used to 'Start/Complete macro').

We can now use this key combination in the Enhanced Editor to sort rows.

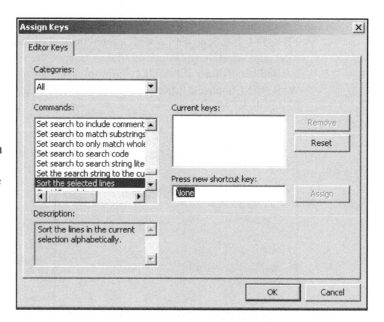

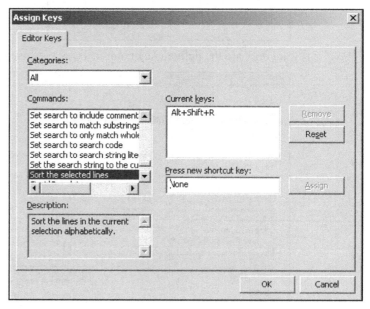

Editor - Untitled1 *
1   subject
2   clinnum
3   lname
4   fname
5   ssn
6   sex
7   dob
8   death
9   race
10 edu
11 wt
12 ht
13 symp
14 death2
15

The list of variables on the left, which have been pasted into the Enhanced Editor, are in the ADVRPT.DEMOG data set and have been written in variable number order. If the first 14 rows are highlighted and we press the Alt+Shift+R keys the rows are reordered. And the list becomes the one shown on the right.

Editor - Untitled1 *
1   clinnum
2   death
3   death2
4   dob
5   edu
6   fname
7   ht
8   lname
9   race
10 sex
11 ssn
12 subject
13 symp
14 wt
15

SEE ALSO

A short write-up on select Enhanced Editor keys can be found on the Tek-Tips Forum http://www.tek-tips.com/faqs.cfm?fid=5140.

The following LinkedIn thread has a number of suggestions http://www.linkedin.com/groupItem?view=&srchtype=discussedNews&gid=70702&item=63659 611&type=member&trk=eml-anet_dig-b_pd-ttl-cn.

Marking a Block of Text

While most applications allow you to hold the left mouse button (LMB) while dragging the mouse to highlight entire lines of text, in the Enhanced Editor you can go a couple of steps further. Dragging while the cursor is in the gray area (left side of the editor), ensures that all the text in the first and last lines will be highlighted.

You can highlight, while controlling for columns and lines, by also pressing the 'Alt' key at the same time as dragging with the LMB depressed. In the image to the right, 'delete' would shift lines 14-23 five columns to the left.

13⊟proc report
14   column
15   define
16   define
17   define
18   define
19   rbreak
20   comput
21   rat
22   endcom
23   **run;**

SEE ALSO

Under SAS 9.1 there was a problem with the feature to mark blocks of text on machines for which SAS Enterprise Guide was also installed. Problem note #30455 shows how to resolve this conflict. This note can be found at http://support.sas.com/kb/30/455.html.

AUTOSAVE – Finding the Backup File

Files being edited by the Enhanced Editor are automatically saved every few minutes (the frequency is set in the DM preferences under the TOOLS → OPTIONS → PREFERENCES → EDIT tab). If you need to recover the saved file the location can be a bit difficult to find and it varies with OS and version of SAS. Usage Note 12392 states: Enhanced Editor Autosave should be consulted to find the location of these backup versions of your program. Under Windows the file extension is .ASV.

SEE ALSO

Usage Note 12392 can be found at http://support.sas.com/kb/12/392.html.

14.4.4 Macro Abbreviations for the Enhanced Editor

The Enhanced Editor enables you to build abbreviations for your editor. Much like abbreviations in other applications, a single word or part of a word can be typed and then other (generally longer) text can be substituted at a keystroke.

For the purposes of this example assume that you want to type the following header block at the top of each of your programs. Typing it once is fine, but more than once becomes tedious. Let's make a macro abbreviation that does the typing for us.

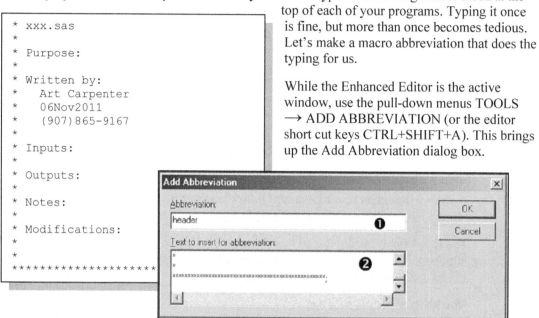

```
*  xxx.sas
*
*  Purpose:
*
*  Written by:
*     Art Carpenter
*     06Nov2011
*     (907)865-9167
*
*  Inputs:
*
*  Outputs:
*
*  Notes:
*
*  Modifications:
*
*
*********************
```

While the Enhanced Editor is the active window, use the pull-down menus TOOLS → ADD ABBREVIATION (or the editor short cut keys CTRL+SHIFT+A). This brings up the Add Abbreviation dialog box.

In the Add Abbreviation dialog box enter a name for the new abbreviation (header) ❶. This becomes a keyboard macro, so you must select a name that has not already been used. Then type (or more practically paste) the substitution text into the 'Text to insert for abbreviation' dialog space ❷. Pressing the OK button creates and stores the abbreviation.

To use the abbreviation simply type in the name of the abbreviation while in the Enhanced Editor. As soon as the last letter of the abbreviation has been entered, a small pop-up 'tip' text box containing the first few characters of the abbreviation is displayed. If at that point you press the TAB or ENTER key, the name of the abbreviation will be replaced by the text that you stored.

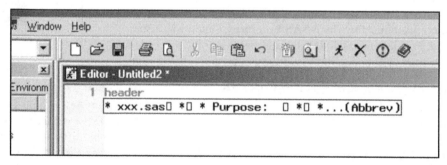

The following screen shot shows that the name of the HEADER abbreviation has been entered in the Enhanced Editor and the first few characters of the text to be substituted is shown in the pop-up 'tip' box. Pressing the TAB or ENTER key causes the abbreviation name to be replaced by the stored text.

Once created macro abbreviations can be edited or deleted just like any other keyboard macro.

Use the pull-down menus TOOLS →
KEYBOARD MACROS →MACROS to bring up
the KEYBOARD MACROS dialog box.

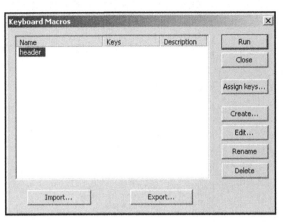

Using this dialog box you may also export/import your macro abbreviations so that they may be standardized across your work group. Use the EXPORT button to create a file with the KMF extension. This file can then be imported by another SAS user by using the IMPORT button. A number of KMF files have been gathered at the sasCommunity.org site http://www.sascommunity.org/wiki/Abbreviations/Macros.

The date in the HEADER abbreviation shown above is static. There are a number of predefined edits that we can apply to a keyboard macro (remember that a macro abbreviation is a special form of a keyboard macro). Several of these predefined edits allow the insertion of date values. Using these we can automatically insert the current date time stamp from when the abbreviation is executed. The following steps reestablish the HEADER abbreviation with the current date time value replacing the static date (06Nov2011).

Edit the HEADER keyboard macro (it has already been established as an abbreviation (TOOLS →KEYBOARD MACROS → MACROS). Select HEADER and the EDIT button.

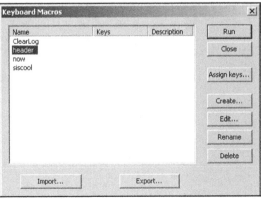

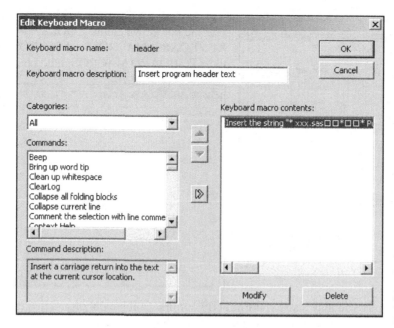

The original HEADER definition is going to be replaced so delete the one line that starts with "Insert the string. . .". This will completely clear the abbreviation definition. That is OK for now. We are about to insert new text. The important thing is that HEADER has already been established as an abbreviation.

The commands on the left of the Edit Keyboard Macro dialog box can be inserted into the macro. Once you have deleted the definition, scroll down the list of commands until you find "Insert the string". Select it and press the double arrow in the middle of the dialog box to move the command to the right-hand box.

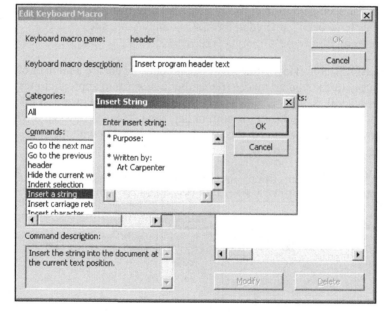

Paste the portion of the header text up to, but not including, the static date into the Insert String dialog box and select OK.

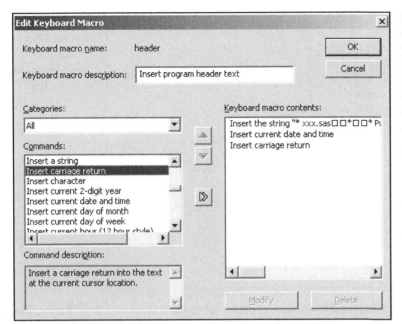

Select the date or date time values of interest. Here the "Insert current date and time" command has been selected. We want the subsequent text to start on the next line so the command to "Insert a carriage return" has also been selected.

The remainder of the header text can now be inserted using the 'Insert the string' command.

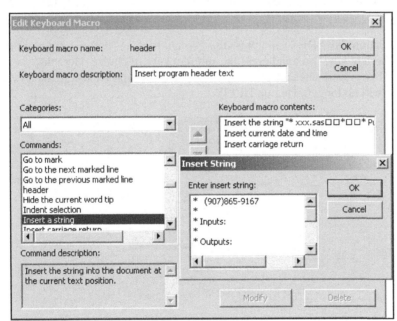

The "Insert a string" command is again selected and the remainder of the text is typed or pasted into the Insert String dialog box.

Specify OK as you exit from each of the dialog boxes. The revised HEADER abbreviation is now ready to use.

```
* xxx.sas
*
* Purpose:
*
* Written by:
*    Art Carpenter
*    Thursday, November 17, 2011 20:49:54
*    (907)865-9167
*
* Inputs:
*
* Outputs:
*
* Notes:
*
* Modifications:
*
*
********************************;
```

The HEADER abbreviation will now insert the current date time stamp into the header text.

The name of the abbreviation will now also appear in the Enhanced Editor Keys dialog box (introduced in Section 14.4.3). You can use this dialog box to assign a set of keys to execute the abbreviation.

Select the abbreviation (HEADER) and press the 'Assign keys. . .' button. Highlight the text in the 'Press new shortcut key:' box and press the shortcut keys of choice. Here the CTRL+SHIFT+H keys were chosen. This key combination will now execute the HEADER abbreviation and will no longer bring up HELP.

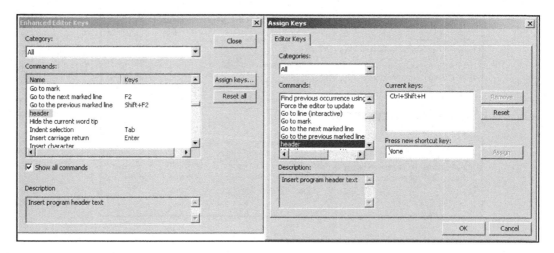

SEE ALSO
Carpenter (2003b) introduces this example for the creation of a macro abbreviation. A collection of abbreviations as well as links to other references with more information can be found on the sasCommunity.org article http://www.sascommunity.org/wiki/Abbreviations/Macros.

14.4.5 Adding Tools to the Application Tool Bar

Like most applications that have pull-down menus and tool bars, it is possible to modify or customize the list of available tools. A common usage is when you have a program or code snippet that you run regularly, and would like to have it readily available. By modifying the tool bar, you can add an icon that will instantly execute your program.

Consider the following program that will delete all the data sets in the work directory. We would

```
proc datasets library=work
              memtype=data
              kill
              nolist;
   quit;
```

like to add an icon on the tool bar associated with the Enhanced Editor that will execute this step. The icon could be placed on any of the tool bars in the DM, this one seems most logical.

With the Enhanced Editor in the active window, use the pull-down menus to select TOOLS → CUSTOMIZE (or right click on the tool bar itself). The CUSTOMIZE TOOLS dialog box for the

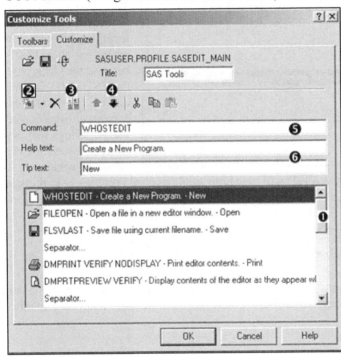

tool bar is shown. From here you can add or remove items on the tool bar. We would like to add an icon that will execute our PROC DATASETS program.

❶ Move the slider to the approximate position of the icon (the final position can be refined later).

❷ Add a blank icon (separators can be useful to make things clearer).

❸ Select an icon for your tool.

❹ Refine the location.

❺ On the command line enter the text that is to be executed. In the example that follows we will be executing a GSUBMIT command.

❻ Help and Tip text should be added.

The tool bar icon used to execute the PROC DATASETS step is shown below. The trash can icon ❼ has been selected for the tool bar from a list of supplied icons ❸.

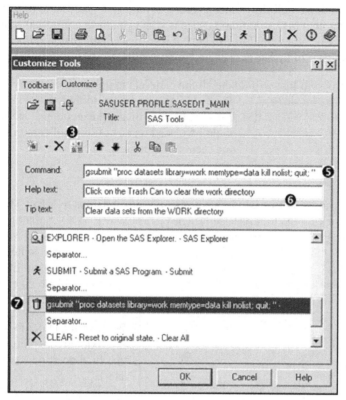

The GSUBMIT command is a corollary of the DM statement. It allows you to insert code where a command is otherwise expected. In this example the entire step has been placed in the GSUBMIT command. This would have been quite inconvenient if the step had been any longer. Actually the GSUBMIT command only allows 500+ characters. For longer steps consider submitting a macro call or a %INCLUDE statement.

The %INCLUDE becomes a bit problematic if the path to the location of the program contains a macro variable. This is an issue because the string associated with the GSUBMIT command must be quoted. However, the %INC also expects either a quoted string or a *fileref*. Strings within strings are often an issue for macro language elements, because at some point the macro variable (*e.g.*, &PATH) will be within single quotes. Fortunately in this case, because of how the line is parsed, the single quotes can be used without masking the macro variable.

```
gsubmit '%inc "&path\sascode\chapter14\e14_4_5.sas";'
```

The GSUBMIT command is only one of a very long list of commands that can be issued from within the Display Manager. Other useful commands include those that bring up other DM windows, such as:

- KEYS
- LIBNAME
- LOG
- FILENAME
- TITLE

SEE ALSO

Howard (2004) shows this and some similar examples. The PMENU procedure can be used to design, build, and save customized pull-down menus and tool bars. Charlie Huang's 9/11/2011 blog entry "Add 10 buttons to enhance SAS 9.3 environment" suggests a number of buttons that could be added to the tool bar http://www.sasanalysis.com/2011/09/10-buttons-to-tweak-sas-93-environmnet.html.

14.4.6 Adding Tools to Pull-Down and Pop-up Menus

Sometimes adding a specialized tool to the pull-down or pop-up menus used in the DM can be very beneficial. In Section 14.4.5 the GSUBMIT command was used to execute a SAS program from a tool bar. You can do the same sort of thing from a pull-down or pop-up menu.

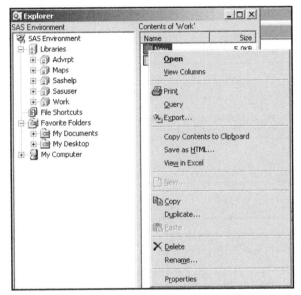

Consider the pop-up menu shown on the left, if you right-click on a data set from within the Explorer window. We would like to have the ability to execute a specialized tool against a SAS data set simply by clicking on a menu item in this pop-up menu.

While the Explorer window is active, this menu is controlled through the TOOLS → OPTIONS → EXPLORER menus. This

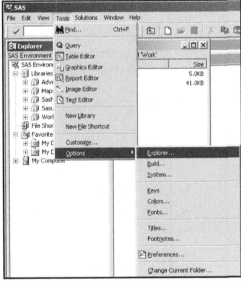

brings up the EXPLORER OPTIONS dialog box. This box is worth exploring just because it is used to control access to a variety of objects from within the SAS Explorer. Since we want to apply our tool to a SAS data table, we select the MEMBERS tab ❶ and then highlight the TABLE (SAS data set) ❷ line. Clicking on the EDIT button ❸ brings up the EXPLORER OPTIONS: TABLE OPTIONS dialog box.

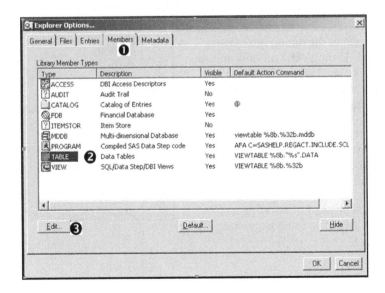

Notice that the EXPLORER OPTIONS: TABLE OPTIONS dialog box is used to form the

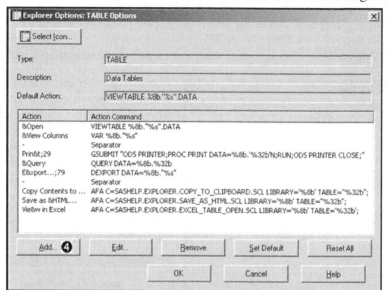

primary pop-up menu that you see when you right-click on a SAS data table (see the first figure in this section). The items in the ACTION COMMANDS section in this dialog box are worth examining. Notice the use of the VIEWTABLE, VAR, GSUBMIT, and QUERY commands. The table name is brought into the script using %8b for the *libref*, and '%32b'N for the data set name.

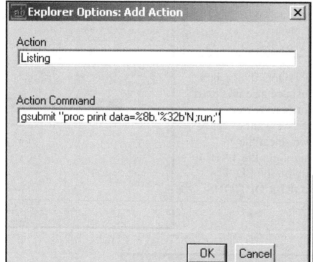

We are going to add a GSUBMIT item to this menu by using the ADD action button ❹. Highlight the item below, which you want to insert the new command, and press the ADD action button ❹. This brings up the ADD ACTION dialog box. Here we enter a name for the action and the action (`gsubmit "proc print data=%8b.'%32b'N;run;"`) that is to

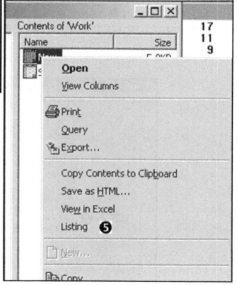

take place. Exit from each of the dialog boxes using OK. A new entry, titled 'Listing' ❺, will now appear on the pop-up menu associated with a SAS data set. Selecting the 'Listing' entry will perform a PROC PRINT on the highlighted data set.

Inserting a PROC PRINT is a bit of a silly thing to do. If you notice the third item in the 'table options' list, you will see that it already contains a GSUBMIT for a PROC PRINT which utilizes ODS.

In the following example instead of inserting a GSUBMIT for a PROC step, we use it to submit macro language elements.

Because the scripting uses the percent sign in the data set name, you must be careful when calling macro language elements from within the GSUBMIT. This is demonstrated by adding a call to the %OBSCNT macro which returns the number of observations in a SAS data set. If we use it in a

```
gsubmit "%%put Obs count is %%obscnt(dsn=%8b.'%32b'N);"
```

%PUT statement the number of observations is written to the LOG.

%OBSCNT (see Section 13.4.2) is an autocall macro and is part of the autocall library that comes with the programs associated with this book. Notice that the percent signs associated with the macro language are doubled. This delays their interpretation until the macro statement has been submitted which takes place after the data set name has been inserted. For the data set WORK.NEW the resultant submitted %PUT statement will become:

```
%put Obs count is %obscnt(dsn=work.new);
```

The TABLE OPTIONS dialog box shows this definition, and the ObsCount entry now appears on the pop-up menu.

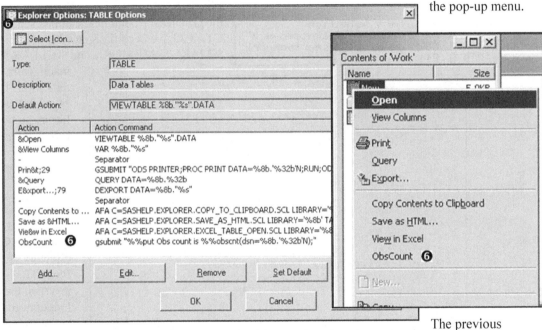

The previous example uses a %PUT statement to write the number of observations to the LOG. You could just

```
gsubmit "%%nrstr(%%printit(dsn=%8b.'%32b'N));"
```

have easily have passed the name of the data set to a macro for execution. The GSUBMIT would be essentially the same. The following GSUBMIT executes the %PRINITIT macro for the displayed data set. The macro quoting function %NRSTR is used to delay the interpretation of the

```
gsubmit "%%nrstr(%%printit(dsn=%8b.%32b));"
```

macro call and may be required when the macro resides in a stored compiled macro library.

For most standard data set names this code can usually be simplified by removing the quotes around the data set name(%32b).

SEE ALSO
Art Trabachneck *et al* (2010) demonstrate additional and more extensive techniques.

14.4.7 Adding Tools to the KEYS List

The KEYS window (TOOLS → OPTIONS → KEYS or F9 or KEYS on the command line) lists DM commands that have been pre-assigned to specific key combinations, including function keys and mouse buttons. The key assignments can be modified by the user and can include Display Manager commands and even macro calls.

```
▦▦▦KEYS <DMKEYS>                    _ □ ×
Key         Definition                     ▲

F1          help
F2          reshow
F3          end; /*gsubmit buffer=def
F4          recall
F5          wpgm
F6          log
F7          output
F8          zoom off;submit
F9          keys
F11         command focus
```

```
SHF F12
CTL F1      sort selection
CTL F2      gsubmit "%maketempwork"
CTL F3      gsubmit "%cleartempwork"
CTL F11
```

The first few key definitions of the KEYS window are shown to the left. Notice that multiple commands can be concatenated with a semicolon.

You can type in a new command, either by overwriting an existing command or by typing it in any available space. Commands can also be inserted into the key definitions using the KEDYDEF command on a DM statement (see Section 14.42).

Like in Section 14.4.6, if you want to submit code, the GSUBMIT command can be used. CTL F2 and CTL F3 have been designated to submit the %MAKETEMPWORK and %CLEARTEMPWORK macro calls. The macro definitions have been saved in the autocall library so that they can be loaded when the appropriate key combinations are selected.

```
%macro MakeTEMPWORK;
%local rc;
%let rc=%sysfunc(fileexist("c:\tempwork"));
%if &rc=0 %then %do;
    %sysexec md "c:\tempwork";
    %let rc=%sysfunc(libname(tempwork,c:\tempwork));
%end;
%mend maketempwork;

%macro ClearTEMPWORK;
%local rc;
%let rc=%sysfunc(fileexist("c:\tempwork"));
%if &rc ne 0 %then %do;
    %let rc=%sysfunc(libname(tempwork));
    %sysexec del /Q "c:\tempwork\*.*";
    %sysexec rd /Q "c:\tempwork";
%end;
%mend cleartempwork;
```

The %MAKETEMPWORK macro creates a directory and assigns the *libref* TEMPWORK to it. When the library is no longer needed, perhaps at the end of the session, the user can press CTL+F3, which executes the macro %CLEARTEMPWORK, which in turn clears the contents of the directory and deletes the directory. Since these macros utilize the %SYSEXEC macro function, the system option NOXWAIT should be declared to prevent prompts from the OS.

Once modified, the new key definitions can be saved using the SAVE command. Key definitions are stored as catalog entries with an entry type of keys. Saved definitions can be recovered through the use of the COPY command.

MORE INFORMATION

Display Manager commands can also be executed through the use of the DM statement (see Sections 14.1.3 and 14.4.2).

SEE ALSO

Rosenbloom and Lafler (2011c) use SUBMIT rather than GSUBMIT in the assignment of several different macros to KEYS.

14.5 Using SAS to Write and Send E-mails

During or after the processing of a program, you can use SAS to generate e-mails. These can be notifications of error conditions or the successful termination of the program. The e-mail can even contain attachments. This simple example will get you started. Read the SEE ALSO references for details that provide refinements of the technique.

```
options emailsys=SMTP emailid="sam@caloxy.com"  ❶
        emailhost="caloxy.com" EMAILPORT=25;

* Define the fileref with the email engine;
FILENAME genmail email  ❷
      subject="Patient 205 ConMeds"  ❸
      to     = "Fred@caloxy.com"  ❹
      from   = "Sam@caloxy.com"  ❺
      attach ="&path\results\E14_5.pdf";  ❻

ods pdf file="&path\results\e14_5.pdf"  ❼
        style=journal2;
proc print data=advrpt.conmed
                      (where=(subject='205'));
   id subject;
   var medstdt medspdt drug;
   run;
ods pdf close;

data _null_;  ❽
   file genmail;  ❾
   put "Here are the ConMeds for Subject 205";  ❿
   run;
```

❶ The attributes of the e-mail server for this e-mail ID are specified in options. These options could also be specified in the configuration file. The EMAILID= and the EMAILHOST= options will take on values specific to your e-mail server.

❷ The FILENAME statement is used to define the e-mail *fileref* (GENMAIL). Notice the use of the EMAIL engine.

❸ The SUBJECT= option defines the subject line.

❹ The recipient's e-mail address is specified using the TO= option.

❺ The FROM= option designates the sender's e-mail. This will generally match the EMAILID ❶.

❻ The ATTACH= option designates the file to be attached.

❼ The file that is to be attached is generated.

❽ A data _NULL_ step is used to generate the e-mail body. The process of starting to generate the e-mail takes place when the DATA step is compiled. Conditional execution of statements within the DATA step can change the body of the e-mail, but remember if the DATA step is compiled, an e-mail will be generated.

❾ The FILE statement points to the *fileref* (GENMAIL) generated using the EMAIL engine.

❿ The text of the e-mail is generated using a PUT statement.

SEE ALSO

A number of papers have been written on using SAS to e-mail results. Hunley (2010) has a number of extended examples including a discussion of texting. Whitworth (2010) has an example that includes zipping the file before e-mailing it. Rosenbloom and Lafler(2011d) have a brief e-mail example.

The SAS Jedi (Mark Jordan) has written a couple of blogs on the use of SAS to e-mail documents http://blogs.sas.com/sastraining/index.php?/archives/81-Jedi-SAS-Tricks-Email-from-the-Front-Part-2.html. He has also created a YouTube video on the topic http://www.youtube.com/watch?v=qPobPZg2osc&feature=related.

Two SAS Forum threads include discussions on the use of e-mail http://communities.sas.com/thread/10467 and http://communities.sas.com/thread/11086.

14.6 Recovering Physical Location Information

Under the WINDOWS OS there are a number of techniques that you can use to find the physical location of a data set or file given the information available to SAS.

SEE ALSO

Carpenter (2008b) discusses these and other techniques in more detail.

14.6.1 Using the PATHNAME Function

The PATHNAME function returns the physical path for a given *fileref* or *libref*. While rarely used in the DATA step it has proven to be invaluable in the macro language. Often times we do not necessarily know the physical path even though we know a *libref* or *fileref*. Here the physical path is loaded into a macro variable where

```
filename saspgm "&path\sascode\e14_6.sas";

%let pgmpath = %sysfunc(pathname(saspgm));
%put &pgmpath;
```

```
C:\InnovativeTechniques\sascode\e14_6.sas
```

it is displayed in the LOG.

You can even use it on concatenated *fileref*s such as the autocall library. To gather the current location of all of the locations in the SASAUTOS *fileref* you could specify:

```
%sysfunc(pathname(sasautos))
```

MORE INFORMATION

One of the examples in Section 13.8.2 recovers some location information using the PATHNAME function.

14.6.2 SASHELP VIEWS and DICTIONARY Tables

The path information for existing *libref*s and *fileref*s can be gathered by examining the SASHELP views and SQL DICTIONARY tables. Here you can find not only the path that would be returned by the PATHNAME function, but other things such as the ENGINE as well.

SASHELP.VLIBNAM and DICTIONARY.LIBNAMES

Each row in the view SASHELP.VLIBNAM (note the spelling) contains the *libref* and path information for each *libref* (more than one row for concatenated *libref*s). A portion of a listing of this view shows the primary variables. Notice that the first few locations of the concatenated SASHELP *libref* are included as well.

```
Returning a Physical Location
14.6.2 SASHELP VIEWS and DICTIONARY Tables


Obs libname engine path

  1 ADVRPT     V9   C:\InnovativeTechniques\Data
  2 SASHELP    V9   C:\Program Files\SAS\SASFoundation\9.2\nls\en\SASCFG
  3 SASHELP    V9   C:\Program Files\SAS\SASFoundation\9.2\core\sashelp
  4 SASHELP    V9   C:\Program Files\SAS\SASFoundation\9.2\inttech\sashelp
  5 SASHELP    V9   C:\Program Files\SAS\SASFoundation\9.2\mddbserv\sashelp
                  .... portions of the table are not shown ....
```

SASHELP.VEXTFL and DICTIONARY.EXTFILES

The location of external files (such as raw data and programs) can also be retrieved from either the SASHELP.VEXTFL view or the SQL dictionary table DICTIONARY.EXTFILES.

```
data _null_;
    set sashelp.vextfl(keep=fileref xpath
                       where=(fileref='SASPGM'));
    call symputx('pgmpath2',xpath,'l');
    run;
%put &pgmpath2;
```

Using the SASHELP.VEXTFL view, the path information can be retrieved through a DATA step and loaded into a macro variable. The view as well as the dictionary table can be accessed from within an SQL step.

```
proc sql noprint;
select xpath into :pgmsqlpath
    from dictionary.extfiles
        where fileref='SASPGM';
quit;
%put &pgmsqlpath;
```

MORE INFORMATION

Section 13.8.1 goes into more detail on the use of these SASHELP views and SQL DICTIONARY tables.

14.6.3   Determining the Executing Program Name and Path

Sometimes we need to be able to automatically detect the name or location of an executing program. This can be especially helpful when we write applications that need to self document, perhaps by placing the name and location of the executing program in a footnote of the generated table.

This is fairly straightforward when the executing program is running in batch mode. In batch mode the name of the executing program is stored in the system option SYSIN, and the value of system options can be retrieved using the GETOPTION function

Under the Windows OS, the name of the executing program and its path is stored in the environmental variables SAS_EXECFILENAME and SAS_EXECFILEPATH. Environmental variables are maintained by the OS; however, SAS can both populate and access their values. Whenever a SAS program is executed, this includes when it is executed through the Display Manager from the Enhanced Editor, these environmental variables are updated.

The values of environmental variables are accessed through the use of the %SYSGET macro function. The returned value can then be loaded into a macro variable or just written to the LOG as is done here.

```
%put %sysget(SAS_EXECFILENAME);
```
E14_6.sas

When you want the name of the program without the SAS extension it can easily be stripped off using the %SCAN or %QSCAN function.

```
%put %qscan(%sysget(SAS_EXECFILENAME),1,.);
```
E14_6

When we need to know not just the name, but the location of the SAS program (when executing from the Enhanced Editor this is the location from where the executing program was retrieved), we can use the SAS_EXECFILEPATH environmental variable. Here the value is retrieved by the macro %GRABPATHNAME and then written to the LOG.

```
%macro grabpathname;
    %sysget(SAS_EXECFILEPATH)
%mend grabpathname;

%put %grabpathname;
```

```
114  %put %grabpathname;
C:\InnovativeTechniques\SASCode\Chapter14\E14_6.sas
```

14.6.4 Retrieving the UNC (Universal Naming Convention) Path

When a program resides on a network server, the server name is generally mapped to a drive letter. Since this drive letter can be user specific, knowing that a program resides on the F:\ drive for one user is not necessarily helpful to someone else. As was shown using all of the previous methods, it is always the mapped drive that is returned; therefore, a different approach is needed to retrieve the actual or UNC path. Although the UNC path information is not stored in a location that is directly available to SAS, it is still possible to get this information - the process is just a bit more challenging.

Certainly we know that the OS has to know the relationship between the mapped drive letter and the actual UNC location. Under Windows this information is stored in a Dynamic Link Library, DLL. Windows has internal tools for accessing the information contained in a DLL and these tools can be accessed from within SAS using the CALL MODULE routine (the MODULEN and MODULEC functions can also be used). To make the tools available to the CALL MODULE routine we must first create a CATALOG SOURCE entry for it to operate against. This entry contains the arguments that are passed to and from the Windows DLL routine, which for retrieving the UNC path, is named WNetGetConnectionA (be careful, this name is case sensitive).

The arguments themselves are specific to each routine. The WNetGetConnectionA routine expects three arguments. Here the SOURCE entry has been written to a catalog in the WORK directory; however, you will generally make this permanent so that you only have to run this step once.

```
filename sascbtbl catalog "work.temp.attrfile.source";  ❶

data _null_;
  file sascbtbl;  ❶

  put "routine WNetGetConnectionA module=mpr minarg=3 maxarg=3  ❷
       stackpop=called returns=long;";
  put "  arg 1 char input byaddr format=$cstr200.;";
  put "  arg 2 char update byaddr format=$cstr200.;";
  put "  arg 3 num update byaddr format=pib4.;";
  run;
```

❶ The attributes needed by the WNetGetConnectionA DLL are specified by writing them to a SOURCE catalog entry using a DATA _NULL_ step. The *fileref* must be SASCBTBL, and the CATALOG engine must be specified.

❷ The routine attributes are written using PUT statements. The DLL name is case sensitive and must be specified exactly as written.

The CALL MODULE routine can then be used to access the WNetGetConnectionA routine and to retrieve the location.

```
%macro grabdrive;  ❸
   %qtrim(%qleft(%qscan(%sysget(SAS_EXECFILEPATH),1,\)))
%mend grabdrive;

%MACRO getUNC;
   %local dir path;
   %* Determine the UNC path for the SAS program being executed.;
   DATA _NULL_;
      length input_dir $200 output_dir $200;

      * The input directory drive letter: ONLY e.g. j: ;
      input_dir = "%grabdrive";❹
      output_dir = ' ';
      output_len = 200;
      call module("WNetGetConnectionA",
                  input_dir,  ❺
                  output_dir,
                  output_len);

      call symputx('dir',input_dir,'l');   ❻
      call symputx('path',output_dir,'l');  ❼
      RUN;

   %* Get the name for the program of execution.;
   %put drive letter is &dir;   ❽
   %put path is &path;  ❾
   %put name is %grabpathname;   ❿
%MEND getunc;
```

❸ The %GRABDRIVE macro returns the drive letter alone.

❹ The %GRABDRIVE macro obtains the mapped drive letter with the colon *e.g.*, F:. This becomes the input for the DLL. The second argument (OUTPUT_DIR) will contain the returned UNC path.

❺ The MODULE routine is called by passing it the mapped drive letter that is contained in the variable INPUT_DIR. The UNC path is passed back by WNetGetConnectionA by storing it in OUTPUT_DIR.

❻ The input drive letter is stored in the macro variable &DIR.

❼ The UNC path is placed into the macro variable &PATH. Notice that this is a local macro variable. Had it been global we would overwrite the &PATH used in most of the examples in this book.

❽ Write the mapped drive letter to the LOG.

```
drive letter is F: ❽
path is \\CALOXYDELL\InnovativeTechniques ❾
name is F:\sascode\Chapter14\e14_6_showUNC.sas ❿
```

❾ Write the UNC path to the LOG. This is the portion of the path that has been mapped to the drive letter (F:\ in this example).

❿ Write the program name to the LOG.

```
%* Place this program in a mapped
%* drive (NOT the C: drive);
%getunc
```

The macro %GETUNC is called by a program that is not on the C: drive. In this case the program resides on the F:\ drive. The UNC path ❾ shows that the F:\ drive letter has been mapped to the \InnovativeTechniques directory on the \\CALOXYDELL server.

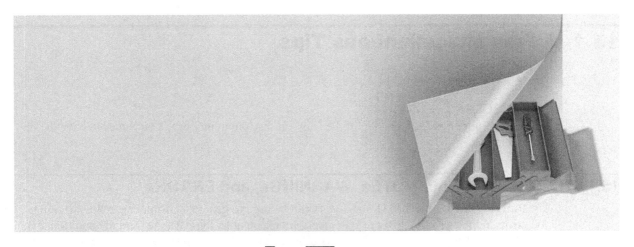

C h a p t e r **15**

Miscellaneous Topics

As if everything in this book was not eclectic enough to be an entire book of miscellaneous topics, the final chapter is the miscellaneous chapter of this eclectically miscellaneous book.

15.1 A Few Miscellaneous Tips

Here are just a few of many tips.

SEE ALSO

If you want more tips read the 'Tip of the Day' on sasCommunity.org. You can even have the tip tweeted to you.

15.1.1 Customizing Your NOTEs, WARNINGs, and ERRORs

The PUT, PUTLOG, and %PUT statements can be used to generate customized notes, warnings, and errors in the LOG. Since these statements are executable, logic can be used to conditionally execute them. The text written to the LOG follows the systems conventions when the statement keyword is immediately followed by one of the following:

- NOTE: The note is written in blue.
- WARNING: Warnings are written in green.
- ERROR: Errors are written in red and are summarized along with other errors.

The keyword must be capitalized and must be immediately followed by a colon.

```
lock sashelp.class;
%put NOTE: class should be locked;
```

```
41   lock sashelp.class;
NOTE: SASHELP.CLASS.DATA is now locked for exclusive access by you.
42   %put NOTE: class should be locked;
NOTE: class should be locked
```

You can also follow the words NOTE, WARNING, or ERROR with a dash instead of a colon. The customized message will still appear in the LOG with the appropriate color; however, the word NOTE, WARNING, or ERROR will not appear.

SEE ALSO

Don Henderson wrote a tip on sasCommunity.org that discusses this topic
http://www.sascommunity.org/wiki/Tips:Using_NOTE,_WARNING,_ERROR_in_Your_Progra
m%27s_Generated_Messages.

15.1.2 Enhancing Titles and Footnotes with the #BYVAL and #BYVAR Options

When your PROC step uses a BY statement the values of the BY variables or even the variable names themselves can be inserted into the title or footnote. Although available for use in a few other locations, the #BYVAL, #BYVAR, and #BYLINE options were designed to be used in the TITLE and FOOTNOTE statements.

The #BYVAR option is used to place the variable name in the title while the #BYVAL option is used to place the value of that BY variable. Both the #BYVAL and #BYVAR options have two ways of addressing a specific BY variable from the list of BY variables. These two forms (implicit and explicit) can be used interchangeably. Implicit naming uses a number that corresponds to the list of BY variables (left to right): #BYVAR1 and #BYVAL1 would both refer to the first variable in the list of variables. Explicit naming uses the variable name in parentheses. The variable must be on the BY list: #BYVAR(RACE) and #BYVAL(RACE). Only a single variable may be used within the parentheses.

Here the #BYVAR and #BYVAL options are used to identify the RACE in a PROC FREQ step.

```
options nobyline;  ❶
title2 'Summary for #byvar1 #byval1';  ❷
proc freq data=demog;
   by race;
   table sex;
   run;
```

❶ Since the BY variable information will appear in the title the procedure's BY line is not needed.

15.1.2 BY Values in Titles
Summary for race 2

The FREQ Procedure

		patient sex		
sex	Frequency	Percent	Cumulative Frequency	Cumulative Percent
F	10	55.56	10	55.56
M	8	44.44	18	100.00

❷ Both the #BYVAR and #BYVAL options are used and the variables are selected implicitly. Notice that these options may be used inside of either single or double quotes.

The same title statement could have been written using these options with explicit specifications.

```
title2 'Summary for #byvar(race) #byval(race)';
```

These options work well with the macro language, but make sure that, unlike the titles above, you use double quotes so that the macro language elements will be resolved correctly.

In the PROC FREQ shown above the procedure automatically creates a separate page for each combination of the BY variables. This ensures that the titles and the BY variable values for any given page will be synchronized. Some procedures do not necessarily generate a separate page for each BY group, when this is the case the titles and the actual BY variable values may appear to be incorrect.

PROC PRINT will only generate a new page for each BY group when the PAGEBY statement is used. Without the PAGEBY statement the output will be separated using the BY group combinations; however, since multiple combinations can appear on the same page, or a given combination can span pages, it is easy to see how the title would not reflect what is really on the page. This is demonstrated in the following example.

```
title2 'BY Information #byline'; ❸
proc print data=demog;
   by race sex;
   var lname fname dob;
   run;
```

❸ The #BYLINE option is used in the title statement to insert each of the BY variables and their values.

❹ The #BYLINE shows both the BY variables and their values in the title and mimics the BY line generated by the procedure.

❺ This line of text is the BY line generated by the procedure, and in this case it has been continued from the previous page and does not match the title.

❻ The BY group that starts on the current page is reflected in the title ❹.

```
15.1.2 BY Values in Titles
BY Information race=2 patient sex=F ❹

race=1 patient sex=M ❺
(continued)

Obs     lname          fname          dob

 42     Thomas         Daniel       23MAY38
 43     Uno            Robert       21MAR44

race=2 patient sex=F ❻

Obs     lname        fname          dob

 44     Adams        Mary         12AUG51
 45     Adamson      Joan            .
 46     Batell       Mary         12JAN37
       . . . . portions of the report not shown . . .
```

MORE INFORMATION
Examples in Section 7.4.3 and also in 11.3.1 use the #BYVAL option on the TITLE statement.

SEE ALSO
Carpenter (1998) discusses these options in detail. A sasCommunity.org tip written by Mary Rosenbloom demonstrates these options http://www.sascommunity.org/wiki/Tips:Use_BYVAL_to_Write_Better_Titles.

15.1.3 Executing OS Commands

There are several ways to execute operating system commands from within SAS. You can execute these OS commands from within a DATA step or through the use of global statements or macro language statements which are essentially global in this context.

Global Execution

The three global statements include:

- X
- %SYSEXEC
- SYSTASK COMMAND

The Windows example that is shown here collects the names of all the SAS controlled files in the C:\TEMP directory. The names are stored in a text file and then read into a SAS data set. This type of operation is common when we want to create a list that is to be stored in macro variables. The full step is first shown using the X statement which is probably the most commonly used statement for executing OS commands and statements.

```
❶x 'dir "c:\temp\*.sas7*" /b/o > c:\temp\SASFiles.txt';
    ❷              ❸        ❹ ❺           ❻

filename flist 'c:\temp\SASFiles.txt'; ❼
data filelist;
infile flist truncover;
input name $20.;
run;
```

❶ The X statement is used to specify and pass the DIR command to the operating system. Notice that the command to be passed is enclosed in quotes.

❷ The DOS DIR command makes a list of files. DOS commands are still available under Windows even though the current OS is not written using DOS.

❸ Only SAS controlled entities are to be selected by the DIR command. Microsoft requires that Windows paths to be enclosed in double quotes.

❹ Switches are used to limit the results to just the names of the files.

❺ The > symbol is used to route the results of the DIR command to a file.

❻ The file containing the list of items is named.

❼ A FILENAME statement is used to point to the text file containing the list.

The %SYSEXEC statement is a macro language statement that can be executed in open code. As such it is global in nature. In the macro language quote marks are not used as parsing characters; consequently, the DIR command is not quoted.

```
%sysexec(dir "c:\temp\*.sas7*" /b/o > c:\temp\SASFiles.txt);
```

The SYSTASK COMMAND statement is also a global statement that can be used to execute OS commands. Its syntax is essentially the same as that used for the X statement.

```
systask command 'dir "c:\temp\*.sas7*" /b/o > c:\temp\SASFiles.txt';
```

When the temporary file (C:\temp\sasfiles.txt) is only a means to an end and is of no lasting value, we can create it virtually and avoid actually creating the physical file. The PIPE device type is used on the FILENAME statement to essentially route the file to memory. Here the DIR command is included on the FILENAME statement. Instead of routing the results to a file, they are directly available to the DATA step.

```
filename flist pipe 'dir "c:\temp\*.sas7*" /b/o';
```

DATA Step Execution

Within the DATA step the SYSTEM function and the CALL SYSTEM routine can be used to execute OS commands. One advantage of this technique is that it is executable, which means that it can be conditionally executed. In this DATA step the CALL SYSTEM routine is used to generate the same list of files, but only after verifying that the directory exists.

```
data _null_;
  if fileexist('c:\temp') then do;
    call system('dir "c:\temp\*.sas7*" /b/o > c:\temp\SASFiles.txt');
    end;
  run;
```

Sub-session Execution Comments

With the exception of the SYSTASK COMMAND statement, when these statements are executed an OS sub-session command window is by default initiated and opened. Under Windows this is seen as a DOS command window. This window must then be closed (the DOS command is EXIT) before SAS can continue with its next statement or operation. This behavior is controlled with the XWAIT system option. Changing this option to NOXWAIT will automatically close this window at the completion of the command.

You will probably notice that even with the NOXWAIT option, the black command window still at least flashes. While this flashing window is not a problem it can be a bit annoying. You can avoid opening the window altogether with the SYSTASK COMMAND statement, but there are other subtle differences. The XMIN option can be used to minimize the command box and a brief command message box. By using the NOXWAIT, NOXSYNC, and XMIN options, the command box does not flash.

```
options noxwait xmin noxsync;
x 'dir *.*';
```

The X statement executes synchronously with SAS (XSYNC system option). This means that the SAS process is suspended until the command generated by the X statement has completed. The NOXSYNC system option can be used to allow SAS and the sub-session to execute asynchronously. When operating asynchronously SAS is not suspended while the sub-session command is completed. This can be an issue if a SAS step that follows the X statement depends on the result of the command before the command is complete. The SYSTASK COMMAND statement is by default executed asynchronously.

SEE ALSO
Walsh (2009) goes into more detail on the differences between the X and SYSTASK COMMAND statements. Varney (2008) discusses a number of DOS commands that can be accessed using PIPES.

Quoting issues within an X statement are discussed in the SAS Forum thread http://communities.sas.com/thread/32486?tstart=0. A sasCommunity.org tip discusses the WAITFOR statement which can be used with the SYSTASK COMMAND statement http://www.sascommunity.org/wiki/Tips:Schedule_SAS_Programs_with_SYSTASK_and_WAIT FOR.

15.2 Creating User-defined Functions Using PROC FCMP

The FCMP procedure allows you to write, compile, and test DATA step functions and CALL routines that you can then use in the DATA step, with the macro language, and within a number of procedures that allow the use of functions.

In the simplest sense creating a function is fairly straightforward, as is shown in the examples in this book. More complex functions are possible. As is the case with so many of the topics in this book, this section is a teaser. The FCMP procedure is very powerful and the concepts are not that difficult, but look deeper than the presentations in this section – there is a lot more.

MORE INFORMATION
Several user-defined functions appear in Section 12.5.5

SEE ALSO
The classic introduction to the FCMP procedure was written by Jasson Secosky (2007). Adams (2010) and Eberhardt (2011) both also provide nice introductions.

15.2.1 Building Your Own Functions
The first version of the QNUM function which is shown here was written by Rick Langston, senior manager of software development at SAS. It is used to convert a SAS date into a quarter (Q1, Q2, etc.) without the year portion that is returned by the YYQ. format. This allows us to consolidate dates into quarters without regard to year.

```
proc fcmp outlib=work.myfuncs.tmp;  ❶
   function qnum(date) $;  ❷
      length yyq4 $4;
      yyq4=put(date,yyq4.);  ❸
      if substr(yyq4,3,1)='Q'  ❹
         then return(substr(yyq4,3,2));  ❺
      else return(yyq4);  ❻
   endsub;  ❼
run;
options cmplib=(work.myfuncs);  ❽
data qlabs;
   set advrpt.lab_chemistry
                  (keep=subject labdt);
   qtr=qnum(labdt);  ❾
   run;

title2 'Quarters without years';
proc freq data=advrpt.lab_chemistry;
   table qtr*visit;  ❿
   run;
```

❶ The compiled function is saved in a special data set that includes a *packet*, which in this case is named TMP.

❷ The FUNCTION statement names the function, lists its arguments in parentheses and, if it is to return a character value, includes the $ before the semicolon.

❸ Use the YYQ. format to translate the date into a quarter.

❹ The third character will be a 'Q' if the date was successfully translated using the YYQ. format.

❺ Use the RETURN statement to specify the value to be returned by the function. The SUBSTR function is used to strip off the year (YY) portion of the formatted value.

❻ The date must have been missing or illegal for the YYQ. format.

❼ The FUNCTION statement always ends with the ENDSUB statement.

❽ The CMPLIB option is used to point to the data set that contains the TMP 'packet' that holds the function definition.

	\multicolumn{10}{c}{Table of qtr by VISIT}
	\multicolumn{10}{c}{VISIT(VISIT NUMBER)}
qtr	1	2	4	5	6	7	8	9	10	Total
Q1	3	3	3	3	3	5	6	5	5	36
	2.21	2.21	2.21	2.21	2.21	3.68	4.41	3.68	3.68	26.47
	8.33	8.33	8.33	8.33	8.33	13.89	16.67	13.89	13.89	
	17.65	18.75	18.75	20.00	18.75	31.25	50.00	35.71	35.71	
Q2	3	2	2	2	2	2	0	2	3	18
	2.21	1.47	1.47	1.47	1.47	1.47	0.00	1.47	2.21	13.24
	16.67	11.11	11.11	11.11	11.11	11.11	0.00	11.11	16.67	
	17.65	12.50	12.50	13.33	12.50	12.50	0.00	14.29	21.43	
Q3	7	7	7	4	5	5	3	6	2	46
	5.15	5.15	5.15	2.94	3.68	3.68	2.21	4.41	1.47	33.82
	15.22	15.22	15.22	8.70	10.87	10.87	6.52	13.04	4.35	
	41.18	43.75	43.75	26.67	31.25	31.25	25.00	42.86	14.29	
Q4	4	4	4	6	6	4	3	1	4	36
	2.94	2.94	2.94	4.41	4.41	2.94	2.21	0.74	2.94	26.47
	11.11	11.11	11.11	16.67	16.67	11.11	8.33	2.78	11.11	
	23.53	25.00	25.00	40.00	37.50	25.00	25.00	7.14	28.57	
Total	17	16	16	15	16	16	12	14	14	136
	12.50	11.76	11.76	11.03	11.76	11.76	8.82	10.29	10.29	100.00

❾ A new character variable (QTR) with a length of $2 is created using the QNUM function. The lab date is passed into the function as the single argument.

❿ The new variable is used in the TABLE statement.

MORE INFORMATION
A variation on the QNUM function is discussed in Section 12.5.5, and a simplified form can be found in Section 15.2.2.

SEE ALSO
An example of a function that calls a macro that contains a PROC FREQ can be found at http://tech.groups.yahoo.com/group/sas_academy/message/438. A FCMP CALL routine is created at

http://tech.groups.yahoo.com/group/sas_academy/message/430. A function is used to calculate a person's societal age at http://support.sas.com/kb/36/788.html. PROC FCMP is used to create an INFORMAT that converts a fraction to a decimal value in a sasCommunity tip by Mike Zdeb http://www.sascommunity.org/wiki/Tips:Create_an_Informat_from_a_User-Defined_Function.

15.2.2 Storing and Accessing Your Functions

The OUTLIB option on the PROC statement is used to name a storage location for your function. The function is stored in a special SAS data set. You cannot use this data set to also store data nor can you store a function in an existing data set.

From an operational perspective it makes sense to organize the storage of your functions. All the functions associated with the ADVRPT project might be stored in the ADVRPT.FUNCTIONS data set. Those functions dealing with dates could be stored in the DATES packet.

The OUTLIB option specifies the *libref* (ADVRPT), data set (FUNCTIONS), and the packet (DATES). Function names are unique within a packet but not necessarily across packets. A given packet can contain multiple function definitions, and a given data set can contain multiple packets.

```
proc fcmp outlib=AdvRpt.functions.Dates;
   function . . . .
```

When you want to use a compiled function, the CMPLIB system option is used. This option specifies one or more SAS data sets that contain the packets defined by PROC FCMP. The packet is not specified, and all packets within the data set are made available.

```
options cmplib=(advrpt.functions);
```

In the example in Section 15.2.1 the data set MYFUNCS is written to the non-permanent work directory. Here we want to create a more permanent version of this simplified version of the QNUM function. The function QNUM is added to the DATES packet in the permanent data set ADVRPT.FUNCTIONS.

```
proc fcmp outlib=advrpt.functions.dates;
   function qnum(d) $;
      return(cats('Q',qtr(d)));
   endsub;
run;
```

We may want to add some other functions to this same data set. Here a new packet (CONVERSIONS) is added to the ADVRPT.FUNCTIONS data set, and we define two functions in one call to the FCMP procedure. These two functions convert from degrees centigrade to Fahrenheit (C2F) and from Fahrenheit to centigrade (F2C).

```
proc fcmp outlib=Advrpt.functions.Conversions;
   function c2f(c);
      return(((9*c)/5)+32);
   endsub;

   function f2c(f);
      return((f-32)*5/9);
   endsub;
run;
```

Later we can add more conversion functions. Here we add functions to calculate the Body Mass Index (BMI) using both Imperial and Metric units.

```
proc fcmp outlib=AdvRpt.functions.Conversions;
   function E_BMI(h,w);
      return((w * 703)/(h*h) );
   endsub;
   function M_BMI(h,w);
      return(w /(h*h));
   endsub;
run;
```

It is likely that you will have a number of functions stored in several data sets. To make all of these functions available each of the data sets must be listed on the CMPLIB= system option.

```
options cmplib=(advrpt.functions work.myfuncs);
```

Here two data sets are listed. All the functions in all the packets in both of these data sets will be available for use. The exception will be for multiple functions with the same name. Unlike most library searches, the search order across multiple function libraries is from *right to left*. In the example shown here functions in WORK.MYFUNCS will be found first. The packet is not named in the CMPLIB option; consequently, if the same function name is used in two different packets within the same data set, it will be harder to anticipate which will be used. I would recommend that function names be unique within data set.

15.2.3 Interaction with the Macro Language

Functions and routines created using PROC FCMP are typically called from within a DATA step; however, they may also be called from within the macro language by using the macro function %SYSFUNC and the macro statement %SYSCALL.

It is also possible to call or execute a macro from within the function or routine by using the RUN_MACRO function. The following rather silly example demonstrates some of the issues when calling a macro from within a function or routine.

```
proc fcmp outlib=advrpt.functions.utilities;
subroutine prntcrit(dsn$,kvar $,cvar $);     ❶
   rc=run_macro('printit',dsn,kvar,cvar);     ❷
endsub;
run;
%macro printit();     ❸
%let dsn  = %sysfunc(dequote(&dsn));     ❹
%let kvar = %sysfunc(dequote(&kvar));
%let cvar = %sysfunc(dequote(&cvar));
title2 "&dsn";     ❺
proc print data=advrpt.&dsn;     ❻
%if &kvar ne %then id &kvar;;
%if &cvar ne %then var &cvar;;
run;
%mend printit;

options cmplib=(advrpt.functions);     ❼
title1 '15.2.3 Macro Language Interface';     ❽
data _null_;
   set advrpt.dsncontrol;
   call prntcrit(dsn,keyvars,critvars);     ❾
   put 'Print ' dsn keyvars critvars;
   run;
proc print data=advrpt.dsncontrol;     ❿
run;
```

❶ The SUBROUTINE statement declares this as the routine's definition. Like with the FUNCTION statement the routine is named. Character argument names are followed by a $.

❷ The first argument of the RUN_MACRO function is the name of the macro to be called. The remaining arguments are the parameters for that macro.

❸ The positional macro parameters are not named on the %MACRO statement. The parameter names will flow from the RUN_MACRO function on through to the macro %PRINTIT ❷.

❹ The values of the macro parameters arrive quoted. Since we need them to be unquoted in the application shown here, the DEQUOTE function is called.

❺ TITLE2 will be displayed when the PROC PRINT executes. Although the TITLE1 statement will have been executed ❽, within the domain available to the compiled routine its definition is not available, and will, therefore, remain undefined.

❻ The PROC PRINT within the macro %PRINTIT is constructed using the macro parameters.

❼ The function library is specified.

❽ TITLE1 is defined before the DATA step that will call the routine; however, this title will not be available to the macro executed by the PRNTCRIT routine.

❾ The PRNTCRIT routine is called using three variables from the ADVRPT.DSNCONTROL data set. The variables in the routine call must be in the same order as they are defined ❷. The PRNTCRIT routine is called for each observation on the incoming data set. The routine calls and executes the macro %PRINTIT. Unlike the CALL EXECUTE routine. which pushes the macro call to a stack, %PRINTIT will be executed immediately.

❿ The TITLE1 will be honored, but the TITLE2 will be undefined.

Because functions and routines are compiled, we cannot use a macro call directly within the function definition. If we had attempted to specify the macro call directly as shown here, %PRINTIT would have executed while the PRNTCRIT routine was being compiled. Macro calls such as this one would be used to generate routine or function code. Macros that are to be executed during function execution should be specified with the RUN_MACRO function as was shown above.

```
proc fcmp outlib=advrpt.functions.utilities;
subroutine prntcrit(dsn$,kvar $,cvar $);
   %printit(dsn,kvar,cvar);
. . . . .
```

MORE INFORMATION
The DEQUOTE function is introduced in Section 3.6.7.

SEE ALSO
Chapter 8 in *Carpenter's Complete Guide to the SAS® Macro Language, 2nd Edition* (Carpenter, 2004) discusses the use of macros and macro variables within the context of compiled code.

15.2.4 Viewing Function Definitions

User-defined functions do not appear in the SASHELP.VFUNC view or in the DICTIONARY.FUNCTIONS table. However, under Windows you can see both the list of available functions and their attributes through the use of the FCMP Function Editor.

To start this editor, go to SOLUTIONS ⟶ ANALYSIS ⟶ FCMP FUNCTION EDITOR. The list of available data sets and the functions that they contain is shown in the left pane. Selecting a function brings up the editor dialog box, which allows you to see the details of the function's definition.

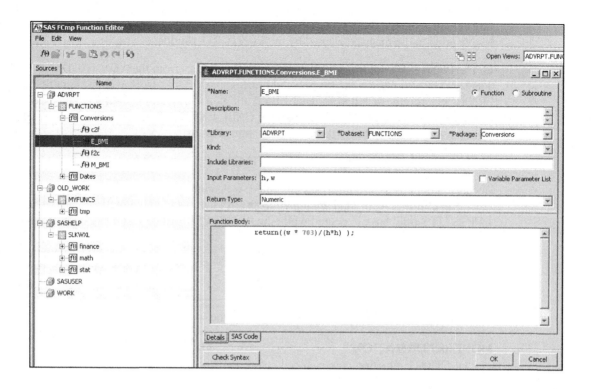

15.2.5 Removing Functions

Although functions and routines are stored in data sets, you cannot use standard data management techniques to delete an instance of a function or a routine. Fortunately the FCMP procedure comes with the DELETEFUNC and DELETESUBR statements that can be used for this purpose.

For the purposes of this example let us assume that a version of the function START has been stored in two different locations (two different data sets). The search order for multiple data sets is from *right to left* (this is the opposite order for searches across libraries, *e.g.,* formats, autocall). The START function is to be written so that it will return the first date of the specified interval type.

```
proc fcmp outlib=work.funcs.dates;  ❶
function start(int$,date);
   return(intck(int,date,0,'b'));
   endsub;
run;
proc fcmp outlib=advrpt.functions.dates;  ❷
function start(int$,date);
   return(intnx(int,date,0,'b'));  ❸
   endsub;
run;
options cmplib=(advrpt.functions work.funcs );  ❹
```

❶ A temporary location is set up to hold the START function. Unfortunately this function has been specified incorrectly using the INTCK function and will fail.

❷ The permanent data set to collect functions in the DATES packet is specified.

❸ The correct version of START, which uses the INTNX function, is stored in the permanent location.

❹ The CMPLIB option specifies that the WORK.FUNCS data set will be searched first, and this means that the bad version of the START function ❶ will be found and used.

We need to have the ability to remove the bad definition of START from WORK.FUNCS.DATES. This can be done using the DELETEFUNC statement.

```
proc fcmp outlib=work.funcs.dates;  ❺
   deletefunc start;  ❻
   run;
```

❺ PROC FCMP is called with the OUTLIB option pointing to the packet containing the bad definition of START.

❻ The DELETEFUNC statement can then be used to delete the specific function.

Using the function libraries defined above ❹, the version of START in ADVRPT.FUNCTIONS.DATES will now be the only version available.

```
data list;
do d = '01jan2010'd to '05feb2010'd;
   styr = start('year',d);  ❼
   stmo = start('month',d);  ❽
   output;
   end;
   format d styr stmo date9.;
   run;
```

❼ The first day of the year that contains the date stored in D is returned by the function START.

❽ The START function returns the first day of the month that contains the date in D.

MORE INFORMATION
You can also delete functions and routines by using the FCMP Function Editor (see Section 15.2.4).

15.3 Reading RTF as Data

RTF is a proprietary document file format developed by Microsoft Corporation in the late 1980s. Unlike an MS Word .DOC binary file, an RTF file can be read by text editors. This means that if we treat an RTF file as text, we can use SAS to read and write the RTF file as data, and this opens the door for the power and flexibility associated with the use of the SAS DATA step and the SAS macro language.

The example shown in this section modifies a CONsolidated Standards Of Reporting Trials, CONSORT, flow diagram by filling in the blanks.

SEE ALSO
These specific examples are presented in more detail in Carpenter and Fisher (2011).

15.3.1 RTF Diagram Completion

The layout of the CONSORT table depends on the study design. This includes the number of ARMS and the phases of the study. The techniques discussed in this section, however, are completely independent of the study design. The first step in this process is to create a template form of the CONSORT table. This RTF table will contain all the needed information with blank fields. This figure shows the "Enrollment" portion of a CONSORT table, which will show the number of subjects and their status relative to the study. Typically the N= values would be filled in by hand once they had been determined.

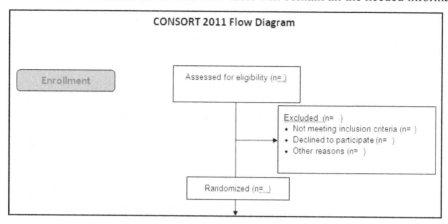

The RTF CONSORT table can easily have over a dozen fields that require completion. In the process described below, each field will be assigned a code unique to the table. The entire table (RTF file) will then be read as data and the codes will be translated into the final values through the use of DATA step functions. The resulting modified table will be rewritten, again as an RTF file, where it will then be available for use by a word processor.

15.3.2 Template Preparation

The template is prepared for use by SAS by filling in each of the individual fields using unique codes. Here the unique codes for the first six fields are TOTASSESSD, TOTEXCL, INELIG, DECLIN, EXCLOTH, and NRAN. For our purposes we are assuming that these names never occur otherwise in the table. Other than being unique, the code that you choose is unimportant, but for a more complicated table the field code names can be used to help make sure that the values are inserted in the correct location.

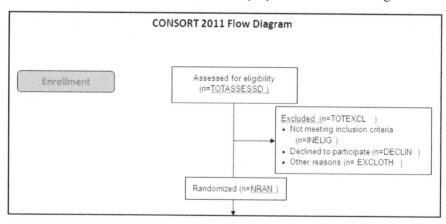

15.3.3 RTF as Data

Fortunately we need to know very little about RTF code in order to work with it using SAS. A quick look at a portion of the RTF code that generated the figure shown in the previous section shows a text language which is mostly not human readable. However, a closer inspection shows one of our designated keywords (DECLIN).

```
\par }{\rtlch\fcs1 \af0\afs16 \ltrch\fcs0 \f3\fs16\lang0\langfe1033\langnp0
\fs16\lang4105\langfe1033\langnp4105\insrsid4260155\charrsid1516310    }{\rt
\f1\fs20\lang4105\langfe1033\langnp4105\insrsid1909421 DECLIN}{\rtlch\fcs1
\par }{\rtlch\fcs1 \af0\afs16 \ltrch\fcs0 \f3\fs16\lang0\langfe1033\langnp0
```

Our approach will be to have SAS read the RTF text strings, find the appropriate codes, replace the codes with the values of interest, and then replace the modified RTF text strings. The search and replace operations will be handled by using the TRANSTRN function, which replaces all occurrences of the second argument with the third argument. For our purposes there should only be one occurrence for each of our codes.

The DATA step used to read and write the RTF CONSORT table is fairly straightforward. RTF

```
filename confile1 "C:\temp\CONSORT_Diagram1.rtf";
filename confile2 "C:\temp\CONSORT_Diagram2.rtf";

data _null_;
infile confile1 lrecl=3000;
input;
_infile_ = transtrn(_infile_,'TOTASSESSED','345');
_infile_ = transtrn(_infile_,'TOTEXCL',    '56');
_infile_ = transtrn(_infile_,'INELIG',     '35');
_infile_ = transtrn(_infile_,'DECLIN',     '17');
_infile_ = transtrn(_infile_,'EXCLOTH',    '4');
_infile_ = transtrn(_infile_,'NRAN',      '289');
file confile2  lrecl=3000;
put _infile_;
run;
```

does not have a fixed maximum record length; however, the length is generally under 500 characters. Here the LRECL is set to 3000 – just in case. The incoming RTF file is designated by the *fileref* CONFFILE1. The new version of the CONSORT table is written to the file named in the CONFILE2 *fileref*. Through the use of the automatic variable

INFILE we read each RTF line as an entire entity. This string is then searched and the appropriate codes are replaced. In this example the TRANSTRN function replaces the text 'TOTASSESSED' with the appropriate number, which we have provided (345). In the figure to

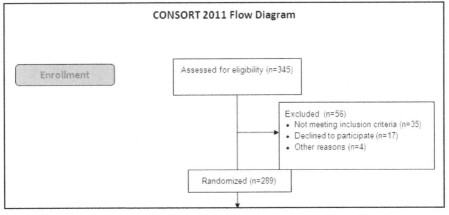

the left we can see that the placeholder codes that we used in the template version of the table have been replaced with the values supplied in the SAS program. In the paper cited above,

(Carpenter and Fisher, 2011) macro code is shown that provides an automated metadata driven coding solution.

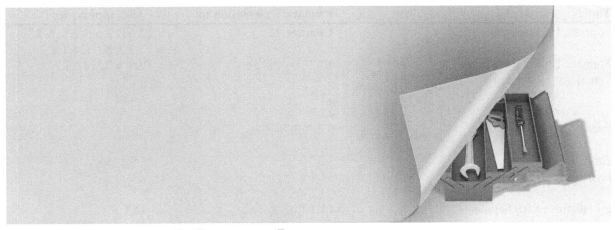

A p p e n d i x   A

Topical Index

(*continued*)

Appendix B

Usage Index

Of all the procedures, statements, and options discussed in this book, only those that are of note specifically to the techniques described in this book are listed in this Appendix.

Global Statements and Options

Statements, Global

Statements, Options	Primary Description in:	Also Appears in:
AXIS	9.3.2	10.1.1, 10.1.2, 10.1.3, 10.2.3
LABEL	9.3.2	10.1.1, 10.1.2, 10.2.3
MAJOR	9.3.2	10.1.1
MINOR	9.3.2	10.1.2, 10.1.3
ORDER	9.3.2	1.2, 10.1.3, 10.2.2
REFLABEL	10.2.2	
VALUE	9.3.2	10.1.2, 10.1.3, 10.2.2
DM	14.4.2	2.6.3, 14.1.1, 14.1.3
FILENAME		2.6.3, 5.4.1, 9.2, 11.4.3, 13.9.1, 4.2, 15.3.3
CATALOG	13.9.1	14.6.4
EMAIL	14.5	
ATTACH	14.5	
FROM	14.5	
SUBJECT	14.5	
TO	14.5	
EXCEL	1.6.2	
PIPE	15.1.3	2.6.3
LEGEND	9.3.2	
ACROSS	9.3.2	
FRAME	9.3.2	
LABEL	9.3.2	
SHAPE	9.3.2	
VALUE	9.3.2	
LIBNAME		1.1.1, 1.1.3, 1.2.5, 10.2.1, 14.2
Concatenated libraries	12.9.3	
SCAN_TEXT	1.1.3	1.1.4
V9		14.2
SYMBOL	9.3.1	10.1.1, 10.1.2, 10.1.3, 10.2.1, 10.2.2
SYSTASK COMMAND	15.1.3	
TITLE	9.1	10.2.2, 10.2.4, 11.3.3, 11.3.5
BOLD	9.1	
BCOLOR	9.1	
#BYLINE	15.1.2	
#BYVAL	15.1.2	7.4.3, 11.3.1
#BYVAR	15.1.2	11.3.1
COLOR	9.1	11.4.4

(continued)

Statements, Global (*continued*)

Statements, Options	Primary Description in:	Also Appears in:
TITLE (*continued*)		
FONT	9.1	10.2.2, 10.2.4
HEIGHT	9.1	10.2.2, 11.3.1
ITALIC	9.1	
JUSTIFY	9.1	10.2.2
LINK	11.4.2	11.4.4
UNDERLINE	9.1	
X	15.1.3	

Macro Language

Statements and Options	Primary Description in:	Also Appears in:
IN operator (#)	13.10	
%LOCAL	13.1	13.7, 13.8.2, 14.6.4
MINDELIMITER	13.10.3	
MINOPERATOR	13.10.2	
SECURE	13.9.1	
SOURCE	13.9.1	
STORE	13.9.1	
%SYSEXEC	15.1.3	14.4.7

GOPTIONS, Graphics

Graphics option	Primary Description in:	Also appears in:
BORDER	9.2	10.2.3, 10.2.4, 11.4.3
DEVICE	8.2.1	9.2, 10.2.3, 10.2.4, 11.4.3
FTEXT	9.2	10.2.3, 10.2.4, 11.4.3
GSFNAME	8.2.1	9.2, 11.4.3
GSFMODE		10.2.3
HTEXT	9.2	11.4.3
RESET	9.2	9.3.1, 10.2.3, 10.2.4, 11.4.3
TARGETDEVICE	9.2	
XPIXELS	10.2.4	
YPIXELS	10.2.4	

Options, System

Option	Primary Description in:	Also Appears in:
in general	14.1	
CENTER		14.2
CMPLIB	12.5.5, 15.2.2	15.2.3
CPUCOUNT	4.5	
DATASTMTCHK	14.1.2	
DATE		14.2
EMAILHOST	14.5	

(*continued*)

Options, System *(continued)*

Option	Primary Description in:	Also Appears in:
EMAILID	14.5	
EMAILPORT	14.5	
EMAILSYS	14.5	
FMTERR		2.3.1
FMTSEARCH	12.9.2	12.9.3
initialization	14.1.1	
ALTLOG	14.1.1	
AUTOEXEC	14.1.1	
AWSCONTROL	14.3.2	
AWSTITLE	14.3.2	
CONFIG	14.1.1	14.3.1
INITCMD	14.3.2	
INITSTMT	14.1.1	
RTFCOLOR	14.1.1	
SASINITIALFOLDER	14.1.1	
SPLASHLOC	14.1.1	
SYSIN	14.1.1	
SYSPARM	14.1.1	
TERMSTMT	14.1.1	
VERBOSE		14.3.1
MAUTOLOCDISPLAY	13.4.1	
MAUTOSOURCE	13.4.1	13.4.2
MCOMPILE	13.9.2	
MERGENOBY	14.1.2	
MFILE	13.11	
MINDELIMITER	13.10.3	
MINOPERATOR	13.10.2	
MISSING	2.10.2	
MLOGIC		13.8.2
MLOGICNEST		13.11
MPRINT		13.8.2, 13.9.1
MPRINTNEST		13.11
MREPLACE	13.9.2	
MSGLEVEL	5.3.2	
NOWORKINIT	14.1.1	
NOWORKTERM	14.1.1	
NUMBER		14.2
SASAUTOS	13.4.1	13.4.2, 14.2, 14.3.1
SORTEQUALS		4.1.6
SYMBOLGEN		13.8.2
THREADS	4.5	
VALIDVARNAME	14.1.2	1.2.3
XMIN	15.13	
XSYNC	15.13	
XWAIT	15.13	

Options, Data Set

Option	Primary Description in:	Also Appears in:
ALTER	2.1.2	5.4.2
DROP	2.1.3	5.1
ENCRYPT	2.1.2	5.4.2
FIRSTOBS	2.1.4	3.1.4, 3.1.5
KEEP	2.1.3	2.1, 2.3.2, 3.1.4, 3.1.5, 4.2, 5.2, 11.1.2, 12.7, 13.5.2
IDXNAME	5.3.4	
IDXWHERE	5.3.4	
IN		3.8.2, 3.7.1
INDEX	5.3.2	
OBS	2.1.4	8.5.2
PW	2.1.2	
PWREQ	2.1.2	5.4.2
READ	2.1.2	5.4.2
RENAME	2.1.3	3.1.4, 3.1.5, 3.7.2, 5.1, 6.4, 12.7
REPEMPTY	2.1.1	
REPLACE	2.1.1	
SORTEDBY	4.4	
WHERE		2.1.3, 2.1.4d, 2.2.2, 3.7.2, 5.1, 7.7, 7.8, 11.5.3, 13.2.3, 13.6, 14.1.1
WRITE	2.1.2	5.4.2

Procedures: Steps, Statements, and Options

Procedures

Statements and Options	Primary Description in:	Also Appears in:
In general (across procedures)		
CLASS	7.1	7.1.2, 8.2.2
DESCENDING	7.1.1	
GROUPINTERNAL	7.1.2	
MISSING	7.1.1	
ORDER	7.1.3	
EXCLUSIVE	12.1	
MISSING	2.10.5, 7.1.1	
ORDER	2.6.2, 8.1.5	2.5.5, 8.4.5
PRELOADFMT	12.1	2.5.5
BOXPLOT	10.1.1	

(continued)

Procedures (*continued*)

Statements and Options	Primary Description in:	Also Appears in:
CATALOG		
CONTENTS		12.9.3
COPY		13.9.2
ENTRYTYPE		13.9.2
DELETE	13.9.2	
FORCE	13.9.2	
SELECT		13.92
COMPARE	5.1	
COMPARE	5.1	
OUT	5.1	
OUTBASE	5.1	
OUTCOMP	5.1	
OUTNOEQUAL	5.1	
CONTENTS		2.6.1, 5.3.1, 13.8.3
ALL	13.8.3	
OUT	13.8.3	
NOPRINT		13.8.3
VARNUM	2.6.1	
DATASETS		2.6.1
AGE	5.6.2	
CHANGE	5.6.2	
CONTENTS		2.6.1
DELETE	5.5	
INDEX CREATE	5.3.2	5.3.3, 6.6
INDEX DELETE	5.3.2	5.3.3
KILL	5.5	14.4.5
MODIFY	5.3.2	2.6.1, 5.3.3, 6.6
NOLIST	5.5	5.6.2, 6.6
DELETE	5.5	
EXPORT	1.2	1.4.2, 11.2.2
SHEET	1.2.3	1.2.1, 11.2.2
FCMP	15.2	12.5.5
DELETEFUNC	15.2.5	
ENDSUB	15.2.1	12.5.5
FUNCTION	15.2.1	12.5.5
OUTLIB	15.2.2	
RETURN	15.2.1	12.5.5
RUN_MACRO	15.2.3	
SUBROUTINE	15.2.3	

(*continued*)

Procedures (*continued*)

Statements and Options	Primary Description in:	Also Appears in:
FORMAT		2.6.2, 3.6.6
CNTLIN	12.7	6.5, 6.7.3, 12.3.2
CNTLOUT	12.7	
DEFAULT		12.5.4, 12.5.5
formats as labels	12.5.1	12.5.2, 12.5.3, 12.8
functions as labels	12.5.5	
INVALUE		12.5.3
LIBRARY	12.9.3	
NOTSORTED	12.4	
MULTILABEL	12.3	
PICTURE	12.2	
DATATYPE	12.2.1	
directives	12.2.1	
MULT	12.2.3	
PREFIX	12.2.3	
ROUND	12.2.2	
REGEXPE	12.5.4	
SAME	12.5.4	
VALUE	12.7	2.6.2, 3.6.6, 7.1.2, 8.1.1, 8.1.5, 11.3.4, 11.3.5, 11.4.4, 11.5.1, 11.5.2, 12.1, 12.8
FREQ	8.3	2.5.6, 2.9.4, 10.3
NLEVELS	8.3.2	
OUTPUT	8.3.1	
ALL	8.3.1	
SPARSE		2.5.6
TABLE		
ALL		8.3.2
CHISQ		8.3.1
PLOTS	10.3	
GFONT	10.2.1	
GKPI	10.2.4	
GPLOT	10.1.2	
IMPORT	1.2	14.1.2
DATAROW	1.2.3	
DBMS		14.1.2
GETNAMES	1.2.3	1.2.5, 14.1.2
GUESSINGROWS	1.2.3	
MIXED	1.2.3	14.1.2
RANGE	1.2.3	1.2.5
SCANTEXT	1.2.3	14.1.2
SCANTIME		14.1.2
SHEET	1.2.3	1.2.1, 14.1.2

(continued)

Procedures (*continued*)

Statements and Options	Primary Description in:	Also Appears in:
IMPORT (*continued*)		
TEXTSIZE	1.2.3	
USEDATE		14.1.2
MEANS/SUMMARY	Chapter 7	2.5.3, 2.5.4, 2.5.5, 13.3.1
CHARTYPE	7.6	
CLASS		
EXCLUSIVE	12.1.3	
MLF		12.3.2
PRELOADFMT	12.1.3	
CLASSDATA	7.9	2.5.4
COMPLETETYPES	7.10	2.5.3, 12.1.3
EXCLUSIVE	7.9	
MISSING	7.1.1	
NWAY		7.6, 7.7
OUTPUT	7.3	7.1
AUTONAME	7.2	5.3.3, 7.4.3
AUTOLABEL	7.2	
GROUPID	7.4.2	
LEVELS	7.2, 7.11	
MAXID	7.4.1	
MINID	7.4.1	
WAYS	7.2, 7.11	
TYPES	7.8	
WAYS	7.7	
OPTIONS		14.3.1
OPTLOAD	14.1.3	
OPTSAVE	14.1.3	
PRINT	8.5	11.5.4, 11.7.1
BY	8.5.1	11.5.4
ID	8.5.1	11.5.4
STYLE	8.5.2	11.5.4
PWENCODE	5.4.1	
REG		9.3.1, 9.3.2, 9.4
REPORT	8.4	2.5.5, 10.2.4, 11.2.3, 11.3.1, 11.3.2, 11.3.3, 11.3.4, 11.3.5, 11.4.1, 11.5.3
absolute column reference	8.4.2	8.4.3, 8.4.5
alias specification	8.4.3	8.4.2, 8.4.5, 8.4.7, 12.1.1
CALL DEFINE	8.4.7	10.2.4, 11.4.1, 11.5.3
COMPLETECOLS	12.1	
COMPLETEROWS	12.1, 12.1.1	2.5.5

(*continued*)

Procedures (*continued*)

Statements and Options	Primary Description in:	Also Appears in:
REPORT (*continued*)		
compute block	8.4.3	
CONTENTS	11.4.3	
DEFINE		
EXCLUSIVE	12.1.1	
NOPRINT	8.4.4	8.4.6
PRELOADFMT	12.1.1	
ORDER	8.4.5	8.4.8
NOZERO	8.4.8	
LINE	8.4.10	8.4.6, 11.3.2, 11.3.3, 11.3.4
STYLE override	8.4.6	8.4.8, 11.5.3
SHEWHART	10.1.3	
SORT	4.1	
DUPOUT	4.1.2	
EQUALS	4.1.6	
FORCE	4.1.5	
NODUPKEY		2.5.1, 6.7.1
NODUPLICATES (NODUPREC)	4.1.1	2.9.1, 4.1.2, 4.4
SORTSEQ	4.1.4	
TAGSORT	4.1.3	
SQL	1.5	2.8.1, 2.9.3, 3.7.4, 5.2, 5.4.2, 11.2.2, 11.2.3, 11.4.1, 11.4.3, 13.5.1, 13.5.2
CONNECT	1.5.1	5.4.2
CREATE INDEX	5.3.2	5.3.3
DESCRIBE	13.8.1	
DISCONNECT	1.5.1	5.4.2
DISTINCT	2.9.3	
DROP TABLE	5.5	
ORDER BY		2.9.3
UNION	2.8.1	
WHERE	2.2.2	
TABULATE	8.1	11.4.4, 11.5.2
CLASS	8.1.1	11.5.2
EXCLUSIVE	12.1.2	
MLF	12.3.1	
PRELOADFMT	12.1.2	
CLASSDATA	8.1.4	
CLASSLEV	8.1.3	11.4.4
EXCLUSIVE	8.1.4	
KEYLABEL	8.1.1	
KEYWORD	8.1.3	
ORDER	8.1.5	
TABLE	8.1.1	11.5.2

(*continued*)

Procedures (*continued*)

Statements and Options	Primary Description in:	Also Appears in:
TABULATE (*continued*)		
BOX	8.1.1	8.1.3, 12.3.1
MISSTEXT	8.1.1	
percentages	8.1.2	
PRINTMISS	12.1.2	
RTS	8.1.3	
VAR	8.1.1	11.5.2
TRANSPOSE	2.4.1	2.5.2, 3.6.5, 5.1
UNIVARIATE	8.2	11.1.1, 11.1.2
CLASS	8.2.2	
KEYLEVEL	8.2.2	
HISTOGRAM	8.2.1	
INSET	8.2.1	
OUTPUT	8.2.4	
PROBPLOT	8.2.1, 8.2.3	
QQPLOT	8.2.1, 8.2.3	

DATA Step: Statements and Options

Statements, DATA Step

Statements and Options	Primary Description in:	Also Appears in:
ARRAY	3.10.1	2.2.3, 2.4.2, 2.10.4, 3.1.7, 3.6.1, 3.6.6, 3.10.3
ARRAY, implicit	3.10.4	
ARRAY, temporary	3.10.2	3.1.2, 3.6.2, 3.6.4, 3.6.5, 3.6.6, 3.8.1, 3.10.3, 6.7.1, 6.7.2, 6.7.3
CLASS		2.6.2
DO	3.9	
FILE		14.5, 15.3.3
INFILE	1.3.3	5.4.1, 15.3.3
DELIMITER	1.3.3	
DLM	1.3.3, 1.4.1	
DLMSTR	1.3.3, 1.4.1	
DSD	1.3.3, 1.4.1	
FLOWOVER	1.3.4	
LRECL		15.3.3
MISSOVER	1.3.4	
TRUNCOVER	1.3.4	5.4.1
INPUT	1.3	5.4.1
See also informat modifiers		
INFILE		15.3.3

(*continued*)

Statements, DATA Step (*continued*)

Statements and Options	Primary Description in:	Also Appears in:
LENGTH		3.7.1, 3.8.2, 3.10.3, 5.2, 5.4.1…
MERGE, one-to-one		3.1.4
PUT, %PUT	15.1.1	
MISSING	2.10.1	
RETAIN		5.2, 6.7.3
SET, options and usage	3.8	
END	3.8.1, 3.83	2.9.5, 3.1.5, 3.6.2, 3.9.1, 3.9.3, 3.10.3, 6.7.2, 6.7.3, 7.4.3, 13.3.2
INDSNAME	3.8.2	
KEY	3.8, 6.6	
NOBS	3.8.1	3.9.3
POINT	3.8.1	3.1.6, 3.9.3
SET, double		3.1.5, 3.1.6, 3.6.2, 6.4
sum	3.1.8	3.1.1, 3.1.6, 3.1.7, 3.3.4, 3.8.1
WHERE	2.7	2.3.1
X		2.6.3

Format Modifiers

Format Modifiers	Primary Description in:	Also Appears in:
?	1.3.1	1.4.5
??	1.3.1	2.3.1, 2.3.2, 3.6.1
&	1.3.2	
:	1.3.2	1.3.3
~	1.3.2	1.3.3

Functions

Functions	Primary Description in:	Also Appears in:
ANYALNUM	3.6.1	
ANYALPHA	3.6.1	3.6.6
ANYDIGIT	3.6.1	3.6.5
ANYPUNCT	3.6.1	
ANYSPACE	3.6.1	
ANYUPPER	3.6.1	
ANYXDIGIT	3.6.1	3.5.2
ARCOS	3.6.6	
ATTRN	13.8.3	13.7
%BQUOTE	13.12	2.6.3, 3.6.7, 13.10.1
CALL DEFINE		2.6.3
CALL EXECUTE		10.2.4, 11.1.2
CALL MISSING	2.10.4	2.9.3, 3.1.2, 3.6.6, 3.10.3
CALL MODULE	14.6.4	
CALL SYMPUTX	13.2	6.7.3, 11.2.2, 13.5.2, 13.8.1, 14.1.1, 14.6.4

(*continued*)

Functions (*continued*)

Functions	Primary Description in:	Also Appears in:
CALL SYSTEM	15.1.3	
CAT	3.6.3	
CATQ	3.6.3	
CATS	3.6.3	10.2.4, 11.1.2
CATT	3.6.3	
CATX	3.6.3	
CEIL		2.2.1, 3.8.1
CLOSE	13.8.3	13.7
COALESCE	3.6.6	2.2.6
COMPARE	3.6.2	
COMPLEV	3.6.2	
COMPGED	3.6.2	
COMPCOST	3.6.2	
COMPRESS	3.6.7	3.6.1
CONSTANT	3.6.6	
COUNT	3.6.6	
COUNTC	3.6.6	
COUNTW	3.6.6	13.7
CMISS	2.10.3	
DAY		3.2.3
DEQUOTE	3.6.7	15.2.3
DIF		3.1.3
DIM	3.10.3	3.6.6
IFC	3.6.6	
IFN	3.6.6	2.2.3, 3.1.4, 3.1.5
INDEX	3.6.6	
INDEXC	3.6.6	
INDEXW	3.6.6	13.8.2
INPUT	3.5.1	2.3.1, 2.3.2, 3.5.2, 6.7.2
INPUTC	3.5.1	
INPUTN	3.5.1	3.4.4
INTCK	3.4	3.2.2, 3.2.3
INTNX	3.4	3.5.1
FILEEXIST		13.8.2, 14.4.7, 15.1.3
FILENAME	13.8.2	
FIND	3.6.6	
FINDC	3.6.6	
FINDW	3.6.6	
FLOOR		3.2.3
GEOMEAN	3.6.6	
GETOPTION	13.8.2	
HBOUND	3.10.3	
LAG	3.1.3	
LARGEST	3.6.4	3.6.6
LBOUND	3.10.3	
LEFT		3.5.1, 3.6.1, 3.7.1, 6.5
%LEFT	13.12	14.6.4

(*continued*)

Functions (*continued*)

Functions	Primary Description in:	Also Appears in:
LENGTH		3.6.7
LIBNAME		5.4.1, 14.4.7
MAX	3.6.4	2.2.5
MIN	3.6.4	2.2.5, 3.6.6
MISSING	2.10.3	2.2.6, 2.3.1, 13.5.2
MOD	3.1.7	
MONTH		2.2.1, 2.2.3
NMISS	2.10.3	
NOTDIGIT	3.6.1	
NOTXDIGIT		3.5.2
%NRSTR	13.12	11.1.2, 14.4.6
OPEN	13.8.3	13.7
ORDINAL	3.6.4	
PATHNAME	14.6.1	13.8.2
PUT	3.5.1	3.5.2, 3.6.3, 6.5, 12.5.2
PUTC	3.5.1	
PUTN	3.5.1	3.4.4, 12.2.1
QTR		8.4.8
RANUNI		3.8.1
RENAME	5.6.1	
ROUND	3.6.6	
SCAN, %SCAN	3.6.6	3.8.2, 13.8.2, 14.6.3
SIGN	2.2.6	
SMALLEST	3.6.4	
%STR	13.12	2.6.3
STRIP	3.7.6	
SUBSTR, %SUBSTR	3.6.6	2.3.1, 3.6.5, 3.6.6, 13.8.2
SUM		2.2.6, 3.1.8
%SYMEXIST	13.7	
%SYSFUNC		3.4.4, 3.5.1, 3.6.6, 12.2.1, 13.7, 13.8.2, 13.12, 14.4.7, 14.6.1, 15.2.3
SYMGET	5.4.2	
%SYSGET	14.6.3	
%SYSRC	6.6.2	
SYSMSG		5.4.1
TRANSLATE	3.6.7	
TRANSTRN	3.6.7	15.3.3
TRANWRD	3.6.6	
TRIM, %TRIM		2.2.2, 3.6.1, 14.6.4
TRIMN		3.6.7
%UNQUOTE	13.12	2.6.3
%UPCASE		13.10.2
VARNAME	13.8.3	
VFORMATX	3.6.5	
VNAME	3.6.5	3.6.1, 3.10.3
VNAMEX	3.6.5	

(*continued*)

Functions (*continued*)

Functions	Primary Description in:	Also Appears in:
VNEXT	3.6.5	
VVALUEX	3.6.5	
WHICHN	3.6.6	2.2.2, 3.10.3
YEAR		2.2.3, 3.2.1, 3.5.1, 3.6.6, 8.4.8
YRDIF	3.2.2	

Hash Object

Statements and Methods	Primary Description in:	Also Appears in:
ADD	3.3.2	3.3.4, 3.3.5, 3.3.7
CHECK	3.3.6	
DECLARE HASH	3.3.1	2.9.5, 3.3.3-3.3.7, 6.8
DECLARE HITER	3.3.4	3.3.5, 3.3.6
DEFINEDATA	3.3.2	2.9.5, 3.3.3-3.3.7, 6.8
DEFINEDONE	3.3.2	2.9.5, 3.3.3-3.3.7, 6.8
DEFINEKEY	3.3.2	2.9.5, 3.3.3-3.3.7, 6.8
DELETE	3.3.5	
FIND	3.3.4	3.3.6, 3.3.7, 6.8
FIRST	3.3.4	3.3.5
NEXT	3.3.4	3.3.5, 3.3.6
OUTPUT	3.3.2	2.9.5, 3.3.6
REPLACE	3.3.2	2.9.5, 3.3.4, 3.3.6

Output Delivery System, ODS

ODS Destinations and Tagsets

ODS Destinations and Tagsets	Primary Description in:	Also Appears in:
CSV	1.4.4	
HTML		10.2.4, 11.4.1, 11.4.3
LISTING		11.1.2
MARKUP		1.6.1, 11.2, 11.2.3
OUTPUT	11.1	8.3.2
PDF		2.5.5, 8.4.9, 10.2.4, 10.3, 11.3.1, 11.3.2, 11.4.2, 11.4.3, 11.5.2, 11.5.4, 11.6, 14.5
RTF		8.4.6, 9.1, 11.3.1, 11.3.5, 11.4.4
TAGSET.EXCELXP	11.2, 11.2.1	11.2.2
OPTIONS	11.2.2	11.2.3

ODS Attributes

Attributes	Primary Description in:	Also Appears in:
ASIS	11.7.1	
BACKGROUND	8.1.3	8.4.7, 8.5.2, 11.5.2, 11.5.3, 11.5.4
CELLWIDTH	8.4.7	
FLYOVER		2.6.3
FONT_FACE	8.1.3	8.4.6
FONT_SIZE	8.1.3	8.4.6
FONT_STYLE	8.1.3	8.4.6
FONT_WEIGHT	8.1.3	8.5.2
FONT_WIDTH	8.1.3	
FOREGROUND	8.1.3	8.5.2, 11.5.2, 11.5.3, 11.5.4
hyperlinks	11.4	
in-line formatting	11.3	8.4.6
attributes	11.3.3, 11.3.4	11.6
DAGGER	11.3.2	
LASTPAGE	11.3.1	
PAGEOF	11.3.1	
raw commands	11.3.5	
subscript	11.3.2	
superscript	11.3.2	
THISPAGE	11.3.1	
JUST		8.1.3, 8.4.6, 8.4.8, 8.4.9, 11.7.1
links, forming	11.4	8.5.3
URL	11.4.4	

ODS Options

Options	Primary Description in:	Also Appears in:
ANCHOR	11.4.3, 11.4.4	
BODYTITLE	9.1	
DELIMITER	1.4.4	
PROCLABEL	11.4.3	11.4.4
STARTPAGE	11.6	
STYLE override		2.6.3
TABULATE	8.1.3	11.4.4, 11.5.2
PRINT	8.5.2	11.5.4, 11.7.1
REPORT	8.4.6	11.4.1

ODS Statements

Statements	Primary Description in:	Also Appears in:
ODS _ALL_ CLOSE		11.4.2, 11.6
ODS ESCAPECHAR	11.3	
ODS GRAPHICS	10.3	
ODS LAYOUT	11.6	
ODS REGION	11.6	
ODS RESULTS	11.7.2	
ODS TRACE	11.1.1	

SAS Display Manager

Display Manager Commands

Command	Primary Description in:	Also Appears in:
AF		14.3.2
CLEAR		14.4.2
COMMAND		14.3.2
DMOPTLOAD	14.1.3	
DMOPTSAVE	14.1.3	
GSUBMIT	14.4.5	14.4.6, 14.4.7
KEYDEF		14.4.2
POST		14.4.2
TOOLCLOSE		14.3.2
VIEWTABLE		14.4.2
WEDIT		14.4.2
WSTATUSLN		14.3.2
window name	14.42	
WWINDOWBAR		14.3.2
ZOOM		14.3.2

References

The links shown in this section are intended to be live links; however, if you are reading this book on the traditional paper, the links of course cannot be live. The links shown below, along with the links shown throughout the book, are available in electronic form on sasCommunity.org, where the links can be live. Look for the category associated with this book's title.

User Publications

Adams, John H., 2010, "The new SAS 9.2 FCMP Procedure, what functions are in your future?", *Proceedings of the Pharmaceutical SAS User Group Conference* (PharmaSUG), 2010, Cary, NC: SAS Institute Inc., paper AD02. http://www.lexjansen.com/pharmasug/2010/ad/ad02.pdf

Adams, Sara, and Chris Colby, 2009, "Age Is Just a Number: Accurately Calculating Integer and Continuous Age", published in the *Proceedings of the Western Users of SAS Software Conference* (WUSS), Cary, NC: SAS Institute Inc., paper COD-Adams. http://www.wuss.org/proceedings09/09WUSSProceedings/papers/cod/COD-Adams.pdf

Aker, Sandra Lynn, 2000, "Using KEY= to Perform Table Look-up", published in the conference *Proceedings of the Twenty-Fifth Annual SAS Users Group International Conference*, SUGI, Cary, NC: SAS Institute Inc., paper 234-25. http://www2.sas.com/proceedings/sugi25/25/po/25p234.pdf

Aker, Sandra Lynn, 2002, "Table Look-up Using Techniques Other Than the Matched Merge DATA Step", published in the conference *Proceedings of the Twenty-Seventh Annual SAS Users Group International Conference*, SUGI, Cary, NC: SAS Institute Inc., paper 195-27. http://www2.sas.com/proceedings/sugi27/p195-27.pdf

Andrews, Rick, 2006, "SAS® to Excel® and Back Again", published on sasCommunity.org. http://www.sascommunity.org/mwiki/images/9/93/CMSSUG-0603-Excel.pdf.

Andrews, Rick and Tom Kress, 2006, "SQL vs SAS®: Clash of the Titans", published on sasCommunity.org. http://www.sascommunity.org/mwiki/images/5/52/CMSSUG-0506-SQL.pdf

Andrews, Rick, 2008, "Printable Spreadsheets Made Easy: Utilizing the SAS® Excel XP Tagset", *Proceedings of the 21st Annual NorthEast SAS Users Group Conference*, NESUG, Cary, NC: SAS Institute Inc. paper AP06. http://www.nesug.org/Proceedings/nesug08/ap/ap06.pdf

Bahler, Caroline, 2001, "Data Cleaning and Base SAS Functions", published in the *Proceedings of the Twenty-Sixth Annual SAS Users Group International Conference* (SUGI), Cary, NC: SAS Institute Inc., paper 56-26. http://www2.sas.com/proceedings/sugi26/p056-26.pdf

Benjamin, Jr., William E., 2007, "Hurry!!!, Hurry!!! Step Right UP. Use The 'Magical Compound Where Clause' to Eliminate Data Steps, Reduce Processing Steps, Speed Job Turnaround, and Mystify Your Friends.", *Proceedings of the SAS Global Forum 2007 Conference*, Cary, NC: SAS Institute Inc., paper 034-2007. http://www2.sas.com/proceedings/forum2007/034-2007.pdf

Bilenas, Jonas V. 2005, *The Power of PROC FORMAT*, Cary, NC: SAS Institute Inc. http://www.sas.com/apps/pubscat/bookdetails.jsp?catid=1&pc=59498

Bryant, Lara, Sally Muller, and Ray Pass, 2000," ODS, YES! Odious, NO! – An Introduction to the SAS Output Delivery System", published in the conference *Proceedings of the Twenty-Fifth Annual SAS Users Group International Conference*, SUGI, Cary, NC: SAS Institute Inc., paper 149-25. http://www2.sas.com/proceedings/sugi25/25/hands/25p149.pdf

Burlew, Michele, 2006, *SAS Macro Programming Made Easy, Second Edition*, Cary, NC: SAS Institute Inc., 426 pp. https://support.sas.com/pubcat/bookdetails.jsp?pc=60560

Carpenter, Arthur L., 1994, "Techniques to Avoid: What Momma Should Have Told You About SAS/GRAPH", published in the *Proceedings of the Nineteenth Annual SAS Users Group International Conference* (SUGI), Cary, NC: SAS Institute Inc. Also published in the *Proceedings of the Second Annual Western Users of SAS Software Conference* (WUSS), 1994, Cary, NC: SAS Institute Inc. http://www.sascommunity.org/sugi/SUGI94/Sugi-94-222%20Carpenter.pdf

Carpenter Arthur L. and Charles E. Shipp, 1995, *Quick Results with SAS/GRAPH® Software*, Cary, NC: SAS Institute Inc., 249 pp. http://www.sas.com/apps/pubscat/bookdetails.jsp?catid=1&pc=55127

Carpenter, Arthur L., 1998, "Better Titles: Using The #BYVAR and #BYVAL Title Options", published in the *Proceedings of the Twenty-Third Annual SAS Users Group International Conference* (SUGI), Cary, NC: SAS Institute Inc. http://www2.sas.com/proceedings/sugi23/Coders/p75.pdf

Carpenter, Arthur L., 1999, *Annotate: Simply the Basics*, SAS Institute, Inc., Cary, NC., 94 pp. http://www.sas.com/apps/pubscat/bookdetails.jsp?catid=1&pc=57320

Carpenter, Arthur L., 2001a, "Building and Using Macro Libraries", *Proceedings of the Ninth Annual Western Users of SAS Software Conference*, Cary, NC: SAS Institute Inc. Also in the *Proceedings of the Twenty-Seventh Annual SAS Users Group International Conference* (SUGI), 2002, Cary, NC: SAS Institute Inc. as well as in the proceedings of MWSUG 2001, PharmaSUG 2002, and PNWSUG 2002 and 2005. http://caloxy.com/papers/45-p17-27.pdf

Carpenter, Arthur L., 2001b, "Table Lookups: From IF-THEN to Key-Indexing," presented at the *Ninth Western Users of SAS Software Conference* (September, 2001) and the *Twenty-Sixth Annual SAS Users Group International Conference*, SUGI, (April, 2001), and the Pacific Northwest SAS Users Group Conference (November, 2005). The paper was published in the proceedings for each of these conferences. http://www2.sas.com/proceedings/sugi26/p158-26.pdf

Carpenter, Arthur L., 2002, "Macro Functions: How to Make Them - How to Use Them", *Proceedings of the Twenty-Seventh Annual SAS® Users Group International Conference*, Cary, NC: SAS Institute Inc., paper 100-27. Also in the Proceedings of the Pharmaceutical SAS® Users Group Conference, Cary, NC: SAS Institute Inc. 2002, paper CC06, pp. 87-91, and in the *Proceedings of the MidWest SAS Users Group Conference* (MWSUG), 2005, Cary, NC: SAS Institute Inc. http://caloxy.com/papers/46-ts200.pdf

Carpenter, Arthur L., 2003a, "Building and Using User Defined Formats", *Proceedings of the Eleventh Annual Western Users of SAS Software Conference*, Cary, NC: SAS Institute Inc. Also in the *Proceedings of the Twenty-Ninth Annual SAS Users Group International Conference* (SUGI), 2004, Cary, NC: SAS Institute Inc., paper 236-29. http://caloxy.com/papers/53-TU02.pdf

Carpenter, Arthur L., 2003b, "Creating Display Manager Abbreviations and Keyboard Macros for the Enhanced Editor", *Proceedings of the Twenty-Eighth Annual SAS® Users Group International Conference*, Cary, NC: SAS Institute Inc., paper 108-28. Also in the *Proceedings of the Pharmaceutical SAS® Users Group Conference* (PharmaSUG), Cary, NC: SAS Institute Inc. (2003), paper CC025, pp. 127-130. http://www2.sas.com/proceedings/sugi28/108-28.pdf

Carpenter, Arthur L., 2004, *Carpenter's Complete Guide to the SAS® Macro Language, 2nd Edition*, Cary, NC: SAS Institute Inc., 476 pp. http://www.sas.com/apps/pubscat/bookdetails.jsp?catid=1&pc=59224

Carpenter, Arthur L., 2005, "Make 'em %LOCAL: Avoiding Macro Variable Collisions", published in the *Proceedings of the Thirteenth Annual Western Users of SAS Software Conference* (WUSS), Cary, NC: SAS Institute Inc., paper sol_make_em_local_avoiding. Also published in the *Proceedings of the Pharmaceutical SAS Users Group Conference* (PharmaSUG), 2005, Cary, NC: SAS Institute Inc., paper TT04. http://caloxy.com/papers/62_TT04.pdf

Carpenter, Arthur L., 2006a, "In The Compute Block: Issues Associated with Using and Naming Variables", published in the *Proceedings of the Fourteenth Annual Western Users of SAS Software Conference* (WUSS), Cary, NC: SAS Institute Inc., paper DPR_Carpenter. Also published in the *Proceedings of the SAS Global Forum 2007 Conference*, Cary, NC: SAS Institute Inc., paper 025-2007 and in the *Proceedings of the Pharmaceutical SAS Users Group Conference* (PharmaSUG), 2007, Cary, NC: SAS Institute Inc., paper CC05. http://caloxy.com/papers/70-DPR.pdf

Carpenter, Arthur L., 2006b, "Advanced PROC REPORT: Traffic Lighting - Controlling Cell Attributes With Your Data", published in the *Proceedings of the Fourteenth Annual Western Users of SAS Software*, Conference (WUSS), Cary, NC: SAS Institute Inc., paper TUT_Carpenter. http://www.caloxy.com/papers/69-TUT.pdf

Carpenter Arthur L., 2007a, *Carpenter's Complete Guide to the SAS® REPORT Procedure*, Cary, NC: SAS Institute Inc., 463 pp. http://www.sas.com/apps/pubscat/bookdetails.jsp?catid=1&pc=60966

Carpenter, Arthur L., 2007b, "Advanced PROC REPORT: Getting Your Tables Connected Using Links", *Proceedings of the Pharmaceutical SAS Users Group Conference* (PharmaSUG), 2007, Cary, NC: SAS Institute Inc., paper HW04. Also presented in 2007 at the *Fifteenth Annual Western Users of SAS Software Conference* (WUSS), San Francisco, CA, in 2008 at MWSUG, and in 2009 at WUSS, SESUG, SCSUG, and PNWSUG. http://caloxy.com/papers/75LinksDrillDown.pdf

Carpenter, Arthur L., 2008, "The MEANS/SUMMARY Procedure: Getting Started and Doing More", presented at the *Sixteenth Annual Western Users of SAS Software Conference* (WUSS), Universal City, CA. Also presented at the PharmaSUG conference, 2009, papers TT05 and TT06. http://caloxy.com/papers/79MeansSummary.pdf

Carpenter, Arthur L., 2008b, "The Path, The Whole Path, And Nothing But the Path, So Help Me Windows", Proceedings of the SAS Global Forum Conference, 2008, NC: SAS Institute Inc., paper 023-2008. http://www2.sas.com/proceedings/forum2008/023-2008.pdf

Carpenter, Arthur L., 2009, "Manual to Automatic: Changing Your Program's Transmission". Presented at the *Seventeenth Annual Western Users of SAS Software Conference*, WUSS, Cary, NC: SAS Institute Inc., paper APP-Carpenter. Also presented at the Vancouver SAS Users Group, 2010, and the PharmaSUG conference, 2010, paper AD25. http://www.sas.com/offices/NA/canada/downloads/presentations/Van10/Manual.pdf

Carpenter, Arthur L., 2010a, "PROC TABULATE: Getting Started and Doing More", presented at the 2010 Pharmaceutical SAS Users Group Conference, PharmaSUG, Cary, NC: SAS Institute Inc., papers HW03 and HW04. http://www.pharmasug.org/cd/papers/HW/HW03.pdf

Carpenter, Arthur L., 2010b, "SAS/GRAPH® Elements You Should Know –Even If You Don't Use SAS/GRAPH", Presented in 2010 at the Western Users of SAS Software Conference , WUSS, and also in 2010 at the Southeast SAS Users Group, SESUG, SAS Global Forum 2010 Conference , and at the 2011 Pharmaceutical SAS Users Group Conference, PharmaSUG. http://analytics.ncsu.edu/sesug/2010/HOW04.Carpenter.pdf

Carpenter, Arthur L. and Dennis G. Fisher, 2011, "Reading and Writing RTF Documents as Data: Automatic Completion of CONSORT Flow Diagrams", presented at the Western Users of SAS Software Conference, WUSS. http://www.wuss.org/proceedings11/Papers_Carpenter_A_74920.pdf

Cassidy, Deb, 2005, "How Old Am I?", published in the *Proceedings of the Thirtieth Annual SAS Users Group International Conference*, SUGI, 2005, Cary, NC: SAS Institute Inc., Paper 060-30. http://www2.sas.com/proceedings/sugi30/060-30.pdf

Cates, Randall, 2001, "MISSOVER, TRUNCOVER, and PAD, OH MY!! or Making Sense of the INFILE and INPUT Statements.", published in the *Proceedings of the Twenty-Sixth Annual SAS Users Group International Conference*, SUGI, 2001, Cary, NC: SAS Institute Inc., Paper 009-26. http://www2.sas.com/proceedings/sugi26/p009-26.pdf

Chapal, Scott E., 2003, "Using SAS® and Other XML Tools Effectively", published in the *Proceedings of the Southeast SAS Users Group Conference*, SESUG, 2003, Cary, NC: SAS Institute Inc., paper TU11-Chapal. http://analytics.ncsu.edu/sesug/2003/TU11-Chapal.pdf

Chapman, David D., 2003, "Using Formats and Other Techniques to Complete PROC REPORT Tables", *Proceedings of the Twenty-Eighth Annual SAS® Users Group International Conference*, Cary, NC: SAS Institute Inc., paper 132-28. http://www2.sas.com/proceedings/sugi28/132-28.pdf

Chen, Ling Y., 2005, "Using V9 ODS LAYOUT to Simplify Generation of Individual Case Summaries", presented at the 2005 Pharmaceutical SAS Users Group Conference, PharmaSUG, Cary, NC: SAS Institute Inc., papers PO02. http://www.lexjansen.com/pharmasug/2005/posters/po02.pdf

Cheng, Alice M., 2011, "Hunting for Columbus' Eggs in the SAS® Programming World: A Guidance to Creative Thinking for SAS® Programmers", published in the *Proceedings of the Western Users of SAS Software Conference* (WUSS), Cary, NC: SAS Institute Inc., paper 74930. http://www.lexjansen.com/wuss/2011/coders/Papers_Cheng_A_74930.pdf

Choate, Paul A. and Carol A. Martell, 2006, "De-Mystifying the SAS® LIBNAME Engine in Microsoft Excel: A Practical Guide", published in the *Proceedings of the Thirty-first Annual SAS Users Group International Conference*, SUGI, 2006, Cary, NC: SAS Institute Inc., Paper 024-31. http://www2.sas.com/proceedings/sugi31/024-31.pdf

Chung, Chang Y. and Ian Whitlock, 2006, "%IFN – A Macro Function", published in the *Proceedings of the Thirty-first Annual SAS Users Group International Conference*, SUGI, 2006, Cary, NC: SAS Institute Inc., Paper 042-31. http://www2.sas.com/proceedings/sugi31/042-31.pdf .

Clifford, Billy, 2005, "Frequently Asked Questions about SAS® Indexes", published in the *Proceedings of the Thirtieth Annual SAS Users Group International Conference*, SUGI, 2005, Cary, NC: SAS Institute Inc., Paper 008-30. http://www2.sas.com/proceedings/sugi30/008-30.pdf

Cody, Ron, 2004, "An Introduction to Perl Regular Expressions in SAS 9", published in the *Proceedings of the Twenty-Ninth Annual SAS Users Group International Conference*, SUGI,2004, Cary, NC: SAS Institute Inc., paper 265-29. http://www2.sas.com/proceedings/sugi29/265-29.pdf

Cody, Ron, 2008a, "Using Advanced Features of User-defined Formats and Informats", published in the *Proceedings of the SAS Global Forum Conference*, 2008, Cary, NC: SAS Institute Inc., paper 041-2008. http://www2.sas.com/proceedings/forum2008/041-2008.pdf

Cody, Ron, 2008b, *Cody's Data Cleaning Techniques Using SAS, Second Edition,* Cary, NC: SAS Institute Inc., 248 pp. https://support.sas.com/pubscat/bookdetails.jsp?catid=1&pc=61703

Cody, Ron, 2010, SAS Functions by Example, 2nd Edition, Cary, NC: SAS Institute Inc., 445 pp. https://support.sas.com/pubscat/bookdetails.jsp?catid=1&pc=62857

Crawford, Peter, 2006a, "List Processing - Make Light Work of List Processing in SAS®", published in the *Proceedings of the Thirty-first Annual SAS Users Group International Conference*, SUGI, 2006, Cary, NC: SAS Institute Inc., Paper 012-31. http://www2.sas.com/proceedings/sugi31/012-31.pdf

Crawford, Peter, 2006b, "The Personal Touch: Control Your Environment as a SAS® User", published in the *Proceedings of the Thirty-first Annual SAS Users Group International Conference*, SUGI, 2006, Cary, NC: SAS Institute Inc., Paper 237-31. http://www2.sas.com/proceedings/sugi31/237-31.pdf

Davison, John W. Jr., 2006, "SAS® by Design – A Disciplined Approach", published in the *Proceedings of the Thirty-first Annual SAS Users Group International Conference*, SUGI, 2006, Cary, NC: SAS Institute Inc., paper 003-31. http://www2.sas.com/proceedings/sugi31/003-31.pdf

DelGobbo, Vincent, 2007, "Creating Multi-Sheet Excel Workbooks the Easy Way with SAS®", *Proceedings of the SAS Global Forum Conference*, 2007, Cary, NC: SAS Institute Inc., paper 120-2007. http://support.sas.com/rnd/papers/sgf07/sgf2007-excel.pdf

DeVenezia, Richard A., 2004, SAS programs originally presented on SAS-L. http://www.devenezia.com/downloads/sas/samples/hash-6.sas

Dorfman, Paul M., 2000a, "Private Detectives In a Data Warehouse: Key-Indexing, Bitmapping, And Hashing", published in *Proceedings of the Twenty-Fifth Annual SAS Users Group International Conference*, SUGI, Cary, NC: SAS Institute Inc, paper 129-25. http://www2.sas.com/proceedings/sugi25/25/dw/25p129.pdf

Dorfman, Paul M., 2000b, "Table Lookup via Direct Addressing: Key-Indexing, Bitmapping, Hashing", published in the *Proceedings of the Southeast SAS Users Group Conference*, SESUG, June, 2000. http://analytics.ncsu.edu/sesug/2000/p-105.pdf

Dorfman, Paul M., 2002, "The Magnificant DO", published in the *Proceedings of the Southeast SAS Users Group Conference*, SESUG, 2002, Cary, NC: SAS Institute Inc., paper TU05. http://www.devenezia.com/papers/other-authors/sesug-2002/TheMagnificentDO.pdf

Dorfman, Paul M. and Gregg P. Snell, 2002, "Hashing Rehashed", published in the *Proceedings of the Twenty-Seventh Annual SAS Users Group International Conference*, SUGI, 2002, Cary, NC: SAS Institute Inc., paper 12-27. http://www2.sas.com/proceedings/sugi27/p012-27.pdf

Dorfman, Paul M. and Gregg P. Snell, 2003, "Hashing: Generations", published in the *Proceedings of the Twenty-Eighth Annual SAS Users Group International Conference*, SUGI, 2003, Cary, NC: SAS Institute Inc., paper 004-28. http://www2.sas.com/proceedings/sugi28/004-28.pdf

Dorfman, Paul M. and Lessia S. Shajenko, 2004a, "Data Step Programming Using the Hash Objects", published in the *Proceedings of the Seventeenth Annual NorthEast SAS Users Group Conference*, NESUG, 2004, Cary, NC: SAS Institute Inc., paper PM06. http://www.nesug.org/Proceedings/nesug04/pm/pm06.pdf

Dorfman, Paul M. and Koen Vyverman, 2004b, "Hash Component Objects: Dynamic Data Storage and Table Look-Up" published in the *Proceedings of the Twenty-Ninth Annual SAS Users Group International Conference*, SUGI, 2004, Cary, NC: SAS Institute Inc., paper 238-29. http://www2.sas.com/proceedings/sugi29/238-29.pdf

Dorfman, Paul M. and Koen Vyverman, 2005, "Data Step Hash Objects as Programming Tools", published in the *Proceedings of the Thirtieth Annual SAS Users Group International Conference*, SUGI, 2005, Cary, NC: SAS Institute Inc., paper 236-30. http://www2.sas.com/proceedings/sugi30/236-30.pdf

Dorfman, Paul M. and Koen Vyverman, 2009, "The DOW-Loop Unrolled", published in the *Proceedings of the SAS Global Forum Conference*, 2009, Cary, NC: SAS Institute Inc., paper 038-2009. http://support.sas.com/resources/papers/proceedings09/038-2009.pdf

Dunn, Toby and Chang Y. Chung, 2005, "Retaining, Lagging, Leading, and Interleaving Data", published in the *Proceedings of the Pharmaceutical SAS Users Group Conference*, PharmaSUG, Cary, NC: SAS Institute Inc., paper TU09. http://www.pharmasug.org/2005/TU09.pdf

Dunn, Toby, 2010, "Efficiency: How Your Data Structure Can Help or Hurt!!!" published in the *Proceedings of the South-Central SAS Users Group Conference*, SCSUG, Cary, NC: SAS Institute Inc. http://www.scsug.org/SCSUGProceedings/2010/Dunn_3/Efficiency-How_Your_Data_Structure.pdf

Eberhardt, Peter, 2010, "The SAS Hash Object: It's Time To .find() Your Way Around", published in the *Proceedings of the Pharmaceutical SAS Users Group Conference*, PharmaSUG, Cary, NC: SAS Institute Inc., paper HW01. http://www.pharmasug.org/cd/papers/HW/HW01.pdf

Eberhardt, Peter, 2011, "A Cup of Coffee and Proc FCMP: I Cannot Function Without Them", published in the *Proceedings of the Pharmaceutical SAS Users Group Conference*, PharmaSUG, Cary, NC: SAS Institute Inc., paper TU07. http://www.pharmasug.org/proceedings/2011/TU/PharmaSUG-2011-TU07.pdf

Edney, Shawn, 2009, "Creating Common Information Structures Using List's Stored in Data Step Hash Objects", *Proceedings of the SAS Global Forum Conference*, 2009, Cary, NC: SAS Institute Inc., paper 011-2009. http://support.sas.com/resources/papers/proceedings09/011-2009.pdf

Fehd, Ronald J., 2007, "Do Which? Loop, Until or While? A Review Of Data Step And Macro Algorithms", *Proceedings of the SAS Global Forum Conference*, 2007, Cary, NC: SAS Institute Inc., paper 067-2007. http://www2.sas.com/proceedings/forum2007/067-2007.pdf

Fehd, Ronald J. and Arthur L. Carpenter, 2007, "List Processing Basics: Creating and Using Lists of Macro Variables", *Proceedings of the SAS Global Forum Conference*, 2007, Cary, NC: SAS Institute Inc., paper 113-2007. http://caloxy.com/papers/72Lists.pdf

Fehd, Ronald J., 2009, "Using Functions SYSFUNC and IFC to Conditionally Execute Statements in Open Code", *Proceedings of the SAS Global Forum Conference*, 2009, Cary, NC: SAS Institute Inc., paper 054-2009. http://support.sas.com/resources/papers/proceedings09/054-2009.pdf A supporting article can be found on sasCommunity.org http://www.sascommunity.org/wiki/Conditionally_Executing_Global_Statements

First, Steven, 2008, "The SAS INFILE and FILE Statements", *Proceedings of the SAS Global Forum Conference*, 2008, Cary, NC: SAS Institute Inc., paper 166-2008. http://www2.sas.com/proceedings/forum2008/166-2008.pdf

Frey, Gerald, 2004, "SAS Excels", Presented at MWSUG in 2004, http://www.sys-seminar.com/pdfs/sas_excels.pdf.

Friendly, Michael, 1991, *SAS® System for Statistical Graphics*, Cary, NC: SAS Institute Inc., 697 pp. http://www.sas.com/apps/pubscat/bookdetails.jsp?catid=1&pc=56143

Gebhart, Eric, 2010," ODS ExcelXP: Tag Attr Is It! Using and Understanding the TAGATTR= Style Attribute with the ExcelXP Tagset", *Proceedings of the SAS Global Forum Conference*, 2010, Cary, NC: SAS Institute Inc., paper 031-2010. http://support.sas.com/resources/papers/proceedings10/031-2010.pdf

Hamilton, Jack, 2001, "How Many Observations Are In My Data Set?", published in the *Proceedings of the Twenty-Sixth Annual SAS Users Group International Conference*, SUGI, 2001, Cary, NC: SAS Institute Inc., paper 095-26. http://www2.sas.com/proceedings/sugi26/p095-26.pdf

Hamilton, Jack, 2007, "Creating Data-Driven Data Set Names in a Single Pass Using Hash Objects", published in the *Proceedings of the SouthEast SAS Users Group Conference*, SESUG, Cary, NC: SAS Institute Inc., paper SD04. http://analytics.ncsu.edu/sesug/2007/SD04.pdf

Haworth, Lauren E., 1999, *PROC TABULATE by Example*, Cary, NC: SAS Institute Inc., 374 pp. http://www.sas.com/apps/pubscat/bookdetails.jsp?catid=1&pc=56514

Haworth, Lauren E., Cynthia L. Zender, and Michele M. Burlew, 2009, *Output Delivery System: The Basics and Beyond*, Cary, NC: SAS Institute Inc., 610 pp. http://www.sas.com/apps/pubscat/bookdetails.jsp?catid=1&pc=61686

Heaton, Ed, 2008, "Many-to-Many Merges in the DATA Step", *Proceedings of the SAS Global Forum Conference*, 2008, NC: SAS Institute Inc., paper 81-2008. http://www2.sas.com/proceedings/forum2008/081-2008.pdf

Heaton, Ed and Sarah Woodruff, 2009, "Implementing User-Friendly Macro Systems", *Proceedings of the SouthEast SAS Users Group Conference*, SESUG, 2009, NC: SAS Institute Inc., paper FF-006. http://analytics.ncsu.edu/sesug/2009/FF006.Heaton.pdf

Hemedinger, Chris, Susan Slaughter, 2011, "Social Networking and SAS®: Running PROCs on Your Facebook Friends", *Proceedings of the SAS Global Forum Conference*, 2011, NC: SAS Institute Inc., paper 315-2011. http://support.sas.com/resources/papers/proceedings11/315-2011.pdf

Howard, Rob, 2004, "GSUBMIT: Simple Customization of your SAS® Application Toolbar in SAS for Windows® ", published in the *Proceedings of the Pharmaceutical SAS Users Group Conference*, PharmaSUG, Cary, NC: SAS Institute Inc., paper CC19. http://www.lexjansen.com/pharmasug/2004/coderscorner/cc19.pdf

Humphreys, Suzanne M., 2006, "MISSING! - Understanding and Making the Most of Missing Data", *Proceedings of the Thirty-first Annual SAS Users Group International Conference*, 2006, NC: SAS Institute Inc., paper 025-31. http://www2.sas.com/proceedings/sugi31/025-31.pdf

Hunley, Chuck, 2010, "SMTP E-Mail Access Method: Hints, Tips, and Tricks", *Proceedings of the SAS Global Forum Conference*, 2010, NC: SAS Institute Inc., paper 060-2010. http://support.sas.com/resources/papers/proceedings10/060-2010.pdf

Hunt, Stephen, 2010, "SAS 1-Liners", *Proceedings of the SAS Global Forum Conference*, 2010, NC: SAS Institute Inc., paper 054-2010. http://support.sas.com/resources/papers/proceedings10/054-2010.pdf

Hurley, George J., 2007, "Customizing Your SAS Initialization", *Proceedings of the SAS Global Forum Conference*, 2007, NC: SAS Institute Inc., paper 063-2007. http://www2.sas.com/proceedings/forum2007/063-2007.pdf

Jolley, Linda and Jane Stroupe, 2007, "Dear Miss SASAnswers: A Guide to SAS® Efficiency", published in the *Proceedings of the SAS Global Forum 2007 Conference*, Cary, NC: SAS Institute Inc., paper 042-2007. http://www2.sas.com/proceedings/forum2007/042-2007.pdf

Keelan, Stephen, 2002, "Off and Running with Arrays in SAS®", published in the *Proceedings of the Twenty-Seventh Annual SAS Users Group International Conference*, SUGI, 2002, Cary, NC: SAS Institute Inc., paper 66-27. http://www2.sas.com/proceedings/sugi27/p066-27.pdf

King, John and Mike Zdeb, 2010, "Transposing Data Using PROC SUMMARY'S IDGROUP Option", *Proceedings of the SAS Global Forum Conference*, 2010, NC: SAS Institute Inc., paper 102-2010. http://support.sas.com/resources/papers/proceedings10/102-2010.pdf

King, John Henry, 2011,"Using a HASH Table to Reference Variables in an Array by Name" published in the *Proceedings of the Pharmaceutical SAS Users Group Conference*, PharmaSUG, Cary, NC: SAS Institute Inc., paper TT04. http://www.pharmasug.org/proceedings/2011/TT/PharmaSUG-2011-TT04.pdf

Kohli, Monal, 2006, "Project Duplication: Eradication Techniques", published in the *Proceedings of the Thirty-first Annual SAS Users Group International Conference*, SUGI, 2006, Cary, NC: SAS Institute Inc., paper 031-31. http://www2.sas.com/proceedings/sugi31/031-31.pdf

Kreuter, William, 2004, "Sample 24808: Accurately Calculating Age with Only One Line of Code", Cary, NC: SAS Institute Inc. http://staff.washington.edu/billyk/TechTips_SC4Q98.pdf and also at http://support.sas.com/kb/24/808.html.

Kuhfeld, Warren F., 2010, *Statistical Graphics in SAS: An Introduction to the Graph Template Language and the Statistical Graphics Procedures*, Cary, NC: SAS Institute Inc., 211 pp. https://support.sas.com/pubscat/bookdetails.jsp?catid=1&pc=63120

Lavery, Russ, 2005, "The SQL Optimizer Project: _Method and _Tree in SAS®9.1", published in the *Proceedings of the Thirtieth Annual SAS Users Group International Conference*, SUGI, 2005, Cary, NC: SAS Institute Inc., Paper 101-30. http://www2.sas.com/proceedings/sugi30/101-30.pdf

Levin, Lois, 2004, "Methods of Storing SAS® Data into Oracle Tables", published in the *Proceedings of the Twenty-Ninth Annual SAS Users Group International Conference*, SUGI, 2004, Cary, NC: SAS Institute Inc., Paper 106-29. http://www2.sas.com/proceedings/sugi29/106-29.pdf

Li, Arthur, 2011, "The Many Ways to Effectively Utilize Array Processing", published in the *Proceedings of the SAS Global Forum Conference*, 2011, Cary, NC: SAS Institute Inc., paper 244-2011. http://support.sas.com/resources/papers/proceedings11/244-2011.pdf

Liu, Ying, 2008, "SAS® Hash Objects: An Efficient Table Look-Up in the Decision Tree", published in the *Proceedings of the SouthEast SAS Users Group Conference*, SESUG, Cary, NC: SAS Institute Inc., paper CS-057. http://analytics.ncsu.edu/sesug/2008/CS-057.pdf

Lund, Pete, 2006, "PDF Can be Pretty Darn Fancy -Tips and Tricks for the ODS PDF Destination", *Proceedings of the Thirty-first Annual SAS Users Group International Conference,* SUGI, 2006, Cary, NC: SAS Institute Inc., Paper 092-31. http://www2.sas.com/proceedings/sugi31/092-31.pdf

Matange, Sanjay and Dan Heath, 2011, *Statistical Graphics Procedures by Example: Effective Graphs Using SAS*, Cary, NC: SAS Institute Inc., 357 pp. https://support.sas.com/pubscat/bookdetails.jsp?catid=1&pc=63855

McQuown, Gary, 2005, "PROC IMPORT with a Twist", *Proceedings of the Thirtieth Annual SAS Users Group International Conference,* SUGI, 2005, Cary, NC: SAS Institute Inc., Paper 038-30. http://www2.sas.com/proceedings/sugi30/038-30.pdf

Miron, Thomas, 1995, *The How-To Book for SAS/GRAPH Software*, Cary, NC: SAS Institute Inc., 286 pp. http://www.sas.com/apps/pubscat/bookdetails.jsp?catid=1&pc=55203

Murphy, William C. , 2006, "Squeezing Information out of Data" *Proceedings of the Thirty-first Annual SAS Users Group International Conference*, SUGI, 2006, Cary, NC: SAS Institute Inc., Paper 028-31. http://www2.sas.com/proceedings/sugi31/028-31.pdf

Nelson, Greg Barnes, Danny Grasse, and Jeff Wright. 2004a. "Automated Testing and Real-time Event Management: An Enterprise Notification System" *Proceedings of the Twenty-ninth Annual SAS Users Group International Conference,* SUGI, 2004, Cary, NC: SAS Institute Inc., Paper 228-29. http://www2.sas.com/proceedings/sugi29/228-29.pdf

Nelson, Greg Barnes. 2004b. "SASUnit: Automated Testing for SAS." *Proceedings of the Pharmaceutical SAS Users Group*, PharmaSUG, Cary, NC: SAS Institute Inc, Paper DM10. http://www.lexjansen.com/pharmasug/2004/datamanagement/dm10.pdf

Nelson, Rob, 2010, "ODS LAYOUT to Create Publication-Quality PDF Reports of STD Surveillance Data", published in the *Proceedings of the SAS Global Forum Conference*,2010, Cary, NC: SAS Institute Inc., paper 216-2010. http://support.sas.com/resources/papers/proceedings10/216-2010.pdf

O'Connor, Daniel and Scott Huntley, 2009, "Breaking New Ground with SAS® 9.2 ODS Layout Enhancements", published in the *Proceedings of the Western Users of SAS Software Conference*, 2009, Cary, NC: SAS Institute Inc., paper DPR-OCONNOR. http://www.lexjansen.com/wuss/2009/dpr/DPR-OConnor.pdf. An excellent related PowerPoint presentation can be found at: http://support.sas.com/rnd/base/early-access/layout.ppt

Palmer, Michael, 2003, "XML in the DATA Step", published in the *Proceedings of the Twenty-Eighth Annual SAS Users Group International Conference*, SUGI, 2003, Cary, NC: SAS Institute Inc., Paper 025-28. http://www2.sas.com/proceedings/sugi28/025-28.pdf

Palmer, Michael, 2004, "XML in the DATA Step", published in the *Proceedings of the Twenty-Ninth Annual SAS Users Group International Conference*, SUGI, 2004, Cary, NC: SAS Institute Inc., Paper 036-29. http://www2.sas.com/proceedings/sugi29/036-29.pdf

Pratter, Frederick, 2008, "XML for SAS® Programmers", published in the *Proceedings of the SAS Global Forum Conference*, 2008, Cary, NC: SAS Institute Inc., paper 042-2008. http://www2.sas.com/proceedings/forum2008/042-2008.pdf

Raithel, Michael A., 2004, "Creating and Exploiting SAS® Indexes", published in the *Proceedings of the Twenty-Ninth Annual SAS Users Group International Conference*, SUGI, 2004, Cary, NC: SAS Institute Inc., Paper 123-29. http://www2.sas.com/proceedings/sugi29/123-29.pdf

Raithel, Michael A., 2006, *The Complete Guide to SAS Indexes*, Cary, NC: SAS Institute Inc., 324 pp. http://www.sas.com/apps/pubscat/bookdetails.jsp?catid=1&pc=60409

Raithel, Michael A., 2009, "Tips:Create_a_PROC_IMPORT_or_PROC_EXPORT_Template_ Program_for_Ease_of_Use", article appearing on sasCommunity.org, http://www.sascommunity.org/wiki/Tips:Create_a_PROC_IMPORT_or_PROC_EXPORT_T emplate_Program_for_Ease_of_Use

Ray, Robert and Jason Secosky, 2008, "Better Hashing in SAS® 9.2", published in the *Proceedings of the SAS Global Forum Conference*, 2008, Cary, NC: SAS Institute Inc., paper 306-2008. http://support.sas.com/rnd/base/datastep/dot/better-hashing-sas92.pdf.

Rhodes, Dianne Louise, 2005, "Speaking Klingon: A Translators guide to PROC TABULATE", published in the *Proceedings of the Thirtieth Annual SAS Users Group International Conference*, SUGI, 2005, Cary, NC: SAS Institute Inc., paper 258-30. http://www2.sas.com/proceedings/sugi30/258-30.pdf

Rosenbloom, Mary F.O., 2011a, "Using PROC CONTENTS and a Macro to Convert Internal Data Values to their Associated Format Values", published in the *Proceedings of the Nineteenth Annual Western Users of SAS Software Conference*, WUSS, Cary, NC, SAS Institute Inc., Paper 74974. http://www.wuss.org/proceedings11/Papers_Rosenbloom_M_74974.pdf

Rosenbloom, Mary F.O., and Art Carpenter, 2011b, "Macro Quoting to the Rescue: Passing Special Characters", published in the *Proceedings of the Nineteenth Annual Western Users of SAS Software Conference*, WUSS, Cary, NC, SAS Institute Inc., Paper 74973. http://www.wuss.org/proceedings11/Papers_Rosenbloom_M_74973.pdf

Rosenbloom, Mary, and Kirk Paul Lafler, 2011c, "Assigning a User-defined Macro to a Function Key", published in the *Proceedings of the Nineteenth Annual Western Users of SAS Software Conference*, WUSS, Cary, NC, SAS Institute Inc., Paper 76113. http://www.wuss.org/proceedings11/Papers_Rosenbloom_M_76113.pdf

Rosenbloom, Mary F.O., and Kirk Paul Lafler, 2011d, "Best Practices: Clean House to Avoid Hangovers", published in the *Proceedings of the Nineteenth Annual Western Users of SAS Software Conference*, WUSS, Cary, NC, SAS Institute Inc., Paper 76114. http://www.wuss.org/proceedings11/Papers_Rosenbloom_M_76114.pdf

Rozhetskin, Dmitry, 2010, "Choosing the Best Way to Store and Manipulate Lists in SAS®" published in the *Proceedings of the Fourteenth Western Users of SAS Software Conference*, WUSS, Cary, NC, SAS Institute Inc., Paper COD-Rozhetskin. http://www.wuss.org/proceedings10/coders/2972_9_COD-Rozhetskin.pdf

Scerbo, Marge, Craig Dickstein, and Alan C. Wilson, 2001, *Health Care Data and the SAS® System*, Cary, NC: SAS Institute Inc., 274 pp. http://www.sas.com/apps/pubscat/bookdetails.jsp?catid=1&pc=57638

Schreier, Howard, 2001, "Now _INFILE_ is an Automatic Variable – So What?", published in the *Proceedings of the Fourteenth Annual NorthEast SAS Users Group Conference*, Cary, NC: SAS Institute Inc., paper cc4018bw. http://www.nesug.org/proceedings/nesug01/cc/cc4018bw.pdf

Schreier, Howard, (2003), "Interleaving a Dataset with Itself: How and Why?" *Proceedings of the Sixteenth Annual NorthEast SAS Users Group* (NESUG) *Conference, 2003.* www.nesug.org/proceedings/nesug03/cc/cc002.pdf

Schreier, Howard, 2007, "Conditional Lags Don't Have to be Treacherous", *Proceedings of the Twentieth Annual NorthEast SAS Users Group Conference*, NESUG, Cary, NC: SAS Institute Inc. paper CC33. http://www.howles.com/saspapers/CC33.pdf

Secosky, Jason and Janice Bloom, 2007, "Getting Started with the DATA Step Hash Object", published in the *Proceedings of the SAS Global Forum 2007 Conference*, Cary, NC: SAS Institute Inc., paper 271-2007. http://www2.sas.com/proceedings/forum2007/271-2007.pdf

Secosky, Jason, 2007, "User-Written DATA Step Functions", published in the *Proceedings of the SAS Global Forum 2007 Conference*, Cary, NC: SAS Institute Inc., paper 008-2007. http://www2.sas.com/proceedings/forum2007/008-2007.pdf

Sherman, Paul D. and Arthur L. Carpenter, 2007, "Secret Sequel: Keeping Your Password Away From the LOG", *Proceedings of the Pharmaceutical SAS Users Group Conference* (PharmaSUG), 2007, Cary, NC: SAS Institute Inc., paper TT07. Also in the *Proceedings of the SAS Global Forum 2009 Conference*, Cary, NC: SAS Institute Inc., Paper 013-2009. http://caloxy.com/papers/74Secret.pdf

Shostak, Jack, 2005, *SAS® Programming in the Pharmaceutical Industry*, Cary, NC: SAS Institute Inc., 332 pp. https://support.sas.com/pubscat/bookdetails.jsp?catid=1&pc=59827

Slaughter, Susan J. and Lora D. Delwiche, *The Little SAS Book for Enterprise Guide 4.2*, Cary, NC: SAS Institute Inc., 371 pp. https://support.sas.com/pubscat/bookdetails.jsp?catid=1&pc=61861

Steven, David C., 2007,"Keep your database passwords out of the clear: Quick and easy tips to protect yourself", published in the *Proceedings of the Pacific Northwest SAS Users Group Conference*, PNWSUG, 2007, Cary, NC: SAS Institute Inc. http://www.lexjansen.com/pnwsug/2007/Dave%20Steven%20-%20Keep%20your%20database%20passwords%20out%20of%20the%20clear.pdf

Stroupe, Jane, 2003,"Nine Steps to Get Started using SAS® Macros", published in the *Proceedings of the Twenty-Eighth Annual SAS Users Group International Conference*, SUGI, 2003, Cary, NC: SAS Institute Inc., paper 56-28. http://www2.sas.com/proceedings/sugi28/056-28.pdf

Stroupe, Jane, 2007, "Adventures in Arrays: A Beginning Tutorial", published in the *Proceedings of the SAS Global Forum 2007 Conference*, Cary, NC: SAS Institute Inc., paper 1780-2007. http://support.sas.com/rnd/papers/sgf07/arrays1780.pdf

Stroupe, Jane and Linda Jolley, 2008, "Using Table Lookup Techniques Efficiently", published in the *Proceedings of the SAS Global Forum 2008 Conference*, Cary, NC: SAS Institute Inc., paper 095-2008. http://www2.sas.com/proceedings/forum2008/095-2008.pdf

Sun, Eric and Arthur L. Carpenter, 2011, "Protecting Macros and Macro Variables: It Is All About Control". *Presented in 2011 at the Pharmaceutical SAS Users Group Conference*, PharmaSUG, paper AD17. http://www.pharmasug.org/proceedings/2011/AD/PharmaSUG-2011-AD17.pdf

Tabachneck, Arthur S., Randy Herbison, Andrew Clapson, John King, Roger DeAngelis, Tom Abernathy, 2010, "Automagically Copying and Pasting Variable Names", published in the *Proceedings of the SAS Global Forum 2010 Conference*, Cary, NC: SAS Institute Inc., paper 046-2010. http://support.sas.com/resources/papers/proceedings10/046-2010.pdf

Tyndall, Russ, 2005,"Give Your Macro Code an Extreme Makeover: Tips for even the most seasoned macro programmer", Technical Support Tip 739, Cary, NC: SAS Institute Inc. http://support.sas.com/techsup/technote/ts739.pdf

Varney, Brian, 2008, "Check out These Pipes: Using Microsoft Windows Commands from SAS®", published in the *Proceedings of the SAS Global Forum 2008 Conference*, Cary, NC: SAS Institute Inc., paper 092-2008. http://www2.sas.com/proceedings/forum2008/092-2008.pdf

Virgile, Robert, 1998, *Efficiency: Improving the Performance of Your SAS Applications,* Cary, NC: SAS Institute Inc., 232 pp. https://support.sas.com/pubscat/bookdetails.jsp?catid=1&pc=55960

Vora, Premal P., 2008, "Easy Rolling Statistics with PROC EXPAND", published in the *Proceedings of the SAS Global Forum 2008 Conference*, Cary, NC: SAS Institute Inc., paper 093-2008. http://www2.sas.com/proceedings/forum2008/093-2008.pdf

Waller, Jennifer L., 2010, "How to Use ARRAYs and DO Loops: Do I DO OVER or Do I DO i?", *Proceedings of the SAS Global Forum 2010 Conference*, Cary, NC: SAS Institute Inc., Paper 158-2010. http://support.sas.com/resources/papers/proceedings10/158-2010.pdf

Walsh, Irina, 2009, "Pros and Cons of X command vs. SYSTASK command", published in the *Proceedings of the Western Users of SAS Software Conference* (WUSS), Cary, NC: SAS Institute Inc., paper COD-Walsh. http://www.wuss.org/proceedings09/09WUSSProceedings/papers/cod/COD-Walsh.pdf

Whitlock, Ian, 2003," A Serious Look Macro Quoting", published in the *Proceedings of the Twenty-Eighth Annual SAS Users Group International Conference*, SUGI, 2003, Cary, NC: SAS Institute Inc., paper 11-28. http://www2.sas.com/proceedings/sugi28/011-28.pdf

Whitlock, Ian, 2008, "The Art of Debugging", *Proceedings of the SAS Global Forum 2009 Conference*, Cary, NC: SAS Institute Inc., Paper 165-2008. http://www2.sas.com/proceedings/forum2008/165-2008.pdf

Whitworth, Ryan, 2010, "Zip and Email Files Using SAS® To Reduce Errors and Make Documentation Easy", *Proceedings of the SAS Global Forum Conference*, 2010, NC: SAS Institute Inc., paper 084-2010. http://support.sas.com/resources/papers/proceedings10/084-2010.pdf

Wright, Jeff, 2006, "Drawkcab Gnimmargorp: Test-Driven Development with FUTS", published in the *Proceedings of the Thirty-first Annual SAS Users Group International Conference*, SUGI, 2006, Cary, NC: SAS Institute Inc, Paper 004-31. http://www2.sas.com/proceedings/sugi31/004-31.pdf

Zender, Cynthia L., 2007, "Funny ^Stuff~ in My Code: Using ODS ESCAPECHAR", *Proceedings of the SAS Global Forum 2007 Conference*, Cary, NC: SAS Institute Inc., Paper 099-2007. http://www2.sas.com/proceedings/forum2007/099-2007.pdf

Zender, Cynthia L., 2008, "Creating Complex Reports", *Proceedings of the SAS Global Forum 2008 Conference*, Cary, NC: SAS Institute Inc., Paper 173-2008. http://www2.sas.com/proceedings/forum2008/173-2008.pdf

Generally Good Reading – Lots More to Learn

Where can you go to get more information? There are a number of sites and opportunities available that have a great variety of types of information. A few of these are collected here. Certainly there are many others including those that have come into being since the publication of this book. These links and others will be published on sasCommunity.org (search for this book's title). There you can add your own favorite links to share with others.

A number of interesting articles can be found under the sasCommunity.org category 'SAS Traps'. http://www.sascommunity.org/wiki/Category:SAS_Traps

A Tips and Tricks thread on SAS-L contains a number of items that are definitely worth knowing. http://listserv.uga.edu/cgi-bin/wa?A2=ind1001d&L=sas-l&F=&S=&P=5105

SAS Documentation
"XML Engine with DATA Step or PROC COPY" http://support.sas.com/documentation/cdl/en/movefile/59598/HTML/default/xmlchap.htm

SAS Usage Notes
Usage Note 15727: Writing PAGE X OF Y in RTF does not work with BODYTITLE http://support.sas.com/kb/15/727.html

Discussion Forums
Discussion forums allow you to not only receive information, but post questions as well. It is this give and take that makes these sites so valuable.

SAS-L is arguably the longest running online help forum. You can participate or just observe http://listserv.uga.edu/archives/sas-l.html.

The SAS sponsored SAS Forums allow you to ask and answer questions http://communities.sas.com.

LinkedIn has over 600 groups that include SAS in their description and a number of these encourage forum-style discussions http://www.linkedin.com/groupsDirectory.

SAS Professionals offers a forum discussion site http://www.sasprofessionals.net/.

Stack Overflow includes discussion forums on virtually all topics related to computing. This includes a number related to SAS http://stackoverflow.com/questions/tagged/sas.

On Google Groups the group comp.soft-sys.sas http://groups.google.com/group/comp.soft-sys.sas/topics?hl=en has a large number of SAS related entries and a large following.

Newsletters, Corporate and Private Sites

Newsletters and corporate sites that regularly include tips and 'how-to' information include:

Amadeus Software Ltd.
http://www.amadeus.co.uk/sas-technical-services/tips-and-techniques/. You can sign up for their newsletter at: http://www.amadeus.co.uk/about-us/newsletter-signup/.

Richard DeVenezia
This Website has links to downloads, papers and other useful information about SAS
http://www.devenezia.com/downloads/sas/actions/.

San Diego SAS Users Group
The SANDS Newsletter contains at least one tip in each issue http://sandsug.org/.

System Seminar Consultants, Inc.
The Missing Semicolon newsletter is loaded with tips and coding techniques
http://www.sys-seminar.com/newsletter.

VIEWS User Group
The newsletter contains tips in addition to information on the use of SAS
http://www.sascommunity.org/wiki/VIEWS_News.

User Communities

Sponsored by the SAS Global User Group the wiki site sasCommunity.org contains thousands of user-supplied articles on all aspects pertaining to SAS http://www.sascommunity.org. This site publishes a daily tip, and current and past tips can be reviewed at
http://www.sascommunity.org/wiki/Tip_of_the_Day. This site can also be searched using a Google appliance.
http://www.sascommunity.org/wiki/Tips:You_can_use_Google_to_search_sascommunity.org_for_tips_and_articles

Publications

Lex Jansen
While user conference proceedings can be found on numerous sites, most of these papers have been indexed at this site http://www.lexjansen.com/.

Blogs about SAS
A number of active SAS blogs can be found on sasCommunity.org
http://www.sascommunity.org/planet/.

SAS Press
Books written about SAS by those who use SAS can be found in the SAS Press catalog
https://support.sas.com/pubscat/complete.jsp. Most of these books include sample programs and data which can be downloaded even if you do not buy the book.

Learning SAS

SAS Institute offers a variety of types of learning opportunities, from instructor led to computer based, and are available here: http://support.sas.com/training/

University at Albany School of Public Health
A collection of links put together by Mike Zdeb can be found here
http://www.albany.edu/~msz03/.

Index

A

B